IMPORTANT THINGS WE DON'T KNOW

ABOUT NEARLY EVERYTHING

JOHN E. BEERBOWER

PUBLISHED BY FASTPENCIL PUBLISHING

Important Things We Don't Know

Third Edition

Copyright © John E. Beerbower 2022

All rights reserved. No part of this publication may be reproduced, stored in a retrieval system, or transmitted, in any form, or by any means, electronic, mechanical, photocopying, recording, or otherwise, without the prior consent of the author.

Sale of this book without a front cover may be unauthorized. If the book is coverless, it may have been reported to the publisher as "unsold or destroyed" and neither the author nor the publisher may have received payment for it.

http://www.fastpencil.com

Printed in the United States of America

Table of Contents

Preface ix
Introduction.. xv
Knowledge and Understanding 21
    The Nature of Comprehension 22
    The Role of Theories and Models 28
    The "Problem" of Induction.. 36
    Prediction versus Explanation.. 44
    Requirements of a Good Theory.. 53
    Explanation Revisited 60
    Biological Limits to Understanding? 65
    Endnotes 68
Mathematics... 75
    What is Mathematics?... 75
    Characteristics of Mathematics 81
    Wigner's Famous Question 84
    Some Comments on Logic 89
    Zero and Infinity 92
    Concepts of Probability... 97
    Laws of Nature 102
    Usefulness Revisited 105
    Endnotes... 113
Social Science...117
    Science? 118
    The Subject of Economics 122
    The Science of Economics 125
    The Theory of the Firm 129
    The Marginalist Debate...134

- The Methodological Issues 138
- Some Concluding Comments 147
- Endnotes 150

Darwinism 153
- The Diversity of Life 154
- Natural Selection 159
- The Origin of Species 182
- The Role of Altruistic Behavior 195
- The Role of Beauty 200
- The "New" Biology 205
- Some Other Matters 215
- The Nature of Neo-Darwinian Explanation 219
- Endnotes 224

Chemistry 231
- Lead to Gold 231
- Atoms and Their Parts 234
- The Elements 245
- Radioactive Decay 248
- Molecules and Compounds 249
- Reactions 256
- Emergence 258
- Endnotes 261

Physics 265
- Action at a Distance 265
- Light 272
- Time 277
- Thermodynamics 283
- Special Relativity 295
- General Relativity 313
- Time Revisited 323
- Endnotes 329

Particle Physics and Quantum Mechanics................................333
    The (Sub) Microscopic World 334
    The Standard Model................................336
    The Higgs Boson 344
    Quantum Mechanics................................348
    Alarming Implications?................................366
    Theories of Everything................................370
    Final Observations................................372
    Endnotes................................378
Cosmology 381
    Setting the Stage................................382
    The Legacy of Edwin Hubble 388
    Origin of the Universe 392
    Black Holes 402
    Inflationary Cosmology................................410
    "Dark" Things 414
    Time (Again) 421
    Endnotes................................428
Consciousness................................433
    A Focus on Consciousness 433
    At the Limits of Science? 436
    The Makeup of the Brain 439
    What the Brain Does................................455
    Computers and Consciousness................................464
    The Source of Consciousness?................................473
    Mind & Cosmos 483
    Endnotes................................488
Some Other Matters................................493
    The Practice of Science................................493
    The "Evolution" of Science 503
    The "Evolution" of Cooperation 509

Religion and Design	518
Endnotes	532
The Big Questions	535
Life	535
Fine Tuning	545
Progress	561
Intelligence	575
And, Us	584
Endnotes	597
Conclusion	601
PostScript	605
References	607

Preface

The perspective of this book differs from that of most books that one would find in the popular science section of a library or bookshop. I do not undertake to surprise or educate the reader with examples of the marvelous discoveries of science and the amazing things that we have come to learn about the world around us. Instead, and of more importance, I try to show the reader what we do not know, illustrating both the scope and the depth of our ignorance.

I have been tempted to think that at least the broad brush strokes of what I have to say here are already largely familiar to or known by most well-educated persons today. Of course, if that were the case, then there would be little point in my writing it all down. However, I am pretty confident that it is not the case. In my youth, I attended three of the English-speaking world's most elite institutions of higher education, but I did not learn these things. Admittedly, I did get an introduction and enough exposure to pique my interest and continuing curiosity (and to bring me back to this undertaking some 40 years later). But, I should have learned it at university. What we do not know ought to be an integral part of a meaningful education.

Given some of the themes that will be developed in the pages that follow, it seems appropriate right up front to address one obvious question: why do I think that I am qualified to address the matters set forth herein? It is certainly a fair question, and it is one on which I have pondered at length, especially when considering alternative and less frustrating ways to spend my time than writing the many words that follow. I have answered the question to my own satisfaction. Let me explain.

As an undergraduate majoring in economics, I took a particular interest in methodological questions arising in the social sciences, leading me deeply into the philosophy of science. I continued that interest as a postgraduate research student at the University of Cambridge. At the Harvard Law School, I studied the concepts of causation, evidence and proof, as well as the nature of laws–subjects just as integral to jurisprudence as to the philosophy of science. Indeed, the historical intellectual interrelationships between law and the relevant areas of philosophy are extensive. In addition, for over 50 years, I have been an avid reader of popularized science (biology, paleontology, physics, astrophysics, cosmology, genetics, evolution, mathematics). Many of the hundreds of books I have read were written by leading scientists in their fields as attempts to explain their science to the intelligent lay reader.

I also spent 35 years as a trial lawyer. Apart from deepening my understanding of the subjects studied in law school, how is that professional experience relevant?

Well, I quickly learned in my practice four things that I think apply very clearly to the subjects that I address herein:

- There are always at least two sides to every issue. Every issue.
- Disputes of fact are very often genuine (there is no clearly correct version).
- "Truth" is relative and a matter of perspective.
- Advocacy and presentation matter.

In addition, as a trial lawyer, I had years of practice interviewing and examining witnesses and extensive experience with the use of experts and of expertise to "prove" factual propositions or to persuade an audience to reach the desired conclusions, with the use of logic to promote or rebut positions and with the assemblage and presentation of evidence. Although it was not always apparent at the time, looking back, I see the direct relationship between my professional experiences and some of the central issues in the philosophy and methodology of science (as well as, of course, in the politics and techniques of scientific advocacy).

As will become clear, my guiding prejudice in undertaking this project, and one of the beliefs to which I hope to convert the reader in the pages that follow, is that non-scientists can achieve valuable insights into and make constructive critiques of the developments that occur in the continuing quest for scientific knowledge, even of developments in the more difficult and technical areas of exploration. Thus, the writing of this book constitutes an effort to establish that my prejudice is, indeed, defensible. In other words, one of the arguments that I am advancing herein is that someone like me is capable of doing something like this.

In part, I am concerned about what I see in our age of specialization as a tendency toward a "tyranny of expertise". Because science has become so complicated and highly technical, various groups of lay persons just presume that their opinions are likely to be wrong or, in any event, irrelevant. One result is apathy and withdrawal. The presumption of one's irrelevance can feed the ever present human temptation to take the easy road, to avoid both confrontation and serious effort.

I do not mean to suggest that this tyranny is caused by the experts. It is just in the air. (My concern is quite different from that of William Easterly, who argues that third-world development efforts have been hindered by a "tyranny of experts", promoted by authoritarian governments and by the experts themselves. *The Tyranny of Experts: Economists, Dictators, and the Forgotten Rights of the Poor* (2013).)

It is important for the lay person to realize how much scientists do not know about even their own areas of expertise and to keep clearly in mind the limited application of what they do know to areas outside their own specialties. Engaged citizens need to recognize that scientists' views on most things are pretty much as likely to be wrong as those of anyone else and that scientists' opinions on, for example, matters of public policy should not matter more than the opinions of persons of reasonable intelligence and common sense. At the same time, the scientist must always be open to the possibility that he or she—as well as the authorities that preceded—is wrong.

Finally, I concluded in the course of my own formal education that both understanding and reason are promoted by an awareness of the methodological issues underlying the physical and social sciences. Asking the questions about what we are trying to do, how we are trying to do it and what we can realistically expect to result from those efforts will bring humility and sensitivity to discussions of science and of policy. I think that many heated confrontations, with abundant self-righteousness on both sides, arise out of what are essentially miscommunications—the proverbial "ships passing in the night."

We need to understand why and how it is that many debates over science, morals and public policy do, in fact, fail to engage meaningfully and seem to generate only heat, not enlightenment. Matters of methodology and the theory of knowledge should be an essential part of a liberal education. It is possible that greater familiarity with methodological issues could substantial improve communication or, at least, reduce glaring instances of miscommunication.

In the end, the average engaged citizen does not need to know the actual science, but he or she would greatly benefit from an understanding of the nature of the limits and uncertainties about what we know and about how it is that we know what we think we know.

So, in pursuit of my long-standing interests and utilizing my experience, I have re-examined the writings that piqued my interest in the first place, then researched and read much more and, finally, tried to put the pieces together. I utilized mainly the more accessible writings of working scientists (and of mathematicians and philosophers of science), many of whose works were clearly intended for a lay audience. To be as current as possible, I have attended lectures and presentations by working scientists and tried to follow the new discoveries reported in the press and presented in leading journals. I have also relied on the works of leading science writers for history and context.

Naturally, I take full responsibility for all of the errors in the pages to follow. I regret the mistakes, but I hope that there are also many comments and conclusions with which knowledgeable readers can and will take issue but must finally concede are worthy of consideration. Otherwise, I did not do adequately the task that I set for myself.

In the opening chapters, I discuss the philosophy of science (focusing on theories of knowledge) and the nature and role of mathematics. Then, I discuss the social sciences (focusing on economics), which present the methodological issues rather clearly. Next, I examine Darwinian theory and its relationship to what we call the scientific method. With this background, I go on to look at major issues in chemistry, physics, particle physics and quantum mechanics, cosmology, and neuroscience, trying to reflect the broad span of the sciences. Thereafter, I address what I consider to be major issues related to science and scientific inquiry. Finally, I discuss what consider to be the most important open questions that we face. Throughout, I stress methodological issues and debates.

We shall see that most areas of science began as what we would generally call philosophy, then, with the introduction of empirical analysis and the extensive use of mathematics, became more and more like the modern concept of science. As noted above, that trend led in the last century to the enthusiastic speculation that someday, even someday soon, "science" might encompass everything and explain it all. But, the mysteries still remain.

I do not purport to provide answers, only to ask questions. Admittedly, I allow myself a fair number of comments and criticisms. Perhaps some of these will be useful in and of themselves; perhaps some will lead to observations by others that will be useful. In all events, I do hope that my work will be a constructive contribution to the search for understanding, whether by adding something to the substantive debate or by inducing others to become participants.

I realize that some (perhaps, many) of my readers will have found the description of what I am trying to do here to sound too analytical and dry and, even, boring. If you say to yourself, I am not going to bother reading any further because this book will not move me, I have a request. Please read the postscript before you give up. Then decide whether to resume this journey. If, instead, you are willing to continue reading already, then feel free to ignore the postscript entirely!

Notes

I have included a large number of endnotes (footnotes having proved too difficult for multi-format publication) and have peppered the text with citations. In part, it is simply the lawyer in me. But, I intend the generous citations to be of assistance to the reader interested in sources. Ample citations are also used because I try hard not to assert on my own authority any facts. (The sources for the facts are clearly identified.)

My task was to marshal evidence: to review, select and organize the "facts". Then I make arguments, draw inferences and ask questions. However, I caution the reader that this approach is not a guarantee of objectivity. The marshaling of evidence is an iterative process and inevitably involves personal prejudices and hunches and wishes. (Thus, I find the statements "the facts speak for themselves" and "we go where the facts lead us" to be naïve, if not downright disingenuous.)

My citations are a modified form of legal citation. For example, where a reference is to the same source just cited I use *id.*, rather than the more conventional *ibid*. And, *cf.* essentially means "compare". The other signs are pretty much self-explanatory. The endnotes are something of a copout. When you cannot figure out how to include information in a coherent narrative, drop a footnote (or endnote)! Yet, they are important. I do recommend them to the reader.

After I was diagnosed with ALS in 2015, I panicked and rushed the first edition of *Limits of Science?* to print in mid-2016. Over the subsequent months, however, I fortunately was able to reread what I had done. That led to the second edition, correcting numerous typographical errors and removing superfluous discussions, shortening the book by 50 pages. I have subsequently added numerous new references and discussions, often of materials that appeared since 2015. I have also made some significant changes in the organization.

Introduction

> " ...[T]he more I learn the more I learn
> how little that I know... ."
>
> Don Williams
> "The Answer"

The human endeavors collectively characterized today as science undoubtedly constitute one of the grandest and most distinctive achievements of our species. Indeed, along with artistic expressions in the visual and musical spheres, science may be deemed a defining characteristic of *Homo sapiens*. In the pages that follow, I shall discuss and explore the nature and the limits of our scientific knowledge.

My focus here is on science—that is, on our attempts to ascertain the truth about our physical environment and to understand the phenomena we observe, directly and indirectly, in the world. I have only occasional things to say about technology—by which I mean the skills to predict, control and manipulate the physical world around us to and for our benefit. I certainly do not mean to suggest that technology is of lesser importance; the practical consequences of our technological endeavors on the lives of human beings have been nothing short of stupendous. Indeed, much of the prestige attached to science and scientists in our modern world is undoubtedly a reflection of the acknowledged significance of man's technological achievements. But, science is distinct from technology; it is different in its objectives and its methodologies. Sometimes, technological advances have been a result of scientific discovery; sometimes, technological advances have preceded and even prompted scientific progress, presenting the challenge of understanding and explaining why the technological achievement actually worked.

The point that I seek to establish in the following pages is that despite the astonishing achievements and practical consequences of technology, science has been remarkably unsuccessful. Upon a close examination, the lack of actual knowledge about and real understanding of the deeper fundamentals of reality are among the most striking and surprising characteristics of modern science.

And, this is not bad news. To the contrary, our ignorance is a source of wonder—one consequence of mystery—and the stimulus of our continuing quest to discover. Science is neither dead nor done, and probably never will be. For me, that is a source of motivation and an inspiration for creativity. How much we do not know is good news.

The reader may immediately wonder whether working scientists share the view here expressed. Numerous important scientists have, in fact, at least paid lip-service to the belief that what is yet to be discovered by science is far greater than what we currently know.

Such an observation was made by one of the fathers of modern science, Galileo Galilei, working in the first half of the seventeenth century. It has been recently paraphrased as follows: "[H]owever much we discover about the building-blocks of life, what we do not know will always be infinitely greater than what we know. All our brilliant detective work amounts to no more than a little gloss on the shining masterwork of creation." Harry Eyres, "How to Cultivate a Growth Industry," FT.com, October 15, 2010.

Over 300 years later, Nobel laureate physicist Robert Laughlin wrote:

> *"Even this room is teeming with things we do not understand. Only people whose common sense has been impaired by too much education cannot see it. The idea that the struggle to understand the natural world has come to an end is not only wrong, it is ludicrously wrong. We are surrounded by mysterious physical miracles, and the continuing, unfinished task of science is to unravel them."*

A *Different Universe* (2005), p.218.

Some leading scientists over the last two centuries, however, have expressed quite a different view, declaring that the tasks of science were soon to be largely completed. In fact, such an attitude was arguably pervasive in the late nineteenth century. At the conclusion of that century, astronomer Simon Newcomb, physicist Albert Michelson and physicist William Thomson (Lord Kelvin) and announced to the world that the end of scientific discovery was sufficiently near that most of the remaining tasks would be more akin to cleanup than revolutionary advance. See, e.g., Rupert Sheldrake, *The Science Delusion: Freeing the Spirit of Inquiry* (2012), p.19. And, curiously, that attitude did not fully disappear despite the dramatic scientific upheavals of the first half of the twentieth century. For example, John Horgan, then a senior science writer at *Scientific American*, based on interviews with many leading scientists, advanced this view: "If one believes in science, one must accept the possibility—even the probability—that the great era of scientific discovery is over. ...Further research may yield no more great revelations or revolutions, but only incremental returns." *The End of Science: Facing the Limits of Knowledge in the Twilight of the Scientific Age* (1977).

Indeed, one of the original celebrity-scientists, Carl Sagan, prophesized in 1979 that the process of discovery was almost complete:

> *"This book is written just before—at most, I believe, a few years or a few decades before—the answers to many of these vexing and awesome questions on origins and fates are pried loose from the cosmos. ...Had we been born fifty years later, the answers would, I think, already have been in. ...In all of the four-billion-year history of life on our planet, in all of the four-million-year history of the human family, there is only one generation privileged to live through that unique transitional moment: that generation is ours."*

Broca's Brain: Reflections on the Romance of Science (1979), p.xv.

I am, in fact, of that very generation of which Sagan wrote and, of course, we are now almost four decades into the future that he was imagining. Yet, this book today discusses most of the same "fundamental" and "awesome questions" to which Sagan had referred. In his words:

> *"questions on the origins of consciousness; life on our planet; the beginnings of the Earth; the formation of the Sun; the possibility of intelligent beings somewhere up there in the depths of the sky; as well as, the grandest inquiry of all—on the advent, nature and ultimate destiny of the universe."*

Id., p.xiii.

As you might have guessed, the answers are not yet in. Not even close. You younger readers were not, in fact, born too late to experience the wonder and mystery or to partake of the process of discovery. Indeed, I believe that the same will be true for your children (and theirs).

Even in the late twentieth century, the attitude of many scientists about the state of science seemed to be to acknowledge the proposition that the search for the truth about the physical world may never result in a declaration of complete victory, but to believe that in general we are continually getting closer and closer. That view is one of the things that I hope to put at issue in the reader's mind.

In recent years, however, some leading scientists (whom I reference repeatedly in the coming chapters) have begun to raise the rather different question of whether science is actually capable of ever explaining everything about our world and, more specifically, whether science is capable of generating an underlying and unifying theory that would combine a theory of gravity (such as General Relativity) and quantum mechanics and, then, all the rest of science:

- Cambridge University Research Professor of Mathematical Sciences John D. Barrow has expressed skepticism about the possibility of achieving a theory of everything: "The process of discovery could continue indefinitely either because the complexity of Nature is truly bottomless or because we have chosen a particular way of describing Nature which, while being as accurate as we desire, is none the less at best always but an asymptotic approximation that only an infinite number of refinements could make correspond exactly to reality. More pessimistically, our human frame and its evolutionary past may have placed real limits upon the concepts that we can accommodate." *Theories of Everything: The Quest for Ultimate Explanation* (2005), p.2.
- Cambridge University Professor of Cosmology Martin Rees, Astronomer Royal and Baron Rees of Ludlow, has speculated that: "some aspects of reality—a unified theory of physics, or a full understanding of consciousness—might elude us simply because they're beyond human brains… Perhaps complex aggregations of atoms, whether brains or machines, can never understand everything about themselves." *From Here to Infinity* (2011), pp.110–11.
- Oxford University Professor of Mathematics Roger Penrose, in speculating about an ultimate "theory of everything," has suggested: "There may be a sense in which the three worlds [mental perceptions, the physical world and the world of mathematical forms] are not separate at all, but merely reflect, individually, aspects of a deeper truth about the world as a whole of which we have little conception at the present time." *The Road to Reality: A Complete Guide to the Laws of the Universe* (2005), pp.22–3.
- Former Cambridge University Lucasian Professor of Mathematics Stephen Hawking, in contrast, concluded that sufficient progress had been made toward a theory of everything in the beginning of the twenty-first century that we are almost there. *The Grand Design* (2010), pp.10, *passim* (with Leonard Mlodinow).

It may simply be that there are things very real about the Universe in which we participate that are not susceptible to explanation by science as we know it. If so, there will be some questions that necessarily elude scientific answers. This is an important issue. I shall return to (but not try to answer) the questions it raises many pages from now, in a discussion of the science of human consciousness and the much and long debated issue of free will.

Irrespective of the answer to this ultimate question, it is important to recognize the fundamental limits of our current understanding and knowledge of the world around us. It is not just the matter of what new discoveries are to be made at the fringes of our existing knowledge, but of how much of what we think we already know will turn out to be wrong. Thus, it is essential–and scientific–to remain constantly open-minded and skeptical, to remember, in the words of the song quoted above, that the "fool" is likely to be the "[ ]one who [b]elieves they have the answer".

- "'Great value of a satisfactory philosophy of ignorance,' [twentieth-century physicist Richard] Feynman jotted on a sheet of notepaper one day, 'teach how doubt is not to be feared but welcomed.'" James Gleick, in the Introduction to Richard Feynman, *The Character of Physical Law* (1994) (originally published by The British Broadcasting Corporation in 1965), pp.viii, x.

- "How about a little agnosticism in our scientific assertions–and even, as with Richard Feynman, a little sense of humor so that we can laugh at our errors and move on? We should all remember that Feynman also said, 'If you think that science is certain–well that's just an error on your part.'" Daniel B. Botkin, "Absolute Certainty is Not Scientific," *The Wall Street Journal*, December 2, 2011.

- "Science is a very human form of knowledge. We are always at the brink of the known, we always feel forward for what is to be hoped. Every judgment in science stands on the edge of error, and is personal. Science is a tribute to what we can know even though we are fallible. [But, heed the words of] Oliver Cromwell: **'I beseech you ...think it possible you may be mistaken.'**" Joseph Bronowski, *The Ascent of Man* (1973), p.374 (emphasis added).

Historian Yuval Noah Harari argues that an awareness of ignorance (of the fact that there were significant things about the world yet to be discovered) that emerged in Western Europe around 1500 was the basis of the Scientific Revolution. It was the factor, he says, that transformed Western Europe from a provincial backwater to a position of global pre-eminence by the late eighteenth century. *Sapiens: A Brief History of Humankind* (2015), pp.247-74.

There has even been some relatively recent academic interest expressed in ignorance. The New York Times reports that a professor of surgery at the University of Arizona, Marlys H. Witte, began teaching, despite some resistance, a class entitled "Introduction to Medical and Other Ignorance" in the mid-1980s. James Holmes, "The Case for Teaching Ignorance," *The New York Times*, August 24, 2015. But, Holmes adds, "[t]he study of ignorance...is in its infancy." Columbia University professor of neuroscience Stuart Firestein describes a course that he began teaching in 2006 entitled "Ignorance". He stresses how what we do not know frames the questions asked by science and that the evidence of good answers or discoveries is the new questions that those discoveries reveal. He even praises the process of writing grant applications because that process focuses on the unknown and how one might go about finding answers. *Ignorance: How It Drives Science* (2012), p.59.

Of course, working scientists are focused on and engaged in, and sometimes obsessed by, their efforts to advance the particular fields that they have chosen. That focus is good. What they do represents some of the best of human activity and aspiration—the quest for knowledge and the desire for understanding. The fact that they still fail to "know," and may be doomed never really to "understand," is irrelevant.

As Lord Rees, referencing Ely Cathedral arising out of the fens of East Anglia, said in tribute to those who "built this cathedral—pushing the boundaries of what was possible. Those who conceived it knew that they wouldn't live to see it finished. Their legacy still elevates our spirit, nearly a millennium later." *From Here to Eternity*, p.150. The dignity and grandeur of the quest matter. "The struggle itself toward the heights is enough to fill a man's heart. One must imagine Sisyphus happy." Albert Camus, (1991), p.123. But, we should also be awed by the creativity, the beauty and the elegance of the theoretical edifices that man has been able to erect in his search to understand the world in which we live. And, we should also marvel at, and be inspired by, how much we simply still do not know about nearly everything that matters.

KNOWLEDGE AND UNDERSTANDING

> "The most incomprehensible thing
> about the physical world
> is that it is comprehensible."
>
> Albert Einstein

Initially, in thinking about how we know things, it is probably common to seize upon the distinction between direct sensory perceptions and knowledge from inference or deduction. We feel a solid surface; we see a star; we hear the waves; we taste and smell the ripe fruit. However, as a result of the discoveries of science, we can all appreciate how tenuous such a distinction really is.

Surfaces are not solid, but composed of molecules the volume of which is only a small part particles, part energy and a lot of nothing; some of the stars we see are no longer there and have not existed for thousands or even millions of years. But, that is not all. "Sound" consists only of vibrations or waves transmitted through the air that have significance because of their impact on our eardrums; taste and smell occur because receptors (proteins) connected to neurons bind with molecules in our mouths or floating through the air. And, sight? Photons activate receptors in our eyes that feed the optical nerve. Color is not part of nor a characteristic of an object, but the result of the frequencies of light that the object reflects (not absorbs), emits (*e.g.*, molten metal or a lightbulb) or transmits (*e.g.*, stained glass or a colored lens), which are physical characteristics of the object.

The physical stimuli of our nerves and receptors are transmitted to our brains where they are, most importantly, "interpreted," a process that actually takes time (as discussed in the chapter on Neuroscience). Thus, for example, different frequencies of light waves are perceived by us as different colors. At the same time, with our senses "[w]e barely see just a tiny window of the vast electromagnetic spectrum emitted by things. We do not see the atomic structure of matter, nor the curvature of space. We see a coherent world that we extrapolate from our interaction with the universe ... ." Carlo Rovelli, *The Order of Time*, p.210 (2018).[1]

So, how do we comprehend or "understand" what the brain receives as inputs from the senses? And, moreover, how do we attempt to understand not just that which we "see", but that which we do not "see"?

The Nature of Comprehension

The role of the brain

The first point here is simple, reflected in the statement "We see with our brains, not with our eyes," a quotation attributed to Paul Bach-y-Rita (a scientist and rehabilitation physician who developed a "tactile-vision device" that would enable a blind person to "see" by providing electrical stimulation to the brain from the device rather than from the impaired eye). Norman Doidge, MD, *The Brain that Changes Itself* (2007), p.15. See also, Joseph Mazur, *The Motion Paradox* (2007), p.27 ("the mind, not the eye, is the seeing organ"). In fact, we not only see, but also hear, taste, smell and feel, with our brain. In short, our perceptions of the world are the result of how our brain interprets the inputs received from our senses. It is the brain, in fact, not any sensory organs, that "perceives" the physical world around us.

(To underscore this point, note that plants appear to "respond" to physical stimuli, including sound; but, that does not mean that plants "hear". Indeed, plants receive and react to stimuli through very different physical mechanism than do animals. See, e.g., David Nield, "Plants May Not Have Ears, But They Can 'Hear' Way Better Than We Thought," *sciencealert.com*, 19 January 2019. There is even some evidence that plants under distress produce "sounds", but at a frequency that humans cannot hear. See, e.g., Katherine J. Wu, "Plants May Let Out Ultrasonic Squeals When Stressed: Human ears can't hear them, but other plants or animals might," *Smithsonian.com*, December 9, 2019.)

Comprehensibility depends first upon the existence of certain regularities in the physical world that allow us to generalize. If knowledge of the physical world required specific or unique information about each atom or constituent piece of the world, then the magnitude of the data required would clearly exceed the capacity of the human brain, and we could never know the world. See Carl Sagan, *Broca's Brain*, p.16. However, it just so happens that there are substantial uniformities, regularities and patterns, so that much of the physical world can be accurately described in a relatively small amount of information.

Some of the more interesting of these regularities are often referred to as "Laws of Nature." We shall talk more about that concept in the next chapter. But, I think that something more subtle and intriguing is encompassed in Einstein's reference to comprehensibility. In part, his comment undoubtedly reflected the surprising ability of mathematics to be used to predict physical events, which I discuss below. In part, the comment also likely captured this other concept of insight or understanding—not just that we have tools that predict accurately, but that we have explanatory paradigms such that the relationships appear to make sense, i.e., be comprehensible.

It is not clear that there is any reason why the ability of the human mind to "understand" physical phenomena would necessarily have any direct correlation with the actual physical relationships that exist, except that the species would probably not have survived if the mental constructs (or models) of which we were capable veered too dramatically away from the actual environmental facts. At the same time, by such reasoning, that apparent conformity need only extend to matters relevant to our existence, specifically to our survival and reproduction. In fact, it may be that human cognition is capable of no more than a relatively superficial comprehension of our Universe, so that the deeper truths (those not directly related to our survival, at least not yet) are simply beyond our ability to comprehend. We shall encounter that possibility from time to time below.

THE CONCEPT OF CAUSALITY

It appears that man's concept of comprehensibility depends fundamentally upon a belief in the existence and pervasiveness of causality or causal relationships.[2] There are some interesting questions about what we mean by "causality."

At a minimum, as we humans understand it, if one event is a cause of another, then the causal event must have preceded the caused event in time. A subsequent event cannot be a cause. Someday, of course, we may discover that temporal relationships are more ambiguous or fluid; but, for now, chronology plays a central role in our understanding of reality. Next, we would expect that the caused event, under the existing circumstances, would not have occurred "but for" the causing event, that is, if the causing event had not occurred, then the caused event would not have occurred either. In addition, one might say that A causes B if, all else equal, the occurrence of A will (or probably will/or in a certain percentage of the time will) be followed by the occurrence of B.

The "existing circumstances", or "all else equal", qualification is important. This qualification is referred to as the *ceteris paribus* condition. It has a rather particular meaning. The "nothing else" that cannot change does not really mean nothing else, which could never be satisfied, but only nothing else relevant to the causal relationship under examination. But, even so, that qualification can still be surprisingly limiting. In certain relationships, the other relevant factors may be quite well-defined, few in number and largely controllable. For example, a hot surface burns a finger, assuming that the finger is more or less at room temperature just before the contact and that the contact is complete enough and long enough. But, in other cases, the other relevant factors may be numerous and not well-understood or, even if understood, not readily controlled for or measured. We shall see later some of the issues that such circumstances present.

We achieve some expectations and some sense of causal relationships through observation and induction—the drawing of inferences from a series of individual observations. For example, there has been extensive use of surveys or experimental collections of data in connection with medical science and the study of health issues. These studies are classic opportunities for inductive and probabilistic reasoning. We find a significant statistical correlation between certain environmental or behavioral characteristics and certain health characteristics in a population. That statistical correlation suggests that there may be a more substantive relationship. It may or may not, in fact, be a causal relationship. To address that underlying issue of causality, one needs an explanatory theory as to how and why the two phenomena relate to one another. We discuss theories shortly.

A PRIORI KNOWLEDGE

But, let's step back for a moment. From where do we derive the sense or awareness of causality? Perhaps, the belief in causality is an empirical derivation, that is, we observe correlations and conclude, from the empirical experience, that there is causation.

Hadley Arkes, a professor of political science, argues that "[t]he notion of causation must be in us, so to speak, as a precondition of experience; it is one of those understandings, built into our natures, which makes it possible for us to 'have experiences.'" *First Things: An Inquiry into the First Principles of Morals and Justice* (1986), p.67. Arkes attributes this conception of causation to Immanuel Kant and Kant's argument that there are necessary, or *a priori*, truths that exist independent of experience.

He paraphrases and quotes Kant as follows:

> "*the nature of the rule becomes more evident to us as we see it manifested in any experience... But unless we understand already the immanent necessity of a rule—unless we presuppose that events are indeed arranged in a temporal order, with certain events A preceding certain events B and determining the position of B in the sequence—unless this elementary point is absorbed as a necessary truth, it becomes impossible to impute any meaning to the succession of perceptions that flash before us. As Kant remarked, 'it is under this supposition only...that an experience of anything that happens becomes possible.' ... [T]he awareness of the rule—the awareness of the very notion of causation—'was nevertheless the foundation of all experience, and consequently preceded it a priori.'"*

Id., p.64, quoting from Kant's *Critique of Pure Reason*.

In short, humans have built-in an understanding or, perhaps, an assumption of causality that leads us to look for and enables us to see causal relationships. In fact, it would be asserted, if we did not have an inherent understanding of causality, we would be unable to detect it.

There are other examples of innate or inherent mental constructions or structures, and the same caveats would appear to apply. As an example, the conception of space and time would seem to precede *a priori* actual experience. Indeed, one could argue that we could not understand our physical experiences without such concepts being given in advance of and as a basis for perceiving those experiences. See Arkes, *First Things*, pp.56-57.[3]

Experiments have demonstrated that our visual perception is facilitated by certain innate "prejudices." For example, the laws of perspective, which were only gradually discovered by Renaissance artists, appear to be embedded in our minds. The point is demonstrated by adjacent images of the Leaning Tower of Pisa which the mind persistently perceives as diverging, with one tower appearing to lean noticeably more even though the two are identical. The reason is that the mind "knows" that two neighboring towers will appear to lean toward each other as they rise from the ground in a single scene (the law of perspective), so if the towers do not in the photo lean toward each other as they rise, the mind concludes—and the eyes see—that one tower tilts more than the other. Stephen L. Macknik and Susana Martinez-Conde, *Sleights of Mind: What the Neuroscience of Magic Reveals About Our Everyday Deceptions* (2010), as reported in *The Wall Street Journal*, November 20-21, 2010, p.C12.

Recent research seems to confirm the suspicions that the human mind is simply incapable of perceiving things objectively—that is, independently of human mental "biases"—or with complete accuracy.

A somewhat different but related point is that the mind necessarily simplifies, restricts and orders the information received—if such restrictions and simplifications did not occur, a person would be reduced to helpless confusion by the unmanageable and overwhelming volume of data being received from the senses.[4]

However, the fact that humans inherently perceive causality (or space and time or perspective or right and wrong) is not itself evidence for the independent truth or accuracy of such perceptions. Arguably, such hard-wiring in the human brain results in the detection of such relationships even where none exists in fact, exhibiting a kind of irrationality. See, e.g., *House MD*, "Small Sacrifices," Season 7, Episode 8, November 2010 (in a typically exasperated reaction to religious belief, when he is treating a man who believes that his promise to reenact the crucifixion explains his daughter's miraculous recovery from cancer, House mutters: "Causal determinism: hard-wired to need to find answers," a legacy of the caveman for whom the belief in causal relationships had a survival benefit, but which House clearly considered to be irrational where religion is concerned, despite his own obsessive drive to find causation in sources that he considers legitimate).

Undoubtedly, the presence of such hard-wiring would be attributed by many to evolution: for example, the ability to appreciate and perceive causation, spatial relationships and the passage of time all would tend to improve the rates survival and reproduction. But, if it is concluded that evolution is the source of such mental constructs, would that be evidence that the constructs reflect reality?

Importantly, no.

That is the case as a matter of logic. And, as we shall see below, twentieth century developments in physics certainly pose challenges to our preconceptions and perceptions of reality, suggesting that the truths about the Universe (especially at very small and very large scales) may be quite different from what our common sense (and senses) tell us about the world on a human scale and from what our evolutionary heritage provides us.

> "Until the advent of modern physics it was generally thought that all knowledge of the world could be obtained through direct observation, that things are what they seem, as perceived through our senses. But the spectacular success of modern physics ... has shown that that is not the case."

Stephen Hawking and Leonard Mlodinow, *The Grand Design* (2010), p.6.

This issue reappears explicitly and implicitly in the following chapters.

Comprehensibility

To return to the subject of comprehensibility or understanding, an important component of comprehension to modern man is the existence of some theory or model that generates an explanation of the causal relationship, i.e., why it is that "if A, then B." A theory will postulate what things are causally related and why (or how).

An attractive theory will generally accomplish two things: it will be consistent with a body of observations or data and it will provide the person contemplating it with a sense of explanation, a feeling of insight or understanding, not held before the theory was propounded. This second achievement can be described as an experience akin to putting on a pair of prescription spectacles—one can see things more clearly or, even, see things that one previously could not.[5]

As we shall see, the role of the concept of understanding (or explanation) in science is subject to some dispute. Various philosophers of science and many working scientists have argued that the accuracy of the predictions that can be made is the real scientific test of a theory. Nonetheless, there has been a persistent feeling among many others that the essence or heart of science involves something more.

Michael Polanyi, in an influential paper from 1966, stressed the central importance of what he called "vision," both in the process of discovery and in the subsequent acceptance of new theories. "The vision which guided the scientist to success lives on in his discovery and is shared by those who recognize it. It is reflected in the confidence they place in the reality of that which has been discovered and in the way in which they sense the depth and fruitfulness of a discovery. Any student of science will understand—must understand—what I mean by these words. But their teachers in philosophy are likely to raise their eyebrows at such a vague emotional description of scientific discovery." "The Creative Imagination," in Michael Kransz (ed.), *The Idea of Creativity* (2009), pp.147-8 (reprinted from *Chemical Engineering News* 44 (1966) p.85).

Polanyi explained that Copernicus had a vision, subsequently shared by Kepler and Galileo and others, that persisted despite the practical problems of connecting the new theory to the substantial data that had been compiled, problems that required Kepler, Galileo and then Newton to resolve. He asserted that they did not believe in the new theory because the theory solved the problems affecting their ability to predict accurately the planetary movements. Instead, the scientists managed to solve the problems because they already believed the theory.

Polanyi also asserted that this vision is the scientists' glimpse of reality: they "believe that it refers to no chance configuration of things, but to a persistent connection of certain features, a connection which, being real, will yet manifest itself in numberless ways, inexhaustibly. It is to believe that it is there, existing independently of us, and that for that reason its consequences can never be fully predicted." *Id.*, p.149.

The Role of Theories and Models

Accepted wisdom is that we understand things about the world only through the use of models and theories. In other words, sensory perceptions or other "information," apart from that which generates an instinctive response, can be processed only through theories we hold about how the world around us functions. We are not objective, neutral observers; instead, we absorb information through the organizing filters of preconceptions about causality and other relationships. Obviously, we can learn and our preconceptions—the models and theories that we utilize—can and do change over time. But, we could not function in a human capacity without such models and theories.

The use of theories would appear to be a significant distinguishing feature of man, setting us apart from the rest of the animal kingdom. All other animals have reactions to stimuli that are apparently genetically determined (and transmitted) and many animals have some ability to "learn" about relationships (regularities or patterns) between various sensory perceptions and likely benefits or detriments to the organism associated with those perceptions (learned either through direct experience or observations of the experiences of other individuals). Relatively recent studies have suggested that some animals are even able to recognize the implications of certain observed phenomena and engage in some form of reasoning. This work is discussed in the chapter on Neuroscience.

With these caveats, it seems fair to conclude that humans have a unique capability of "deducing" broader conclusions or chains of implications from particular sensory perceptions. And, humans do so through the power of explanatory theories or models. (To the extent that the theories or models are more than mere analogies or metaphors based upon actual experience, we must wonder how they are generated and then used by the mind to reason.) But, the concept and role of "theories" are much broader than the mechanisms by which we make inferences.

Perhaps, the most important point to be made about theories is their ubiquitousness. Theories shape the questions we ask and the observations we make, as well as the conclusions we draw. Indeed, it is the creation and then communication of theories that enable humans to attempt to make sense of and adapt to the world around them. All of our observations about the world are based upon theories; we do not receive any meaningful information directly from our senses. All knowledge is a result of interpretation. "[W]e perceive nothing as what it really is. It is all theoretical interpretation: conjecture." John D. Barrow, *Theories of Everything: The Quest for Ultimate Explanation* (1991), p.10. Theories shape the questions we ask; theories enable the observations and measurements we make; theories structure the explanations we derive to make sense of those observations.

One of the clearest illustrations of the nature and role of theories is the basic human enterprise of categorization. We have a long history of organizing the world we observe into categories. Our choices of categories reflect implicit and explicit assumptions about what things "matter," what similarities are relevant. These assumptions are the results of "theories." We have grouped all living things into kingdoms and then into phyla or divisions. We group animals by species. The groupings are based upon a variety of assumptions about what things are similar. For example, we identify animals with backbones as a meaningful group ("vertebrates"). We could have decided that relevant groupings were by color (everything that is red), by weight or by height. Indeed, for other purposes, color, weight and height are relevant criteria for groupings.

A similar example is our choice of scale for the measurement of temperatures.[6] The Celsius scale is based upon the temperatures at which water freezes (defined as $0°$ C) and boils (defined as $100°$ C), with the interval being divided into 100 "degrees." The Fahrenheit scale takes the boiling point of salt water (defined as $0°$ F) and the temperature of the human body (defined as about $100°$ F) and divides the interval into 100 "degrees." See, e.g., Tim James, *Elemental: How the Periodic Table Can Now Explain (Nearly) Everything* (2019), p.39. So each degree Centigrade is equal to 1.8 degrees Fahrenheit. The particular relevance of these ranges is seen in the search for "Goldilocks" plants, discussed in the last chapter.

All of these activities are based upon models or theories about the world, but they are not predictive, do not test hypotheses and are not constructed of inferences.

As recently expressed by physicist David Deutsch:

> "*Observations are theory-laden. Given an experimental oddity, we have no way of predicting whether it will eventually be explained merely by correcting a minor parochial assumption or by revolutionizing entire sciences. We can know that only after we have seen it in the light of a new explanation. In the meantime we have no option but to see the world through our best existing explanations—which include our existing misconceptions. And that biases our intuition.*"

The Beginning of Infinity (2011), p.119.

Deutsch explains that "[o]ne cannot make even the simplest prediction without invoking quite a sophisticated explanatory framework." *Id.*, p.15. Testing and measurement techniques themselves depend upon theories of testing and measurement. *Id.*, p.317. "The very idea that an experience has been repeated is not itself a sensory experience, but a theory" because "we are tacitly relying on explanatory theories to tell us which combinations of variables in our experience we should interpret as being 'repeated' phenomena in the underlying reality, and which are local or irrelevant." *Id.*, p.7.

One consequence of the pervasive and decisive nature of theories can even be the inability to communicate. In a lecture given in 1960, philosopher of science Stephen Toulmin noted that "Men who accept different ideals and paradigms have no common theoretical terms in which to discuss theory problems fruitfully. They will not even have the same problems: events which are 'phenomena' in one man's eyes will be passed over by the other as 'perfectly natural.'" *Foresight and Understanding* (1961), p.57. Moreover, we all face a challenge in our efforts to understand science, because the theories that constitute an integral part of that science shape not just our understanding of the science, but also the vocabulary with which we discuss it. See Deutsch, *The Beginning of Infinity*, p.199.

These issues have been extensively addressed by philosophers of science. For example, Thomas Kuhn has argued that new paradigms in science are not directly comparable to prior theories because both the concepts and the vocabulary are different. *The Structure of Scientific Revolutions* (1962), pp.147-50, 170-73. And, Paul Feyerabend has asserted that this conceptual incompatibility of successive theories creates a conservative bias favoring existing theories, partly because the development of a new theory generally requires a new way of thinking and of seeing. "Explanation, Reduction and Empiricism," in H. Feigl and G. Maxwell (eds.), *Scientific Explanation, Space and Time* (1962) at pp.28-9.[7]

The second most important point to be made about theories is that they are always likely to be wrong. As stated by Cox and Forshaw in 2009: "In science, there are no universal truths, just views of the world that have yet to be shown to be false." Brian Cox and Jeff Forshaw, *Why Does $E=mc^2$? (And why should we care?)*, p.xi. Or, as Deutsch put it:

> "[T]he nature of science would be better understood if we called theories '**misconceptions**' from the outset, instead of only after we have discovered their successors. Thus we could say that Einstein's Misconception of Gravity was an improvement on Newton's Misconception, which was an improvement on Kepler's. The neo-Darwinian Misconception of Evolution is an improvement on Darwin's Misconception, and his on Lamarck's. If people thought of it like that, perhaps no one would need to be reminded that science claims neither infallibility nor finality."

The Beginning of Infinity, p.446 (emphasis added).

In all events, the source of man's creative actions lies in the construction of theories, which start initially with speculation and the formulation of hypotheticals that are then "tested" against observations of phenomena in the external world. The remarkable history of the development of human culture and learning reflects the power of the theory-building mind to test, improve, reject and then create ever-more extensive theories to explain the human situation and the external world.

DEDUCTIVE THEORIES

"Scientific" theories generally fall into one of four major and ostensibly different patterns of explanation: (i) the deductive model in which the event to be explained is a logically (mathematically) necessary consequence of the explanatory premises; (ii) probabilistic relationships, which often resemble deductive models but include statistical premises; (iii) teleological or functional explanations, which identify the functions or role played by various agents in events and (iv) "genetic" explanations in which phenomena emerge from the events preceding them. See Ernest Nagel, *The Structure of Science* (1961), pp.20-26.

Most of our discussion in what follows will address the first pattern, with references to the. second. They dominant the "hard" sciences. The fourth pattern of explanation characterizes Darwinism. As for the status of the third, teleology, one is reminded of the once popular quotation: "Teleology is a mistress without whom no biologist can live, but with whom none wishes to be seen in public."[8] It lurks behind many scientific efforts at explanation but must be denounced for fear of giving comfort to religion.

It is commonly accepted that the most powerful and versatile of theories are deductive in structure. They contain a set of assumptions or premises (or axioms) followed by a set of rules for derivation or logical manipulation (sometimes called rules of transformation), so that one can draw conclusions of the form if A, then B, but often in far more complex structures with many variables.

Such deductive theories can allow a variety of more remote identification (or prediction) of the consequences of specified starting conditions. However, nothing "new" is introduced—the subsequent implications are all contained in the initial axioms or assumptions, given the rules governing the manipulation of the axioms and the derivation of additional propositions. Deductive theories of this type generally lend themselves to formalization, specifically often to mathematical representation. Indeed, such theories can be viewed, and could have started life, as purely logical constructs. In fact, a good deductive theory will necessarily be a good logical construct. The question that arises is whether or how a deductive system bears any relationship with the external, real world.

Of interest here are theories that tie to or incorporate real, observable data or events—theories about the world in which we live. Such theories must contain what are called "contingent propositions"—statements about the world that are subject to verification or observation or, in other words, statements that can be "true" or "false". See Richard Bevan Braithwaite, *Scientific Explanation* (1968), p.24. Thus, a theory would include contingent propositions, and an observer could attempt to ascertain whether those contingent propositions are true in the circumstances at hand. If so, then one could presumably make predictions (deductive inferences) that also included contingent propositions. Such inferences would be logically necessary within the theory (or deductive system). But, such predictions could also be "tested" against real world data or used (relied upon) in real world applications. (Generally, an "hypothesis" is contingent, subject to testing. One asks, are the conclusions from it, derived through a model, meaningful, reflect reality or otherwise make sense? In contrast, an assumption is a definition or a contingent statement generally believed likely to be true.)

In all events, it is crucial to be able to recognize when terms used in a theory merely appear to reference actual things (or are carelessly assumed to reference actual things) and when the theories contain real contingent propositions. A formal deductive system can be expressed in symbols. The symbols could be given names that are used in normal conversation to refer to physical objects.[10] However, to tie the deductive system to the physical world, one needs also to introduce into the system logical propositions utilizing the "things" to which the names refer and that contain some factual (non-circular and not purely definitional) relationships that are subject to empirical testing—in other words, assertions that have some real content concerning the physical world.

There are terms that have meaning only in the context of the theory or the logical structure. An illustration set out by philosopher Gilbert Ryle is from the vocabulary of games. The word "trump" has no meaning out of the context of bridge, and one cannot explain its meaning without some explanation of the rules of that game. *Dilemmas* (1954), p.33. Another matter that deserves attention, to avoid misunderstandings, is the use in theories of terms that have an ordinary usage but technical meanings in the context of the theory, such as the "mass" or the "temperature" of an object. The technical meanings are defined by the theory. As Ryle explained, "technical terms ... are theory-laden, laden, that is, not just with theoretical luggage of some sort or other but with the luggage of...theory. Their meanings change with changes in the theory. Knowing their meanings requires some grasp of the theory." *Id.*, p.90.

At some level, all words are "theory-laden," that is based upon or reflecting models of the physical world that people use to comprehend their environment. Obviously, it is important at least to identify and communicate the particular models involved in statements made, in order to avoid misunderstandings and miscommunications.

Modern scientific theories regularly make use of terms or concepts that are not subject to direct sensory observation. A frequently cited example is the "electron." It is an assumed or hypothesized "object" defined by several characteristics, including a negative electrical charge. The catalogue of characteristics defines the term functionally and enables it to be utilized in theories of atomic and molecular structures. Obviously, the related theories have been very successful and powerful. But, does an electron actually "exist"? See, *e.g.*, Brian Greene, *The Elegant Universe* (2005), pp.108–16. (This issue will be discussed further in the chapter on Particle Physics and Quantum Mechanics.)

An example: The "gene"

Perhaps a more vivid and suggestive example is the concept of the "gene." In the 1860s, the Moravian monk Gregor Mendel, in his famous experiments with hybrid peas, crossing round with wrinkly and yellow with green peas, discovered patterns of heredity. Subsequently, the experimental data were expressed in a theory of "genes" expressing recessive and dominant traits, with which we are all at least passingly familiar.[11] The second half of the twentieth century brought the emergence of molecular biology, with the "discovery" of DNA in 1953, and then RNA and so on.

One might have expected that these breakthroughs would lead to a physical identification of the "gene" within the cell structure, that is, that we would be able to identify objects (by chemical composition and structural location) that constitute the gene that is the transmitter of heritable traits. Instead, biologists seem to have discovered that the gene, as previously conceived, does not exist. Instead, heritable traits are transmitted by a variety of types of quite complex processes and elements contained within the cell. See, *e.g.*, Rheinberger, Müller-Wille and Meunier, "Gene," *The Stanford Encyclopedia of Philosophy*, Spring 2015.

Interestingly, however, the Mendelian genetic theory is still a good—indeed, perhaps the most effective—method of predicting the inheritance of traits. In addition, we have been able to ascertain that the human genome consists of only about 20,000 genes, not much more than many other life forms of complexity greater than the bacteria (and far fewer than had been expected when the human genome project began). We can also identify numerous genes that a variety of quite different animals have in common. See Frank Ryan, *Virolution* (2009), p.2.

Thus, the theory is eminently useful. It appears that the concept of a gene reflects various bundles of processes and cell constituents that as a bundle function just as the supposed gene. In addition, different traits are transmitted by considerably different bundles—some much more complicated than others.

So, is the theory based upon the concept of genes true or false?

One answer is to say that the Mendelian theory reflects a higher level or macro view of the phenomena while molecular biology is dealing with a more granular or micro level of activity. The assumption would be that when the theories have been perfected, they would both be capable of generating the same predictions or inferences at the level of the organism, but that the micro level theories would be able to explain many details that the macro theory would not even perceive as existing. It would then be the case that the macro theory (based on genes) would be reducible to the molecular theories. Even if that reduction were to be achieved, the question of whether the genetic theory is true would still present itself.

One might say that the question is not important or that it is not really meaningful. The macro theory "explains" what happens (that is, what occurs) and enables action based upon reliable predictions, even if it does not "describe" the details of how, or even why, it occurs.

MORE ON DEDUCTION

The common view has been that a theory of scope can achieve something that mere inductive inferences cannot.[11] Such a theory can enable one with logical certainty to make predictions about or to foresee consequences of certain actions beyond the immediate relationship that gives rise to the inductive inference of causality. That ability would seem to offer the promise of a statement of causality that has real content, even if certain aspects of the theory may more appropriately be called fictions (like the "gene"). For example, instead of the relative primitive understanding that fire burns (or is hot), one could achieve a theory of what causes combustion, what the process of fire actually is, what the impact on living cells caused by fire is and why we feel pain following contact with fire.

Of course, one would not need or expect to generate theories sufficient to explain the entire chain of events at one time. Pieces of the chain can be examined and theories propounded for each piece. In part, the success of each such theory will ultimately be judged by whether that theory fits into the theories that are tentatively accepted for the other pieces. If it all works when the pieces are put together, one feels more confident that each piece is right.

Great excitement is generated in the broader scientific community when a researcher is able to identify an empirically testable proposition and then find that the prediction of the theory is "confirmed" by the empirical test. The excitement arises even when the testable hypothesis is in no sense central or at the core of the theory; it can be a remote and very technical implication of the theoretical construct. The scientific community views this result as a partial confirmation that the theory is true; but, what does true mean in this context?

Do we expect that the theory might, can or even should correspond to reality in the sense of resembling or modeling the details of reality? Can say that a theory can ever be expected to be true or to reflect reality. Cf. Paul Feyerabend, *The Tyranny of Science* (2011) (originall published in Italian in 1996), p.66. Many scientific theories consist of a substantial edifice of theoretical (or mathematical) propositions that touch the physical world only occasionally and may, in some of those places, do so only incidentally. As we shall see later, much of modern physics consists of abstract mathematical models, many parts of which have no currently known or even imagined counterparts in the physical world.

The "Problem" of Induction

Philosophers of science have devoted considerable attention to what has been called the "problem of induction." The problem is interesting, because the process which philosophers call induction reflects the principle means by which man has made sense of the external world.

The process of induction essentially assumes that apparent patterns that are observed in the course of natural events are likely to reoccur. At a minimum, induction assumes that there are patterns or regularities in the physical world. The issue arises from the difficulties in articulating a reason why it is logical to conclude that event B will occur subsequent to an occurrence of event A, based upon the fact that on occasions in the past, B has been observed to follow A? Or, in another formulation, how we can justify inferring from a particular event or series of events that occur, a rule that applies generally in other cases or to future events?

The question for philosophers is whether there is a logical or philosophical "principle of induction." To take the classic example, just because the Sun has come up every morning for all of recorded history (as far as we know), is there a sound logical or philosophical basis (or argument) to believe that it will come up tomorrow morning? Clearly, the fact of regular occurrences in the past in and of itself does not prove that the same thing will occur in the future; indeed, one could speculate that in some contexts it could be more likely that something different will happen, at least, eventually.

I can grant the philosopher these points, on their own limited terms; but, I would not be willing to conclude that we are not justified in believing that the Sun will come up tomorrow. The logical fact that historical patterns do not guarantee future repetition does not make the expectation of such repetition unreasonable (or, unscientific).

INFERENCES AND PREDICTIONS

Philosophers have struggled with the question of why such an assumption is philosophically justifiable (as opposed to practically or pragmatically justifiable: *e.g.*, the fact that the assumption has worked in the past has proven to be a good practical reason to assume that it will work again). The question is underscored, or perhaps generated, by the comparison of induction with deductive reasoning, in which the conclusions are compelled as a matter of logical necessity given the premises and rules of deduction. The simple philosophical argument is that if induction could be reformulated so that you could deduce the conclusions being drawn, then it would no longer have the characteristics of induction (*i.e.*, it would have become deduction), but if the demonstration of validity of inductive reasoning is through some other level of inductive reasoning, then the proof is necessarily circular—you cannot prove that induction is valid by means of induction.

I have two concerns about this argument.

First, it is constructed on the assumption that there are two and only two types of inference: deductive and inductive. If that assumption is correct, then the dilemma inevitably arises—if you can justify induction by deduction, then it is no longer induction; if you can only justify it by induction, then your proof is necessarily circular. So, one must necessarily look beyond deduction and induction to find a principle of induction. (Interestingly, the same point can be made about deduction. If deduction can be justified only by deductive inference, then deduction is also circular. Of course, as previously discussed, we already recognized that deduction essentially is circular, that is, valid by definition.)

Second, the "problem" arises because of the application of the standards of deduction, that is, of logical necessity—the way in which a conclusion is "guaranteed" or mandated in a deductive system. Induction cannot and should not be judged by those standards. In fact, it is interesting to ask what it means to say that some conclusion from actual observations could ever be "guaranteed".

As we discussed above, there are serious limitations in how deductive reasoning relates to real world events. The reason that deduction is so compelling is that the conclusions are necessarily incorporated in the premises. As a result, deductive reasoning cannot lead to conclusions that contain anything "new." Now, such reasoning may (and often does) lead to conclusions that had not previously been noticed or "seen." They may not have been obvious, but they are not "new." (This subject is discussed further in the next chapter on Mathematics.) Deduction, unlike induction, is not ampliative—it does not and cannot lead to something new and additional.

In contrast, the power of induction is that it necessarily leads to conclusions or predictions that extend beyond the known or existing premises and observations. That is why induction is not, and inevitably cannot be, as logically "certain" as deduction. That is also why induction is far more interesting and relevant to our day-to-day lives. If one rejects the idea that the external world is made up of elements that have direct correspondence to innate mental images (that real world relationships are accurately reflected by human mental processes), so that man can know the world through *a priori* reasoning; then, for any understanding of the external world, there must be some means of access by the mind to the external phenomena. Such access is presumably gained through the senses.

One can speculate that the initial tool that the mind uses to gain understanding from the sensory perceptions is induction, based upon the identification of apparent patterns. It also seems that mankind made an important leap when man began to formulate models and theories that provided explanations for the predictions that were suggested by induction. One could also conceive of induction as generating "rules of thumb" for selecting likely appropriate responses to external stimuli, and which rules often assisted in the preservation of the individuals and the species.

When theory building began, induction is likely to have been an important source of conjectures, as well as a means of identifying problems, which new deductive theories could be propounded to solve.

DAVID HUME

The original identification of the problem of induction has traditionally been attributed to David Hume. See, e.g., Ian Hacking, The Emergence of Probability (1975), p.176. Hume's A Treatise on Human Nature was published in 1739. A more focused (or, at least, more concise) statement of the arguments relevant here was set out in An Enquiry concerning Human Understanding, An Enquiry concerning Human Understanding and Other Writings (2007) (edited by Stephen Buckle), p.6. A more focused (or, at least, more concise) statement of the arguments relevant here was set out in An Enquiry concerning Human Understanding, published in 1748 (originally entitled Philosophical Essays concerning Human Understanding).

Hume set the stage for his argument as follows: "Man is a reasonable being; and as such, receives from science his proper food and nourishment: But so narrow are the bounds of human understanding, that little satisfaction can be hoped for in this particular, either from the extent or security of his acquisitions." Hume, An Enquiry concerning Human Understanding and Other Writings (2007) (edited by Stephen Buckle), p.6.[12] He furthered stressed the inherent limits on man's ability to know and understand the real world. Then, he presented the following thought: "It may, therefore, be subject worthy of curiosity, to inquire what is the nature of the evidence which assures us of any real existence and matter of fact, beyond the present testimony of our senses, or the records of our memory." Id., p.29.

Hume effectively rejected the applicability of deduction to the external world.[13] He expressly distinguished between "relations of ideas" and "matters of fact", the former including mathematics and logic, subject to intuition and deductive reasoning with the consequent necessity of the conclusions (because contrary conclusions would be logically contradictory and, therefore, inconceivable).

In contrast,

> "[m]atters of fact...are not ascertained in the same manner [through intuition and deductive reasoning]; nor is our evidence of their truth, however great, of a like nature.... The contrary of every matter of fact is still possible; because it never implies a [logical] contradiction.... Were it demonstratively false, it would imply a contradiction, and could never be distinctly conceived by the mind."

Id., pp.28-29; see also, id., p.36.

Hume argued that everything that man knew was a result of sensory perceptions combined with the search for patterns or "associations" of events and facts. The most important (or only) such association for matters of fact was "cause and effect." Id., pp.20-21. ("All reasonings concerning matters of fact seem to be founded on the relation of *cause* and *effect*. By means of that relation alone we can go beyond the evidence of our memory and senses." Id., p.29). In other words, man continually scans the available sensory inputs and attempts to identify relationships of cause and effect.

Such activity was essential to man's survival, but was also the heart of scientific knowledge. Hume believed that induction was inherent and essential—there is no alternative available. He recognized the circularity involved in efforts to justify inductive reasoning. ("It is impossible, therefore, that any arguments from experience can prove this resemblance of the past to the future, since all these arguments are founded on the supposition of that resemblance." Id., p.38.)

Hume also highlighted the distinction between practical conduct and philosophical inquiry. "My practice, you say, refutes my doubts. But you mistake the purport of my question. As an agent, I am quite satisfied in the point; but as a philosopher, who has some share of curiosity, I will not say skepticism, I want to learn the foundation of this inference."[14] Subsequent philosophical inquiries have focused on the alleged logical shortcomings of induction or on solving the "problem" of induction.

There is something more to say about the alleged problem of induction.

Probabilities

It is obvious that the fact two events happen in sequence once does not mean that they will or are even likely to do so again. Even the fact that the two events happen in sequence twice (or one hundred times) does not guarantee that they will do so again. Even if there has never been a recorded instance in which A was not followed by B, one cannot say that it is inevitable that the next time A occurs, B will follow. We might say that we can at least assert that it is likely.[15]

Philosophers supposedly have shown that the characterization of induction as a matter of probability does not solve the "problem". But, regardless, I do not think that our understanding of induction is simply probabilistic.

More repeated observations of a pattern can give rise to a belief that it is possible (or more probable) that there is a causal relationship between them. But, what gives rise to a reasonable and justifiable belief that B will follow A is a theory of the relationship between them that explains why B will occur when A has occurred.[16] This element is, again, the concept of causation and the belief in the necessity of causal relationships. Man's apparently inherent belief in the existence of causation was discussed above. The nature of the causal relationship is set out in the theory. The theory explains why the observed pattern is expected to repeat itself. That theory will also often identify factors that could cause the pattern not to be repeated in a given instance.

Commonly, a theory will contain (explicitly or implicitly) the *ceteris paribus* condition, saying that if A occurs then B will necessarily follow if, but only if, all relevant things remain the same. The use of "an unspecifiable *ceteris paribus* (all else being equal) clause" is a technique used in efforts to "connect" deductive models to the real world. Peter Lipton, *Inference to the Best Explanation* (Routledge, 2004) (Second Edition), p.64. But, one would hope that the real life variations in the initial conditions that can be identified when the predictions do not materialize will make sense in terms of the relationship being presented by the theory. If so, then the theory provides a basis for understanding the events that occur, including the failure of a predicted B to follow A.

EXAMPLES: BELLS, BEER AND RAVENS

Suppose that over three years as an undergraduate, one observed that whenever the bells of St. John's College chapel rang, the bells of Trinity College chapel rang almost immediately thereafter. Inductively, one might infer (correctly) that there was a relationship between the two events. However, one might also (incorrectly) infer that the ringing of the John's bells "caused" the ringing of the Trinity bells. In fact, presumably, the cause of the regular occurrence was that both bells were set to chime on the hour but the Trinity clock was slightly slower than the John's clock. The causal relationship is thereby described by a more complex and extended theory.

The example that I recall being used in my statistics class 45 years ago was the close negative correlation between infant mortality and beer consumption (or, perhaps, it was a positive correlation between life expectancy and beer consumption). The misleading inference that could be drawn was a result of the selection of the data used. It reflected experiences across several countries in different stages of economic development. What was apparently behind the results was that economic development caused increased per capita beer consumption and also caused greater life expectancies (or lower infant mortality). There was no causal relationship between the two variables that were being correlated, only correlations of each with something else. But, those other correlations may well have reflected causal relationships.

Finally, there are many examples of empirical observations that would satisfy the description of "successful predictions" that could not meaningfully be considered as confirmation of the theory. One example is the "raven problem." The proposition that "all ravens are black" logically entails the proposition that "All non-black objects are not ravens." Yet, the observation of an object of a color other than black that is not a raven (*e.g.*, a white swan) can hardly be deemed inductive support for the initial proposition about ravens. See, *e.g.*, Peter Lipton, *Inference to the Best Explanation* (2004), pp.14–6.

To the non-philosopher, these problems raised by these examples may seem rather contrived. It is tempting simply to say "I know valid inductive support when I see it" and, I think, one would fare quite adequately. But, the philosophers want to define precisely both how one would "know it" and how one would "see it".

Some Observations

In short, while we might support efforts better to understand and to systematize how we make valid inductive inferences, non-philosophers may have trouble taking too seriously Hume's circularity argument with respect to the justification for induction.

Lipton stated the issue as follows: "In short, induction will work because it has worked. This seems the only justification our inductive ways could ever have or require. Hume's disturbing observation was that this justification appears circular, no better than trying to convince someone that you are honest by saying that you are." *Inference to the Best Explanation*, p.10. But, is that really a fair analogy? Is it not a more apt comparison to look at someone trying to convince another that he is honest and will continue to be so by pointing to a track record of honest behavior?

As discussed below, Sir Karl Popper criticized the notion that successful predictions prove the correctness of a theory. He allowed that success constituted corroboration (a kind of confirmation), but he emphasized that there could be an abundance of rather trivial corroborations that should give rise to little confidence in the theory and that, more importantly, no amount of corroboration could ever conclusively establish that the theory was an accurate reflection of reality.

Thus, he asserted that the real essence of the scientific method was (or should be) the seeking of the refutation of theories (disconfirmation) through falsification. Then, it is the success of a theory relative to its competitors in that process of attempted refutation that constitutes a basis for belief in the theory. One consequence of this reformulation of the process of scientific inquiry was believed by Popper to be the elimination of the "problem" of induction by the elimination of inductive inference in the process of testing theories. Karl Popper, *The Logic of Scientific Inquiry* (1959), pp.27-34. The questions become how to test or assess theories and how to choose among them. Thus, we are simply back to the discussion of theories and how they can be tested that we were discussing above. The problem of induction is the problem of scientific inquiry and understanding.

Hume argued that simple induction was the only way in which man interpreted and understood the world. I am suggesting a somewhat different view.

I can assume that early man's efforts to comprehend the world and survive were based upon simple induction–the observation of apparent patterns and the conclusion that they were likely to be repeated. That conclusion is reasonable when choice or action is required. For example, take the mouse that twice discovers food in a particular place. Next time the mouse is hungry should it go back to that place or try someplace new? Is it not "rational" to try first the place of previous success? Of course, if that place is not productive, then the mouse presumably would try elsewhere, perhaps at random.

But, mankind has been functioning in a much more complicated fashion for millennia. Apparent patterns are observed. Efforts to construct theories of cause and effect are made. Those theories then are tested and, thereafter, eventually revised or replaced. Certainly, scientists today do not simply look for patterns and then make inferences; although, that is undoubtedly part of what they do. I suspect that many discoveries are the result of manipulating existing theories and playing with new ones, or simply indulging guesses and imaginative speculation.

Inductive reasoning has and does play a role in the generation of discoveries, but it cannot be the basis of scientific explanation or understanding. It is a tool for making educated guesses–sometimes for pragmatic reasons (like in the search for food), sometimes for intellectual ones (like in the formulation of new theories or revision of old ones). Its record of success in this process is a justification for its continued use.

Prediction versus Explanation

At this point, it seems appropriate to address explicitly one debate in the philosophy of science–that is, whether science can, or should try to, do more than predict consequences.

The issue

One view that held considerable influence during the first half of the twentieth century is called the predictivist thesis: that the purpose of science is to enable accurate predictions and that, in fact, science cannot (or need not) actually achieve more than that. The test of an explanatory theory, therefore, is its success at prediction.

This view need not be limited to actual predictions of future, yet to happen, events; it can accommodate theories that are able to generate results that have already been observed or, if not observed, have already occurred. Of course, in such cases, care must be taken that the theory has not simply been retrofitted to the observations that have already been made—it must have some reach beyond the data used to construct it.

In 1960, Stephen Toulmin attacked the "predictivist thesis," a philosophical approach that he claims he once shared. *Foresight and Understanding*, pp.22-3. He asserted:

> "Forecasting...is a craft or technology, an application of science rather than the kernel of science itself. If a technique of forecasting is successful, that is only one more fact, which scientists must try to explain, and may succeed in explaining.... [A] novel and successful theory may lead to no increase in our forecasting skill; while, alternatively, a successful forecasting-technique may remain for centuries without any scientific basis. In the first case, the scientific theory will not be necessarily any the worse; and, in the second, the forecasting-technique will not necessarily become scientific ... ."

Id., p.36.

Toulmin explained how historically several theories that have been rejected were capable of more precise predictions than the theories that superseded them, such as Babylonian astronomy and tables showing the timing of the tides Id., pp.27-34. Now, those methods of forecasting were the result of the compilation of data reflecting apparent patterns or regularities in the phenomena under examination combined with the assumption that the patterns would repeat themselves, which they did. The techniques were mechanical (or, as Toulmin characterized them, merely "arithmetical"); they did not otherwise explain the successes or the failures of the forecasts. Id., pp.28, 29.

"The Babylonians acquired great *forecasting-power*, but they conspicuously lacked *understanding*. ... [Newton gave us] a number of general notions and principles which *make sense* of the observed regularities, and in terms of which they all hang together." Id., pp.30, 33 (emphasis in original). As discussed later, one might question the extent to which Newton's theories actually gave us explanations; but, there clearly is a rather profound sense in which Newton's theories were an improvement over prior theories. It is the nature of that "sense" that is of interest here.[17]

Some scientists do take an interest in these issues, at least when writing for the general public. For example, contemporary physicist David Deutsch recently set out an attack on the philosophical view that "denies that what I have been calling 'explanation' can exist at all." He claims that "during the twentieth century, most philosophers, and many scientists, took the view that science is incapable of discovering anything about reality. Starting from empiricism, they drew the conclusion ... that science cannot validly do more than predict the outcome of observations, and that it should never purport to describe the reality that brings those outcomes about." *The Beginning of Infinity*, p.15. Deutsch asserts, to the contrary, that the physical world—reality—really does exist and that it is accessible to rational inquiry. *Id.* Furthermore, he says that the aim of science is explanation which enables us to understand the external world and that the use of theories permeates man's mental activities.

THE NATURE OF UNDERSTANDING

Columbia University Professor of Physics and Mathematics Brian Greene recently wrote:

> "There's a difference between making predictions and understanding them. The beauty of physics, its raison d'être, is that it offers insights into why things in the Universe behave the way they do. The ability to predict behavior is a big part of physics' power, but the heart of physics would be lost if it didn't give us a deep understanding of the hidden reality underlying what we observe."

The Hidden Reality (2011), p.271.

Albert Einstein, we can assume, would have wholly agreed.

Interestingly, there is a similar contrast or tension in the philosophy of mathematics among types of proof. Ian Hacking, *Why Is There Philosophy of Mathematics at All?* (2014), pp.21-40. At one end, there are long, detailed proofs that appear to establish a proposition and can be carefully checked step by step. These proofs are often identified with Leibniz (1646-1716). At the other end, there are proofs that can be absorbed as a whole and which are compelling in their simplicity. This type of proof is associated with Descartes. The point of interest here is that the Cartesian proofs are thought to impart understanding—"they enable one to see not only that something is true, but why it is true. They give a feeling of understanding of the fact proven." *Id.*, p.32.

But, what does it mean to "make sense" of the observed regularities, to create a sense of insight or understanding? Natural scientists and many social scientists seem to aspire to the ideal of the deductive theory, and much of modern science appears in the mathematical forms to which a deductive theory lends itself. However, what we mean by an explanatory theory in the sense here is not necessarily or even primarily a mathematical model of the phenomena under investigation.

Toulmin argues that the notion of explanation "...involves appeal to some principle of regularity or ideal of natural uniformity." *Foresight and Understanding*, pp.41-42. And, he acknowledges that those "notions" will reflect the paradigms and ideals of the persons involved. Toulmin attributes to Copernicus the objective of a scientific theory as "consistent with the numerical data" while being "absolute" and "pleasing to the mind." *Id.*, pp.41, 115. In more contemporary language, a satisfactory explanatory theory will be based upon circumstances that the observer considers to be natural or self-evident, circumstances that require no further explanations or raise no questions.*See, id.*, p.41.

Toulmin refers to "ideals of natural order," which are perceptions of circumstances or events—such as motion or change—that are in the natural order of things and that, therefore, require no explanation. The goal is to explain new data or observations of events that differ from the natural order in terms that make sense of the differences. *Id.*, pp.44-82.

Of course, man's perception of what is the natural order of things is a historical/cultural phenomenon that can and does vary among persons and over time. Again, those perceptions necessarily shape the questions that are asked (i.e., the observations that "need" explanation) and the nature and content of the theories that are developed to provide the explanations. This notion clearly introduces a significant subjective element into a field of intellectual inquiry (or "creation") that we like to think of as "objective."

Ian Hacking says, "[O]thers talk of the proof explaining the fact that is proved. Such words—*understand, why, explain*—are sound, but do little more than point to a satisfying phenomenon that is experienced, rather than one that can be well defined." *Why Is There Philosophy of Mathematics at All?*, p.32.

(As we shall discuss later, developments in quantum theories in the second half of the twentieth century gave rise to an increasing emphasis on the predictive successes and testability of the mathematical models being employed, as opposed to the concepts of explanation and understanding. However, it would appear that there are among the current generation of leading physicists some who seem still to see the possibility of explanatory models that can be grasped by the intelligent layman—or, at least, the marketability of books that purport to present such explanations.)

An example: Gravity

I would like to investigate further the concept of an explanation that provides us with understanding with an example in the natural sciences. So, let us consider gravity. (We will have much more to say about gravity in later chapters.)

Newtonian mechanics, a model the fundamental elements of which are some 400 years old, has often been heralded as the paradigm of the successful scientific theory and has indisputably yielded some of the most spectacular and useful practical applications of modern society. Yet, the fundamental elements of that model fail to achieve a meaningful theory of causation or a satisfying explanation of the how and why of the observed phenomena. See, e.g., Joseph Mazur, *The Motion Paradox* (2007), p.181. The situation has been described by a mathematician as follows:

> "Contrary to popular belief, no one ever discovered gravity, for the physical reality of this force has never been demonstrated. However, the mathematical deductions from the quantitative law proved so effective that the phenomenon has been accepted as an integral part of physical science. What science has done, then, is to sacrifice physical intelligibility for the sake of mathematical description and mathematical prediction. **This basic concept of physical science is a complete mystery, and all we know about it is a mathematical law describing the action of a force as though it were real.**"

Morris Kline, *Mathematics for the Nonmathematician* (1967), p.361 (emphasis added).

In other words, we do not know how gravity actually works and, therefore, cannot even say that such a thing clearly exists (even though the consequences of what we call gravity clearly exist and can be accurately predicted by us).[18] By "how it works" I am referring to what we have described as an explanatory paradigm or model—a conceptual structure that is consistent with the observed events and gives us a feeling of understanding of the type of causality that is at work. As discussed above, we tend to look for an analogy, something with which we are familiar and think we understand, that displays at least some of the characteristics of the phenomenon that we are attempting to explain.

With respect to gravity, there are two things in particular that need to be explained. First, how does the attractive force manage to operate over distances of empty space? Second, what is the reason that the force is inversely proportional to the square of the distance between two objects? The first question arises because of our ordinary understanding that things interact through direct or indirect contact. The second question is really the question of why there is such a simple mathematical relationship between gravitational force and distance. It makes sense that distance matters, but why is the relationship based upon the square of the distance?

One way to highlight what we are missing is to discuss examples of what an explanatory theory might look like.

A Newtonian version

Richard Feynman set out a hypothesized theory of gravity in his 1964 lectures at Cornell University. He said, suppose that all bodies are subject to continuous bombardment of particles from all directions, which particles would "push" those bodies except that they are exerting pressure from all directions at once. Then imagine two bodies close together. Each would block particles coming from behind it toward the other body, with the result that the "normal" balance of pressures would be disrupted so that the two bodies would be pushed toward each other by the particles that are not being blocked.

One interesting prediction from this scenario is that the forces pushing the bodies toward each other would be inversely related to the distance squared (as in the mathematical formula). The reason is that the areas being blocked (where no particles reach each of the bodies to push it) would get smaller the farther away the two bodies are from each other and the diminution in areas would be a function of the distance squared. Id., pp.31–33.

Now, we have not said anything about what the particles are, where they come from or how they are generated. But, assuming such particles, we have set out a model or theory of gravity that in some important way "explains" it. The model speaks to our intuition and is capable of visualization, like a good analogy.

Feynman goes on to note that this model does not work in fact, because it leads to predictions that are incorrect. For example, a moving body would be impacted by more particles on its front, from the direction toward which it is moving, than on its back, where the particles would be "chasing" it, which phenomenon should tend to slow down the forward movement of the body, contrary to the law of inertia. Id., p.33.[19] (In addition, the mathematical formula relates gravity to the masses of the objects, not to their sizes or diameters. Therefore, we would need to say that the blockage of particles is a function of the mass, not the volume, of each body. That refinement undermines the rather neat geometrical calculation supporting the inverse squared relationship.)

Einstein's version

Newton's theory has been subsumed into (or refined by) Einstein's theory of gravity, which will be discussed below in the chapter on Physics.

Einstein's theory proposes that the observed phenomenon occurred as a result of "curvatures" or distortions to the fabric of space itself or, more precisely, of spacetime.[20] The object with greater mass creates greater warp in spacetime, causing objects with lesser mass to move towards it. Thus, gravity is not a force that operates across empty space; it is a phenomenon that occurs in spacetime and is part of the nature of spacetime. There is no need for a hypothetical particle or force to convey gravitational attraction; the "fabric" of spacetime is the medium through which gravity operates. Greene, *The Hidden Reality*, pp.14-15.

The theory was, of course, propounded in an elaborate and innovative mathematical structure, what have been called "Einstein Field Equations". Id., p.16. Einstein's General Theory of Relativity has enabled highly accurate calculations of the motions of planets and other heavenly bodies by calculating the presumed curvature of spacetime caused by matter (mass plus energy).

However, in order to try to convey a feeling of what is occurring, Einstein and subsequent physicists have used an analogy to a trampoline (or tightly stretched sheet) with two heavy balls placed on it, giving us another example of an "explanatory" model of gravity. See, e.g., Greene, *The Elegant Universe*, pp.67-71. Each ball would tend to create a depression in the surface. At a sufficient distance, each depression would have no discernible effect on the other ball. But, as the balls were placed closer together, the depression caused by one would "reach" toward the other, tending to cause the surface to tilt toward the first ball, and vice versa. The expected consequence would be that the balls would begin to move toward each other, with more force, the closer together they were. Also, the mass of each body would determine the degree of depression and, therefore, the amount of "tilt" that would occur on the surface near the other body. Thus, we have a "model," in the form of a physical analogy that may be said to allow us to visualize and "understand" the force that we have called gravity.[21]

The physicists and mathematicians can undertake to explore whether the observed phenomena can be generated from mathematical representations of the assumed "surface" (the trampoline or sheet) and the effects of bodies of different masses. If they are successful, we would conclude that we have a working model, at least temporarily.

But, we have said nothing about the nature of the surface that we have used as a central feature of our explanation. In addition, we have no explanation of why the curvature or indentation occurs—it cannot be gravity "pulling" downward on the body on the surface, causing the surface to sag, since it is gravity that we are trying to explain. Id., p.71. Similarly, our initial reaction that the indentation of the surface would explain the tendency of the two balls to move towards each other in a manner consistent with Newton's formula is incorrect. In the analogy, it is the force of gravity that causes a ball to roll down the slope (like a ball on an inclined plane)—absent gravity, the ball would just "float" where it is, whether the surface on which it appears to rest (absent gravity, it would not actually be resting on the surface) is level or sloping.

Also, this analogy is in only two dimensions, while gravity obviously operates in (at least) three dimensions. One can clearly conceptualize a three-dimensional space, but then what does it mean to "curve" or "warp"? Id., pp.72–3. The sheet or trampoline analogy is easy. In contrast, imagine a bowling ball dropped into a tank of water. The ball displaces water, pushing water in all directions. But, how could the ball cause a curvature of the three dimensional body of water? In short, into what dimension does the distortion caused by the object of mass occur?

There is finally the matter of the "physical mechanism" by which the phenomena is caused, calling again for an analogy to things familiar to us in ordinary life. See id., p.71. Newton's theory calls up the image of a force of attraction, like magnetism; Einstein's theory of gravity pictures objects taking the shortest paths through a curved or warped spacetime, in which, because of the curvature, the shortest path will be a curve rather than a straight line—yet, we have no identification of the mechanism by which the curvature is caused by the object with mass or, in fact, the mechanism causing the object to take the "shortest" route.[22]

These are significant questions about the analogy.

AND SO...?

Nonetheless, we may still tend to believe that we had achieved something more than we had with just the mathematical formula. The mathematics seems to be clear, internally consistent and powerful. Perhaps the problem is simply the limits of the human imagination, that we just cannot "see" what is taking place. Perhaps the analogy brings us closer to understanding what it is. Alternatively, one could say that these types of explanation are really just appeals to familiar phenomena, which do not really "explain" anything either; they just make the observed relationships seem familiar or normal to us, so that we feel that there is no pressing need for further explanation. But, how satisfactory is that?

Before being too critical, we should ask, what is a good explanation of a causal relationship? Do we not always face the question of what caused that which we have identified as a cause?

Richard Feynman has claimed to paraphrase Newton's reported response to the accusation that his theory did not really tell us anything as follows: "'It tells you *how* it moves. That should be enough. I have told you how it moves, not why.'" *The Character of Physical Law*, p.31.

Should it be enough? I do not have a good or satisfying response.

In part, there is a factual question of whether a particular theory is just an analogy or, in some way, actually reflects the mechanism by which the phenomenon occurs. If the latter, then I think it is fair to say that we have at least a first step closer to an explanatory theory than we would have simply with a mathematical formula.

On the other hand, in much of current physics, especially with respect to the very small (quantum mechanics) and the very large (cosmology), many of the explanations offered are little more than mathematics. There seem to be obstacles, at least for now, to our ability even to conceive of or visualize the phenomena described by the mathematical theories. Often, the implications of the theories simply defy our common sense and sharply conflict with our perception of reality.

Hawking and Mlodinow assert:

> "To understand the universe at the deepest level, we need to know not only how the universe behaves, but why.
> Why is there something rather than nothing?
> Why do we exist?
> Why this particular set of laws and not some other?
> This is the Ultimate Question of Life, the Universe, and Everything"

The Grand Design, pp.9-10 (emphasis removed).

So, "why" is the big question—the "ultimate question."

Except, scientists (including Hawking and Mlodinow, as we shall see) seem to feel that the creation of an elegant mathematical model that is consistent with observed empirical phenomena and that is testable empirically, at least in theory (that is, one could imagine an empirical test of the theory or model even if current technology does not permit one to perform it), constitutes an satisfactory "explanation" of the phenomena under investigation. In other words, these scientists appear to believe that mathematical representation constitutes understanding. However, as discussed below in our examination of mathematics, we do not have a very satisfactory theory of the relationship between mathematics and reality.

REQUIREMENTS OF A GOOD THEORY

There have been serious debates as to whether a theory that provides a sense or feeling of understanding is adequate to qualify as scientific or whether other criteria must also be met.

Of course, the theory must be internally consistent (not lead to two conflicting predictions or conclusions). See, e.g., Alfred Tarski, *Introduction to Logic* (1946), pp.108-9. More precisely, the axioms or initial premises must be logically consistent and the rules of manipulation must be constructed so as not to allow the introduction or creation of inconsistencies in the propositions derived from the axioms. See, e.g., Barrow, *Theories of Everything*, p.32. Clearly, a useful theory will have the quality of generality, that is, it can extend to a wide variety of observations and consequences. And, there are often suggestions that certain aesthetic criteria be met concerning generality, simplicity and elegance. These criteria are not strictly requirements for achieving the status of a scientific theory; they are standards for preferring one theory over another.

Philosophers of science and many scientists themselves consider simplicity to be not just a virtue but an inherent quality of a good theory. Simplicity may be hard to define and it would certainly seem to have some cultural basis (or bias), but it seems that it is often possible and even easy, when confronted with two alternative theories, to identify the one that is more simple. Finally, there is elegance: "closely allied to simplicity, this regulative maxim separates what is ugly and cumbersome from sweeping ideas that carry élan and give pleasure upon comprehension."[23] Henry Morgenau, "What is a Theory?," in Sherman R. Krupp (ed.), *The Structure of Economic Science* (1966), p.26.

VERIFIABILITY

Probably the most frequent requirement for a scientific theory has been that the theory be testable by experimentation or observation. The idea that theories should be subject to assessment and testing should not seem strange. Otherwise, one could hardly claim that theories are relevant to knowledge or understanding. Indeed, it would seem unscientific for a proponent of a theory to declare that it is unfair to require that the theory be subject to testing and possible disconfirmation.

A theory or hypothesis should enable one to make predictions about observable facts or events beyond the elements contained in the hypothesis itself. If one successfully observes the occurrence of such predicted consequences, then there is a basis for considering that hypothesis or theory to be supported.

One might consider that the successful predictions constitute a "verification" of the theory involved; but, during the second half of the twentieth century, the concept of verification or verifiability came under increasing attack as a matter of logic. The issue was not about how scientists engaged in the process of scientific discovery, but the philosophical issue of what was logically defensible.[24]

Verification by successful prediction has the same logical problems as inference by induction (or, for that matter, deduction): there are many reasons why a prediction might come true that are not supportive of the theory by which the prediction was made (coincidence, a common underlying cause, etc.). So, generally, a successful empirical test simply will not provide substantial confirmation of the theory, because there are likely to be multiple alternative theories that could generate the same empirical prediction.

"Falsifiability"

A formal challenge to the concept of verification was presented by Sir Karl Popper. The position that Popper took was that the theory had to be capable of "falsification" for it to be a true scientific theory. If, over time, experimental efforts to falsify the theory fail, one can increasingly become more confident in the theory. See Popper, *The Logic of Scientific Discovery*; *Objective Knowledge: An Evolutionary Approach* (1972). This philosophical approach stresses the central role of criticism of and challenge to established theories and portrays the proper scientist as fully objective and psychologically prepared to reject his or her own theory in the face of falsification.

The testability of a theory has come to mean not that the theory can be confirmed by successful predictions, but that the theory is susceptible to being disproved by an imaginable and identifiable test or set of observations—in other words, that it is falsifiable. That means that the testing of a theory for the philosophers of science is formulated in terms of a possible experimental result that would be viewed as inconsistent with the theory and, if achieved, would demand that the theory be rejected.

This may seem to be a rather odd concept of proof, but this characterization of science and scientific methodology has become widespread among philosophers of science and at least some scientists, especially physicists who have expressly addressed these methodological questions while touting their open-mindedness and objectivity. It is much less clear the extent to which working scientists (or, perhaps, it is more accurate to say scientists while working as scientists) really embrace and incorporate this methodological approach in the process of their work.

A strict Popperian approach might suggest that the scientist must be prepared to commit to reject a theory based upon obtaining an empirical result inconsistent with the theory's prediction. In real life, of course, most scientists will be reluctant to quickly abandon an accepted theory, and rightly so. At a minimum, they would want to check carefully for errors in the experiment and would attempt to reproduce the results. If the inconsistent result still obtained, they would want to examine the theory with care to see if the result was truly inconsistent. Then, they would consider whether there was some minor adjustment that could be made in the theory, or in the theories upon which the test instrumentation or the test methodology were constructed, that would accommodate the inconsistent result.

Moreover, as a practical matter of application, the efforts to find testable, empirical propositions that are predicted by the theories increasingly have led, especially in modern physics and related subjects, to hypotheses that are distant, minor logical implications of the theory, not matters at the core, or that involve observations or testing procedures that are themselves heavily theory-laden. Thus, a failure of the test is susceptible to many potential explanations that preserve the central core of the theory, in other words, the central theory is not really falsifiable by that test. See, e.g., Imre Lakatos, "Science and Pseudoscience," Philosophical Papers, vol. 1 (1977), pp.1–7.

PROBLEMS AND CRITICS

Philosopher of science Imre Lakatos was sharply critical of Popper's definition of (the methodology of) science as based upon the sole use of propositions that are falsifiable.

> "...Popper's criterion ignores the remarkable tenacity of scientific theories. Scientists have thick skins. They do not abandon a theory merely because facts contradict it. They normally either invent some rescue hypothesis ...[or] they ignore [the anomalous results]. ...[S]cientists talk about anomalies, recalcitrant instances, not refutations. History of science, of course, is full of accounts of how crucial experiments allegedly killed theories. But such accounts are fabricated long after the theory had been abandoned."

Lakatos, "Science and Pseudoscience," Philosophical Papers, vol. 1 (1977), p.1.

This criticism, as expressed, is based upon his view of how scientists act; but, Lakatos also argues, in effect, that the test of falsification is not actually implementable.[25]

There's another type of critique. Many theories are not susceptible to the formulation of a conclusive test. One cause is the *ceteris paribus* condition, the satisfaction of which often cannot be guaranteed in advance. Another problem is "underdetermination," where the theory does not provide a unique explanation of causation. Peter Lipton, Inference to the Best Explanation (2004), pp.5–7.

We know that economists regularly make predictions based upon their theories, and economists seem always to have explanations for why their predictions did not come true–explanations that often are quite consistent with the theory. The typical explanation will invoke the *ceteris paribus* condition or assumption. Not everything that was relevant stayed the same.[26]

In some cases, we may have a "closed system"–all relevant variables within the parameters of current measurement technology, i.e., within experimental error, are included in the theory. But, in "open systems," like in economics and biology, it is almost inconceivable that the *ceteris paribus* condition could be actually satisfied in the real world. Such theories will never be able to incorporate and control for or measure all relevant factors, and there will always be relevant factors that have changed. These types of theories have been called "explanation sketches," since a complete explanation can never be achieved, only more and more complete explanations as additional variables are incorporated into the theory. Carl G. Hempel and Paul Oppenheim, "Studies in the Logic of Explanation," *Philosophy of Science*, XV (1948), pp.130–39.

Models (theories) of science have been characterized as descriptions of how things change or how circumstances evolve over time. See, e.g., Barrow, *Theories of Everything*, p.3. However, even if one were to have a theory of change that faithfully reflected the reality of change, the accuracy of the predictions will necessarily depend upon the initial conditions from which the process of change occurs. Many theories simply do not have an adequate specification of the initial conditions to enable unique predictions.

One might mistakenly assume that minor errors in the specification of the initial conditions would result only in minor errors in the predictions; however, it is clear that there are systems in which very small, apparently trivial differences in the initial conditions will result in enormous differences in the final results over time. *Id.*, p.41.[27] Perhaps the most familiar example is the weather (the notorious hypothetical flutter of a butterfly's wings "causing" a devastating storm half a world away). See, e.g., Rees, *From Here to Infinity*, p.91.

Finally, systems that may be described as chaotic will appear not to be deterministic, because they seem to be unpredictable. In fact, however, chaotic systems may be deterministic, at least in theory, in the sense that the same result will always be obtained from the same initial conditions.

Of course, the specification of those conditions may clearly exceed current and, perhaps, foreseeable human capabilities, and we may lack a theory capable of predicting the results.

The question then arises as to whether any such theories could ever satisfy (or purport to satisfy) the falsifiability requirement. Yet, with such models, a simulation of a representative or completely plausible outcome is achievable, and such analyses may be of considerable value in the understanding of the nature of the phenomena at issue even if falsifiable predictions cannot practically be generated. See, e.g., Roger Penrose, Shadows of the Mind, pp.21–3. In short, there are useful theories that have been deemed to be very respectable in the scientific community but that do not and cannot generate specific predictions. These theories have achieved their success largely through their apparent power to explain the past.

AN EXAMPLE: DARWINIAN THEORY

Probably the most widely known example is Darwin's theory of natural selection, which has been an impressive tool in generating explanations for the emergence and success of various characteristics among living organisms. The theory of evolution by natural selection does not even purport to make predictions in the normal sense of the word—for example, that such a characteristic will appear in a particular type of organism and will do so in a particular period of time. See, e.g., Toulmin, Foresight and Understanding, pp.24–5. That "flaw" is explicitly embedded in Darwinian theory itself, which assumes random mutations or alternations as the source of variation. Random change by definition, is not predictable.[28]

Related and similar theories are utilized in the social sciences, particularly in micro-economics. Of course, the parallels between the models of competition in the marketplace and competition in the natural world are obvious (the differences are discussed in some detail in the chapter on Darwinism). Indeed, the underlying model or explanatory paradigm emerged almost simultaneously in biology, with Darwin and Wallace, and in economics, with Adam Smith. All three were certainly influenced by Thomas Malthus.[29]

ANOTHER EXAMPLE: MEDICAL SCIENCE

Medical science is generally probabilistic—that is, the predictions are set out in terms of statistical likelihood rather than absolute predictions. However, modern medical science tends to incorporate statistical correlations with theoretical structures into which other empirical evidence can be folded for some greater degree of confirmation.

As just one illustration, after the identification in 1989 of an abnormality of the prostate labeled "prostatic intraepithelial neoplasa" or PIN and the observation that the condition PIN occurred with great frequency with prostatic cancer, efforts were made to understand the relationship between PIN and cancer. For example, was PIN a precursor to cancer in the sense that the same cells displaying the abnormality would likely become cancerous (in a direct causal pathway)? Did PIN tend to occur simultaneously with cancer in different cells as a result of some other underlying causal factor that would lead to both conditions? Or, was PIN a likely event on a progression that led to cancer appearing in other cells?

The analysis involved not only sampling of a statistical nature but also the examination of the cell structures arising in both conditions, the types of genetic alterations that occurred and the progression of both conditions. The studies also utilized animal models tested through experimentation. See David G. Bostwick and Isabelle Meiers, "Neoplasms of the Prostate", in Bostwick and Liang Cheng, *Urologic Surgical Pathology* (2008), pp.443–580 (Chapter 9).

Again, the predictions could not strictly be disconfirmed, since probabilistic outcomes can be consistent with very unrepresentative results (like finding the toss of a coin coming up heads 10 times in a row), but it would seem that this approach to science is clearly appropriate and meaningful. Indeed, the methodological approaches in medicine are largely different from the deductive theories we normally equate with the physical sciences.[30]

Of course, one might argue that the methodological approaches in medical science reflect only the limitations of our current knowledge, rather than a different ultimate conception of what the science should and, hopefully, someday could be. It is conceivable that with a sufficient understanding of all of the relevant initial conditions (the physiology of the patient) and of the pathogens to which the patient has been exposed, we may someday be able to make predictions that are not just probabilistic.[31]

Even if that turns out to be the case, the methodological approaches that make use of the knowledge that we do have and can hope to achieve in the shorter term are entirely justifiable.

And So...

What should we make of these deviations from the asserted norm for scientific theories? These alternative approaches to knowledge and understanding can be very instructive and productive. Indeed, a persuasive case can be made that our classically-derived Western view of science and scientific method, as reflected in classical mechanics as well as modern physics, threatens actually to be a serious limitation on or hindrance to comprehensive knowledge of the world. It has been asserted that this traditional scientific view is too narrow and restrictive, ignoring (or excluding) major aspects of human experience (including art, music, many emotions and, perhaps, even consciousness). See, e.g., Feyerabend, *The Tyranny of Science*, pp.6–12, 54–6, 68–9. Arguably, the dominance of a restrictive view of science based upon the model of classical physics may not just hinder the appreciation of other types of knowledge, it may also hinder the continued development of physics itself.

Explanation Revisited

These arguments lead me to add some elaboration to the second of the two characteristics I noted above as generally true of a useful theory—that it causes one to feel that understanding or insight has been achieved by virtue of the theory—and what that characteristic might mean. A person might manage the external world based only upon theories or models that accurately predicted the consequences of particular actions or events. But, many of us would not necessarily say that that person really comprehends or understands the world. We want more than accurate predictions; we want understanding.

Revelations?

The feeling may be quite powerful and exciting, as a sense of illumination or revelation: a "Eureka" moment. New connections are revealed. A feeling of comprehension is achieved. At least a partial glimpse of some underlying truth has been obtained.

Obviously, such experiences are inherently and necessarily subjective. "Whatever we mean by illumination, it is intimately connected to the mind, and therefore to intelligence." William Byers, *The Blind Spot*, p.41. Indeed, "'understanding' ... means developing a subjective feeling for the subject." *Id.*, p.103. This undeniable fact poses a problem for the classical view of science as objective, that is, as independent of the observer.[32] While I doubt that many would deny that the experience of revelation does occur, what is the source of the feeling of insight and what does the feeling indicate with respect to the question of whether the theory is true, that is, the issue of whether it conforms with or at least partially depicts reality?

Two rather different possibilities jump to mind.

Perhaps the feeling occurs because, in fact, we have achieved at least some glimpse of reality, which seems to be what Polanyi considers scientific "vision" to be.

The other possibility is that the sense of comprehension or understanding is achieved not because of a new theory's closer relationship to reality, but because of factors inherent in the human mind or arising through society and culture or when a proffered theory happens to support or vindicate a prejudice or bias that one already has based upon emotions or the subconscious awareness of self-interest.[33]

Paradigms

Significant contributions to our understanding of science were made during the twentieth century through efforts to examine the ways in which science was actually practiced, how creative discoveries have occurred and how, in fact, scientific knowledge has grown. The examination of the history (as opposed to the philosophy) of science, in terms of the emergence and acceptance of new ideas, provides useful insights into the methodological questions.

One of the more important influences on the developments in the history of science was Thomas Kuhn's concept of paradigm shifts. Thomas S. Kuhn, *The Structure of Scientific Revolutions* (1962). In simple terms, Kuhn explained how significant changes in science occur with a fundamental shift in the paradigm by which man understands the phenomena at issue. Thus, we have the shift from the pre-Copernicus world to the post-Galileo world (whatever debates there may be about what happened in between) or from the Newtonian world to the world of Einstein. These shifts suggest the interrelationships between science and culture generally, with influences going both ways.

The example of the developments leading to Newton's theories is relatively easy to understand. The change from an Earth-centered to a Sun-centered then a no-centered worldview is widely understood and accepted in modern society. There was a paradigm shift that was far broader than the scientific theories directly involved. It is likely that certain societal and cultural developments were useful, if not necessary, to the acceptance of the new theories and, perhaps, even to the conception of them.

What about Einstein and his theories of relativity? At a superficially appealing level, one can note that societal developments concurrent to the scientific theory included the breakdown of established orders and universal truths and the emergence of cultural and moral relativity. Are these all elements of a paradigm shift? Can one say that the Special Theory of Relativity is a cousin to moral relativity, cultural relativity and then to political correctness?

I think not. The relativity of the theories in physics does not in any way suggest that there are no universal truths or bases from making judgments of the correctness or value of various positions. Indeed, as Lord Rees has observed, "It's a pity, in retrospect, that he called his theory 'relativity'. Its essence is that the local laws are *just the same* in different frames of reference. 'Theory of invariance' might have been a more apt choice, and would have staunched the misleading analogies with relativism in human contexts." *From Here to Infinity*, p.137.

REVOLUTIONS?

Kuhn viewed those shifts as revolutions. Before and after the shift, scientist will attempt to resolve experimental and observational problems by modifications and amendments to the dominant theory. At some point, the scientific community, in part influenced by general societal developments, becomes able to see the relationships under study in a new light, in and through a different paradigm. With time, the new worldview then becomes widespread in the educated community.

Stephen Toulmin disputed Kuhn's formulation, describing the changes in worldview that ultimately occur as gradual developments with multiple inputs from many sources. The groundwork is laid over time by new insights and new discoveries in a variety of scientific fields. Toulmin was a contrarian in several ways, but his disagreement on this point does not detract from the observations being offered here. Both philosophers of science agree that paradigm shifts result from the confluence of numerous developments in the scientific community and the society in which the scientists function.

Professor Barrow argues that contemporary physics is defined or constrained by the successes of existing established theories. The established theories would constitute special limiting cases of the yet to be discovered underlying comprehensive theory. The underlying theory, if discovered, would both open new insights and explain the interconnections of the existing theories. He argues that "[r]eal advances in our understanding of the physical world always seem to involve" one or more of six types of development, ranging from "revelation" to "enumeration". *The Constants of Nature*, pp.60–66. Professor Barrow disputes the term "revolution" as an accurate description of the progress of the sciences for seeming to diminish the interrelatedness and interdependency of the theories and ideas.

Philosopher of science Peter Lipton, as indicated above, defended a type of inductive reasoning. He used "inductive" to refer to any inferences other than those that are deductive, *i.e.*, those inference that are the logically necessary conclusions from the premises. Thus, "[i]nductive inference is ... a matter of weighing evidence and judging probability, not of proof." *Inference to the Best Explanation* (Second Edition), p.5. "If an inference is inductive, then by definition it is underdetermined by the evidence and the rules of deduction. ...By definition, even a good inductive argument is one where it is possible for there to be true premises but a false conclusion." Id., p.7.

In this formulation, the problem of induction is the question of "[h]ow ... we go about making these judgments, and why [we] should ... believe they are reliable," questions of description and justification. Id., p.5. In other words, the "problem" is an invitation to try to understand how we actually make sense of the world and successfully make predictions about it. As his title suggests, Lipton described the process of science as the continual replacement of one theory by another, where the successor appears to present the "better" explanation.

A somewhat different approach was taken by Professor Imre Lakatos, who described science as a "research programme." See "Science and Pseudoscience," *Philosophical Papers*, vol. 1, pp.1-7. Lakatos argued that it is misleading to focus on particular hypotheses in isolation. A scientific theory will have a "hard core" of propositions surrounded by auxiliary hypotheses that are dispensable. The research programme is "a powerful problem solving machine" that incorporates anomalies as it evolves. The most important feature to Lakatos is that the effective and successful research programs "predict novel facts, facts which had been either undreamt of, or have indeed been contradict by previous or rival programmes."

Also, such programs will be either "progressing" or "degenerating." The degenerating programs will eventually be abandoned and the progressing programs will attract proponents and young talent, because they will be perceived as where the action is. "Criticism is not a Popperian quick kill, by refutation: there is no refutation without a better theory. Kuhn is wrong in thinking that scientific revolutions are sudden, irrational changes in vision." Id.

We should note in passing that these criticisms, while punchy, are certainly overstated. Popper contemplated a process of competition among theories, not just the rejection of any theory that failed the falsification test; Kuhn contemplated more than an irrational change in worldview. Similarly, the disputes over whether the paradigm shifts are revolutions are not really important. Scientific advances involve significant and fundamental changes in world views and in the accepted and acceptable (dominant) explanatory paradigms.

COMPLEXITY AND PREJUDICE

The process of science is more nuanced, complex and haphazard than suggested by our discussion above of explanatory models and falsifiability. One important factor is the asserted connection between theories in science and the concepts or structures that we use more broadly to understand and explain things within our society and culture.

Here is an example that should be accessible to everyone. To the surprise of many, we were recently told that the evils of fat and cholesterol in our diet have been seriously overstated for decades, based upon faulty science. The real culprits instead are sugars and carbohydrates. The challenges to the accepted wisdom arose out of studies that began appearing in 2010. See, e.g., Joanna Blythman, "Butter is bad—a myth we've been fed by the 'healthy eating' industry," *The Guardian*, 23 October 2012; Matt Ridley, "Experts have been feeding us a big fat myth," *The Times*, 30 June 2014. Finally, the U.S. Dietary Guidelines Advisory Committee has made it official. Anahad O'Connor, "Nutritional Panel Calls for Less Sugar and Eases Cholesterol and Fat Restrictions," *The New York Times*, February 19, 2015. In other words, the government may have actively sponsored health advice and instructions that contributed to the creation of a serious new health threat in the form of obesity and associated disorders like diabetes. (However, a large new study suggests that the concerns about cholesterol were, in fact, justified. See, e.g., Betsy McKay, "Study Links Eggs to Higher Cholesterol and Risk of Heart Disease," WSJ.com, March 15, 2019.)

In response to the question of how this could have happened, some have suggested that it is the result of methodological flaws: the use of observational rather than experimental data and the confusion of correlation with causation. See, e.g., Nina Teicholz, "The Government's Bad Diet Advice," The New York Times, February 20, 2015. I think that the commentators have been overlooking one of the principle reasons that bad advice becomes accepted dogma—it often makes sense, based upon our (incorrect or, at least, incomplete) understanding of how things work in the body (eating high cholesterol food results in high cholesterol in the blood). The observational studies of groups of people over time throw up apparent associations between types of diets and health issues; our common sense of how food affects our bodies gives rise to conclusions about causality. Those conclusions may be wrong. (Several similar examples of how scientific mistakes arose out of common sense can be found in David Sloan Wilson, This View of Life: Completing the Darwinian Revolution (2019), pp. 51-74.)

The obvious obstacles to the performance of long-term clinical or experimental trials on humans prevent the most direct testing of our hypotheses. Only now are we glimpsing how much more complicated the relationship is between food and health. The mistake was the overconfidence placed by the public advocates and policy makers on the correctness of the scientific conclusions. Models are inevitable, models matter and models are likely to be wrong.

BIOLOGICAL LIMITS TO UNDERSTANDING?

A somewhat different theory about the nature and source of scientific insight postulates that the human mind is wired to respond to certain types of logic and structure, so that some of secrets of the Universe are part of our very physical being. Therefore, as scientific thinking more closely approaches the underlying structures of the Universe, we respond with recognition. This type of explanation has been used by some mathematicians to explain the appeal and forcefulness of mathematics.

One interesting inference from this line of thought is that if some part of a reality in the Universe is reflected or embodied in the human mind, then our search for understanding can and does or, at least, tends to lead us toward the truth and that that process can be pursued through thought as opposed to (or, in addition to) observation and experimentation. This embodiment could be the result of the inherent structure of the mind, as part of the Universe, or as an adaptive development, occurring by chance (or otherwise) and enhancing the survival of the species.

It is also possible that the pervasive role of the workings of the mind and mental constructs in our efforts to comprehend our physical environment inevitably intrude into or "taint" the process so that we will always see some of ourselves in any structure that we can create to model the Universe, even if such forms do not really exist in the Universe being modeled. The Kuhnian observation that new theories depend upon contemporaneous developments in other areas of science and, perhaps, in other areas of learning and culture is relevant. But, that does not tell us whether the experience of improved understanding reflects the discovery of universal truths or simply the consequences of either innate or societally-derived biases of the human who achieves the feeling of understanding.

An equally speculative argument would assert that the human mind evolved to meet the needs presented by its environment so one cannot, therefore, expect man to have the capabilities to understand phenomena that were completely irrelevant to man's ability to reproduce and survive. Thus, it may be that there are inherent limitations on our ability to comprehend the Universe beyond our parochial niche. I quote various well-known scientists speculating as to some such limitations in the Introduction. Similarly, professor of mathematics William Byers asserts that there are "fundamental limits" to reason and human understanding. He writes, "The discovery of ... 'limits to reason' is in many ways the key scientific discovery of the twentieth century, one that our society has still not fully assimilated." *The Blind Spot: Science and the Crisis of Uncertainty* (2011), p.4.

This somewhat pessimistic view is strongly opposed by David Deutsch, who asserts, with no actual support, that man's ability to continue to expand his knowledge of the broader Universe is unlimited. He responds to the supposed evolutionist's reasoning by arguing that "the reach of human adaptations does have a different character from that of all other adaptations in the biosphere. The ability to create and use explanatory knowledge gives people a power to transform nature which is ultimately not limited by parochial factors ... ." *The Beginning of Infinity*, p.56. A good illustration of man's ability to transcend the formative influences of his biosphere and biochemistry is his demonstrated ability to survive in environments that are inhospitable to human life (or, even, to any life as we know it) such as the Moon. The evolution of culture and knowledge certainly occurs much more rapidly and with much broader reach than biological evolution.

Of course, Deutsch cannot prove that man's mind has the capability ultimately to understand everything in the Universe. There could be relationships or facts that simply cannot be grasped by mankind because of physical characteristics of the brain. *Cf. id.*, pp.60–61. Perhaps, we do not and cannot know. After all, as Deutsch repeatedly asserts, that yet to be discovered is currently unknowable. *See, e.g., id.*, pp.193, 197.

ENDNOTES

[1] We are limited in what we can detect with our senses. Although, sound waves range from almost 0 to 100,000 hertz, the average person can hear only waves from 20 to 20,000 hertz. Similarly, electromagnetic waves range in frequency from less than 1 to more than 10^{25} hertz (cycles per second, but the human eye only responds to frequencies in the range of 430-770 terahertz (10^{12} hertz)–which are wavelengths of about 380 to 740 nanometers ("visible light"). See, e.g., Daniel J. Levitin, *This is Your Brain on Music: The Science of a Human Obsession* (2006), pp.24-5; Theodore Gray, *Reactions: An Illustrated Exploration of Elements, Molecules, and Change in the Universe* (2017), pp.118-9. Our eyes are much less discriminating than our ears or noses, able to respond to only three frequencies in that narrow range, the ones that roughly equate to the colors red, blue and green. Gray, *Reactions*, p.120. And, yet, we can distinguish millions of different colors. Also, people differ as to what they see, hear, smell and taste. See, e.g., Heather Murphy, "You Will Never Smell My World the Way I Do: Scientists find that whiskey's smokiness, the smell of beets and lily of the valley perfume can be utterly different depending on your genetic wiring," NYTimes.com, May 3, 2019. But, there are much greater differences between people and other animals (and among animals), especially with respect to sight and smell. These differences reflect the differing receptors involved. As to sight, most mammals have two receptors (for blue and green), humans have three (adding red) and certain birds have four, enabling them to see colors we cannot. Virginia Morell, "Hummingbirds see colors we can't even imagine," *National Geographic*, June 15, 2020. I have a note on color in the chapter on Chemistry.

[2] Perhaps one could fashion an "understanding" of the natural world based upon the existence of patterns and regularities, without introducing causality. At a minimum, it seems clear that science could not have arisen if the Universe had not been characterized by regularities and constants. (Actually, forget science. Life as we know it could not have existed without such regularities.) And, perhaps mere recognition of the existence of regularities or patterns would have sufficed for purposes of survival. But, I think that normal conception of understanding implies causal explanations—e.g., at least, why there are regularities that make certain events predictable.

[3] Some current research has indicated that the human mind is "hardwired" with modules that are suitable for quite specific purposes. We may

be born with "domains of intuitive knowledge" that govern particular areas of cognition. Both the facility for and the resulting structure of language seem to be of such a nature. Certain models that arise in psychology, biology and physics may also be intuitive. Adam Frank, About Time (2012), pp.7–8. It has even been suggested that the evolution of the processes for moving information among the modules gave rise to analogy and metaphor, thereby boosting human creativity. Id.

[4] As Professor Barrow wrote, "The mind is the most effective algorithmic compressor of information that we have so far encountered in Nature. It reduces complex sequences of sense data to simple abbreviated forms which permit the existence of thought and memory. The natural limits that nature imposes upon the sensitivity of our eyes and ears prevent us from being overloaded with information about the world." Theories of Everything, p.11.

[5] This analogy of a theory being like a pair of spectacles suggests another point of relevance to the discussion here. As with a pair of spectacles, it is difficult to look at a theory (the spectacles) and through it at the same time. See Stephen Toulmin, Foresight and Understanding (1961), p.101.

[6] For a detailed discussion of the impressive scientific achievements made and their contribution to the advancement of science involved in developing accurate measurement techniques in one particular area (thermometry: the measurement of temperature), see Hasok Chang, Inventing Temperature: Measurement and Scientific Progress (2004).

[7] This debate among philosophers of science has been labelled the issue of "incommensurability." See Oberheim and Hoyningen-Huene, "The Incommensurability of Scientific Theories." The Stanford Encyclopedia of Philosophy, Fall 2018.

[8] The quotation is regularly attributed to J. B. S. Haldane in the 1930s, but it has been traced to Ernst Wilhelm von Brücke, written in German in the mid-19th century. See John S. Wilkins, "Teleology as a mistress," evolvingthoughts.net, 8 September 2009.

[9] Rudolph Carnap described the deductive construction of the logician as "a skeleton of a language rather than a language proper, i.e., one capable of describing facts. It becomes a factual language only if supplemented by descriptive signs." "Foundations of Logic and Mathematics," International Encyclopedia of Unified Science (1939), No. 3, p.32. He also quoted Einstein

as saying, "So far as the theorems of mathematics are about reality they are not certain; and so far as they are certain they are not about reality." Id., p.56.

[10] The term "gene" was not used by Mendel. The concept was set out by Dutch botanist Hugo de Vries in 1889 (an inheritable unit that caused a specific trait, which unit he called a panagene). The word "gene" was introduced in 1909 by Danish botanist Wilhelm Johannsen. See Rheinberger, Müller-Wille and Meunier, "Gene," *The Stanford Encyclopedia of Philosophy*.

[11] I use "inductive inference" here in the common, lay sense of drawing at least tentative inferences from the observation of apparent patterns or regularities in events. The philosopher's "problem of induction" is discussed at some length in the next section of this chapter.

[12] The editor added a footnote there explaining that Hume used the word "science" in the pre-nineteenth century sense of knowledge or learning generally.

[13] Hume used "knowledge" to refer to pure mathematics or pure deduction, things known from "first principles." Understanding of the physical world came from (and with) probabilities and was referred to by Hume as "opinion." Any theories about cause and effect in the physical world were necessarily of the latter type. See Ian Hacking, *The Emergence of Probability*, pp.180–1.

[14] Hume goes on to characterize inductive reasoning in a rather pejorative manner. He asserts that the inductive inferences from experience are the mere result of "custom or habit." "There is some other principle which determines him to form such a conclusion [that the future will resemble the past]. This principle is *custom or habit.*" Id., p.43. "All inferences from experience, therefore, are effects of custom, not of reasoning." Id., p.44. "[A]fter a repetition of similar instances, the mind is carried by habit, upon the appearance of one event, to expect its usual attendant [the apparently related event], and to believe that it will exist." Id., p.69. "Our idea, therefore, of necessity and causation arises entirely from the uniformity observable in nature, which similar objects are constantly cojoined together, and the mind is determined by custom to infer the one from the appearance of the other." Id., p.75.

[15] Probability is not an unambiguous concept. Rudolf Carnap argued that there is concept of "logical probability" or "inductive probability" that reflects the confidence we can have in a particular theory or proposed law of nature in light of certain evidence–i.e., that reflects the likelihood that the law is true or the degree of confirmation provided by the evidence. He contrasted this concept with the more familiar concept of "statistical probability." See Rudolf Carnap, *Philosophical Foundations of Physics* (1966), pp.20, 34-5. This subject is discussed further in the chapter Mathematics.

[16] John Stuart Mill used the concept of causality or causal inference to explain the essence of inductive reasoning in his book *A System of Logic* (1904). See Peter Lipton, *Inference to the Best Explanation*, pp.18-19. An inference of causality was justified where all known cases of B were preceded by A or where there is only one identifiable difference between similar situations in which B occurs and those in which it does not (e.g., the presence of A). Id.

[17] I use the terms prediction and explanation, words that I think are more modest than foresight and understanding. Toulmin's concepts seem to imply somewhat greater grasps of the causal relationships involved.

[18] If this statement seems strange, I refer the reader back to the discussion of the concept of the "gene." To summarize, the theory of genes provides a highly accurate basis for predicting the heritability of traits, with actual hands-on applications, e.g., in the breeding of animals. But, modern biology strongly suggests that there is no such thing as a "gene," at least not in the sense that the theory was traditionally understood.

[19] Of course, a characteristic of light, which we discuss below, is that it appears to travel at the same speed whether approaching the front or the back of a moving body, that is, that its speed relative to a body is independent of the speed or direction of the motion of the observing body, contrary to "normal" experience. Thus, light as particles do not display the characteristics that Feynman says would be expected of his imagined gravity-inducing particles.

[20] Einstein's theory included the distortion of time as well as space, so that gravity also operates through the warping of time, which is a concept harder to visualize than the warping of space. *Id.*, p.15n.

[21] There is a gap between intuition and our pure sensory perceptions. Take the airplane. Our common sensory experiences probably tell us that

the airplane flies because the force of the air on the underside of the moving wing pushes the airplane upwards. The aeronautical engineers tell us that, instead, the movement of the wing through the air creates a vacuum on the top of the wings which pulls the airplane upwards. The result is the same. The explanatory model is different. I think that our reasoning abilities allow us to incorporate the engineers' model into our intuitive experience, even if our initial physical sensation would have suggested to us that the first model was correct. When you stick your arm out of the window of a moving car, you probably would say that you feel the air pushing your arm upwards and backwards. But, if you ask yourself is it possible that, in fact, a vacuum is pulling your arm upwards and backwards, you can conclude that the sensation is consistent with either explanation.

[22] Lord Rees says that Einstein transcended Newton by offering insights into gravity that made it seem more "'natural' and linked it to the nature of space and time, and the Universe itself." *From Here to Infinity*, p.81. I, like many (I would think), find the hypothesized curvature of space-time to be a concept that is far from normal. However, what Lord Rees probably means by "natural" is that the theory presents gravity as an integral part of the natural order, even if we find that natural order to seem contrary to the experiences of our sense or to our common sense.

[23] Elegance is undoubtedly a culturally-influenced criterion—certainly, what it is that constitutes elegance will be culturally determined; perhaps the relevance of the criterion itself is also a cultural phenomenon.

[24] It can be argued that a history of successful predictions, especially predictions of things not already known, justifies a belief in the theory, at least until a better one comes along. This approach to science was formalized as the hypothetico-deductive model by Carl Hempel in *Philosophy of Natural Science* (1966). Professor Lipton observed that "the hypothetico-deductive model seems genuinely to reflect scientific practice, which is perhaps why it has become the scientists' philosophy of science." *Inference to the Best Explanation*, p.15.

[25] *See also*, Polanyi, "The Creative Imagination," *Chemical and Engineering News* 44 (1966), p.85 ("Verification and falsification are *both formally indeterminate* procedures").

[26] As we shall discuss further, all predictions will be subject to the *ceteris paribus* condition: the prediction of the theory will be realized on the assumption that nothing else material to the result, beyond what has been

hypothesized in the making of the prediction, occurs in the process. Many scientific predictions can be tested in a laboratory where extraneous factors can be controlled or, at least, measured.

[27] Professor Barrow attributes the realization of the crucial significance of the initial conditions to two of the "deepest thinkers of the nineteenth century," James Clerk Maxwell and Henri Poincare, quoting Poincare's statement that "a very small cause which escapes our notice determines a considerable effect that we cannot fail to see, and then we say that the effect is due to chance. ... [I]t may happen that small differences in the initial conditions produce very great ones in the final phenomena. A small error in the former will produce an enormous error in the latter. Prediction becomes impossible... ." *Id.*

[28] Let me add, and explain, a caveat to this assertion. There has been some empirical evidence of regularity or predictability in the observed, laboratory evolution of bacteria. See Carl Zimmer, "Watching Bacteria Evolve, With Predictable Results," *The New York Times*, August 15, 2013. Research conducted under the direction of Joao Xavier at Memorial Sloan-Kettering Cancer Center, the results of which were published in the journal *Cell Reports*, demonstrated that a common species of bacteria called *Pseudomonas aeruginosa* repeatedly experienced mutations in the same gene that produced bacteria that could travel faster through the medium in which they were raised so as to obtain more food. This study dealt with only a specific adaptation in a specific environment. Dr. Xavier is quoted as saying: "In this case, it could be that there are only a few solutions in the evolutionary space." *Id.* Given that the particular mutation was capable of producing a physical characteristic that was clearly useful, if there were no other viable mutations that could be as useful, then it was predictable with large numbers and repeated reproduction that such a mutation would be likely to reappear in new samples. But, it is hard to imagine the state of knowledge that would be required for scientists to have predicted such a mutation if they had never seen it before.

[29] Of course, one could undertake an interesting exploration in the history of ideas to ascertain the actual origins of the root concept and the influences of various intellectuals on others. One could also speculate on the apparent concurrent emergence of ideas (whose "time has come") that seems to appear in the history of ideas. Naturally, the concepts of competition and of survival of the fittest also can be—and have been—applied to the "evolution" of various bodies of thought, including the natural sciences.

[30] The practice of diagnostic medicine may more accurately be characterized as intuitive than probabilistic. The distinction I am making can be captured in the question "Could a computer be programmed (at least, theoretically) to be as effective a diagnostician as a human being is?" To the extent that the successful practice of diagnosis and treatment depends upon intuitive and creative insights, then that level of achievement will be beyond the capability of the computer because such insights are not computable.

[31] For example, concerns have been expressed about the meaning of certain clinical trials that fail to detect any statistically meaningful effects of a drug across the sample group, despite individual stories of almost miraculous effects of the same drug on particular individuals. The explanation may be in our failure to understand "just how individualized human physiology and human pathology really are." Clinton Leaf, "Do Clinical Trails Work," *The New York Times*, July 14, 2013. In other words, there may be quite specific factors in particular individuals that make a drug highly effective, and many others with the "same" disease may not have those factors.

[32] Byers discusses the different uses of the words "subjective" and "objective." Citing the *New Oxford American Dictionary*, he notes that "objective" can be used to mean free of personal opinion or of a personal (biased) perspective but can also used to mean "not dependent on the mind for existence." Byers suggests that science is objective in the first sense, but not necessarily so in the second. *Id.*, pp.92, 98.

[33] I admit that part of the wonder I felt as I began to study neoclassical value theory, and then more traditional industrial organization economics, was that it seemed to provide an intellectual framework for the 1950s midwestern Republican values on which I had been raised. I then saw that one of the reasons for the almost violent antipathy toward the Economics Department being expressed by most of my college contemporaries was the very fact that it did indeed seem to provide support for free markets, inequality in the distribution of wealth and income and *laissez faire* economic policies.

Mathematics

> "[M]athematics is about everything and nothing.
> [I]t is the root answer to all questions,
> but the actual answer to hardly any ... ."
>
> Theodore Gray
> *Reactions*, p.104

Mathematics is encountered everywhere in today's world. Its usefulness (even, indispensability) as a tool in applications in almost all fields is beyond dispute. Moreover, pure mathematics may be the leading example of the unique capabilities of the human mind, reflecting both astonishing creativity and imagination coupled with powerful reasoning.

The specific question that I address here is: why is it that mathematics is so useful? Or, in other words, what is the relationship between mathematics and the physical world? The question is a methodological one, similar to the types of issues that have already been introduced. See, e.g., Horsten, "Philosophy of Mathematics," *The Stanford Encyclopedia of Philosophy*, Spring 2019; Ian Hacking, *Why Is There Philosophy of Mathematics at All?* (2014); Mark Colyvan, *An Introduction to the Philosophy of Mathematics* (2012).

What is Mathematics?

The term mathematics refers to a rather wide range of intellectual methods and activities that can look quite different and can serve very different methodological functions in science. Think of just the basic academic subjects: arithmetic, geometry, algebra, trigonometry and calculus. Many of the symbols used are different, the types of statements presented look different and the uses of the various disciplines are different. Yet, one recognizes intuitively that there is something fundamentally similar among all of these activities, even though the computations, if set forth on the page, would look very different one from another, both in form and function. So, what is mathematics?

An abstract science?

The physicist Richard Feynman said that "Mathematics is a language plus reasoning; it is a language plus logic. Mathematics is a tool for reasoning. ...Mathematics, then, is a way of going from one set of statements to another." *The Character of Physical Law* (1994) (originally published by The British Broadcasting Corporation in 1965, based upon a series of lectures given at Cornell University in 1964), pp.34, 39. Similarly, mathematician Morris Kline states: "Mathematics is concerned primarily with what can be accomplished by reasoning". Kline, *Mathematics*, pp.1-5. (For a long list of answers to the question, see Hacking, *Why Is There Philosophy of Mathematics at All?*, pp.41-7.)

Alfred North Whitehead, in an important essay on the nature of mathematics, wrote:

> "[A]rithmetic ... applies to everything, to tastes and to sounds, to apples and to angels, to the ideas of the mind and to the bones of the body. The nature of the things is perfectly indifferent, of all things it is true ... . Thus we write down as the leading characteristic of mathematics that it deals with **properties and ideas which are applicable to things just because they are things.**"

An Introduction to Mathematics (1911), p.9 (emphasis added).

In this sense, mathematics is an "abstract science"—that is, it is independent of all of the physical characteristics, the location, the feelings or emotions and the surrounding circumstances or environment of the "things" to which it might be applied. As a result, the concept of mathematics is particularly relevant to the expression of scientific laws. The mathematics can capture the permanent, determinative relationship among things. Whitehead declared:

> "To see what is general in what is particular and what is permanent in what is transitory is the aim of scientific thought. ...This possibility of disentangling the most complex evanescent circumstances into various examples of permanent laws is the

> *controlling idea of modern thought. ...[A]ll science as it grows towards perfection becomes mathematical in its ideas."*

Id., pp.11, 14.

Since the basis of mathematics is the systematic use of logic or reasoning or, worded differently, the creation of elaborate systems based upon rules of logic; mathematics always appears as a system of definitions (or concepts) and rules of transformation. Various criteria can be set out as to what constitutes a sound or valid logical system. Characteristics like internal consistency would seem to be of central importance. (Although, the need for that criterion has been challenged by Wittgenstein. Id., p.25.)

Yet, such logical systems cannot in and of themselves be said to be true or false as we normally think of the concept of truth, *i.e.*, as accurately or inaccurately reflecting some aspect of objective reality. That is the case because the logical or mathematical system as postulated has no factual content, no connection to objective reality external to the logical system itself.

Rudolf Carnap expressed this thought in a more sweeping form:

> *"It is true that the laws of logic and pure mathematics (not physical geometry, which is something else) are universal, but they tell us nothing whatever about the world. They merely state relations that hold between certain concepts, not because the world has such and such a structure, but only because those concepts are defined in certain ways."*

Philosophical Foundations of Physics: An Introduction to the Philosophy of Science (1966), p.9.

Or, as science writer Theodore Gray stated, just before the language quoted to begin this chapter, "Mathematics ... is the field that transcends all individual, specific issues and speaks only in universal and absolute truths." As already explored in connection with the use of models and theories, deductive logical systems, which most pure mathematics is, derive "true" conclusions from the stated premises as a matter of logical necessity, often as a matter merely of the definition of the terms. Such systems are self-contained and self-consistent. They simply do not touch the physical world.

In the *Principia Mathematica*, published in three volumes in 1910, 1912 and 1913, Alfred North Whitehead and Bertrand Russell sought to demonstrate that all of mathematics could be set forth as a system of logic that was complete and consistent. The general consensus seems to be that the effort was not successful.[1]

A PART OF REALITY?

One question is whether mathematics is created or discovered. One f the preeminent mathematicians of the twentieth century, G.H. Harding, maintained that mathematical advances were "discovered" or "observed". A *Mathematician's Apology* (1940), pp.123-4 (citing 26th edition). Does that mean that mathematics "exists" independent of humans?

Oxford professor of mathematics (and theoretical physicist) Roger Penrose endorses a version of Plato's mathematical forms, where mathematics exists in a perfect or idealized form external to the world of normal human experience. See *The Road to Reality: A Complete Guide to the Laws of the Universe* (2005), pp.11–7. Penrose does not claim that mathematical forms "exist" physically, in space or in time, or in some other reality, but that they embody a type of objectivity that is true across cultures and, perhaps, even across all intelligent beings of various kinds (if there are other such). *Id.*, pp.15, 17. Penrose makes several interesting points about why mathematics may be relevant to the physical world, some of which I discuss below; but he readily acknowledges that "only a small part of the world of mathematics need have relevance to the workings of the physical world" and that "the vast preponderance of the activities of pure mathematics today has no obvious connection with physics, nor any other science ..., although we may frequently be surprised by unexpected important applications." *Id.*, p.18.

There is a potentially significant difference between the conception of Plato and that of Penrose, however. Plato's mathematical forms were conceived as the pure and perfect forms of things that, in a less perfect form, existed in our world. The reflections of the forms with which we interact are only approximations of the pure forms, tainted or distorted by the imperfections of the physical world in which man and they exist. Thus, we would expect to find that our reality would be similar to and suggested, but not precisely described or defined, by the mathematical forms. Plato's forms could be thought of as the ideal of, or as epitomizing, the forms we experience.

In Penrose's conception, however, as we shall later see, mathematics, where it appears to be applicable to the physical world, is treated as the ultimate truth about or answer to how things actually are. Where our empirical examinations and observations do not provide the complete story, scientists turn to mathematics to derive the conclusions about what was, is and will be.

Penrose has also referred to "the mystery of why such precise and profoundly mathematical laws play such an important role in the behaviour of the physical world. Somehow the very world of physical reality seems almost mysteriously to emerge out of the Platonic world of mathematics." *Shadows of the Mind* (1995), p.413. He does not expressly address in *The Road to Reality* the possibility that the apparent truth of mathematics—and of logic in general—might be a reflection of the structure and nature of the human mind. In fact, in that work, he expressly treats mathematics as having an existence or reality, and a truthfulness, independent of man's mind (*see, e.g., id.*, p.17) and poses as one of the mysteries to be explained how it is that the mind can comprehend mathematical truth (*id.*, p.21).[2]

In an earlier work, Penrose identified what he considered to be three deep mysteries that relate one to another the world of our consciousness, the physical world and the world of mathematical forms. *Shadows of the Mind*, pp.412-4. The first mystery is why mathematical laws play such significant roles in the behavior of the physical world. The second is how the physical world can give rise to entities that are conscious and able to perceive it. The third is that that the human mind is able to develop mathematical models that seem so accurately to describe the physical world.

THE ONLY REALITY?

A curious argument has been recently proffered by physicist Max Tegmark. He suggests that there is a physical reality independent of the human mind (or other intelligence) and that it is mathematical. *Our Mathematical Universe: My Quest for the Ultimate Nature of Reality* (2014). In other words, the "real" reality is mathematics or, more precisely, a mathematical structure. *Id.*, p.254. "Everything in our world is completely mathematical—including you." *Id.*, p.260.

His argument is facially simple: An external physical reality independent of humans (or of the human mind) must be completely free of all "human baggage." A mathematical structure is free of all human baggage. *Id.*, pp.255-60. Thus, if there is "an external physical reality completely independent of us humans," then it "is a mathematical structure." *Id.* Finally, if that description (or "theory of everything") is truly complete, then the description and the physical reality must be equivalent and correspond to one another in all aspects. If they are so equivalent, then they are the same thing (the "principle of equivalency"). *Id.*, pp.262, 280-1. *Ergo*, reality is a mathematical structure.

By mathematical structure, Tegmark means something that has no intrinsic physical properties. It is purely relational, reflecting the structural relationships among abstract entities, themselves without intrinsic properties. One example is space, which one can argue is nothing in itself, only positional relationships among things. Another example is the electromagnetic force identified by fields. (Indeed, our discussion of fields in the chapter Physics illustrates the concept of mathematical structures. In the chapter Particle Physics and Quantum Mechanics, we will see that quantum mechanics views point particles as having no intrinsic properties themselves, only mathematical characteristics. *See id.*, pp.253-4, 274, 278-9.)

I have some comments about his assumptions and inferences:[3]

- There is a serious question as to whether any mathematical structures are or could indeed be free of "all human baggage." It is hard to see how that could be true of any mathematics that has already been developed or "discovered" by man. Even in the Plato/Penrose conception of mathematics as existing independent of man's discovery of it, the perception or discovery of it necessarily introduces "human baggage."
- To the extent that modern science based upon mathematics has reduced space, particles, motion and most everything else to mathematical structures, it is circular reasoning to conclude that that "fact" demonstrates that everything is, in reality, only a mathematical structure. The exercise begins with the assumption that mathematics reflects reality. The important questions of whether that is true and if so, why it is true, remain unanswered.
- Of course, we do not now (at least, not yet) and may never have a "theory of everything".
- The principle of equivalence evades the question. Where is the proof that any mathematical structure embodies everything about reality? Indeed, is there not something fundamentally different between a mathematical structure and a stone, at least when your foot strikes the stone?[4]
- And, there is a serious question, as discussed in the chapter Consciousness, as to whether consciousness can be explained by any reasonable extensions of existing scientific theories or by mathematics at all.

So, to me, Tegmark does not make his case with respect to a truly "objective" physical reality, whether mathematical or something else.

CHARACTERISTICS OF MATHEMATICS

USEFUL?

Arithmetic and geometry are recognized as highly useful tools for dealing with the world around us, whether in our daily errands, our work or our play. Indeed, many of the applications of the more elementary branches of mathematics as a tool seem almost second nature. Counting, adding, subtracting, multiplying are functions that tie directly to our actual experience of the things of the external world. Geometry also appears to be an idealized conceptualization of relationships that exist and can be observed in the physical world, such as the volume of a cube or the area of a circle.[5] We could say that these branches of mathematics may have been initially developed in order to represent relationships that existed.

But, science makes use of many forms of mathematics and does so in many different ways, beyond the construction of theories. Statistical analyses and related quantitative methods are used to organize data and to extract information from it, such as correlations between or among variables reflected in the data. This branch of mathematics, including probability theory, has found wide application in biology and the social sciences, as well as in astronomy and other natural sciences. Kline, *Mathematics for the Nonmathematician* (1967), p.499. In addition, "game theory" models are used in economics and evolutionary biology. See, e.g., Martin A. Nowak (with Roger Highfield), *SuperCooperators* (2011).

Many modern sciences make use of large simulations in which values for certain variables are placed into sets of equations and operations are performed hundreds, thousands or millions of times by computer. An intriguing use of such simulations is to perform hypothetical tests for potential causality. The idea is to make discreet changes in specific assumptions, run the model and then see how the resulting scenario compares to our observations of the existing world. If the constructed results resemble what has in fact happened, one has some support for the validity of the new assumption. See, e.g., Jonathan Webb, "Evolution 'favours bigger sea creatures,'" *BBC News*, 19 February 2015.[6]

True?

"Nature speaks in equations. It is an odd coincidence. The rules of mathematics were built around counting sheep and surveying property, yet these very rules govern the way the universe works." Charles Seife, *Zero: The Biography of a Dangerous Idea* (2000), p.117. Are the propositions derived by mathematics true as a factual matter (assuming that the original axioms or propositions are factually accurate)? In other words, does the world conform to the logic of mathematics? For this question, we need to focus on the distinction between mathematics derived (or derivable) from observations of the physical world and mathematics that has been constructed based upon concepts that are not part of our physical experiences (infinity, complex numbers, *etc.*).

There is the possibility that man has been able to extract causal relationships from observable experience and incorporate the observed logic of those relationships into a mathematical system. To the extent that the observations have indeed captured the truth of the underlying physical relationships, then the mathematics should enable us to extrapolate to phenomena yet to be observed and to make predictions with accuracy. The idea here is not that the constructed theory necessarily corresponds to reality in fact, but that the structure of the relationships corresponds to what does happen—*i.e.*, our theory is incorrect as an explanation but the relationships hypothesized in the theory do occur in fact.

The historians of science can debate, and maybe establish, the extent to which mathematics developed through an analysis of real world applications, such as counting things or describing the physical features of our three dimensional space, or as an independent, logical exercise that to the surprise of the creators has some real world application. Undoubtedly, there has been some of each. *See, e.g.*, Kline, *Mathematics*, pp.30-7, 51-4; Wigner, "The Unreasonable Effectiveness," pp.2-3.

Universal and timeless?

Some rather different sounding claims are often made about mathematics, along the lines of this recent statement: "Mathematical knowledge is unlike any other knowledge. Its truths are objective, necessary and timeless." Edward Freckle "Is the Universe a Simulation?" *The New York Times*, February 14, 2014. But, what does this claim mean? Is this the Platonic notion that mathematics has an existence separate from and independent of the world of objects that we inhabit or something else? G.H. Hardy asserted that mathematics is more "real" than our physical "reality", because mathematics, unlike physical objects, is just what it seems or appears to be. He explained: "317 is a prime, not because we think so, or because our minds are shaped in one way rather than another, but *because it is so...* ." A *Mathematician's Apology*, p.130.

Certainly, some of man's efforts to communicate with potential extraterrestrial intelligent life have assumed that certain of the mathematical relationships we observe in our physical world (such as the value of Pi) would be recognizable to other intelligent beings. It may be that the validity of that assumption depends upon the extent to which the other world inhabited by such beings resembles our world in terms of the geometry of its space. Perhaps it does, perhaps it does not. We may wonder further whether there may be differences in the mental processes of other beings that we would concede to be "intelligent" and "self-conscious" that would interfere with our ability to communicate using the pure forms of mathematics.

One can say that Pi reflects inherent, objective relationships embodied in the geometry of a circle, that is, numerical relationships between diameter and circumference and between radius (squared) and area. Those are relationships that exists independent of the human mind.

Or, are they?

Or, human-centric?

Is it not possible, on the contrary, that the concepts of diameter, circumference and area, as well as that of the circle, are human constructs and that an alien intelligence might simply not recognize Pi, because it does not recognize the circle? In fact, one could posit another, different sort of relationship between logic and truth. To the extent that logic and mathematics reflect the inherent characteristics of the human mind, they may not be truly objective nor universal nor, in that sense, independent of the physical world (that is, of the human mind). Apparent consistency between mathematical models and our understanding of the external world could simply be the result of both types of models being human creations.

If mathematics is structured around concepts of purely of human invention, then it would seem that we might find some coincidences of correspondence to the physical world and many examples of mathematics as simply an intellectual exercise. From this perspective, mathematics may not actually be "objective and concrete" nor mean the same thing to other intelligent life (if such aliens exist) as to human beings. Cf., e.g., Nowak, *SuperCooperators*, p.2 ("universal logic acting on universal rules").

It is suggested that both mathematics and human reasoning are the results of evolution and that, as such, we would expect them to reflect reality. But, such an argument would only apply to aspects of the world around us that have been directly relevant to our survival and reproduction. Modern science, and its mathematical models, go far, far beyond such matters, into realms of knowledge that have only recent or expected future relevance to our physical existence and other realms of knowledge that appear to have no currently foreseeable practical application.

However, the question, then, is why does human logic seem able to capture natural phenomena?

WIGNER'S FAMOUS QUESTION

Nobel-prize winning physicist Eugene P. Wigner gave a famous lecture in 1959 entitled "The Unreasonable Effectiveness of Mathematics in the Natural Sciences," *Communications of Pure and Applied Mathematics*, Vol. XIII (1960), pp.1–14. The title seems to have captured the philosophical puzzle in an imaginative way, since the lecture is frequently cited, and there are numerous articles published utilizing easily recognizable variations on Wigner's title. I do not cover all of the issues illuminated in that lecture. I simply note here that Wigner did not, in my view, present a very satisfactory answer to the question he asked.

Nonetheless, many scientists since have limited their discussion of the issue to a citation to that lecture, as if it held the dispositive answer. That is not to suggest that there have not been any more detailed explorations of the question. See, e.g., Colyvan, *An Introduction to the Philosophy of Mathematics*, pp.99–117. One suggestion is that the question focuses on the notable successes and fails to note the instances of failure in the application of mathematics.[7]

Of course, as already suggested, it may be a mistake to talk about mathematics as if it were all the same thing epistemologically. One might observe that arithmetic, at least in its simple form, reflects one aspect of material things, captured in counting, addition and subtraction. (Kant argued that arithmetic reflected the structure of time, based upon counting which takes place in time.) Geometry reflects the structure of space. One might speculate that each of these branches of mathematics developed simply as an accurate representation of relationships that exist in various aspects of the real world.

THE USEFULNESS OF MATHEMATICS

The impressive success of modern technology is a result of the inexplicable fact that the physical world can be simulated by computational models and that greater and greater accuracy can be (has been) achieved through more and more sophisticated mathematics. See Penrose, *Shadows of the Mind*, p.203. It is the "fruitfulness" of mathematics that is, perhaps, the most surprising fact. Mathematics has been used not only to capture the patterns of observed phenomena, but as a guide to the discovery of new theories and models that have then led to the discovery of other empirical patterns and relationships that had not previously been noticed. See *id.*, pp.416-7.

This capability may seem particularly surprising if one accepts what has been said so far about mathematics being just one or more deductive system in which the implications of the initial assumptions are all necessarily contained in those assumptions. In that sense, all mathematics is circular. There can be nothing new derived by manipulating the logical system, only different ways of expressing the same thing. So, how can that be such a powerful tool?

Where a set of relationships in the observed world has been defined (or hypothesized) logically as a deductive theory, it may be the case that the logical rules of the theory can be represented mathematically, using existing mathematics or even creating mathematics to fit the situation. The mathematical representation of the theory often provides very significant pragmatic benefits.

Mathematics is based upon a system of notation using symbols. The use of symbols in place of words and phrases forces precision and clarity, as well as providing great economy and simplicity in the formulation of propositions or assertions. The simplicity itself is important. As Whitehead observed, "[b]y relieving the brain of all unnecessary work, a good notation sets it free to concentrate on more advanced problems, and in effect increases the mental power." *An Introduction to Mathematics*, p.59.

Moreover, the process of breaking logical relationships down into many small steps makes it easier to avoid errors in constructing predictions and theories or to check for errors after the fact. These are very significant practical benefits attainable by setting forth theories in mathematical form. Finally, the mathematical rules allow extensive manipulation by hand and, for calculations, by machine (including the computer).

It also appears that the use of the mathematics may even enable the development of concepts that have useful analogs for or provide surprising insights consistent with the non-mathematical theory. Indeed, physicists Cox and Forshaw assert that "[e]quations are the most powerful tools available to physicists in their quest to understand nature." *Why $E=mc^2$*, p.21.

Roger Penrose writes:

> *"Computation can supply extremely valuable aid to understanding, but it never supplies actual understanding itself. However, mathematical understanding is often directed towards the finding of algorithmic procedures for solving problems. In this way, algorithmic procedures can take over and leave the mind free to address other issues."*

Shadows of the Mind, p.198.

Let us be more specific.

An equation sets out on each side of the "equals sign" two formulations that, when reduced to numerical values, will be equal in quantity. Sometimes the equality is "by definition," in which case the equation can be considered to be a definition of one (or both) of the terms. In other cases, the equation expresses an empirical relationship. It may be hard to tell the two apart. For example:

Distance (traveled) = Time (in transit) × Speed.

Is that an empirical statement or a definition? Are these equations "factual"–do they make contingent statements about the physical world? You might immediately say "of course, it is obvious that if one travels for T minutes at S speed, one will in fact travel D distance." Agreed. But, is that a factual statement? Or, is it merely a definition of D (and of T and S) and, as a definition, is true because it is defined so?

Either way, however, one can manipulate the formula and derive other relationships that are "necessary," that is, are compelled by or implicit in the original proposition, *e.g.*, $S = D/T$ and $T = D/S$. It is apparent that these simple manipulations of the formula generate new statements of the relationship that can be useful.

Similarly, mathematical logic tells us that if you do the same thing to both sides of an equation, then the two sides will still be equal. So, you can add the same value to both sides or subtract the same value, or multiple or divide by the same value and the equality will still hold. Why is this proposition "true"? Because it is inherent in the definition of "equals." But, that fact does not make the proposition vacuous. Surprisingly, there are many examples where this process of performing the same operation on both sides of an equation, followed by manipulations of the resulting equation according to the established rules, results in a statement (in the form of something "=" something else) that seems to provide new, unexpected information about the physical world. We can generate propositions that were not intuitively obvious before the manipulation was performed. We will come later to some examples. For now, suffice it to say that this application of simple mathematical methods appears to yield many quite striking results.

Mathematics can be utilized to explore various possible characteristics of a theory and make "predictions" of results that would occur if those characteristics obtained in the real world. In other words, the theories we construct about relationships in the world often lend themselves to mathematical representation. In those cases, we can use the power and versatility of mathematics to investigate those relationships and to make predictions about events, as well as to gain insights into the relationships.

But, is there something more? Can we actually answer Wigner's question: why is it that we find that many branches of mathematics have real world applications?

Unexpected Uses of Mathematics

One might approach the matter with a more limited question. Why is it the case that one branch of mathematics so often can be applied to or is useful in another branch of mathematics? See Hacking, *Why Is There Philosophy of Mathematics at All?*, pp.3, 8–11. It is a curious fact that algebra (derived from arithmetic) can be successfully applied to geometry and geometry to algebra, each being used to simplify or solve problems arising in the other. See *Id.*, pp.20–21. This fact may seem particularly strange given that the two branches of mathematics developed historically in different geographical regions and different cultures (geometry from the Greeks and algebra from India through Persia and Islam). In fact, sometimes a form of proof in one branch can be applied to create a proof in another branch. Perhaps, this applicability evidences some underlying logic in the nature of reality. (However, as we have come to learn, Euclidean geometry is apparently not a totally accurate representation of space.)

The possible answers to this question include the assertion that (i) there exists a mathematics that is true, parts of which are reflected in various branches, (ii) it is mere coincidence or (iii) the relationships are really more of analogies between the branches than applications of one to the other. Id., pp.11–4, 16–21.

A similarly curious phenomenon occurs where it is "discovered" that a branch of mathematics just happens to seem to reflect other relationships that exist in the physical world.

An interesting example is that of Boolean algebra. In 1854, George Boole published a book entitled *An Investigation of the Laws of Thought on Which are Founded the Mathematical Theories of Logic and Probability*. In that work, he undertook to set out in symbolic form Aristotelian logic. Every symbol had only two possible values: 0 for false and 1 for truth. He constructed rules of manipulation for the system. Multiplication is equivalent to "and"; addition is equivalent to "or". So, for example, a false statement **and** a true statement is false ($0 \times 1 = 0$); but a true statement **and** a true statement is true ($1 \times 1 = 1$). In addition, a true statement **or** a false statement would be true ($1 + 1 = 1$), and a true statement **or** a true statement would be true ($1 + 1 = 1$). This last proposition would not be true for integers; but, for statements that can be either true or false, the logic is understandable and correct. Interestingly and importantly (for practical purposes),

it happens that alternating current circuits can be analyzed with Boolean algebra, which captures the dynamics of two-phase or on/off circuitry. So, one can specify a result that one wishes to achieve through a series of switches, perform computations with Boolean algebra to design a simplified circuitry and find that the new design performs the desired function.[8]

Why? Or, perhaps, the question is, how?

The point for here is that this example and many others like it suggest that man has not always "forced" his own constructs on to nature (or nature into his own constructs) but, at least sometimes, has "found" an apparent good fit. Interestingly, Boole created the algebra in 1854, but it was not until 1937 that a young graduate student named Claude Shannon wrote a master's thesis in which he combined Boolean logic with electrical engineering, demonstrating that the logic could be implemented electronically. "Thus is borne the electronic 'logic gate'–and soon enough, the [computer] processor." Brian Christian, *The Most Human Human* (2011), pp.49–9.

Whatever the answer, it is clear that mathematics has led to scientific discoveries and scientific insights have led to new mathematics. There are numerous examples of what seems to be an interactive relationship between mathematics and the physical sciences. For example, Shing-Tung Yau has described the human process leading to his proof of the Calabi conjecture and the subsequent developments in string theory that flowed from or followed that proof. See Shing-Tung Yau and Steve Nadis, *The Shape of Inner Space: String Theory and the Geometry of the Universe's Hidden Dimensions* (2010).

Professor Barrow describes instances where scientific exploration found existing mathematics available to fill the scientists' needs and examples where the work of scientists identified areas of new mathematics that were needed for the science. These needs, in some cases, have been filled by mathematicians and, in some cases, are still awaiting solutions. *Theories of Everything*, pp.188–193. For example, when Max Born worked to explore the apparently new mathematical system that Heisenberg had utilized for the multiplication of two lists of quantities, representing frequencies and amplitudes in a system representing the behavior of electrons in an atom, he discovered that a whole branch of mathematics had already been developed that would serve for this new quantum mechanics—matrix algebra. David Lindley, *Uncertainty* (2007), pp.123, 113–4.

At the same time, it is also true that much of mathematics has no scientific application, at least not yet. Wigner, "The Unreasonable Effectiveness," p.7.

Some Comments on Logic

At this stage, it may be be instructive to take another look at logic. Let's use a well-known example: "All men are mortal. A is a man. Therefore, A is mortal."

The conclusion seems beyond dispute. In fact, it is hard to imagine how it could not be true. But, what does true mean in this context? Suppose that we discover an apparent "man" B who appears to be immortal because he has substantially outlived any prior example of a man. There are two clear possibilities. Obviously, B may eventually turn out to be mortal—that is, B may die. Alternatively, we may conclude that B is not actually a "man." The explanation may be that mortality is part of the definition of "man": if B is not mortal, then he is not a "man" (he may be an "angel" or a "god"). But, when worded this way, the logic is not very satisfying. There are other possibilities. We might find an ambiguity in the meaning of mortal or, at least, in the empirical observation of mortality (such as the length of life). Another possibility is in the meaning of the word "all." If it has a probabilistic meaning, then occasional counter-examples are compatible.

In any event, we would hope that our logical argument could go beyond merely asserting, just in different words, a definition that we have adopted. To have some meat in the theory, we would want the definition of man to be in some sense independent of the assertion that man is mortal. Clearly, it cannot be completely independent. There could be aspects of man, defined without explicit reference to mortality, that would still mean that man was, in fact, necessarily mortal.

EXAMPLE: ONE PLUS ONE

Take a simple arithmetical example. One plus one equals two. How could it be otherwise? If the proposition means no more than if you have one item and obtain another item, then you have two items; then it seems necessarily true—that is, true by reason of the meanings of the terms. By that, I do not mean just a matter of mere definition, but as a result of the structure of the proposition.

To illustrate, are there situations where one plus one does not "equal" two? If the word "equals" in the question is limited to its definition in logic or mathematics, then the answer must be "no" because of how the term "equals" is defined. In contrast, if "equals" is taken to mean "result in" or "yield," then there are some other interesting possible answers. For example, one male plus one female could turn out to result in three (or four or five, *etc.*). Or, one particle of matter plus one particle of anti-matter supposedly results in nothing or zero (ignoring the energy that assuredly would be released). One would say that these examples are fundamentally different from the original arithmetical proposition. If so, the difference is because the statements consist of more than arithmetic—they are statements that incorporate facts about the physical world and relationships in that world.

So, the arithmetical proposition is a statement about certain characteristics of things (closely connected to the concept of counting). But, it abstracts from or ignores other characteristics (such as the potential for biological, chemical or physical interactions). The logical proposition is clear and unambiguous, but trivial. The possibilities in real world applications are regularly much more nuanced, complex and interesting.

EXAMPLE: PROOF BY CONTRADICTION

Take the assertion that two propositions that contradict each other cannot both be true. A corollary is the idea that the proposition that the demonstration that a statement logically leads to a contradiction constitutes a "proof" that the statement is not true (or *vice versa*). Are these concepts empirical, that is, do they reflect things that are true (or untrue) about the physical world?

Clearly, the propositions seem to be integral to the notion of rational thought and logic and, probably, inherent in the human mental processes. Professor Arkes states that "[t]he law of contradiction expresses a necessary truth, and all efforts to refute it will fall into the embarrassment of self-contradiction." *First Things*, p.51. Following a well-established philosophical tradition, Arkes identifies "necessary truths" that can be known independent of experience or as *a priori* facts. These propositions are things that must be so, because they are innate to the human cognitive process. (They include space, time, causality and "the law of contradiction".) *First Things*, pp.51-84. This "necessary truth," like the similar concept of causality (discussed above), may be "necessary" in terms of the functioning of the human mind, but is it necessarily "true"? It may be that we are incapable of rationally conceiving of two contradictory propositions both being true, but our reason does not control the actual physical relationships of the world. In fact, the concept of inconsistency itself seems necessarily to depend upon human theories or models.

We might say that we have never observed in nature an entity that is simultaneously dead and alive, as evidence that nature does not tolerate inconsistent or contradictory states. But, that claim requires definitions of "dead" and "alive". One could imagine plausible definitions that overlap, allowing the alleged inconsistency to exist in the external world (say a definition of death that turns on brain activity and a definition of life that is based upon cell reproduction). If the definitions are structured to be mutually exclusive, then no entity could satisfy both—by definition. Take a different type of example. We recognize that a person could simultaneously be both "short" and "tall", since those characteristics are relative and we could have two different points of reference.

I noted that scientists often use inconsistencies as a form of proof. A typical "proof" using this type of argument would be as follows: Proposition A is either true or false. I can deduce (using deduction) that if A is false, then both B and not-B are true. Since that conclusion presents a contradiction that cannot be accepted, I conclude that A is true. Actually, the logic tells us only that A is not false, since that is the assumption that led to the contradiction. The conclusion that A is, therefore, true depends upon the initial assertion or assumption that A is either true of false. If A could be something else, then the conclusion that A is true is not established.

Zero and Infinity

Charles Seife has written: "Zero...is infinity's twin. ...The biggest questions in science and religion are about nothingness and eternity, the void and the infinite, zero and infinity." *Zero*, p.2. I think that the statement is a serious oversimplification, and a misleading one at that.[9] But, it is correct to observe that both of these concepts give rise to problems in scientific applications. I think it is useful to explore why. As we shall see, the situations are quite different with respect to zero and infinity. And, the differences are important.

The Construct Called Zero

Mathematics certainly had problems with zero,[10] but at least zero was a concept that can be related to the actual experience of counting (sort of, anyway). Contrary to some popular perceptions, one does not need zero for purposes of counting, i.e., the absence of any As ("no As") does not have to be characterized as "zero As" in the sense that the presence of A and A is clearly "2 As."

On the other hand, the use of zero as a "place holder" in larger numbers does make intuitive sense. For example, a number 100 represents one hundreds, zero tens and zero ones. It is clearly a convenient – indeed, a highly useful – convention. In fact, the first use of zero, Seife says, was not as a number but was as a place holder to distinguish 1 from 10 or 100. *Zero*, pp.26–53. See also, Deutsch, *The Beginning of Infinity*, p.131.[11] Zero also enabled the development of the algebraic form where a series of values can be expressed as equaling zero. Whitehead, *An Introduction to Mathematics*, pp.66–8. The first use of zero as another number is attributed to mathematicians in India. Seife, *Zero*, pp.66–71; Deutsch, *The Beginning of Infinity*, p.131.

Zero may not have direct obvious applications in daily life, but it is not in concept really different from the other cardinal numbers. See Whitehead, *An Introduction to Mathematics*, p.63. While there may not have been many pressing applications for the addition to or subtraction of zero from a positive quantity, the concept does not seem foreign to our physical experiences. Even the notion of zero times a positive number (say, zero groups of five apples) has an intuitive meaning.

The mathematical solution for zero has actually been effective in application to the physical world, with one important exception–division. The concept of division by zero seems of no use in our daily activities. As mathematics developed, however, it was deemed that a positive number divided by zero was infinity. See *id.*, p.71. I say "deemed" to identify this usage as a convention adopted by man. I do not think that it is logically necessary or even logically persuasive to assert that a positive number "contains" an infinite number of zeros (or any zeros, for that matter) or that zero "can go into" a positive number an infinite number of times.

Take the well-publicized issue of "zero divided by zero." (To get an idea of the extensiveness of the discussions, simply enter the phrase in Google and search.) If we address the question using mathematical conventions, we discover that there are multiple possible answers. For example, zero divided by any number is zero, so the answer is zero. But, any number divided by itself equals one, so the answer is one. Or, if you view division as the corollary of multiplication (*e.g.*, 6 divided by 3 equals 2, because 2 times 3 equals 6), then zero divided by zero can equal any number, since zero times any number equals zero. One might simply say that the answer is indeterminate. However, we are told that any number divided by zero is infinity.

Seife asserts that "if you wantonly divide by zero, you can destroy the entire foundation of logic and mathematics. Dividing by zero once–just one time–allows you to prove, mathematically, anything at all in the universe." *Id.*, p.23. This statement is certainly at least an exaggeration, but it does identify a problem with the mathematical treatment of zero. The structure of mathematics required a convention for the treatment of zero in mathematical operations; however, the convention adopted for division by zero created problems–the problems of infinity.

The Construct Called Infinity

A mathematical convention was also adopted for the concept of infinity. But, I think that infinity is quite a different type of concept.

Infinity is simply not a factor or characteristic that we experience in our physical world. We can conceive of a series of numbers going on forever, with the sum of which series continuing to become ever larger and always capable of being greater than any number one would propound. Indeed, it is very difficult to imagine how such a string could ever end—one can always add one more number. Although, such an endless string can be imagined; it is not something that we actually experience. We can talk about something as being "unlimited" or "unbounded". See, Deutsch, *The Beginning of Infinity*, at pp.164-5, 181. Such language is meaningful, even if the underlying concepts have no actual analog in the physical world.[12] But, there is a gap between the conceptualization of such an endless stream of numbers and the concept of the something we have called infinity.

How should infinity be defined in mathematics? For example, is two times infinity greater than infinity? Obviously, it must be, in some sense; but, the concept of infinity would seem to indicate that the common sense answer is no. When we add or subtract a positive number to or from infinity, the answer is unchanged—it is still infinity. How can one infinity be larger than another?

The nineteenth century mathematician Georg Cantor "founded the modern study of infinity." Id, p.166. The modern approach is based upon "infinite sets"–sets with an infinite number of members. "The defining property of an infinite set [or of an infinity] is that some part of it has as many elements as the whole thing." Id., p.167. However, at the same time, Cantor demonstrated that some infinities are larger than others, by showing that there is no one-to-one correspondence between certain pairs of infinite sets (specifically, for example, the number of points on a finite continuous line is greater than the number of whole numbers, since for any rule of correspondence between the points and the whole numbers, there are always points between the points that correspond to the whole numbers). Id., pp.170-71. The conclusion is that one infinity may be larger than another.

One consequence of the mathematical definition of infinity is that probability has no meaning in the context of infinite sets. One simply cannot say that one outcome is more likely than another. Id., p.177. See also, William Byers, *The Blind Spot*, pp.24-8. Another consequence is the "ability" of mathematics to model the performance of an infinite number of tasks in a finite amount of time. Here we have another insight into Zeno's paradox of the tortoise and the hare. Mazur, *The Motion Paradox*, pp.38-9. "Mathematics tells us that it happens without explaining why [or, in this case, how]." *Id.*, p.25. The infinite steps of the hare, each halving the distance between him and the tortoise, take place without the passage of time. If time does not pass, then it is not surprising that the hare fails to overtake the tortoise. But, if time does pass, then the hare will reach the finish line first, despite the mathematical model.

When the mathematical concept of infinity appears in our physical theories, the mathematics often breaks down. *See, e.g.*, Greene, *The Hidden Reality* (2011), pp.208-11; *see also*, Deutsch, *The Beginning of Infinity*, pp.258-62.[13] And, of course, by convention, infinity is introduced whenever one divides by zero. Indeed, it seems that "[e]very time mathematicians tried to deal with the infinite or with zero, they encountered trouble with illogic." Seife, *Zero*, p.113.

We shall return to this subject in the context of scientific theory where it becomes apparent that many of the developments of twentieth century physics, from quantum mechanics to superstring theory, were outgrowths of efforts to cope with the mathematical absurdities that arose because of infinity. *See, e.g.*, Greene, *The Hidden Reality*; Seife, *Zero*, pp.157-209. The use of mathematical tricks to eliminate infinities is widespread. *See, e.g.*, Penrose, *Cycles of Time*, p.210; Hawking, *A Brief History of Time*, pp.157, 162. Indeed, infinities posed problems even for Newton in the development of the calculus, requiring him to engage in a notational trick to get the proper results. Seife, *Zero*, pp.114-7. ("Quantum electrodynamics, it turned out, is a well-behaved theory in which the infinites can be unambiguously removed by introducing two parameters that have to be determined experimentally: the mass and charge of electrons." Sabine Hossenfelder, *Lost in Math: How Beauty Leads Physics Astray* (2018), p.32).

In fact, several important developments in the mathematical models being used to explain the physical world arose as creative means of avoiding the appearance of infinity in the calculations. String theory is a relatively recent example. Vibrating strings solved part of the mathematical problem; the introduction of many additional dimensions (up to 26) finished the job. Yet, no one has any confidence that the real world contains "vibrating strings" or even more than four dimensions (three for space plus time). *See, e.g.*, Krauss, *A Universe from Nothing*, pp.130-31. (String theory replaces particles or points with loops. One benefit is that the points would reduce to "singularities" or zeros, whereas the loops have added dimensions that avoid the unfortunate consequences for the mathematics of the zeros–and the resulting infinities. See Seife, Zero, pp.194-5.[14])

There are many other new conventions required for the further development of pure mathematics, like imaginary numbers. Even though the subjects of some of these conventions had no apparent analogue in the physical world, the resulting mathematics did often have applications. For example, imaginary numbers were integral to the development of vector analyses. See Whitehead, *An Introduction to Mathematics*, pp.87-111. And, the concept of the infinitesimal was at the heart of the development of calculus. *Id.*, pp.217-35. But, the concept of infinity seems to invite all sorts of silly assertions. (For example, utilizing the mathematical construct of infinity, physicist David Deutsch reasons that human knowledge is necessarily at the beginning of its achievements, since any specified point is, by definition, "at the beginning of infinity." See, e.g., *The Beginning of Infinity*, pp.164-95, 196.)

Indeed, twentieth century mathematician David Hilbert delivered a lecture in 1925, declaring:

> "[T]he universe is finite in two respects, i.e., as regards the infinitely small and the infinitely large. ...[T]**he infinite is nowhere to be found in reality**, no matter what experiences, observations, and knowledge are appealed to. Can thought about things be so much different from things? Can thinking processes be so unlike the actual processes of things? In short, can thought be so far removed from reality? ... [T]**he infinite ... neither exists in nature nor provides a legitimate basis for rational thought** – a remarkable harmony between being and thought. ...The role that remains for the infinite to play is solely that of an idea ... ."

"On the Infinite," in *Philosophy of mathematics: Selected readings* (Second edition, 1983) edited by Paul Benacerraf and Hilary Putnam (emphasis added).

An Alternative?

An interesting case can be made for the exclusion of infinity and infinitesimal from our efforts to understand the world or the Universe, as opposed to applications of pure mathematics. We could still think of quantities too small to measure or to notice and of quantities too large to imagine. And, the boundaries of each category can change, and have changed, with increases in our technology and knowledge—we are now aware of things formerly too small to measure and of expanses of space and time far beyond our former ability to contemplate.

Under a philosophy of mathematics that arose in the late nineteenth century, the scope and content of mathematics would be substantially reduced. See Barrow, *Theories of Everything*, pp.186-88. Such a philosophy would limit mathematics to those statements that can be deduced from the natural numbers.[15] Thus, it would exclude the introduction of human constructs that are not based upon our actual experience, such as infinity, imaginary numbers and complex numbers, which "turns out to have the most dramatic consequences for the whole scope and meaning of mathematics." It would even eliminate the use of "such familiar devices as the argument from contradiction (the so-called *reductio ad absurdum*), wherein one assumes some statement to be true and from that assumption proceeds to deduce a logical contraction and hence to a conclusion that the original assumption must have been false." As examples, the General Theory of Relativity and current cosmological science depend at various points upon the use of contradiction. *Id.*

CONCEPTS OF PROBABILITY

The concept of probability will appear frequently in subsequent chapters. Therefore, I think that it is useful to say a few things about the concept and the branch of mathematics that is often referred to as "probability and statistics."

We all have some intuitive notion of probabilities. Many things in our lives are, at any given time, perceived as more or less likely. We feel with virtual certainty that the Sun will rise tomorrow and can predict with confidence the precise time at which it will do so. Whether we will see the Sun is less certain. We might estimate the likelihood based upon the time of year, this evening's weather and the forecast for the morning. We also have an intuitive feeling for the concept of chance (and luck and risk). Indeed, the use of presumably random events to learn the will of the gods or to divine the future is of ancient origin. See, e.g., Hacking, *The Emergence of Probability* (1975), pp.1-3.

LUCK, RISK AND CHANCE

Not surprisingly, the rigorous investigation and analysis of probabilities arose through the mercenary interest in winning at games of chance.

Games of chance have a very long history in human society, perhaps predating history itself. See Amir D. Aczel, *Chance* (2004), pp.viii–ix. And, that is still likely the context in which probability theory has the most obvious meaning to many of us.

Elaborate models were developed to calculate the probabilities of various complicated outcomes in such games in the seventeenth century by Galileo, Blaise Pascal, Pierre de Fermat and Abraham de Moivre and in the eighteenth century by Jacob Bernoulli, Thomas Bayes and Pierre Simon de Laplace. See, *e.g.*, Aczel, *Chance*, p.xii; Carnap, *Philosophical Foundations of Physics*, pp.23–4; Hacking, *The Emergence of Probability*, pp.11–2, 143–53. The rules students learn today about the calculations of probabilities of various permutations and combinations derive from these developments.

As in other areas of mathematics, the creation of a rigorous logical system of probability theory that began with propositions that seemed obvious and natural led to formulae and results that were surprising and even counter-intuitive. For example, the probability that at least one of several possible things might happen is properly calculated not by adding the probabilities of the various possibilities but by subtracting from 1.0. (The probabilities of all possible occurrences must add up to 1.0 or certainty.) Thus, the probability that heads will appear at least once in three tosses of a coin is not three times 1/2 (the odds of getting heads on any one toss); instead, it equals 1.0 minus the probability that the coin toss results in three tails. The probability of tossing three tails is .125 (the result of $0.5 \times 0.5 \times 0.5$), so the probability of getting at least one heads is .875 (or 87.5%). The reason for approaching the question from the opposite end is the need to account for and eliminate the overlaps (*e.g.*, getting a heads on any one or all of the tosses meets the criteria of at least one heads). This rule is referred to as the Law of Unions of Independent Events. See Aczel, *Chance*, pp.25–39.

It may happen that the casual practitioner of such calculations will discover that the theory does not apply very well in many real world situations. The reason is that games of chance have some rather special characteristics. The unusual feature of the classical view of probability is that one would (in an honest game, at least) confront a variety of outcomes each of which is equally likely to occur—such as in the flip of a coin, the roll of dice, the hand of cards. Situations in which alternative outcomes could be intuitively perceived as equally likely, that is, where no reason could be ascertained why one number or card would be more likely to appear than another, are, however, rather limited. Most of our real world experiences are much more complicated and ambiguous.

A subsequent and more general theory of probability focused on the expected frequency of occurrences, generally based upon or support by experience or experimentation. If one did something 100 times, a particular outcome would tend to occur some particular number of times out of the 100 (or 1000, etc.). But, in the real world, we encounter single events where the notion of frequency of outcomes is not directly applicable. What is the meaning of probabilities in such situations? Id., pp.25-8.

Actually, we use a concept of probability that is much more nuanced than the comparisons of relative frequencies. We regularly conclude that one outcome or state of affairs is more probable than another. We may say that something is possible, likely, very likely or, even, almost certain. Such usage often cannot be restated in terms of relative frequencies. Indeed, in many such cases we would not entertain the prospect of putting a numerical value on the probability we have expressed. We may say that it is probable (or very probable) that our child is sick. That is a meaningful statement. Yet, it would seem quite strange to try to assess whether that condition is 70% or 80% probable.

Carnap proposes the phrase "logical probability" or "inductive probability" to refer to this type of usage. We derive logically a sense of the likelihood of something based upon a set of facts or evidence, drawing inferences from the evidence (facts such as a lack of appetite, a fever, a cough, etc.). See Carnap, *Philosophical Foundations of Physics*, pp.20-22, 34-5. See also, Hacking, *The Emergence of Probability*, pp.13-5.

In all events, probability theories developed out of the particular interests and mind-sets of the theories' creators. Yet, such theories are often put to use in contexts far different from those in which they emerged—for example, in statistical inferences and in quantum mechanics. See Hacking, *The Emergence of Probability*, p.9.

Bayes' Theorem

There is another type of probability that I want to discuss. It is based on Bayes' Theorem, a formula discovered by the Reverend Thomas Bayes in the eighteenth century, and is one that can generate some rather surprising conclusions. See, e.g., Aczel, *Chance*, pp.95-6.

It can best be explained by a famous example. Suppose there are three people, one of whom has won a prize through a random drawing, the result of which is still undisclosed. Each had an equal 1/3 chance of winning, so each assumes that the probability that his name was drawn is 1/3 or 33.33%. But, A then says to the moderator, "Obviously, at least one of B and C cannot have won. Tell us the identity of one of the two who did not win, so that that person can leave." The moderator announces that B did not win. What now, with that additional information, is the probability that A is the winner?

Has it become 1/2? Actually, Bayes Theorem tells us that the probability that A won is still 1/3. Perhaps more surprisingly, however, according to the Theorem, the probability that C has won has become 2/3. See *id.*, pp.95–103. How is that possible? The Theorem states that the probability that A wins given that B was excluded equals the probability that B is excluded if A won times the probability that A won divided by the sum of that number plus the results of two other multiplications: the probability that B is excluded, even though B won, times the probability that B won and the probability that B is excluded, even if C had won, times the probability that C won. The key to the result is that the last element, the probability that B is excluded if C won, is 1.0, because the moderator is choosing between B and C. If C had won, then the moderator would have to have excluded B. (The probability that B is excluded if B won is zero.)

What does this really mean and does it matter? Well, before the moderator provided the additional information, the probability that either B or C had won was 2/3 (1.0 minus the 1/3 probability that A won). With the information that B had not won, the probability that one of the two of them had won did not change, but the only one that now could have won was C. So C enjoys the 2/3s chance that B and C had together. A's chance has not changed.

Note that if the request to the moderator had simply been to identify one of the three who had not won, the identification of B as a loser would have improved the chances of A and C (both would then have a 1/2 chance of being the winner). The information contained in the responses to the two requests is different, affecting the calculation of the probabilities with the benefit of the additional information. Note further that this analysis depends upon the fact that the winner had already been determined but was not disclosed (otherwise, the moderator could not have granted the request). That situation is different from the prospective flip of a coin. In the case of the coin, the question is: What is the likelihood that the toss will yield heads (or tails)?

In the game described, we are not predicting the future. Indeed, the probabilities with respect to who won are no longer even probabilities.

Someone did win; two others lost. No additional information or event will change the outcome. It is 100% certain that one of the three did win and 100% certain who it is. We just do not yet know who it is. Instead, the real question being presented is: What is the probability that the person we pick is the person who actually did win? With respect to that question, the additional information is relevant. In other words, the probability that C would become the winner was 1/3, and that probability was not affected by the subsequent events. However, the probability that B was the winner (after the winner had been determined) did change with additional information.

Of course, we are now using probability in a somewhat different sense than we were using it when this discussion started. It is not the probability that something will happen in the future, like with the flip of a coin or roll of the dice, but the probability of making the correct choice. The probability that C was the winner was either 0% or 100%, once the winner had been determined. The probability of one identifying the actual winner changes with the increase of information.

This distinction is worth remembering. Later we will talk about theories that involve the probabilities that a particle **is** in a particular place or **has** particular characteristics and the probabilities that a particle **will be** in a particular place or **will have** certain characteristics. As we have just seen, those two similar-sounding propositions are actually rather different. The answers to both questions depend upon information. In both cases, one needs information relevant to the likelihoods of different future outcomes in order to calculate probabilities; but, in only one case, does information about what can be observed to have already happened become relevant.

THE TOSS OF A COIN

There is one additional aspect of probabilities that I want to mention. There is a very great difference between the probability that out of 100 tosses of a coin, 50 will come up heads, and the probability that one will observe in the 100 tosses, 5 heads followed by 5 tails followed by 5 heads and so on (or that heads and tails will alternate perfectly or that the first 50 will be heads).

The statement is obvious. But, to generalize, any specific sequence of results will be highly improbable in advance. Yet, 100 tosses of a coin will necessarily result in some specific sequence of heads and tails, every time. Thus, whatever actually happens will have been highly unlikely to occur before it happens. Yet, it happens nonetheless, every time. That is just the way it is.

Laws of Nature

> "It is not a cause;
> yet it is more than merely a description.
> It is true
> because it is beautiful and simple;
> yet it is never quite true at all."
>
> James Gleick,
> "Introduction"
> Feynman, *The Character of Physical Law*, p.ix.

The so-called Laws of Nature are generally identified and expressed in the language of mathematics. See Wigner, "*The Unreasonable Effectiveness*," pp.4–6. These Laws, if they exist, have been generally assumed by scientists and philosophers of science to be constant and unchanging both in space (throughout the Universe) and in time (over billions of years).[16] In addition, the Laws of Nature are assumed to incorporate or reflect certain "Constants" of nature that also do not change in space or in time. See Barrow, *The Constants of Nature* (2003). Indeed, the assertion is made that "[i]n fact, it is not logically possible for all the Laws of Nature to be changing." Barrow, *Theories of Everything*, p.27. This last statement, however, turns on the definition of Law, so it is essentially circular and almost vacuous. The relevant issue is an empirical, not a logical, one.

Not all scientists accept the assumption that the Laws are invariable and eternal. Theoretical physicist Lee Smolin unambiguously asserts (with no evidence) that physical Laws have evolved and will continue to evolve through time. *Time Reborn: From the Crisis of Physics to the Future of the Universe* (2013), p.xxv. He claims a long tradition for this view, providing quotations from Paul Dirac, John Archibald Wheeler and Richard Feynman that at least acknowledge the possibility that the Laws of Nature change and, perhaps, evolve.[17]

Similar points can be made about the supposed "Constants" of nature, the numerical values contained in many Natural Laws which we shall discuss below. The Laws of Nature often set forth an algebraic relationship between two variables that includes a Constant. Examples are G (the gravitational constant), c (the speed of light) and h (the Planck constant). The

Constants are assumed to be the same, regardless of when or where they might be found.

Importantly, theories do not generate the numerical values of the Constants; the values exist as factual matters:

> "The laws of science, as we know them at present, contain many fundamental numbers, like the size of the electric charge of the electron and the ratio of the masses of the proton and the electron. We cannot, at the moment at least, predict the values of these numbers from theory—we have to find them by observation...."

Hawking, A Brief History of Time, p.125.[18]

The derivation of definitive values for these Constants has not always been easy; but, even where the empirical results are very robust, they are necessarily limited to the values of those Constants in modern times. It is simply assumed that the values have always been and always will be the same. Moreover, it is certainly not obvious that the Laws and Constants that we have observed on Earth are equally applicable throughout the Universe. Many of these Laws and Constants have been extensively tested and measured empirically only on Earth and within our Solar System, but the theories of which they are part are normally presumed to apply everywhere. Similarly, as we shall see, we currently know that many Laws of our observable macro-world (like Newton's Laws of Motion), as well as many of our maxims of common sense, do not apply at very small scales, the so-called quantum world. Why do we assume that the Laws apply to very large scale phenomena?[19]

"[F]ar out in space lie environments differing hugely from our own. We should not be surprised that commonsense notions break down over vast cosmic distances, or at high speeds [approaching the speed of light], or when gravity is strong." Martin Rees, *Just Six Numbers: The Deep Forces That Shape The Universe* (2000), p.37.

THE CETERIS PARIBUS ASSUMPTION (AGAIN)

We have already discussed the *ceteris paribus* condition. The use of the condition might sound a bit strange in this context. That is because the theories of the physical sciences are not normally set forth in such language.

Most of our theories in the physical sciences are presented (often implicitly) in terms of subsystems, in the context of which everything else is assumed to be irrelevant. This fact is apparent whenever one simply looks at a mathematical model of a theory; there are only a finite, even if sometimes very large (and sometimes very small), number of variables utilized in the model. For essentially all practical purposes, everything else is deemed irrelevant (or, in some cases, can be controlled for). But, such an approach implicitly contains the *ceteris paribus* condition. For example, the Newtonian Laws of Motion focus on the particular bodies of interest. In doing so, they effectively assume that, for example, the gravitational forces arising from all other matter in the Universe do not exist. Where those other forces are very, very small or do not have a systematic influence on the matter under examination (do not consistently bias the results one way or another), the assumption is satisfactory, because the resulting predictions are sufficiently accurate for the purposes at hand.

But, what needs to remain the same are only those characteristics relevant to or that have an effect upon the causal relationship being examined. Laws of Nature are relationships as to which the relevant *ceteris paribus* factors exclude much of the aspects of the natural world that our senses are capable of detecting. If that were not so, then our inherent limitations would prevent us from even identifying those Laws. Simplification and selectivity are necessities for the human mind to cope with our exceedingly rich and complex physical environment.

LAWS AND REALITY

The models incorporating these Laws are simply not full or, therefore, accurate representations of reality. All models are inevitably only approximations, since they apply only to subsystems and exclude the influences of everything else. Smolin, *Time Reborn*, pp.38-39. In addition, many theories are also only approximations even within the subsystems to which they apply, because they do not fully or accurately capture even the relationships that exist within the subsystem. Over time we find that our approximations become better and also, often, more inclusive (such as General Relativity compared to Newtonian mechanics), but even if they come actually to reflect the real relationships within the subsystems, they are still approximations from the standpoint of the Universe as a whole.

Thus, one could argue that such Laws cannot, by definition, be truly universal or timeless. One could then conclude that this type of limitation to our scientific theories will necessarily continue to exist until we have a complete theory of everything, a theory that is able to incorporate everything in and about the Universe, including time. *Id.*, p.40. We will address several times below issues that arise with respect to a potential theory of everything.

Of course, one might say that if nature did not obey certain simple Natural Laws, we would not be able to understand it. See Barrow, *Theories of Everything*, pp.10–11. Of course, such an observation proves very little: we think we understand some natural phenomena, so those phenomena must be obeying Natural Laws; but, if we do not really understand those phenomena (but just think that we do), we can draw no relevant conclusions. Can the idea of Laws be just a human construct that has fooled us into thinking we understand the world?

Such "rules" might not be "Laws," because they might not apply to the very beginning and end of the Universe or across the vastness of space. But, whether one uses the word Law, there are interesting questions as to whether our perception of regularities and rules may simply be an illusion (that possibility does not seem very likely, since the existence of regularities within the portions of the Universe we can observe seems to be a prerequisite to our being here) or whether our speculation that there are Laws reflects a misinterpretation of the evidence (for example, these regularities could be a mere coincidence of our rather small part of the Universe). *See, e.g.*, Barrow, *Theories of Everything*, pp.24–26. These issues obviously become important when one gets to astrophysics and cosmology.

Much of the rest of this book is about issues concerning these supposed Laws.

USEFULNESS REVISITED

As we said above, mathematics reflects the formalization and extension of human logic or the process of rational thought. These extensions have become quite elaborate.

Initially, mathematics reflected the analysis of "things" that people encounter in the physical world—"things" people experience through their senses. These "things" include numbers (as applied to quantities, counting and measurements) and shapes. Arithmetic and geometry became pretty sophisticated over time, revealing relationships and techniques that are not directly obvious. However, mathematics evolved well beyond the practical or even the applicable. Wigner, with some exaggeration, stated in his famous lecture that "Most advanced mathematical concepts, such as complex numbers, algebras, linear operations, Borel sets[,]...were so devised that they are apt subjects on which the mathematician can demonstrate his ingenuity and sense of formal beauty." "The Unreasonable Effectiveness," p.3.

Perhaps, we can say that mathematics is a human creation, that it has largely developed as an exercise in human reasoning and creativity and that it has often been pursued for the aesthetic and intellectual pleasures of the mathematician. *See, e.g.*, Kline, *Mathematics*, pp.3–5; Barrow, *Theories of Everything*, pp.175–76; Wigner, "The Unreasonable Effectiveness," p.7. As noted above, it is possible not only that mathematics is a creation of the mind, but of the human mind, with whatever lack of correspondence with or inability to see reality is inherent in the human mind.

Penrose explains that the use of mathematical models in science is necessary to obtain the precision required in the formulation of questions and "well-defined" answers, treating mathematics as a particularly precise or unambiguous language for communication. *The Road to Reality*, p.12. However, Barrow also asserts that "mathematics is a language that possesses a built-in logic which is unexpectedly attuned to reality." *Theories of Everything*, p.174. I think that this last statement is quite misleading.

THE MIND AND THE PHYSICAL WORLD

These observations bring us back to the question of whether the human mind and its deductive theories do or can accurately reflect the structure of the physical world and, if so, why that is the case. In the end, it is physical reality, not mathematics, that determines what is true in fact.

> "[T]he mistake is to confuse an abstract attribute with a physical one of the same name. Since it is possible to prove theorems about the mathematical attribute, which have the state of absolutely necessary truths, one is then misled into assuming that one possesses a priori knowledge about what the laws of physics must say about the physical attribute."

Deutsch, *The Beginning of Infinity*, p.183.

Deutsch further asserts:

> "a computation of a proof is a physical process in which objects such as computers or brains physically model or instantiate abstract entities like numbers or equations, and mimic their prop-

erties. ...It works **because we use such theories only in situations where we have good explanations** saying that the relevant physical variables in those objects do indeed instantiate those abstract properties."

Id., p.188 (emphasis added).

In addition, Deutsch argues that the conclusions of Hilbert, Turing and Gödel about the limits of computing or proving mathematical statements are the results of the existing laws of physics, reflected in the operations of our brains and our computers:

> "Different physical laws would make different things infinite, different things computable, different truths—both mathematical and scientific—knowable. ...[I]f the laws of physics were in fact different from what we currently think they are, then so might be the set of mathematical truths that we would then be able to prove, and so might the operations that would be available to prove them with."

Id., p.186.

Thus, mathematics will not always reflect the existing physical relationships, so it will not necessarily provide good predictions, let alone good explanations, of the physical world. In addition, there may be significant aspects of our physical reality that are not susceptible to mathematical modeling, aspects that are "non-computable."

DEDUCTIVE THEORIES (AGAIN)

In addition to the use of mathematical logic to explore the implications of the axioms of a deductive theory, there is "the interesting possibility that mathematical consistency might be used to guide us, along with experimental observation, to the laws that describe physical reality." Cox and Forshaw, Why $E=mc^2$, p.25. In other words, in the development of a theory, one might find ideas by considering the assumption that the correct theory will be internally consistent from a mathematical standpoint.

Thus, one can do theory construction by a combination of inductive empirical reasoning and abstract mathematical reasoning, adding to the equations suggested by the data factors suggested by the mathematics. Of course, the prudent course is to conclude that the propositions are, at least, worth considering and then look for experimental methods to attempt to support or disprove the conclusion. One would still want to test the resulting theory against observation.

Deductive theories, as previously discussed, will have a set of axioms or assumptions and rules for manipulation. In some cases, one might be able to test the assumptions to determine whether they are empirically sound. If they are, then the results of the theory (the predictions) would be expected also to be empirically sound, if the rules of manipulation (the mathematics) correspond to the Laws of Nature, that is, to the actual causal relationships at work. In that situation, the "validity" of the theory becomes effectively an empirical question.

In contrast, one might have a deductive theory where the axioms are either not verifiable or are not really empirical, that is, they are either merely assumptions or are essentially definitions. For a theory to have scientific meaning, there will need to be some variables that have empirical content. So, essentially, one inputs certain observations and the theory generates a predication of certain other potential observations. If the theory "works," one will find that the predicted observations occur. In this type of deductive theory, the assumptions or definitions need not have empirical content (or be "true") and the mathematics (rules of transformation) need not mimic the actual causal relationships at work; the theory just gives the right answers. All that one can say is that it works.

THE METHODS OF NEWTON AND GALILEO

This characterization of a theory might sound strange or contrived, but it is essentially a description of Newtonian or classical mechanics and is an apt characterization of much of modern science. Modern science has been attributed to the development of a new scientific method that, with a bit of overstatement, has been described as "fashioned almost entirely by Galileo Galilei" in the early seventeenth century. Kline, *Mathematics*, p.284. See, *id.*, pp.284-290, 337-51.

Galileo concluded that matter (shape and size) and motion were phenomena of the natural world that existed independent of man's perception, so they were suitable subjects for science. *Id.*, p.284. He was interested in quantitative relationships based upon experimentation and observations (rather than introspection and contemplation). So, he focused on concepts that were quantifiable or measurable, like mass, speed, time, distance, *etc.*–concepts fundamentally different than the qualities that were the focus of Aristotelian philosophy, like fluidity, potentiality and purpose.

Id., p.288. And Galileo believed that the quantitative relationships in Nature could be expressed mathematically.

Galileo's interest was essentially descriptive. The question was not why or how, but what. Thus, mathematical formulas were perfectly suited to the pursuit: "formulas do not explain; they describe." Id. "[T]hrough Galileo ... the connection between math and the physical world became solidified." Mazur, The Motion Paradox, p.7; see also, pp.62-5.

Of course, it was Newton, building on the work of Galileo and other Italian scientists of the fifteenth and sixteenth centuries, who put together the pieces and created the science of Dynamics. See, e.g., Whitehead, An Introduction to Mathematics, pp.30-31. ("Mechanics" evolved from the Greek analyses of the mechanical advantage obtained through the use of lever and similar problems concerning the weights of bodies. Whitehead, An Introduction to Mathematics, pp.46-7. Later, the emphasis shifted to the motion of bodies, like planets or cannon balls, leading to "Dynamics."

Newton similarly believed that mathematics underlay the design of the natural world, and he believed that the quantitative principles of that design should be explored through observation and experimentation. He also focused on matter and motion. To the work of Galileo, Newton added the Second Law of Motion and the more general law of gravity. Kline, Mathematics, pp.337, 359. The result was a set of principles that enabled the description of all motion of all matter. Like Galileo, Newton utilized axioms and, through mathematics, deduced the Laws of Motion.

The important point here is the nature of the resulting science, which is mathematics with great predictive and descriptive power, but no explanation of the underlying phenomenon itself.

A Hostage to Mathematics?

It is often observed that contemporary science is almost all mathematics. See, e.g., Barrow, Theories of Everything, p.174 ("Modern science is founded almost entirely upon mathematics."); Kline, Mathematics, p.361 ("...our best knowledge of the physical world is mathematical knowledge"), p.555 ("[i]n all of these and in other significant and powerful bodies of science, mathematics, as we now know, is the method of construction, the framework, and indeed the essence"); Wigner, "The Unreasonable Effectiveness," p.1.

In fact, that is an overstatement, for several reasons. For example, it is not true of modern biology. However, as discussed further below, it does seem accurately to reflect the state of modern physics, including particle physics and cosmology. (Leonard Susskind and George Hrabovsky wrote a book that purports to set out the minimum amount of mathematics that one needs to know in order to do physics. *The Theoretical Minimum* (2013). The mathematics they include goes well beyond trigonometry and calculus.)

We have found that many physical phenomena are susceptible to mathematical modeling. It is certainly not obvious that it had to be so. The apparent congruence between mathematics and the physical world is characterized by Cox and Forshaw as "one of the deepest and in some ways most mysterious insights into the workings of modern science. Physical objects out there in the real world behave in predictable ways, using little more than the same basic laws of mathematics Pythagoras probably knew about when he set about to calculate the properties of triangles." *Why E=mc²*, pp.24-25. (The authors probably should have said "basic laws of mathematics that the Pythagorean School probably knew about when it set out... .") They, like many others, cite to the lecture by Wigner, discussed above, relying mainly upon the title, not the content.

Theoretical physicist Lee Smolin expresses some frustration with the emphasis placed upon mathematics in the development of scientific theories. He asserts that it is absurd to think that "mathematics is prior to nature. Math in reality comes after nature. It has no generative power. ...[I]n mathematics conclusions are forced by logical implication, whereas in nature events are generated by causal processes operating in time. ...[L]ogical implications can model aspects of causal processes, but they're not identical to causal processes. ...Logic and mathematics capture aspects of nature, but never the whole of nature." *Time Reborn*, p.246.

Smolin, however, does not actually to address the question raised here of why mathematics seems to be able to "capture aspects of nature" and whether it is the aspects captured by the mathematics or the aspects that are left out that are the things that really matter. Nonetheless, as we shall see in the later chapter Particle Physics and Quantum Mechanics, there are a host of examples where mathematics has in a sense taken the science hostage and has led us in directions that, at least, merit re-examination.

The curiosity is not that certain specific observed relationships can be depicted by a mathematical formula, but that a few of such formulae corresponding to observed physical relationships can be incorporated into a mathematical system that will enable the user to compute other relationships that can then be observed in fact. In other words, whole complex structures in mathematics can have their analog in the physical world.

One could assert that some regularity and causal relationships are necessary features of the physical world, otherwise we would not find a structured Universe. But, that observation does not explain very much. It simply says that if it were not so, this world would not be (and we would not be either}. Maybe that is all that we can say, but it is not very satisfying nor, one suspects, the whole story. (Moreover, that argument does not establish that such regularities must be or are the same everywhere or for all times, only that existed for a region and a time sufficient to allow the emergence of a hospitable environment and the development of life.)

Thus, we are left with no real answer to the question posed of why mathematics is so useful or to the derivative question of what are the limits, if any, to the use of mathematics to describe the physical world. It seems that we are just left with the fact that relative to much of the physical world, mathematics seems to work. But, to be consistent with what we have said about induction, past success cannot be taken as a guarantee of future success.

Moreover, even if mathematics reflects many important relationships in the physical world, it may not capture the relationships of many other aspects of our "natural world" that we seek to understand. There may be very significant areas that may come someday to be called science but that are not capable of representation by equations or other tools of mathematics.

Nonetheless, it is easy to note that there may be, or even probably are, limitations to knowledge based upon mathematics. What is hard is to figure out what those limitations are and what could supplement or replace our knowledge based upon mathematics. Kline concludes his book with these observations:

> "One should also question the extent to which mathematics really represents the physical world. ...It treats those physical concepts which can be represented by numbers or geometrical figures. But physical objects possess other properties as well. ...[The use of mathematics] may cause us to look at the world with blinders. ...It may be that man has introduced some limited and even artificial concepts and only in this way has managed to institute some order in nature. ...[But, i]n those domains where it

is effective it is all we have; if it is not reality itself, it is the closest to reality we can get."

Mathematics, pp.554, 555.

Concluding comments

Cambridge paleobiologist Simon Conway Morris has asserted that "being a product of evolution gives no warrant at all that what we perceive as rationality, and indeed one that science and mathematics employ with almost dizzying success, has as its basis anything more than sheer whimsy." Conway Morris, "Darwin was right. Up to a point," The Guardian, 12 February 2009. For the reasons already expressed, I think that he overstates his case here. But, he goes on, suggesting a possibility that many scientists would strongly resist:

> *"If, however, the universe is actually the product of a rational Mind and evolution is simply the search engine that in leading to sentience and consciousness allows us to discover the fundamental architecture of the universe—a point many mathematicians intuitively sense when they speak of The Unreasonable Effectiveness of mathematics—then things not only start to make much better sense, but they are also much more interesting."*

Apart from the possibility that there may be many aspects of our world that do not follow the "logic of mathematics," there are two other cautions I would suggest:

- First, the fact that mathematics does not provide actual causal explanations clearly indicates that there may be more fundamental theories, even of the laws of motion and matter.
- Second, where science becomes largely mathematics, with only few and tenuous contacts with observable phenomena, there is the risk that it is only an intellectual game that can create an artificial and erroneous belief that we know something when, in fact, we do not.

Thus, it would seem that we need to remain cautious and be ready to test empirically the applicability and accuracy of mathematical predictions continuously as advances in, extensions to or replacements of theories based upon mathematics are proposed. In other words, the assertion that "this is what the mathematics establishes" should be met with some skepticism: the assertion may be viewed as interesting, maybe insightful, but not dispositive.

Endnotes

[1] "If the amazing three volumes of Whitehead and Russell's *Principia Mathematica* ... (1910-13) had wholly succeeded, our seemingly naïve question would have a direct answer. Something is mathematics if it is logic!" Hacking, *Why Is There Philosophy of Mathematics At All?*, p.54. See also, Douglas Hofstadter, *Gödel, Escher, Bach: An Eternal Golden Braid* (Vintage Books Edition, September 1980). pp.15-24; Linsk and Irvine, "Principia Mathematica", *The Stanford Encyclopedia of Philosophy* (Summer 2019 Edition). Subsequently, Kurt Gödel presented his incompleteness theorems, which "are among the most important results in modern logic, and have deep implications for various issues. They concern the limits of provability in formal axiomatic theories." Paru Raatikainen, "Gödel's Incompleteness Theorems", *The Stanford Encyclopedia of Philosophy* (Spring 2022 Edition),

[2] Roger Penrose does elsewhere write extensively about the relationships between mathematics and the human mind. E.g., *Shadows of the Mind: A Search for the Missing Science of Human Consciousness* (1994); *The Emperor's New Mind: Concerning Computers, Minds, and the Laws of Physics* (1989). Some of his insights are discussed in the chapter on Consciousness. However, Penrose continues to express the view that "major revolutions are required in our physical understanding ... [before] much real progress can be made in understanding the actual nature of mental processes." *The Road to Reality*, p.21.

[3] One possibility, of course, is that we are living in an elaborate computer simulation constructed using the mathematics that we have been in the process of discovering. Thus, perhaps we are simply uncovering the elements of the programming code used in creating the simulation. If so, then it is easy to understand why the mathematics that we develop so accurately reflects the structure of the world in which we think that we live. See Edward Frenkel, "Is the Universe a Simulation?" *The New York Times*, February 14, 2014.

[4] "After we came out of the church, we stood talking for some time together of Bishop Berkeley's ingenious sophistry to prove the nonexistence of matter, and that every thing in the universe is merely ideal. I observed, that though we are satisfied his doctrine is not true, it is impossible to refute it. I never shall forget the alacrity with which Johnson answered, striking his foot with mighty force against a large stone, till he rebounded

from it—'I refute it *thus*.'" James Boswell, *Life of Samuel Johnson*, "Refutation of Bishop Berkeley" (1791).

[5] The discovery of the Pythagorean Theorem presented the problem of "incommensurables," numbers that cannot be written as the ratio of two whole numbers (like 1/2, 1/3, 2/5). An example is the diagonal of a triangle with two other sides of the length of 1, which would be the square root of two. See Joseph Mazur, *The Motion Paradox*, p.18. Geometry happens to present such incommensurables with some frequency, such as the very important number Pi. These "incommensurables" are now referred to as irrational numbers.

[6] For example, a study of marine fossils suggested that marine life have in general gotten larger since the Cambrian period; the question asked was whether the larger body size could have been the result of evolution in which size had adaptational value. In some of the computer simulations, the researchers gave larger size a beneficial effect. When the program with that adjustment was run, the tendency was for the marine life to get larger as appears to have happened. This mathematical model confirms our common sense instinct. The historical study of the fossil record is clearly interesting. The computer simulation is less impressive. Where no advantage is granted for larger size, random variations appear not to lead to size increases; where there are advantages, the advantages tend to affect the evolution of the species. One wonders that computer simulations were necessary to generate that result.

[7] Similarly, it is possible that "we tackle only the physical problems that are amenable to the mathematical methods we have at our disposal." Id., p.105. These propositions are, at best, only partial answers.

[8] An open circuit and a closed circuit in sequence ("×" or "and") will be open because the current will not flow ($1 \times 0 = 0$). A closed circuit and a closed circuit in sequence will be closed because the current will flow ($1 \times 1 = 1$). A closed circuit and an open circuit in parallel ("+" or "or") will be closed because the current will flow through the closed alternative path (i.e., $1 + 0 = 1$), and a closed circuit or a closed circuit will also be closed ($1 + 1 = 1$). Using Boolean algebra, complicated circuits can be symbolically depicted and through the manipulation of the symbols, comparably operating circuits can be devised.

[9] Several reviews of Seife's book were quite complimentary, referring to his evident enthusiasm and the potential of his style of writing to engage

and interest the lay person. *See, e.g.*, Nicholas Lezard, "Explaining nothing, brilliantly," *The Guardian*, 22 March 2003 ("one of the best-written popular science books to have come this way for quite a while..."). A review published by the American Mathematical Society, however, was sharply critical of his generalizations (and some of his history), asking "whether the assertions stem from misunderstanding or simply constitute absurd hyperbole"? Jeremy Gray, "Book Review," Notices of the AMS, Vol. 47, No. 9, October 2000, p.1080.

[10] Seife describes how the concept of zero was absent from the classical Western world—unacceptable to the Greek philosophy and not included in Roman numerals. *Zero*, pp.6-19.

[11] The introduction of zero was of great practical significance in that it enabled Arabic numerical notation (as a place holder so that the same number could be used repeatedly in multiples of 10, depending on the column in which it was placed, as in 1, 10, 100, 1000, etc.). See Whitehead, *An Introduction to Mathematics*, pp.63-51. *Id.*, pp.66-8.

[12] Aristotle's view of infinity was actually somewhat similar to that expressed above. He distinguished between two types of infinity: potential infinity and actual infinity. Potential infinity is the mathematical concept of a series of numbers that never ends. That was acceptable. Actual infinity would exist in the physical world and that Aristotle believed was an impossibility. *See, e.g.*, John D. Barrow, "Beyond numbers—the role of infinity in the understanding of the universe," *University of Cambridge Alumni News*, May 2015.

[13] Lee Smolin observes, "There is no more romantic notion than infinity, but in science the concept can easily lead to confusion." *Time Reborn*, p.227.

[14] Seife asserts that the extra dimensions mean "[no]thing, really. ... They are simply mathematical constructs that make the mathematical operations in string theory work in the manner that they have to." *Id.* p.197.

[15] Barrows refers to this school as "Constructionism." *Id.*, pp.181-86. His usage, however, does not seem to conform to the generally recognized categorizations of Platonism, Logicism, Intuitionism, Formalism and Predictivism. See Horsten, "Philosophy of Mathematics," *The Stanford Encyclopedia of Philosophy*. What Professor Barrows calls Constructionism appears to be more widely called Intuitionism.

[16] Physicist Paul Davies asserts that it "is generally agreed" among scientists that the Laws of Nature have four characteristics: they are universal, they are absolute, they are eternal and they are omnipotent ("all powerful," meaning that they cannot be avoided or evaded). *The Mind of God*, pp.82-3. He observes that these qualities "that were formerly attributed to the God from which [the Laws] were once supposed to have come." *Id.*, p.82.

[17] And, certainly, theoretical work does go on exploring the possible consequences to our understanding of cosmology, for example, if one permits (or assumes) changes in the Laws or the Constants. *See, e.g.,* Smolin, *Time Reborn* (2013); Joao Mangueijo, *Faster Than the Speed of Light* (2003).

[18] Arthur Stanley Eddington, a contemporary and colleague of Albert Einstein, believed that fundamental science describing the physical world could be developed by pure thought. Observation and experimentation were useful tools that could greatly speed the process of discovery but were ultimately unnecessary. He declared that,

> *"My conclusion is that not only the laws of nature but numerical values of the fundamental relationships or forces that are presumed to be constant, such as the speed of light and the gravitational constant, can be deduced from epistemological considerations, so that we can have a prior knowledge of them."*

The Philosophy of Physical Science (1939), p.58, quoted in Barrow, *The Constants of Nature* (2003), p.83.

[19] Lord Rees explains that if one takes a few meters as a normal distance for man, that distance would have to be increased by twenty five factors of ten to reach the observable limits of the Universe. *Just Six Numbers*, pp.5-6. Our smallest measurable size (using electron microscopes and particle accelerators) would be about seventeen negative factors of ten smaller than the meter. Physicists speculate that the smallest structures of nature, like the proposed superstrings, would be smaller by another seventeen negative factors of ten. *Id.*, pp.6-7. Similarly, the world of the atom is some ten orders of magnitude smaller than our observable world. The Universe is estimated to be some 14 orders of magnitude larger than our Solar System. *See, e.g.,* Alexander Unzicker and Sheilla Jones, *Bankrupting Physics: How Today's Top Scientists Are Gambling Away Their Credibility* (2013), pp.57, 60-68.

Social Science

> "It is vain to speak of the higher authority
> of a unified social science.
> ...[I]it does not exist;
> it shews no sign of coming into existence."
>
> Alfred Marshall
> Quoted by John Neville Keynes,
> *The Scope and Method of Political Economy*
> (1891), p.110

In this day and age, it will seem strange to many people for me even to include a discussion of the social sciences in a book on science. If the status of a discipline as a science is to be determined by the ability of the discipline to make concrete and accurate projections about the consequences of various actions, then the social sciences in the early twenty-first century should strive to keep a low profile.

I became interested in the matters discussed herein during my university study of economics. I encountered and investigated a variety of methodological issues concerning the construction and testing of theories, the use of mathematics (and statistics) and the relevance of empirical evidence. Much of the discussion that interested me, and that was clearly secondary to the mainstream work of economists, essentially addressed the question of whether economics is a "science." Of course, had I been a student of physics first and come to economics later, I might be expressing surprise that the methodological issues of physics have parallels in the field of economics. However, it seems to me that, independent of my experiences, there are potential benefits for the lay reader in confronting the methodological issues in the context of economics rather than biology or physics, because the concepts and subjects of economics generally feel more approachable for the non-scientist.

So, we explore these issues here using economics as an illustration.

Science?

"Physics envy"

In September 2010, The New York Times labeled as the "Idea of the Day" the views of columnist Gideon Rachman of *The Financial Times* that: "Economists would do well to learn the modesty of historians...." September 9, 2010. Rachman warned about economists (pointing to Nobel Prize winner Joseph Stiglitz) who, in the face of recent failures, would "simply search for 'new paradigms'—and then presumably go back into the business of scientific prediction." He noted:

> *"With the exception of a few deluded Marxists, historians know that their work cannot be used to predict the future. History can suggest lessons and parallels and provide wisdom—but what it cannot do is provide a sociological equivalent of the laws of physics. Yet this seems to be the aspiration of many economists,* **who notoriously suffer from 'physics envy.'"**

FT.com, 6 September 2010.

The same theme appeared in an opinion column by political scientists Kevin A. Clarke and David M. Primo a year and a half later. "Overcoming 'Physics Envy,'" *The New York Times*, April 1, 2012. They asserted that the social sciences are suffering from the efforts to use the hypothetico-deductive method that has been idealized as "the" method of the hard sciences (formulate a theory, deduce testable empirical hypotheses from the theory, then test the hypotheses—if the evidence contradicts the prediction, then the theory needs to be rejected or revised; if it is consistent with the hypothesis, then the theory is deemed tentatively confirmed, or, perhaps more accurately, not yet disconfirmed). They argued that this model does not accurately represent even how the hard sciences are practiced, ignoring "everything messy and chaotic" that inevitably accompanies scientific inquiry.

More importantly, Clarke and Primo argued that empirical testing and even empirical predictions are not necessary for "good science," noting that many useful theoretical models help explain and enable us better to understand phenomena, even though they do not give rise to testable predictions or even, sometimes, where they are obviously false. (They also asserted that the analysis of empirical data can be helpful even in the absence of a grand theory.) "To borrow a metaphor from the philosopher of science Ronald Giere, theories are like maps: the test of a map lies not in arbitrarily checking random points but in whether people find it useful to get somewhere." Id.

Interestingly, during the late nineteenth and early twentieth centuries, successes at incorporating mathematical methods of analysis into increasingly elaborate deductive theories, in models that bore striking similarities to classical mechanics, led economists increasingly to identify themselves as different from and superior to the other social sciences.[1] There were critics, of course, and not just among the progressives. For example, Frederick Hayek warned in his Noble Prize acceptance speech in 1974 that:

> "[The] failure of the economists to guide policy more successfully is closely connected with their propensity to imitate as closely as possible the procedures of the brilliantly successful physical sciences – an attempt which in our field may lead to outright error."

"The Pretence of Knowledge," Lecture to the memory of Alfred Nobel, December 11, 1974. Facts: NobelPrize.org.

Other social sciences have been increasingly incorporating mathematical methods, as illustrated by various examples set out elsewhere. These efforts have been successful in terms of their own goals. But, it is not physics.

HISTORY AS SCIENCE

Whatever may be said about the "humility" of the historian in our contemporary world, there are indications that at least some historians, as well as other social scientists, have aspirations to achieve the same status as the natural sciences.

For example, Jared Diamond and James A. Robinson (eds.) in *Natural Experiments of History* (2010) discuss history as a science.[2] In their prologue, the editors complain that "laboratory scientists ...look[ ] down on fields of science that cannot employ manipulative experiments" and observe that, nonetheless, "there are many fields widely admitted to be sciences" in which it is not possible to conduct experiments, citing evolutionary biology, paleontology, epidemiology, historical geology and astronomy. Id., p.1.

(In this context, it is worth noting that some practitioners of these other sciences, such as evolutionary biology, have sought to emulate, as far as possible, the methodological approaches of physics in the use of laws and mathematical techniques. See, e.g., Nowak, *Super Cooperators*, pp.1-2, et seq.) Diamond and Robinson go on to discuss the use in history of "natural experiments" or comparative methods "preferably quantitative[ ] and aided by statistical analyses." *Natural Experiments of History*, pp.1-2.[3] The book then describes several such studies or experiments.

Two things are likely to strike the reader of *Natural Experiments of History*. First, it is surprising that the application of comparative analyses and statistical methods in historical studies would be viewed as innovative or controversial. One would have thought that the potential benefits of such techniques would have been obvious and that it was eminently sensible to see what results could be achieved. ("History is full of such potential experiments; it is just that historians have not yet thought of them in these terms." *Id.*, p.249.) Second, it is surprising that such highly respected and accomplished historians as the editors would take such an apologetic and defensive approach to the subject—the useful question is not how history compares to the natural sciences, but whether its techniques of analysis can further our understanding and knowledge.

Recently, historian of science Steven Shapin reviewed a new book by Nobel prize-winning physicist Steven Weinberg entitled *To Explain the World* (2014). "Why Scientists Shouldn't Write History," *The Wall Street Journal*, February 13, 2015. Shapin describes how much of Weinberg's book is about the development of physics from the ancient Greeks through the seventeenth century. Shapin is not impressed. He explains that historians actually use their specialized skills and training to try to understand events of the past in light of the full context of the world at the time in question, including the participants' concept of their place and role in the world. It is not meaningful history just to try to identify who (coincidentally) made discoveries or had insights that seem to be important in today, that is, from the standpoint of the modern worldview. In Shapin's view, Weinberg's attempt to write history falls short.

A Note on Expert Arrogance

Shapin also sets out the following version of a well-known story:

"There's a story told about a distinguished cardiac surgeon who, about to retire, decided he'd like to take up the history of medicine. He sought out a historian friend and asked her if she had any tips for him. The historian said she'd be happy to help but first asked the surgeon a reciprocal favor: 'As it happens, I'm about to retire too, and I'm thinking of taking up heart surgery. Do you have any tips for me?'"

What is the point?

Shapin explains that the story demonstrates "a real asymmetry between two expert practices"—one expert (the cardiac surgeon) has what are recognized to be real, hard to acquire skills, the other (the historian) is thought to do things that can be done by most anyone.

Fair enough, but the telling of the story says something more. First, it indicates that the arrogance of certain experts (and not just physicists) can extend well beyond the areas of their expertise. Second, it shows how the defensiveness toward certain other experts can generate an arrogance toward non-scientists or other sciences in turn (like the "physics envy" displayed by economists). And, third, this example also demonstrates how both arrogance and defensiveness can get in the way of the business at hand.

Shapin acknowledges that there may be a real and legitimate interest is what a leading physicist of the twentieth century thinks about the historical development of his science, but then spends more time objecting to the lack of humility in the undertaking than in demonstrating the alleged errors in the book that are supposedly apparent to the trained historian. It would be far more constructive to let insights and analyses compete on the merits.

THE IMPORTANCE OF METHODOLOGY

I believe that the study of the philosophical underpinnings of the methods and methodologies employed by natural and social scientists can provide significant insights into the state, direction and uses of the scientific knowledge being pursued. Thus, I include this chapter not so much to identify important things we do not know about economics, many of which are believed to be painfully obvious by most students of politics and public policy, as to help identify and elucidate the methodological problems and issues that arise in all efforts to understand, and make predictions about, our world through the construction of theories.

I begin with a brief introduction to the historical roots and general content of the discipline now called economics.[4] Then I discuss some of the early efforts by economists to understand and articulate what they were doing as social scientists.

The issues involved are then illustrated by an examination of the theoretical debate within the profession and among some close observers about the traditional "theory of the firm" and the fundamental assumption of profit maximization, a debate that spilled over to challenge the more pervasive and fundamental concept of "maximization" itself.[5] The example is of special interest because several traditional economists used in the debate some of the concepts of philosophy of science, attempting to demonstrate that their critics simply did not understand the nature of scientific theories and were advancing arguments based upon mere misunderstandings and mischaracterizations.

I conclude by offering some observations about economics as a science, with some implications about scientific knowledge more generally.

THE SUBJECT OF ECONOMICS

The branch of human "knowledge" known as economics has ancient roots, reflecting man's long-standing interest in trade and money (media of exchange). But, then, we see in the eighteenth century a new focus on two rather distinct questions:

- Why are certain nations richer than others, even when one controls for natural resources?
- Why are diamonds—intrinsically useless—worth more than water—an essential of life?

The first question has begged a convincing full answer for almost 250 years (during which time, the answer has undoubtedly changed for different places and different stages of development); although, some striking insights were achieved into causes of increased productivity and growth.

Adam Smith identified the importance of the division of labor and specialization, attributing much of the progress and promise of industrialized economies to the benefits that were derived from the increasing division of labor that greater scale and free trade would enable. The benefits of the specialization that is engendered by the division of labor include improved skill at the task at hand resulting from repetition, improved techniques that may be discovered as a result of the greater experience gained with specialization, new technology made feasible as a result of the increases scale of the process at issue, reduced time and resources spent in frequently changing tasks, efficiencies in training for specialized jobs and, even, the opportunities to capture differences in relative differential abilities in performing particular tasks through trade; although, the full theory of "comparative advantage" awaited David Ricardo, 50 years later. See, e.g., Matt Ridley, *The Origins of Virtue*, pp.42-4.

Ricardo logically demonstrated that in a world of two manufactured commodities and two countries, if one country had a comparative advantage in the production of one good over the other, relative to the other country (e.g., Country A can produce 3 widgets if it forgoes the production of 1 blodget, and Country B can produce only 2 widgets in place of one blodget), then if each country produced the item as to which it was most productive and the two traded to satisfy the demands of each, then the result would be in total more of at least one good and often more of both. Consequently, trade necessarily would make one or both countries better off than they would be without trade. *On the Principles of Political Economy and Taxation* (1817), Chapter 7, "On Foreign Trade." Of course, the model is based upon very restrictive assumptions. Its relevance to the real world is debatable.

Interestingly, these perceptive analytical insights were fruitful in a somewhat indirect way. They led to a recognition of the importance to economic development of institutional features like population density, infrastructure (physical, legal and, even, cultural), political stability, basic security of persons and property, a monetary currency and a banking system to enable trade and specialization..

The second question, in contrast, was answered at the end of the nineteenth century by the theory of the intersection of supply and demand. (Adam Smith and, even, David Ricardo had sought to find value in the "labor content" of goods, as—of course—did Karl Marx.) Thus, while water is essential for life, its relative abundance and low cost of "production" cause water generally to be cheap or, in some places, free. Diamonds (gem quality) have no practical utility as such, but they are scarce and costly to obtain. When sufficient demand arose for diamonds, presumably for aesthetic and status-securing purposes, the value became high. People were prepared to pay large sums because of the scarcity.

Subsequent development of the theory of supply and demand resulted in marginal analysis, based on the hypothesis that decisions were made and that prices (and many other things) were determined at the margin. Such analyses lent themselves to mathematical modeling, which, with the application of the calculus, generated a wide range of intriguing concepts (*e.g.*, elasticity of demand and supply). The graphical representation of demand and supply curves suggested related areas of investigation through graphical and algebraic presentation of the nature of demand and the nature of supply, including analyses of production functions.

The result, as seen in Alfred Marshall's *Principles of Economics* (Ninth Edition, 1961; initially published in 1890), was an elaborate, elegant and almost comprehensive theory of the derivation and satisfaction of human material wants and needs. Because of the use of mathematics and mathematical reasoning, the neo-classical theory also had the trappings of a science.

As efforts were made to make neo-classical theory more comprehensive, economics began to disintegrate into multiple parts or sub-disciplines. The terms micro-economics and macro-economics describe fields focused on value theory and industrial organization, on the one hand, and trends in the economy and the role of money, on the other. More intensive work in each specialty yielded fruit, as one would expect. Economists generated theories of consumer behavior, theories of the firm, theories of markets, theories of international trade, theories of economic development, game theory and theories of money.

Of course, many issues are decidedly outside of the defined scope of economics, but others have been excluded by choice. For example, traditional economics consciously refuses to rank or value individual choices. Instead, not only are people presumed to be the best (or only) judge of what pleases them, but no judgment is made about the relative merits of such pleasures (at least as long as the indulgence does not hurt anyone else). For this reason, the gross domestic product of a nation reflects the sum of the market prices of all of the goods and services produced. No differentiation is made between products that promote health and those that undermine it; no different weighting is made for expenditures that add to wealth or productivity and current consumption of luxuries. Thus, comparable GDPs can represent very different levels of economic wellbeing.[6]

Curiously, progress in the sub-categories of economics appears not to be bringing economists closer to a unified theory but, instead, to be leaving them with relatively unconnected theories for the individual person or decision-maker, for individual markets and for economies as a whole. We do not have is a theory of everything economic. That, however, is not a lacking unique to economics or, even, to the social sciences. As we shall discuss, it is clearly true of the state of all branches of the natural sciences, as well. See, e.g., John D. Barrow, *Theories of Everything* (2005).

THE SCIENCE OF ECONOMICS

The "proper" methodological approach to economics as a branch of study was not obvious at the beginning (and may still not be). In the last part of the nineteenth century, the impressive progress that was apparently being made in the field we now call micro-economics stimulated various economists to discuss the methodological foundations of economics.

John Neville Keynes, the father of John Maynard Keynes (in turn, the father of "Keynesian economics"), undertook an analysis of these issues in *The Scope and Method of Political Economy* (1891). He distinguished between two schools: the theoretical, abstract and deductive and the ethical, realistic and inductive. *Id.*, pp.9–10.

Adam Smith had engaged in both approaches. Thomas Malthus had continued the inductive tendencies of Smith, while David Ricardo had developed the deductive approach. Then, Alfred Marshall stressed the need for inductive as well as deductive analysis and rejected the idea of a comprehensive, axiom-based theory: "The function then of analysis and deduction in economics is not to forge a few long chains of reasoning, but to forge rightly many short chains and single connecting links." Marshall, *Principles of Economics*, p.781.

DEDUCTION VERSES INDUCTION

Deductive systems have strong appeal, presumably in part because they seem scientific. It is also clear that such theories can be immensely flexible, general and powerful. The use of deductive systems also has the advantage of allowing work to be done in pieces. Certain aspects of what might become an all-inclusive theory could be developed independently, with the expectation that the pieces could ultimately be put together into a sweeping whole.

As noted above, J.N. Keynes' prejudice was clear. He admired abstract reasoning. Keynes was impressed by the scientific method of economic theory and believed that, because of it, economics was decidedly superior to the other social sciences, especially sociology. Other economists, in the following decades, were to espouse the same view. The belief was that economists were developing "'a body of 'pure theory' the possession of which distinguishes Economics from the other social sciences." T.W. Hutchinson, *The Significance and Basic Postulates of Economic Theory* (1938), p.3 (economic theory was constructed on the model of mechanical equilibrium, which generated "a body of 'pure theory' the possession of which distinguishes Economics from the other social sciences").

Part of the differences in the approaches of practicing economists reflected different views of what it was that economics was supposed to explain. What became the marginal utility school (the deductive theorists) asserted that the goal was to understand—or, the subject of economics was—the allocation of scarce resources among competing ends. Another school focused on the historical and institutional underpinnings of economic development. See, Sherman R. Krupp, "Types of Controversy in Economics," in Krupp (ed.), *The Structure of Economic Science* (1966), p.42.

In the English-speaking world, in the end, mainstream economics came to focus on explanatory models based upon the model of classical mechanics.[7] As classical theory evolved into neo-classical theory, economics assumed the mantel (almost) of a real science. Its proponents and admirers marveled at the elaborate graphical presentations that appeared to generate definitive and clear results, putting to shame the subjective narrative methodologies of the other social science.

For example, once value (or price) is defined as the intersection of supply and demand and one hypothesizes that both supply and demand over various reasonable quantities and prices are functions that can be depicted as curves (sloping opposite directions), then mathematical techniques can be utilized to explore various possible characteristics of those curves and make "predictions" of the results that would occur if those characteristics obtained in the real world.

In other words, the theories constructed about relationships in the world lent themselves to mathematical representation so that the power and versatility of mathematics can be used.

The Development of Deductive Systems

The first effort to state economic theory in the form of a system of axioms and deductions is attributed to Nassau William Senior, writing in the 1830s:

> "To Senior belongs the signal honor of having been the first to make the attempt to state, consciously and explicitly, the postulates that are necessary and sufficient in order to build up—it is misleading to say to 'deduce'—that little analytical apparatus commonly known as economic theory, or, to put it differently, to provide for it an axiomatic basis."

Schumpeter, A History of Economic Analysis, pp.575-76.

Senior described the analytical work of the economist as follows:

> "his premises consist of a very few general propositions, the result of observation, or consciousness, and scarcely requiring proof, or even formal statement, which almost every man, as soon as he hears them, admits as familiar to his thoughts, or at least as included in his previous knowledge; and his inferences are nearly as general, and, if he has reasoned correctly, as certain, as his premises."

Nassau William Senior, Political Economy (1836), pp.2-3.

A century after Senior published his treatise, Lord Robbins described economics in very similar terms:

> "The propositions of economic theory, like all scientific theory, are obviously deductions from a series of postulates. And the chief of these postulates are all **assumptions involving in some way simple and indisputable facts of experience** relative to the way in which the scarcity of goods...actually shows itself in the world of reality.... But while it is important to realize how many are the subsidiary assumptions which necessarily arise as our theory becomes more and more complicated (i.e., market structure, etc.), it is equally important to realize how widely applicable are the main assumptions on which it rests."

The Nature and Significance of Economic Science (1935), p.158 (emphasis added).[8]

Robbins was not alone in these views. There were other rather elaborate and excited tributes written about the scientific status of economics. *See, e.g.*, T.W. Hutchinson, *The Significance and Basic Postulates of Economic Theory* (1938) and, to be discussed in some detail below, Milton Friedman, "The Methodology of Positive Economics," in *Essays in Positive Economics* (1953). The approach was methodologically sound: if the premises are true, then logical deductions from them will also be true.[9] Today, of course, there is considerable and appropriate skepticism about the obviousness and, indeed, the "truth" of those initial premises.

PROBLEMS AND CONTROVERSIES

The Great Depression raised significant questions about economists' understanding of the macro-operations of economies and of the role of money, but the theoretical model of the determination of price and output came to be subject to sharp criticisms as well. Part of the reason for some of the passion reflected in the attacks were the apparent normative implications of the theory (some of which are discussed below in the subsection Economic 'morality'), coupled with the claims that it was a science resembling Newtonian physics.

Micro-economics appeared to demonstrate that the most efficient allocation of scarce resources would occur in conditions of perfect competition. When equilibrium was reached, the allocation would be such that any change would result in a reduction in total output. (The particular equilibrium in question would depend upon the initial conditions, including the initial distribution of wealth and resources. Different initial conditions would lead to different results.) Thus, economists would argue that free exchange and open markets, with certain caveats developed over time, would lead to the most efficient allocation of resources.

However, while some would call the caveats "minor" qualifications; others would say that the caveats reflect almost everything that matters, rendering the general theory almost or totally useless or, even, dangerous.

This theoretical structure was built upon a large number of propositions or postulates that appeared to have factual content (that is, they could be true or false). The results are deduced from those postulates. Attacks on the theory often challenged specific postulates. Such issues involve the realism of the conditions of perfect competition; the assumptions about (perfect) knowledge and foresight; the silence about the presumed process of adjustment from disequilibrium to equilibrium, including the simple question of whether there are mechanisms that would be expected to cause a movement toward equilibrium as opposed to creating disequilibrium or cycling fluctuations among states of disequilibrium.

THE THEORY OF THE FIRM

The nature of the methodological issues that arose can be illustrated better–or, at least, certainly more concretely–by an example. The example I use here is the controversy over the theory of the firm and the assumption of profit maximization.

In neoclassical theory, relatively simple mathematics demonstrates that for any market with a downward sloping demand curve (people wanting more of a good at a lower price per unit) and an upward sloping supply curve (at some relevant point, as production increases, because of raising costs of inputs or technological limitations, the cost of each additional unit will rise), production at the intersection of the marginal cost curve and the demand curve will result in the most profits for the industry–the production of either more or less would cause the total profits to fall.

The unit or entity of production was assumed to be the "firm", and the firm was assumed to act so as to maximize profits. Therefore, the firm would increase or decrease production of the good it was producing until the marginal cost of production equaled the price of the good. From the standpoint of the individual firm, the price was given by the market (the intersection of industry-wide supply and demand curves). All the firm could do was decide how much to produce.

In a monopoly, the rather simple and elegant analysis of the determination of price by supply and demand for a market applies to the monopolist, with the single firm being the sole source of supply and its cost functions determining the supply curve (so the model for the market and the firm are one and the same).The profit maximizing output in either case is where the marginal cost of production per unit equals the price, where the demand and supply curves intersect.

At some level, early economists probably understood profit-maximization to be a reasonable approximation of what firms actually try to do. The sole proprietorship would seek to make money for the owner; the corporation would seek to make money for their shareholders. However, did these economists believe that firms knew how to maximize profits by comparing marginal costs to price? Did the firms even know what their marginal costs were over any relevant range of output?

The "Firm"

A little history is probably useful. The theory of the firm emerged in the mid-nineteenth century with Augustin Cournot. He began his analysis with a single firm monopolist, then added additional sellers until the point was reached where no one seller could influence the price of the product in the market. Cournot thereby developed a theory of duopoly that was widely ignored until the 1930s. Until then, economists had generally presumed that the economy consisted of monopolists and, elsewhere, highly competitive markets. These two "extremes" present few interesting questions about the behavior of firms, because it was perceived that there was little that a firm could elect to do. The market determined the price; all that any seller could decide was the quantity to produce.

Marshall subsequently focused on the firm. He assumed that the competitive scheme was the most important. Therefore, he believed that the interesting questions arose at the level of the industry or market, not at the level of the individual firm. (I should repeat that the theory assumed that there was nothing that a single buyer could do to influence price.) Thus, it was perceived that the functioning of the market could be analyzed independently from the actions of any of the individual market participants on either the sell or buy side.

In these analyses, no attention was paid to what the firm is, whether the legal entity, the chief executive, the board of directors, the shareholders, *etc*. It did not really matter. One might imagine business school courses in how to maximize profits (or how to survive) in a competitive market, but such instruction would be addressed to the businessman making daily decisions and not have much interest to the theorist analyzing the behavior of markets.

In perfect competition

In a fully competitive market, firms compete away supra-normal profits (returns in excess of the cost of capital), so that the firms earn the normal rate of return. (Additional capital investment will flow to the firms earning above normal returns, causing them to increase production until their returns fall to normal.) Any firm that fails to earn that return will eventually go out of business.

Thus, firms will maximize profits, because any who do not will not be around in the long run. If the market compels profit maximization, because of competition, then it does not matter what the managers or owners are attempting to do. In other words, we do not even need to infer from the profit-maximizing assumption any particular intention or motivation on behalf of the firm. It would not matter what one's intentions or goals were; the end behavior would be the same because, in the end, one did not really have a choice. Profit maximization is a survival condition for a firm: it happens as a consequence of competition (at which point in the discussion, the "assumption" becomes not one of the initial axioms, but a conclusion or prediction of the theory). However, this argument ties the theory of the firm deeply into (makes it dependent upon) the theory of perfect competition. But, the full ramifications of the assumption of a competitive market were not well understood.

As an aside, the assumption or conclusion of profit maximization was indispensable to the concept of the equilibrium position for individual firms in a competitive market. From that model, a host of conclusions were derived, ranging from the optimum allocation of resources, to the appropriate return on capital, the implications for employment of a minimum wage or of collective bargaining by unions.

IN "IMPERFECT" COMPETITION

In 1921, Frank Knight published *Risk, Uncertainty, and Profit*, in which he set out clearly the concept of pure or perfect competition. It was becoming clear that the necessary conditions were not likely often to be realized, at least in the more industrialized economies. In the absence of perfect competition, or something close to it, profit maximization was not necessarily a survival condition. If many, if not most, markets did not satisfy the characteristics of perfect competition, then it would be of much more interest to assess whether firms did, in fact, maximize profits.

In 1933, two important works appeared focused on the theoretical implications of the fact that most markets were not actually fully competitive:

- Joan Robinson, in *The Economics of Imperfect Competition*, argued that the theory of monopoly is the more relevant theory for studies of the real economy, because the conditions for perfect competition (large numbers of sellers, large numbers of buyers and identical products) are not regularly met in the industrialized economy.

- E.H. Chamberlin, in *The Theory of Monopolistic Competition*, was, perhaps, the more revolutionary. He argued that there were important phenomena in the industrialized economies, like differentiated products and selling costs (including advertising and promotion), that simply were outside of the scope of the traditional theories of monopoly and pure competition. Important aspects of modern economic activity were simply ignored by the neo-classical theory and, as a result, the neo-classical theory was not a very useful tool for understanding industrialized economies. Chamberlin did not, however, successfully formulate an alternative theory.

Subsequently, J.R. Hicks explained how barriers to pure competition (such as product differentiation and barriers to entry) allowed wide margins of discretion to allow firms to fail to maximize profits or, even to pursue other goals. "Annual Survey of Economic Theory: The Theory of Monopoly," *Econometrica*, Volume 3, Issue 1 (Jan. 1935), pp.1-20. Indeed, Hicks famously observed that "the best of all monopoly profits is a quiet life." Id., p.8. In other words, a monopolist or other firm with substantial market power might "choose" not to try to maximize profits and to forgo profits in exchange for other benefits.

The development of models of oligopoly and imperfect competition fostered a new interest in the theory of the firm. If some reasonable approximation of perfect competition were not the prevailing condition in many markets, then the assumption of profit maximization clearly merited scrutiny, because it could not be assumed to be a market-imposed result. Instead, the actions of the firm would depend upon the motivations and skills of the decision-makers. Firms are controlled by people and people have a variety of motives.

Is it realistic to think that all business actions will be motivated by the goal of profit maximization? Owners might pursue size for its own sake or for the prestige and power size might bring. Other managers (perhaps in dereliction of their duties to the owners) might pursue leisure and lack of stress or various material "perks." These types of criticisms led to extensive efforts to investigate and understand how business decisions were actually made and how managers behaved.

BEHAVIORAL MODELS

Students of business and the corporate structure suggested that modern business generally was run by salaried managers who, even with bonuses, would have different economic interests than the "owners" of the capital. See, e.g., Berle and Means, *The Modern Corporation and Private Property* (1932).

These commentators questioned how realistic it was to characterize the behavior of a firm as that of a collective entrepreneur. In addition, efforts were also being made to investigate empirically how levels of prices and output were actually being determined in real-life business entities. These studies suggested that business managers did not attempt to engage in marginal cost pricing, as the theory seemed to suggest, but instead relied upon various forms of full-cost pricing, with a target mark-up of profit margin on top of the fully allocated variable costs per unit. See, e.g., R.L. Hall and C.J. Hitch, *Oxford Economic Papers* (1939).

Also, by the late 1930s, another type of issue gained attention. Economists began to struggle with the questions of uncertainty and risk. See, e.g., J.R. Hicks, *Value and Capital* (1941). The introduction of these "facts of life" into the theoretical models of the business decision-making process presented significant complications. These problems were different from and independent of the issues of imperfect competition. They would exist even in a purely competitive industry. Thus, even if the manager were acting solely to maximize the benefits of the shareholders, understood as the market value of the stock; because of uncertainty, he would not act in a profit-maximizing manner as "predicted" in the neo-classical theory.

Combining the questions of the goals of business managers with the complications of uncertainty, Modigliani and Miller argued that profit maximization by a firm is not equal to market value maximization (the firm's valuation in the stock market) when uncertainty is included in the theory. Franco Modigliani and Merton H. Miler, "The Cost of Capital, Corporation Finance, and the Theory of Investment," *American Economic Review*, XLVIII (June 1958), pp.278–85. Work on, and debate over, the theory of the firm continued, with many significant contributions. Some analyses and arguments have concerned what it is that firms attempt to maximize (profits, revenues, market capitalization). See, e.g., W.J. Baumol, *Business Behavior, Value and Growth* (1959).

Other economists have focused on whose interests it is that the firm attempts to maximize or promote (the interests of the managers, the employees, the shareholders). See, e.g., C.E. Williamson, *The Economics of Discretionary Behavior* (1964). And yet others have questioned whether the concept of maximization itself is valid (either as an intentional goal or as an achievable goal). See, e.g., J. Margolis, *Journal of Business* (July 1958);

Kenneth Boulding, A *Reconstruction of Economics* (1950); Neil Chamberlain, A *General Theory of Economic Process* (1950).

THE MARGINALIST DEBATE

In the mid-1940s, the underlying methodological issues were explored in a heated exchange between some of the leading neo-classical economists and dissenters, referred to as the Marginalist Controversy. The argument went beyond the profit-maximizing assumption to challenge the validity of marginal analysis in general and its underlying maximization principle.

THE CHALLENGE

Richard Lester sought to investigate empirically the decision-making process of a group of actual business executives. He concluded that they did not engage in a marginalist approach to decisions, particularly not in marginal variable cost analysis. He also challenged the assumption of a U-shaped cost curve for most businesses within the relevant output range. Lester intimated, but did not expressly state, that these conclusions destroyed the persuasive force of marginal economic analysis. (His political purpose was to challenge the prevailing view of many prominent economists on the minimum wage.) Richard A. Lester, "Shortcomings of Marginal Analyses for Wage-Employment Problems," *American Economic Review*, XXXVI (March 1946), pp.63–82.

THE RESPONSE

Lester's article prompted a response from Fritz Machlup, who asserted that "...the alleged 'inapplicability' of marginal analysis is often due to a failure to understand it, to faulty research techniques, or to mistaken interpretations of 'findings.'" Fritz Machlup, "Marginal Analysis and Empirical Research," *American Economic Review*, XXXVI (September 1946), pp.519–54. Machlup aptly substantiated each point.

With respect to the last two points, he highlighted the inherent problems of empirical research in the social sciences, *e.g.*, the dangers of reliance on oral and written responses, including the likelihood of *ex post facto* "rationalizations" by the participants, the risk of misinterpretation where the same terms have both a lay and scientific meaning, the problems in capturing non-conscious processes, the influence of personal prioritization of causes or events, *etc.* As to the first point, he argued that economic theory "is essentially a theory of adjustment to change," which approximates and idealizes the process by which decisions are made:

> "The explanation of an action must often include steps of reasoning which the acting individual himself does not consciously perform...and which perhaps he would never be able to perform in scientific exactness...To call, on these grounds, the theory 'invalid'...is to reveal failure to understand the most basic methodological constitution of most social science."

Id., p.535.

FURTHER EXCHANGES

Lester responded; Machlup published a rejoinder; and future Nobel Laureate George Stigler added his own comments criticizing Lester's obvious political agenda. Richard A. Lester, "Marginalism, Minimum Wages and Labor Markets," *American Economic Review*, XXXVII (March 1947), pp.135-48; Fritz Machlup, "Rejoinder to an Anti-Marginalist," *American Economic Review*, XXXVII (March 1947), pp.148-54; George J. Stigler, "Professor Lester and the Marginalists," *American Economic Review*, XXXVII (March 1947), pp.154-57. Perhaps the most telling and convincing part of this debate were the cries from each side that their views were being misunderstood and mischaracterized by the other.

MILTON FRIEDMAN'S "POSITIVE ECONOMICS"

An elegant defense of the neo-classical theory of the firm was presented by Milton Friedman in an article called "The Methodology of Positive Economics," which reflected much of the current orthodoxy of the philosophy of science focused on the natural sciences. *Essays in Positive Economics* (1953), pp.3-43.

He quoted J.N. Keynes with respect to the distinction between positive and normative economics, asserting that positive economics is concerned with what is, not what ought to be, and must be free from any ethical positions. Its goal is that of an objective science: to formulate a system of generalizations with which to make accurate predictions about the results of changes in various economic variables. "Its performance is to be judged by the precision, scope, and conformity with experience of the predictions it yields. ...The ultimate goal of a positive science is the development of a 'theory' or 'hypothesis' that yields valid and meaningful (i.e., not truistic) predictions about phenomena not yet observed." Id., p.7.

Friedman asserted that many of the criticisms of neo-classical theory were inappropriate because they focused on the "realism" of the assumptions. That is not the proper standard to apply.

As he explained:

> "A hypothesis is important if it 'explains' much by little, that is, if it abstracts the common and crucial elements from the mass of complex and detailed circumstances surrounding the phenomena to be explained and permits valid predictions on the basis of them alone. **To be important, therefore, a hypothesis must be descriptively false in its assumptions....**"

Id., pp.14–15 (emphasis added).

Sherman R. Krupp illustrated the propositions asserted by Milton Friedman with examples. "Theoretical Explanation and the Nature of the Firm," *Western Economic Journal*, I (1963), pp.191–204. He explained that the concepts of a "firm" and "profit maximization" are involved in three quite different types of inquiry, asking different questions and addressing different concerns:

- In micro-economic analysis, the concern is with the allocation of resources by an economic system (in which process price plays a crucial role), and the theory of the firm assumes profit-maximizing behavior.
- The behavioral or empirical inquiry is concerned with how and to what extent firms do maximize profits.
- The managerial concern is with how firms can maximize profits.

Friedman, in his response, made the additional, derivative assertion that no theory will predict all characteristics of an event or make predictions for all types of phenomena. Therefore, the validity of a theory must be judged in relationship to the particular purpose for which it is to be used and whether its predictions are sufficiently accurate for that purpose. Now this observation may seem to be applicable to the social sciences, but not to the natural sciences; yet, in fact, theories in the natural sciences do not attempt to predict all events or characteristics, only the specific phenomena that the theory had singled out for treatment. The basic principles are the same.

Some Observations

It may seem odd to some to say that a theory should be judged relative to the purpose for which it is being used. Yet, in the sense that a theory will often seek to "explain" only selected phenomena, it seems like an innocuous statement. But, the observation also has meaning in a different and more important sense.

The failure to differentiate the nature of the inquiry at hand leads to confusion. In neo-classical price theory, the firm is part of the explanation (and its function is postulated or assumed); in other types of studies, the firm is the thing to be explained. "Ambiguity in these meanings and activities as well as the frequent lack of a clear distinction between what is to be explained and how the explanation is to be composed, account for substantial confusion in the theory of the firm. ...What economic theories are, how they do their work, and what kinds of work economic theory may or may not do—the answers to these questions are essential to clarity of thought. This kind of knowledge is found in the logic of science and the history of ideas." Id., p.204.

Friedman had defended the profit-maximizing assumption on the assertion that businessmen behave "as if" they were maximizing profits, regardless whether they consciously believed themselves to be doing so or even whether they were even capable of consciously doing so. Friedman sought to exemplify his analysis with the following analogy. One could theorize that the leaves on a tree grew "as if" they maximized the sunlight received by the tree as a whole, considering all of the leaf surfaces. As he explained, that theory would explain with considerable accuracy the actual observable growth patterns of most trees. However, the accuracy of the explanation does not mean that the tree actually intended any such thing or was capable of doing any of the calculations necessary to achieve that result by design. Similarly, he claimed, firms behaved in a manner that could be "explained" by the profit maximization principle even though the managers did not intend that consequence or were incapable of determining how to achieve it, even if they wanted to do so.

Despite the superficial appeal of the analogy, it should be apparent that in the case of the tree, we have an understanding of a biological process that would tend to generate the result described—the response of living material to sources of nutrients. Therefore, this analogy does not really meet the objections of the critics who claim that profit maximization is not, in fact, the objective of real-world business entities and that such entities would not know how to achieve that objective even if they embraced it.[10]

What is it, if anything, that connects the actual operations of business to the model?

Note that the different types of arguments made about or explanations given with respect to the theory of the firm lead to somewhat different conclusions as to how the theory connects to reality. Some approaches suggest that the theory could generate predictions or hypotheses about real world firm behavior that could, at least in principle, be tested. Others suggest that one would not necessarily expect that accurate predictions about the conduct of particular business entities could be made, but that the broader micro-economic theory of which the theory of the firm is part would enable such predictions. The appropriate approach that would be taken to test the theory would be quite different depending on which of the two different conceptualizations of what the theorist is doing applies.

However, Friedman staked out a rather categorical position in his defense of neo-classical price theory: the real test is in the accuracy of its predictions. He wrote: "An even more important body of evidence for the maximization-of-returns hypothesis is the experience from countless applications of the hypothesis to specific problems and the repeated failure of its implications to be contradicted. ...Yet the continued use and acceptance of the hypothesis over a long period, and the failure of any coherent self-consistent alternative to be developed and widely accepted, is strong indirect testimony to its worth. The evidence for a hypothesis always consists of its repeated failure to be contradicted... ." Id., pp.22–23.

Was this a wise position to take? We have already alluded to the dubious record of economics in making successful predictions. We shall consider further what that means. We shall also look more closely at Friedman's assertion that positive economics is objective, free of value judgments or ethical positions.

The Methodological Issues

Prediction and explanation (again)

We have discussed different explanations given for the relationship between the concept in the theories and the real economic world. Previously, we discussed at some length the differences between prediction and explanation. Let's review the argumrnts.

One argument is that value theory and the theory of the firm represent reality, with firms, subject to inevitable distractions and interferences, behaving so as to maximize profits. The disruptions will obscure the underlying process, but it is there and can be extracted through careful analysis of the facts of particular situations. This approach is both predictive and explanatory.

Another argument is that the theories are based upon necessary roles or functions in an economy, which must be performed but may not be tied to particular existing legal entities (like the example of the "gene", discussed above). Thus, an entrepreneurial function could be performed by a person or a sub-group separate from the formal firm. Subject to what could be serious problems of connecting obtainable data in the real world with the terms in the theory, this type of theory could be predictive as well as explanatory.

Yet another argument is that the theories abstract from reality, presenting simplified versions of reality and that, as a result, there will be no direct correspondence between a concept in the theory (like the concept of the firm) and entities identified in the real world. It must be possible to have contingent propositions somewhere in the theory, so that there could be verification in the real world, but, even so, such a theory would be potentially predictive and not explanatory.

And, a somewhat similar argument might be derived from the assertion that firms act to maximize profits because they have to do so or they will not survive. It is common in economics to excuse short run abnormalities as the results transient disruptions, looking to the long run to reveal the important underlying tendencies. Of course, the importance of the long-run tendencies, even if they do exist, may be debated. As (John Maynard) Keynes said, "In the long run, we are all dead."

This line of thought brings us back to the fact that the adequacy of the type of theory depends upon what we intend to use the theory to do. We may be uncomfortable with that standard when it comes to the natural sciences, but that feeling makes little sense. If the theory explains the phenomenon in which one is interested sufficiently for the purposes at hand (whether technological, policy making or future scientific investigation), then it has served a purpose, even if there are many other things it does not explain.

Some Caveats

Despite the frequent pronouncements that economics is a science and can be subjected to the rigorous test of disconfirmability, it seems quite clear that certain characteristics of the theories are simply not compatible with, at least, that latter claim.

Ceteris paribus (again)

Almost all of economics deals with open systems (where not all relevant variables that can affect the result are included in the theory) and, therefore, is constructed based firmly upon the *ceteris paribus* assumption, that is, "assuming all else is the same." The importance of that assumption is that in the actual world, all things are never the same or, at least, it is always possible that they are not. Thus, when a prediction based upon the theory is made and subsequent observations are inconsistent with the prediction, there is always an available answer or explanation–something else that was assumed to remain constant changed and that is the reason the prediction was not fulfilled. In other words, after the fact, it is almost always possible to justify the failure of the test or, at a minimum, to cast doubt on the failure by reason of the necessary assumption of *ceteris paribus*.

The *ceteris paribus* condition applies essentially to all testing of theories making predictions. However, its importance and impact varies greatly. Although, the assumption is that "all else" is the same; the condition is not really that strict. All that needs to remain unchanged are the things that are relevant–that is, that could matter to the result that is being predicted. It is obvious that with respect to various physical science theories, the range of conditions that could matter may be relatively small and may also be controllable. In those cases, meaningful experiments can be conducted, and the *ceteris paribus* condition can be satisfied.

Initial conditions (again)

A similar type of challenge is presented by the identification or specification of the initial conditions or state of affairs from which the change predicted by the theory is to occur. If those conditions are not or cannot be known, then the predictions of the best theory will necessarily be speculative and conditional. See, Barrow, *Theories of Everything*, pp.38–42; Nagel, *The Structure of Science*, p.32.

Again, the initial conditions at issue are those factors that will affect the results of the process of change. Initial conditions could be essentially irrelevant if the process at work leads to a particular result regardless of the starting conditions or, at the other end, could be overwhelmingly important. Where small differences in initial conditions may result in very large, significant differences in the end result, the ability to specify the initial conditions becomes crucial. See Barrow, *id.* In cases leaning toward this later category, the ability to identify those initial conditions fully and accurately will largely determine the usefulness of the theory, at least for purposes of prediction and forecasting actual events.

Some consequences

As a result of both of these problems (the *ceteris paribus* condition and the difficulty of specifying the initial conditions), economic theories are not, in practice, disconfirmable or subject to falsification. Instead, failures of prediction are routinely explained away after the event, and skill in constructing such explanations has been the hallmark of a good economist. Indeed, almost 50 years ago, it was possible to assert, without surprising anyone knowledgeable in the discipline, that "It has been said that a good economist is one who will be able to explain next year why his predictions of last year have not come true." Fritz Machlup, "Operationalism and Pure Theory in Economics," in Krupp (ed.), *The Structure of Economic Science* (1966), pp.66, 74. Of course, there may be circumstances where even the most imaginative economist would be unable to find an *ex post facto* explanation. In such a circumstance, would the theory be abandoned as falsified? Perhaps, but I think it not likely.

AN ELEMENT OF CIRCULARITY

A separate methodological issue has to do with potential circularity, not just the lack of disconfirmability. Much of micro-economics is based upon assumptions that are either true by definition or are identified as beyond the proper scope of the discipline.

For example, it is assumed that whenever a choice is made between two alternatives, the more beneficial option is selected. But, generally no investigation is made as to whether there are objective bases for doubting whether that assumption is factually correct. Actually, efforts have been made to explore that question, but they were often beyond the mainstream of economics and have been sometimes viewed as heretical. It is easier and theoretically more pure to say simply that the option selected is the more beneficial, by definition, because it was the one that was selected. Thus, consumer decisions are taken as demonstration of value. If a person selects one of a set of alternatives, then the person is deriving greater utility or value from the item selected, by definition, or he would not have selected it. Rationality is assumed to prevail and to be the determinant of all decisions. The problem extends to more than the assumptions. The market dynamics are not needless details. The "how" is a necessary part of understanding.

Economists have recognized that lack of information could taint the selection made, and various caveats have been offered to reflect that reality. At the same time, it is often assumed that such problems correct themselves over time and, therefore, can largely be ignored. But, the assumption that I am focused on here is that the choice made reflects the alternative that provides more utility or benefit to the person making the choice, even with perfect information. We do not know what a person's actual motivations are; we do not even know what a person's set of values or priorities is. So we assume that the person's actions and choices must reveal or reflect those values and priorities. In other words, people would not make choices that diminish their welfare as they understand it.

Unfortunately, there are many examples of instances where people seem actually to do just that. This central assumption of economics is that the actors are rational and make rational choices. To what extent and in what ways do the theories fall short because that assumption is simply not valid?

A note on food

> *Although it was published in 2008, I did not read Michael Polan's* In Defense of Food *until 2018, well after the publication of the second edition of this book. I was a bit surprised to see that many of the themes that I set forth (and now repeat) were either expressly stated by Polan or, in other cases, nicely illustrated by his argument. For example, in his discussion of the consequences of nutritional science or what he refers to as "nutritionism"—a focus on nutrients rather than on foods—from the 1970s onward, he provides powerful illustrations of the fallacy of the economist's assumption of informed rational consumer behavior. Whether as a result of lack of information, lack of willpower or susceptibility to manipulation; it appears that the American consumer has been unable to make choices that are actually in his or her long-term best health interests.*

THE RELEVANCE OF EQUILIBRIUM

There is another issue lurking in the background. Traditional economic theory is based upon hypothetical positions of equilibrium–situations in which no one has any incentive to make any change. External events will disrupt that equilibrium and create opportunities for persons or entities to make changes in pursuit of their own self-interests. The theories predict what type of adjustments will take place and, with luck, what the new equilibrium will look like.

Some of the shortcomings of this model are pretty obvious. For example, although it may be the case that in the equilibrium state, no one has any incentive to make a change; if the participants have insufficient information to realize that that is the case, then they may adjust their conduct in the mistaken belief that it will make them better off. And, if they do not have sufficient information to assess the actual results of the changes, then they may be unable to decide even whether the change was beneficial or not. The same observations apply to the assumed responses to external changes.

More importantly, the concept of equilibrium has limited usefulness unless there are strong reasons to believe that there are forces pushing toward equilibrium. Thus, we need a theory of the process of adjustment that says that a state of disequilibrium will move toward, not away from, the state of equilibrium. For example, in perfect competition, a change in demand provides an incentive for producers to change their output. However, changes take time. If a sufficient number of producers increase their output in response to a higher price, then there will be an oversupply at that higher price. So, the price falls. If a sufficient number of producers decrease their output in response to the lower price, then there will be an undersupply at that lower price. So, the price rises. Do the swings get smaller or larger? Will the market move toward or further away from equilibrium?

THE "INVISIBLE HAND"

One distinctive image emerging from the classical theory of Adam Smith is that of the "invisible hand". The proposition is that individuals each working in their own self-interest would produce a result that was in the best interests of the community. The mechanism by which this result occurs (through competition and free exchange) was referred to as the "invisible hand."

Initially, the use of the metaphor was based, in part, upon empirical observation of the economic development in the early industrial age; in part, upon personal belief and, in part, as the necessary implication of the emerging theoretical model. In addition, over the next 150 years, the empirical evidence of the benefits of private property and free markets grew in strength and breadth, as did the theoretical model. The incidental benefits of decentralized decision-making were elaborated by economic theory (less risk of catastrophic mistakes, ability to "utilize" widely dispersed information, greater innovation and creativity, more opportunities for social mobility, etc.), as well as the conclusion that the result could be the optimum allocation of the resources and maximization of the output of the economy.

The importance of the image of the "invisible hand" is its focus on change and adjustment. Neo-classical theory present elaborate models for representing states of economic affairs. But, the real problems in economic policy relate to the processes of adjustment and the acquisition and use of information.

> "*If we possess all the relevant information, if we can start out from a given system of preferences, and if we command complete knowledge of available means, the problem which remains is purely one of logic. That is, **the answer to the question of what is the best use of the available means is implicit in our assumptions**. The conditions which the solution of this optimum problem must satisfy have been fully worked out and can be stated best in mathematical form: put at their briefest, they are that the marginal rates of substitution between any two commodities or factors must be the same in all their different uses. **This, however, is emphatically not the economic problem which society faces**.*"

Friedrich A. Hayek, "The Use of Knowledge in Society," *American Economic Review*, XXXV, No. 4, p. 519, September, 1945 (emphasis added).[12]

He gos on:

> "*...the 'data' from which the economic calculus starts are never for the whole society 'given' to a single mind which could work out the implications and can never be so given. ...The peculiar character of the problem of a rational economic order is determined precisely by the fact that **the knowledge of the circumstances of which we must make use never exists in concentrated*

or integrated form but solely as the dispersed bits of incomplete and frequently contradictory knowledge which all the separate individuals possess."

The metaphor of the "invisible hand" describes the expected beneficial results of thousands of individual actions motivated by self-interest. Many of the actions may be misguided or random. But, extraordinary profits will be eroded by others competing to enjoy some part of them, and bad decisions will be punished by loss of sales or bankruptcy of firms. Productive behavior will be rewarded and then mimicked. Supposedly, it all happens because of the relentless pressure of people pursuing their own self-interest, guided by the "invisible hand". In the free market economy, "[t]he continuous flow of goods and services is maintained by constant deliberate adjustments, by new dispositions made every day in the light of circumstances not known the day before, by B stepping in at once when A fails to deliver." Id.

Hayek argued that the decentralized economy is superior, because decision-making incorporates the details of many different situations and circumstances:

> *"But the 'man on the spot' cannot decide solely on the basis of his limited but intimate knowledge of the facts of his immediate surroundings. There still remains the problem of communicating to him such further information as he needs to fit his decisions into the whole pattern of changes of the larger economic system."*

Id.

The problem of the collection and dissemination of the relevant information is central. As Hayek observed, "in a system in which the knowledge of the relevant facts is dispersed among many people, prices can act to coördinate the separate actions of different people in the same way as subjective values help the individual to coördinate the parts of his plan." Id. Indeed, the emergence of an accepted monetary currency and supporting financial infrastructure has created a common "language" that enabled communication and "coordination" among hundreds of thousands of decision-makers spread over wide geographical areas. The process is not coordinated or directed activity; it is commercial interaction (trading, buying and selling, investing in capital) based upon trust (in the currency) and communication through prices set out in the common currency. See Harari, *Sapiens*, pp.173-87.

Thus, the price system is the "invisible hand," and money makes possible its operation.

THE "GOOD" THEORY (AGAIN)

Part of the goal of a good theory is to abstract from the innumerable details reality and seize upon the key or central factors at operation, so as to illuminate what is really happening. The process of abstraction to identify the essence requires simplification. Thus, the traditional theory assumes that the firm acts to maximize its economic profits. That assumption does not say that the managers of the firm necessarily "intend" that result. Instead, the theory simply says that that is what happens.

Perhaps, economic theory is a little like geometry, where the concepts (points, lines, planes with only one or two dimensions) represent idealized, theoretical constructs that objects in the real physical world can only approximate. In geometry, manipulation of and calculations concerning the idealized structures provide good approximations of the physical world when one accounts for the inevitable and necessary deviations from the ideal as well as the limitations of physical measurement. Is the same true for economics or other social sciences? It certainly does not appear to be so.

Some early economists thought that the proper approach to constructing deductive theories was to identify assumptions that were accurate reflections of factual circumstances and then use the rules of logic to manipulate the assumptions to generate conclusions or predictions. If the assumptions were "true" and the rules of manipulation consistent and logical, then the conclusions derived should also be "true." Over time, increasing debate occurred over the realism or truth of the assumptions. Propositions that were essentially taken for granted by earlier generations appeared more and more questionable under critical scrutiny, especially in the context of changing societal norms and values.

Separately, it is apparent that a useful theory has to simplify reality and extract from the welter of facts the factors and relationships that really mattered. In other words, some abstraction from reality is essential. The goal is to identify the fundamental forces at work.

One example would be making the assumption that there are characteristics of group behavior that exist but that would be hard or impossible to detect by looking at the actions of an individual. If that is correct, then one could produce theories of group actions that are independent of and not informative with respect to individuals. One might speculate that large numbers of individual decisions will lead to certain consequences about which predictions can be made, even though the predictions would tell you little or nothing about any individual decision. (In fact, we see something very similar in physics.[11])

But, Hayek, again in "The Use of Knowledge in Society," disputed the applicability of this "law of large numbers" to economic phenomena:

> *"The comparative stability of the aggregates cannot, however, be accounted for—as the statisticians occasionally seem to be inclined to do—by the "law of large numbers" or the mutual compensation of random changes.* **The number** *of elements with which we have to deal* **is not large enough** *for such accidental forces to produce stability [and the individual movements* **are not random***]."*

So, back to Friedman's analogy to the leaves on a tree. For insight into the causation of the observed phenomena, one needs theories concerning the growth of cells, the assembly of cells into branches and leaves and the development of plants over their life span. Such theories might provide the basis for an understanding of how it happens that leaves become so positioned. What is the relationship between those theories and the more abstract statement that the position of the leaves on the tree will happen to be such as to maximize the amount of sunlight received. Is that latter proposition, even with its misleading anthropomorphic overtones, good science?

Some Concluding Comments

I want to comment on two other topics and offer a final observation.

Pareto optimality

Assume that an economy is in the position in which no changes could be made that would result in making any person better off without making at least someone else worse off. Such a situation became known as Pareto optimality, named after the Italian engineer and economist Vilfredo Pareto (1848–1923), who published his *Manual of Political Economy* in 1906. This definition of an optimum seems "scientific" or, at least, objective. It avoids any attempt at the comparative valuing of benefits for different people, so one could not assert that a change would benefit society, as a matter of economics, if one person sacrificed so that 100 people would benefit. Similarly, the satisfaction of a human "vice" is not assigned less weight than the satisfaction of a fundamental human need. These two features, however, are of debatable value.

But, how could anyone object to a reallocation of resources that made one person better off without disadvantaging anyone else? Unfortunately, this standard is not a practical tool for economic policy-making. We can not know whether any particular circumstance meets the standard, and we cannot know what changes would get us there.[13] Moreover, this standard is utterly irrelevant to the significant and pressing policy issues facing most societies that involve questions of justice and fairness, the establishment of priorities among or weighting of competing goals, the value of human live (or of animal lives), *etc.*

Economic "morality"

One concern that arose, however, was that the phraseology of the theory appeared to carry normative implications. The "fittest" survive; the more "efficient" firms prosper; the more "productive" factors of production get the greater return (at the margin), the allocation of resources was "optimum". Of course, one can argue that the derivation of such normative implications represents a misuse or distortion of the theory. If so, then careful thinking should eliminate any problem. Perhaps one could say that the vocabulary utilized is simply unfortunate in its susceptibility to misuse. But, I think that such responses are unduly facile. The implications were not careless mistakes. They reflected the intention and beliefs of the creators of the theory.

As we look more closely at the methodological elements of neo-classical economics, it becomes clear that the theory is not value-free. The theory contains important assumptions about human beings and society. Perhaps, it is not surprising that much of the early development of the theory was the work of persons from the eighteenth-century English world, where private property, a reasonably high degree of personal freedom and autonomy, a predictable legal system and a history of trade were fundamental.

Strong negative reactions to neo-classical theory arose from multiple sources, often because the theory was perceived to have normative implications. Obviously, the theory appeared to proclaim the superiority of free markets. It also seemed to support minimal roles for government and collective action. Moreover, the analysis of the returns to the factors of production in the process of allocating resources (in the optimum fashion, of course) could be interpreted to suggest that the returns received by each factor were "earned". That implication would morally justify the allocation of income to capitalists, since capital "deserves" the return at the rate provided by the market. It would also morally justify the poverty of the working poor, since the low wages earned were the result of the low comparative marginal productivity of the labor relative to the other factors of production (capital and natural resources or materials).

Of course, economic theory does not formally purport to address the morality of any of the results—as economists have routinely and smugly proclaimed. That answer, while technically true, is surely not sufficient. First, the impressive appearance of the theory, about which much boasting has occurred, has an impact that understandably does lead to misunderstanding and misuse despite disclaimers. Second, the questions of public policy and morality matter a lot, and there is something disingenuous about simply ignoring them. These are serious issues. Nonetheless, it is also true that objections to the societal or moral implications of a theory are not proper grounds for rejecting a theory as a matter of science.

AND SO...

There is broad agreement today that economics, whether micro or macro, falls far short of providing a basis for successful predictions about or effective management of an economy. Yet, this does not mean that economics is not a successful science. There are concepts, structures and methods of analysis that are exceptionally productive tools. Economics provides models, paradigms and theories that assist in understanding various phenomena (at least in isolation) and in identifying significant questions. These methods are now applied throughout the social sciences.[14] Indeed, I believe that an understanding of the major concepts of economics is essential to an intelligent and informed approach to most significant policies issues.[15] However, at the same time, the usefulness of these tools does not establish the validity of the broader theory of which these elements are part.

ENDNOTES

[1] In 1891, J. N. Keynes quoted Alfred Marshall as saying: "It is vain to speak of the higher authority of a unified social science. No doubt if that existed, economics would gladly find shelter under its wing. But it does not exist; it shews no sign of coming into existence. There is no use waiting idly for it; we must do what we can do with our present resources." *The Scope and Method of Political Economy* (1891), p.110. Keynes observed: "Students of economics may, moreover, naturally and fairly ask to have the province of sociology itself more explicitly defined, and to see its own fundamental doctrines more clearly formulated, before they can be expected to show a willingness to have political economy sub-sumed under and absorbed into it." *Id.*, p.134.

[2] The aspiration to mimic the natural sciences or, perhaps more accurately, to be a "science" seems not to be universal in the social sciences. Some recent controversy attended the decision by the American Anthropology Association to change the statement of the Association's purpose in its statement of its long range plan by replacing the word "science" with "public understanding" in the description of what the Association would attempt to advance. See Nicholas Wade, "Anthropology a Science? Statement Deepens a Rift," *The New York Times*, December 9, 2010.

[3] The methodological approach essentially is to posit a hypothesis, make a prediction and then examine historical events to see if the prediction came true in circumstances with sufficient similarities to the hypothesis. An obvious problem is that the construction of the hypothesis or theory may well or often be directly influenced by the historian's existing understanding of the historical events, so that the resulting prediction is almost guaranteed to be confirmed. It is difficult to describe the prediction under those circumstances as a "test" of the hypothesis.

[4] For some of the historical review, I have relied on secondary sources. My primary secondary sources were Joseph A. Schumpeter, *History of Economic Analysis* (1954); Kaman J. Cohen and Richard M. Cyert, *Theory of the Firm* (1965) and Joseph W. McGuire, *Theories of Business Behavior* (1964).

[5] I discovered that in the mid-twentieth century, some leading economists had become engaged in a heated debate over methodology. The heart of the debate—what are we trying to accomplish with our theories

and how do we judge or assess our degree of success—has application in a wide range of intellectual and policy activities.

6 At a more prosaic level, as has been frequently observed, if two caregivers simply take care of each other's wards in exchange for payment (in amounts that can even cancel each other out), rather than stay at home caring for their own, GDP is increased. Yet, there is no actual increase in real output (indeed, there may be time lost in commuting, additional expenses and more taxes paid because of the creation of recognizable income).

7 The paradigm of the scientific method in the nineteenth century was Newtonian mechanics.

8 These "truths" were apparently viewed as self-evident; they were not defended with evidence.

9 This characterization of the process of model-building was widely accepted. For example, J.N. Keynes wrote in 1891, "We may, at any rate in certain departments of enquiry, determine the more constant and permanent tendencies in operation, and hence reach a first approximation towards the truth." *The Scope and Method of Political Economy*, p.115. The Nobel Prize-winning economist George Stigler wrote more than half a century later: "the role of description is to particularize, while the role of theory is to generalize—to disregard an infinite number of differences and capture the important common element in different phenomena." Stigler, *Five Lectures on Economic Problems* (1950), p.23.

10 The sunlight-maximizing principle may be an accurate description of the positioning of leaves in response to changes or the arrangement of the leaves or, even, of the growth of new twigs and small branches. In that sense, it might be offered as an "explanation." But, it does not say anything useful in connection with alternative shapes of trees. The basic structure of the trunk and branches will be determined by the genetics of the species and various external factors.

11 Newtonian mechanics provides a highly reliable model of the movements of visible objects without attempting to say anything about atoms or the sub-atomic world (*see* the chapter Physics). Indeed, an examination of the actions of individual atoms could not generate accurate predictions about the behavior of the aggregate (*see* the chapter Chemistry).

[12] In the mid-twentieth century, there were serious debates about the feasibility of the centralized planning and management of the economy (e.g., socialism). Hayek's article was part of this debate, with him arguing that the free market is the only effective type of economic organization. "We need decentralization because only thus can we insure that the knowledge of the particular circumstances of time and place will be promptly used. A central planner will never have the information about the localized time and place to make the immediate adjustments necessary for an optimal result." Hayek, "The Use of Knowledge in Society."

[13] See, e.g., Hayek, "The Use of Knowledge in Society," n1:

> There is a "myth that Pareto and Barone have 'solved' the problem of socialist calculation. What they, and many others, did was merely to state the conditions which a rational allocation of resources would have to satisfy and to point out that these were essentially the same as the conditions of equilibrium of a competitive market. This is something altogether different from knowing how the allocation of resources satisfying these conditions can be found in practice. Pareto himself (from whom Barone has taken practically everything he has to say), far from claiming to have solved the practical problem, in fact explicitly denies that it can be solved without the help of the market."

[14] Noble Laureate Gary Stanley Becker of the Chicago School of Economics pioneered the application of economic methods to subjects traditionally thought to belong to sociology.

[15] Marginal analysis, the concepts of sunk costs and opportunity costs, and basic cost/benefit analysis to assess the desirability of proposed actions are all fundamental to a modern understanding of the world and are necessary tools for participation in meaningful debate.

Darwinism

> "[T]he single best idea anyone has ever had
> ...unif[ying] the realm of life,
> meaning, and purpose... ."
>
> Daniel Dennett
> *Darwin's Dangerous Idea*, p.21

> "One intuitively senses that
> it is an inherently feeble response
> to an extraordinarily rich history that
> has brought forth an immense
> coruscation of form and diversity."
>
> Simon Conway Morris
> *Life's Solution*, p.1.

Up front, I want to say that I believe that it is established with virtual certainty that humans share significant genetic material with other animals. Moreover, it is highly unlikely that the extent of that common genetic material could have arisen purely by chance. The odds of such common genetic sequences just happening are staggeringly enormous. I am also convinced that *Homo sapiens* descended or derived from ancestors that included other primates and that much of the shared genetic material exists because of that descent. To the extent that the fossil record may be ambiguous, the sequencing of the human genome and others provides pretty clear evidence. See, *e.g.*, Francis S. Collins, *The Language of God: A Scientist Presents Evidence for Belief* (2006), pp.109-42.

But, as we shall see, there are possible explanations other than Darwinism for at least some of the common or shared "genes". (For those of you who wonder why the much publicized DNA analysis does not definitively answer these questions, it is because DNA has only been found dating back 700,000 years. New technology is now giving the possibility of looking back 1.7 million years.[1]) We shall discuss in later chapters the apparent inability of natural selection (and, therefore, neo-Darwinism) to explain the appearance of two of the more important aspects of our reality: life itself and consciousness.

The Diversity of Life

The extent of the genetic relationship is secondary to what I am interested in addressing here. The question I pose for examination is whether traditional Darwinism (including neo-Darwinism, described below), which in recent decades has seemed almost immune from scientific criticism, is an adequate explanation for the diversity of biological forms that exist on Earth today, including the forms that have displayed intelligence and consciousness.

The theory at issue is "natural selection": (i) that individuals of a species experience random heritable changes; (i) that certain of those changes fortuitously improve the ability of those individuals to survive and reproduce; (iv) that, by virtue of "natural selection" or, in the words of Herbert Spencer (in 1864), "survival of the fittest",[2] certain resulting variations in physical characteristics become dominant within a population; and, finally, (v) that over time new species are created by or emerge from this process.

In a recent book, *What Darwin Got Wrong* (2011), Jerry Fodor and Massimo Piattelli-Palmarini distinguish two branches of Darwin's theory, the "taxonomic tree" based upon the asserted fact of descent from common ancestors—the explanation for the existence of phenotypic and genotypic similarities—and the theory of natural selection which attempts to provide the why or how of such descent. See also, Dawkins, *The Greatest Show on Earth*, pp.17-8 (explaining that Darwin thought of both the fact of common descent and the theory of what drives evolution as tentative hypotheses). This distinction is not new. The Marquis of Salisbury, in his presidential address to the British Association for the Advancement of Science at Oxford in 1894, praised Darwin for making a convincing case for evolution but heavily criticized Darwin's proffered mechanism, natural selection. See Frank Ryan, *Virolution* (2009), pp.34-5.

At the turn of the twentieth century, Darwin's theory of natural selection was under serious attack. A fundamental problem was posed by the conclusion by Lord Kelvin that the age of the Earth could be no more than one million years, based upon the physics of the cooling of a hot body. *Id.*, p.35. A span of one million years was simply inadequate for the development of the diversity of life observed on Earth by the process of natural selection. Of course, Lord Kelvin was wrong in his calculation of the age of the Earth, because he (like everyone else) did not know about nuclear fusion.

Among the general public, however, what was most controversial about Darwin's ideas was the assertion that the shared characteristics of various animals are the result of descent, *i.e.*, of genetic derivation. *See, e.g.*, Fodor and Piattelli-Palmarini, *What Darwin Got Wrong*, p.2. Over time, that aspect of Darwin's contribution has been supported by significant scientific discoveries and is considered by many to be established with virtual certainty.

As put by Simon Conway Morris, there is now little question about "the fact of evolution, at least as a historical narrative: very crudely, first bacteria, then dinosaurs, now humans." *Life's Solution: Inevitable Humans in a Lonely Universe* (2003), p.1.

Importantly, however, it is not the genealogical conclusion that is the heart of the neo-Darwinism being promoted by its outspoken advocates, like Daniel Dennett, Richard Dawkins and Jerry Coyne. It is specifically the theory of natural selection. For example, Dennett asserts that it is "the single best idea anyone has ever had.... [I]n a single stroke, the idea of evolution by natural selection unifies the realm of life, meaning, and purpose with the realm of space and time, cause and effect, mechanism and physical law." *Darwin's Dangerous Idea* (1995), p.21. Dawkins characterizes "evolution by non-random natural selection" as "the only game in town, the greatest show on Earth." *The Greatest Show on Earth* (2009), p.426. It was "Darwin's greatest discovery." *Id.*, p.62. With similar effusiveness, biologist Jerry Coyne claims natural selection to be "a mechanism of staggering simplicity and beauty." *Why Evolution Is True* (2009), p.xvi.

Such tributes are not for the conclusion that living things are genetically related; they are expressly focused on the underlying explanatory theory of how those relationships came about.

ULTERIOR MOTIVES?

Another retired lawyer, a full generation ahead of me, found himself drawn into a study of Darwinian theories. (Coincidentally, his daughter, also a lawyer, worked briefly for me a few years after his book was published, leading me to buy and read it.) Beginning in 1959, Norman Macbeth spent years reading the works of the major participants in the theoretical developments of the first half of the twentieth century. In 1971, he published *Darwin Retried*. Macbeth's professed motivation was to inform the lay world that, contrary to general public perception, the experts in the field had effectively concluded that Darwinism was flawed and was "going to pieces." Id., p.2. He supported his conclusion with abundant quotations from the leading evolutionary biologists of the century.

He concluded that the reason that the general public had been left to believe that, like it or not, Darwinism was impeccable scientific dogma was that scientists simply were not prepared explicitly to abandon Darwinian theories because there was not a satisfactory scientific alternative available.[3]

I believe that the ardor of those advocates is driven by a powerful need to challenge Creationism and similar religious explanations of life in general and of *Homo sapiens* in particular. The worry is about the "camel's nose". Thus, the enthusiastic promotion of natural selection may be regarded as a pre-emptive strike. Coyne freely admits his motivation. *Why Evolution is True*, pp.xi–xiv.

Dawkins, in his 2006 best seller *The God Delusion*, constructs much of his argument against the traditional concept of a supernatural Creator upon the proposition that with natural selection, we have a credible–indeed, persuasive–alternative to God, mooting the many originally convincing arguments advanced over the centuries that the world we see and our presence in it are inexplicable absent a Creator and Designer. Id, pp.85, 137–69. Despite the length of the book, Dawkins' entire logical argument consists of the propositions that (i) natural selection is a persuasive explanation for the complexity and diversity of our current biological world, so a Designer is not necessary, and (ii) the hypothesis of a God inevitably results in an infinite regress (e.g., what is the cause or source of God, etc.), involving a highly improbable complexity. In other words, it is not a "simple" explanation (since, e.g., one must then explain what God is and how He came to be). Id., pp.175–80, 188.[4]

Coyne's argument, in comparison, is not phrased in terms of natural selection being a better explanatory theory than Intelligent Design. Instead, he claims that natural selection is **the** answer.

Is Darwinism a "fact"?

Coyne claims that evolution by natural selection is "true," is "as solidly established as any scientific fact" and is "more than 'just a theory.'" *Why Evolution is True,*, pp.xvii, xviii. Apparently, almost the identical words were used by Sir Julian Huxley in a television appearance some 65 years ago. See Macbeth, *Darwin Retried*, p.147. Is it really indisputable that evolution by non-random natural selection is the accurate depiction of what has actually occurred in producing the tremendous variety of life forms we have?

The answer to that question is clearly no. These claims by Coyne are ridiculous. I do not here assert that the theory is demonstrably wrong, but only that it is at least subject to doubt, actually, to significant doubt. Evolution by natural selection is indisputably a theory. And, as we have discussed at length, a theory is not a fact. Moreover, a theory cannot ever properly be claimed to be true or a fact.[5]

Yet, the concept of Darwinian evolution by natural selection has achieved a pervasive presence in modern scientific thought. Indeed, it has become, for many, the hallmark or symbol of objective scientific materialism, one which must be fully accepted by anyone who hopes to be recognized as a respected member of the scientific community. In other words, "allegiance to Darwinism has become a litmus test for deciding who does, and who does not, hold a 'properly scientific' world view." Fodor and Piattelli-Palmarini, *What Darwin Got Wrong*, p.xv. (The wide acceptance of the theory in the scientific world is evidenced by the number of scientists from other disciplines who seem to feel quite comfortable opining about evolution and natural selection.) This development, I think, is most unfortunate. It has suppressed critical analyses of the theory in its many varied applications and has added considerable unnecessary emotion to public debate.

Is it just "tuning the piano"?

Despite the extensive attention that has been paid to the application of the theory of evolution to primates and *Homo sapiens*, there are still highly significant aspects of human evolution that are not explained.

> "That we evolved is considered by the scientific community to be established beyond reasonable doubt, yet at the same time there remain enormous gaps in our knowledge of how it happened. Some of the most fundamental details of our development—why

our ancestors became bipedal, or why their brains tripled in size—remain unexplained."

Stephen Cave, "Planet of the Apes," *Financial Times*, August 18–19, 2012, p.7.

The gradual elimination of those gaps in our knowledge is a necessary process in the effort to satisfy ourselves that the theory is defensible. But, the process of closing gaps and resolving apparent inconsistencies can take a long time. It is always possible that some insurmountable issue will be encountered along the way that may require a radical revision of the theoretical structure that will have been built over time. But, that possibility itself is necessarily highly speculative.

In any event, the discussion here is not focused on those gaps, but on more fundamental issues of the underlying theory that are ripe for analysis. The underlying question in this chapter is whether natural selection operating on random variations is a convincing explanation of what has occurred over time. There are many issues involved.

In a recent book, philosopher Thomas Nagel summarized the challenges he thinks face Darwinian theories as the explanation for life as we observe it today. *Mind & Cosmos: Why the Materialist Neo-Darwinian Conception of Nature is Almost Certainly False* (2012), p.6. He asks:

- First, how did life itself appear? (Clearly, not through natural selection as such, since the process of natural selection assumes the existence of life.)
- Second, has there been sufficient time since the appearance of life on this planet for random mutations and natural selection to have resulted in the sophisticated and varied life forms that we observe and the appearance of intelligence and consciousness?
- And, were the various distinctive human characteristics (like consciousness, intelligence and rational thinking) actually adaptive when they arose (as predicted by natural selection)?

The arguments of Fodor and Piattelli-Palmarini against natural selection as the principle mechanism of evolution, discussed below ("The New Biology"), are based primarily upon criticisms of two fundamental aspects of neo-Darwinian theory:

- The assumption of gradualism, that is, the view that evolution has occurred primarily through small adjustments or changes that gradually accumulate to produce new species.
- The proposition that the changes in populations that led to the emergence of new species was caused by mutations occurring in genes. (Their comments on this assumption are based largely upon relatively recent developments in biology.)

In the end, the question is whether neo-Darwinism provides an adequate explanation of the evolution of existing life forms; in particular, whether natural selection acting on random variations is the primary means by which species originated and current life forms evolved, including intelligent and conscious life. Our observations indicate that life has evolved to become more sophisticated, more adaptive and, in various senses, better. If that perception reflects reality, then one should conclude that there is something more to evolution than suggested by the "selfish gene" theory, described below.

Perhaps such improvements are consistent with Darwin's theory, but are they compelled by it and, if they are, to what extent? This question I discuss in my last chapter. I address the important issue of evolution and the emergence of consciousness at length in later chapters. Also, in the last chapter, I turn to what might seem to be the initial question in the whole inquiry: where did life come from in the first place?

Natural selection, as Darwin used the concept, certainly occurs. But, is it but one of the mechanisms underlying evolution, one of the more important mechanisms or, as neo-Darwinism asserts, *the* mechanism? Fodor and Piattelli-Palmarini argue that "natural selection among traits generated at random cannot by itself be the basic principle of evolution. ...We think of natural selection as tuning the piano, not as composing the melodies." *What Darwin Got Wrong*, p.21. Largely just "fine-tuning"?

NATURAL SELECTION

At a very general level, Darwin's theory of evolution by natural selection assumes the existence of something that can replicate and, in the process, transmit physical (phenotypic) characteristics. That something must also be capable of some degree of variation, at least by accident. Then, natural selection acts on those phenotypic traits, indirectly responding to the underlying changes. It has been claimed that all that is required for evolution is "replication, variation and selection. If these [three elements are present, then evolution seems bound to happen." Kate Distin, *The Selfish Meme* (2005), p.2. See also, e.g., Dawkins, *The Selfish Gene* (1976), pp.13-21; *The Blind Watchmaker*, pp.128-9. This paradigm, however, is far too simplistic and incomplete, as will become apparent as we discuss each of these three necessary and, allegedly, sufficient elements in turn.

HERITABILITY

Darwin simply assumed that the necessary mechanism for replication with variations existed. His theory was given a significant boost by the rediscovery and subsequent elaborations of the work of Mendel at the beginning of the twentieth century, which introduced the concept of the gene and, thereby, provided a mechanism by which traits could replicate, variation could occur and new traits could replace previously prevalent traits in a population.

Some background

Gregor Mendel was an Austrian monk who performed the famous experiments with tens of thousands of pea plants in 1856 through 1863, crossing round with wrinkly and yellow with green (and other peas with different discrete traits). His experiments demonstrated that at least certain hereditable traits appeared to occur in offspring as selections of one or the other of the traits of the parents and not as "blends" of the two. See Nowak, *SuperCooperators*, p.15. He also discovered the concepts of recessive and dominant traits. However, his results, published in 1865 and 1866, appeared in journals for audiences interested in the breeding of plants and he did not speculate on the broader potential applications of what he had discovered. The publications attracted little attention. Mendel did not use the word "gene" or even employ the concept.

Dutch botanist de Vries rediscovered the findings of Mendel. He postulated that specific inheritable traits were the result of discrete units, which he called panagenes in the book *Intercellular Pangenesis*, published in 1889. He duplicated Mendel's work, including the concepts of dominance and recessiveness. He further presented a theory of evolution by mutation of the panagenes in *The Mutation Theory* (1900-1903), supported by experiments he had conducted with plants. See Dawkins, *The Greatest Show on Earth*, p.31n. In 1909, Danish botanist Wilhelm Johannsen coined the term "gene" (perhaps as an abbreviation of "panagene") and also introduced the concepts of phenotype and genotype. Johannsen conceived of the gene, the phenotype and the genotype as abstract constructs not necessarily related to any particular or specific physical elements. See Matt Ridley, The Agile Gene (2004), pp.231-2; Rheinberger, Müller-Wille and Meunier, "Gene," The Stanford Encyclopedia of Philosophy.

The theory that genetic materials could mutate, and supporting evidence that de Vries believed he had found in experiments with evening primroses, completed the framework necessary for the development of a satisfactory theory. Cambridge mathematician G.H. Hardy developed a simple equation for the transmission of traits through "genes"; and Sir Ronald Fisher, J.B.S. Haldane and Sewall Wright, in the 1920s and 1930s, put Darwin's natural selection and Mendelian genetics into a "mathematical framework" creating "the modern synthesis, or neo-Darwinism." See Ridley, The Agile Gene, p.16; Dawkins, The Greatest Show on Earth, pp.28-9; Dawkins, The Blind Watchmaker, pp.114-5; Ryan, Virolution, pp.36-7; Shermer, Why Darwin Matters (2006), pp.145-6.

There was a significant further boost with the discovery of the chemical composition and structure of DNA in the middle of the last century. The substance now called DNA was known to exist by scientists since 1869, but DNA (along with RNA) was first mentioned in The New York Times on July 15, 1947. Nicholas Bakalar, "First Mention/DNA, 1947," The New York Times, February 28, 2012. By 1951, the abbreviation was being widely used and the substance was being identified as the likely means by which heredity is transmitted. Id. The discovery of the structure of the DNA molecule, made by Francis Crick and James D. Watson, was revealed to the world in June 1953. Id.

The "gene" as the relevant unit

In the second half of the twentieth century, a related restatement of Darwinism theory of evolution by natural selection was forged focusing on genes rather than organisms, proposing that evolution is driven by the successful reproduction of particular genes in a population, not of organisms themselves. See Deutsch, The Beginning of Infinity, p.89. That restatement began in the mid-1960s with the work of George Williams and William Hamilton. This work caused a "revolution in biology." Matt Ridley, The Origins of Virtue: Human Instincts and the Evolution of Cooperation (1996), p.17. These advances were partly theoretical and partly experimental, that is, generated partly by logical analysis and partly by experimentation and observation. Id., pp.17-20.

A focus on the gene also rather dramatically expanded the explanations of various phenomena that had previously seemed to be in tension, if not inconsistent, with Darwinian evolution, including various types of cooperative and apparently altruistic behavior. In the simplest and perhaps most striking example, the successful reproduction of genes could readily be promoted by the self-sacrifice by a parent to save its offspring, whereas the survival of the individual organism would not be promoted by its own demise. The work of Williams and Hamilton resulted in related theories of "kin selection" and suggested theories of "group selection", subjects discussed further below.

Reflecting these scientific developments, Richard Dawkins invigorated the public appreciation of Darwin by vividly presenting evolution through natural selection operating at the level of genes, rather than of individuals. *The Selfish Gene* (1976). The gene is characterized as "selfish" because successful or adaptive variations are ones that enable the gene to survive and reproduce at the expense of or instead of the other genes in the gene pool. A gene's ability to out-reproduce its competitor genes does not inevitably depend upon enhancing the survival prospects of its host organism; although, that would often be the case. The theory based upon the "selfish gene" that occupies a place in a "survival machine," like the human body, was very effective. The resulting form of the theory recognized, for example, that there were versions of Darwin's theory that need not be so crude as to have evolution premised solely upon changes in a living organism that enhanced the organism's chances of reproduction.

(Dawkins proposes a technical definition of evolution as "a systematic increase or decrease in the frequency with which we see a particular gene in a gene pool." *The Greatest Show on Earth*, p.33. As a definition, his proposal is rather unsatisfying. Perhaps, it could be used as a mechanical test for whether evolution has occurred; although, the characterization as "systematic" may be less than rigorously objective.)

Now, the apparent subsequent recognition that there is no such thing as a "gene" (*see* discussion above in the chapter on Foresight and Understanding) does not necessarily invalidate the insights obtained through this analytical approach (although, the phrase "selfish gene" is much catcher than the more precise "selfish 'variety of types of quite complex processes and elements contained within the cell' that transmit phenotypic characteristics). One might think of the "gene" as a bit "the firm" in economic theory, discussed above in the previous chapter—a useful analytic tool. (However, several of the scientific discoveries that gave rise to that debate are relevant.) So, I shall continue to use it in the following paragraphs.

Phenotypic traits

Evolution, however, appears to us in the form of changes in the phenotypic characteristics of a species. Indeed, the process of natural selection would seem to require the existence of phenotypic variation; otherwise, there could be no difference in "fitness." Those phenotypic changes may originate in mutations in individual genes, but they will show up as phenotypic changes in individual entities. The process of evolution implies that the individual-level changes have the consequence of changing the characteristics of a gene pool, because the individual members sharing the mutated gene will out-reproduce the individuals that do not share the genetic change until the entities that contribute to the gene pool consist largely or exclusively of individuals with the mutation. (This statement is an over-simplification[6] and overstatement; there can be stable populations with some significant genetic diversity.)

This process over very long stretches of time has, it is claimed, resulted in the appearance of complex, even intelligent organisms out of an initial pool of simple life forms consisting of a handful of cells. As Matt Ridley has noted, apart from the theories of natural selection among genes, "[t]he organism itself needs explaining." *The Rational Optimist: How Prosperity Evolves* (2010), p.16. In other words, from a world of independent, self-sufficient cells, how did we get large complex organisms consisting of billions of cells? Ridley asserts that the first fully formed awareness of this issue was set forth by Richard Dawkins in *The Extended Phenotype* (1982):

> *"[E]ven cells are collectives. They are formed from the symbiotic collaboration between bacteria, or so most biologists believe. Every cell in your body is home to mitochondria, tiny bacteria so specialized as energy-producing batteries that about seven or eight hundred million years ago they surrendered their independence in exchange for a comfortable life inside the cells of your ancestors. ... And chromosomes also are collaborations, not individuals: collaborations of genes."*

Id., pp.16-7.

Of course, what gives rise to traits in the organism is the genetic information contained in the genes. It is the instructions for the creation and operation of proteins that matters, not the nature, structure or physical appearance of what we call the gene. Our genetic code consists of the detailed instructions that direct the formation and subsequent activity of thousands of proteins that actually do the work of constructing the organism with all of its traits. That information is very detailed and specific.

Further developments

Interestingly, while the initial discoveries in genetics provided a physical framework by which Darwinian evolution could be explained, as eloquently done by Dawkins, further advances in the second half of the twentieth and beginning of the twenty-first centuries have very much muddied the waters. Fodor and Piattelli-Palmarini have summarized many of the types of research and discoveries appearing over the last 10 years, which they call the "new biology" (discussed below), that raise issues for the neo-Darwinian model. *What Darwin Got Wrong*, pp.19-27.

The simple framework in which genetic mutations are expressed as traits is a serious over-simplification. There is not a straight-line arrow from a gene to a trait. The interrelationships among genes, including the existence of common control or regulation, mean that various genes are connected or packaged. Those packages or complexes often have self-regulating or correcting mechanisms. As a result, the packages or complexes can be quite resistant to change.

The significant differences in adult forms of related organisms occur because of differing growth rates among different groups of cells constituting different parts of the body. All animal life develops from single cells that are virtually identical. The striking differences that appear occur in the development process, as some differentiated cells grow faster than others. The process of differentiation and specialization of the cells as the growth process occurs generates the dramatically different looking adults. Indeed, there are even great similarities among the embryos of all mammals.

Many phenotypic traits are also bound together. Thus, for example, the same phenotypic traits may be the result of different genes or gene combinations, and the same genes or gene combinations may result in different phenotypic traits due to differences in the gene regulations and expression that occurs. *Id.*, p.45. Furthermore, there are heritable traits that come from things other than the traditional genes. Moreover, developmental plasticity, epigenetics, internal constraints, horizontal gene transfers are among the types of things that are being investigated on the forefront of molecular biology, all of which pose challenges to neo-Darwinism.

RANDOM VARIATION

In the theory, natural selection is defined to be the non-random selection from random variations. We will consider at length the concept of non-random selection in the next section. But, first, we look at the assumption that the variations on which the selection operates are random. To dispose of the need for a Divine Designer, one needs to remove intentionality from the occurrence of the variations on which selection works.

As discussed, variations in genes occur through mutations and through recombination. As to the first source, there are undoubtedly many more failures than successes; that is, it must be the case that most mutations of genes have no discernible effect (such mutations may persist in future generations). Next, many mutations will interfere with the functioning necessary for survival and reproduction. These mutations are likely to die out, because the organism dies out or, as will be discussed below in the section on The New Biology, they are counteracted by the organism's internal controls, repair and regulatory functions. Finally, some mutations presumably alter functions in a way that is or turns out eventually to be useful. These mutations are likely to spread. While the potential types of mutations may be essentially infinite, it is not the case that every conceivable outcome could come about as a result of a mutation or, even, a series of mutations. There are certain boundaries that cannot and will not be crossed. With respect to the recombination of genes from different sources (*e.g.,* the male and the female), the possible outcomes are necessarily even much more limited.

The issue to be addressed here is whether the changes reflected in the phenotype are actually random?

In this context, Dawkins has asserted that "random" means that there is in the genetic mutation "no general bias towards bodily improvement." Richard Dawkins, *The Blind Watchmaker* (1986), p.307. In other words, for his purposes, the mutations need not be truly random but cannot be driven or directed by anything that would make the changes that occur for the specific gene at issue more likely to be beneficial or useful. The point is that any tendency toward improvement must be coming from the process of natural selection, not the source or nature of the variations that arise.

There is an ambiguity in these statements, in light of the discussion above. One normally thinks of the mutations being random. But, that does not mean that the effects of the random genetic mutation will be random. To the extent that the internal processes and capabilities of the organism alter or direct the phenotypic manifestations of the mutations, then there will be less phenotypic variation than genetic variation. This factor imposes at least a caveat on the assumption of randomness, but it does not resolve the issue of interest here. The internal controls may effectuate a built-in conservatism or inertia tending to preserve the outward manifestations of the complex and highly coordinated system that constitutes the organism, but we have identified no reason to conclude that the controls can direct whatever change does occur (for better or for worse).

The significant fact with respect to the assumption of randomness is that, at least to date, there is no means of testing whether, or proving that, the significant mutations or variations leading to complex organisms have arisen randomly or, instead, were subject to some systematic mechanism of creation.[8]

NATURAL (NON-RANDOM) SELECTION

For the informed public, the discovery of Crick and Watson resulted in a very plausible theory of evolution: Physical characteristics are determined by genes. Changes in genes occur all the time due to random mutations. Such changes in genes (sometimes) result in changes in physical characteristics or traits. Natural selection preserves the changes in traits that are useful to survival by allowing the mutated genes responsible for the beneficial phenotypic changes to become established. The physical concept of DNA and the sequence of chromosomes, with the simplistic view of genes, gave the theory considerable appeal. It even appears to be understandable. However, the theory of natural selection exists largely independent of the specific mechanics of heritability. The point of natural selection is that the process of selection is *not* random. It is assumed to be discriminatory or "selective." Thus, evolution by natural selection in the original biological sense occurs when random variation is followed by non-random selection—or, through "the non-random survival of randomly varying hereditary equipment." Dawkins, *The Greatest Show on Earth*, p.63.

Origin of the concept

It seems to be pretty clear how the concept of natural selection arose. Darwin was clearly influenced by his awareness of the power of conscious selection in breeding, both with plants and with animals. The fact that traits could be developed through controlled breeding so as to promote particular characteristics or to suppress others was common knowledge and extensively used in practice.

What was notable about domestication were "its astonishing power to change the shape and behaviour of wild animals, and the speed with which it does so." Dawkins, *The Greatest Show on Earth*, p.28. Indeed, "[i]f so much evolutionary change can be achieved in just a few centuries or even decades, just think what might be achieved in ten or a hundred million years." *Id.*, p.37. "If human breeders can transform a wolf into a Pekinese, or a wild cabbage into a cauliflower, in just a few centuries or millennia, why shouldn't the non-random survival of wild animals and plants do the same thing over millions of years?" *Id.*, p.42.[7]

Clearly, selection in breeding can have dramatic results. However, one could rightly wonder whether selective or intentional breeding is, in fact, a very useful analogy for evolution.

Dawkins asserts that "[n]atural selection is the same [as domestication], with one minor detail changed." *Id.*, p.28. In fact, however, the differences are in more than a single "detail" and they hardly seem to me to be "minor." In the domestication of wild animals and in the subsequent breeding of desirable examples of the species, there is a breeder who selects the entities (and, thereby, the genes) that breed, who makes the selections with particular goals in mind and who controls or prevents the interbreeding with the general population. In evolution through natural selection, the variations supposedly arise by random mutation, the multiplication of certain new genes occurs by the more successful reproduction given the serendipitous characteristics of the environment, which enables the new genes to be established rather than being reabsorbed into the general population of the initial species.

Fodor and Piattelli-Palmarini argue that the prevalence of the intentional analogy has actually done a serious disservice to the advancement of scientific knowledge in this area. *What Darwin Got Wrong*, pp.95–116. They call it "the problem of 'selection-for.'" With an intelligent breeder, acting with intention and design, the concept of "selection-for" makes perfect sense. In a mindless and purposeless process, the concept is dangerously misleading. They argue that the confusion between the concept that a creature that has a trait making it more adaptive to the environment is selected and the concept that a creature is selected for its adaptive trait reflects what philosophers call an "intensional fallacy" in logic. *Id.*, at pp.xvii–xix.

Darwin and his followers caused themselves these problems with the use of the phrase "natural selection." The phrase "natural selection" was reportedly used commonly among breeders to make the distinction between changes for which their efforts were responsible and changes that occurred for other, unknown reasons. See Robert M. Young, *Darwin's Metaphor: Nature's Place in Victorian Culture* (1985), p.95. The analogy to intentional breeding invited a continuing debate about what factors or forces might account for the changes or variations that appeared.

A process for selection?

It is obvious that the fact (as well as the nature) of selection can dramatically impact the pace and content of development.

Michael Shermer references the hypothetical scenario of a monkey at a typewriter generating Hamlet's soliloquy by chance. To guarantee that one successfully obtains even the first thirteen letters ("to be or not to be"), even without the spacing between the words, would require a number of trials that is equal to the number of seconds that have elapsed in the life of our Solar System. Shermer, *Why Darwin Matters*, pp.82-3. Shermer cites a computer experiment by Richard Hardison involving a process by which the randomly generated letters are selected for or against, that is, where the correct letter is preserved and all incorrect letters are rejected, the number of trials required to produce the correct thirteen letters in sequence is quite manageable, taking the computer less than ninety seconds. *Id.* He also reports that Richard Dawkins had performed almost the identical computer simulation at the same time (1984-85), which Dawkins reported in his 1986 book *The Blind Watchmaker*. *Id.*, pp.178-9, n51.

This illustration underscores the significance of selection or discrimination, but it also suggests the importance of the basis for selection that is being utilized.

For example, if the computer generates a random letter and then accepts or rejects it based upon whether it fits the template of "tobeornottobe," then the process of obtaining the "correct" result could be quite speedy. Alternatively, suppose that the computer incorporates a dictionary and the selection criterion is to allow a letter if, in sequence, that letter is consistent with a known word and to reject the letter if it is not. Also, this type of criterion would need to provide for a space whenever the sequence achieves a recognizable word (because we are now trying to construct a properly segmented sentence of six words). Again, this program should be able to produce the sentence far more rapidly than mere random selection, but certainly not as quickly as the first example. (But, of course, this hypothetical process would be relatively effective in the example used here, because the words to be created are only two letters long. If the desired words are longer and of varying lengths, then there may often be known words selected that are not the ones wanted, leading to am impasse.)

What would be a reasonable analogy to the process that is contemplated to occur in natural selection? Certainly, the principle proponents of natural selection would reject the idea of some template or ultimate objective against which the changes are to be evaluated, even though the evaluation would be by natural selection. (If there were some such template, then we need to explain where it came from and how it is implemented.) Instead, natural selection is premised upon an unplanned and unguided series of small steps, each of which is selected or rejected based upon its impact on the ability of the gene (or organism) to survive and reproduce successfully relative to its immediate competitors. This model is quite a bit different from the examples of computer simulations referenced above.

A hypothetical illustration: Early nesting

One point that can often be missed is that evolution by natural selection does not imply that the changes that do occur will necessarily enhance the good of the species or the population. *See, e.g.*, Deutsch, *The Beginning of Infinity*, p.89.

An illustration set forth by Deutsch hypothesizes that the conditions on an island are such that the total number of birds of a particular species will be maximized if the birds nest at the beginning of April. *Id.*, pp.89–91. He hypothesizes further that all of the birds in the population have genes that cause them to nest on April 1. If a mutation occurs that causes one bird to nest on March 31 each year, that bird will enjoy an advantage over all of the other birds in the selection of a nesting place. If the benefits of that advantage outweigh the costs of earlier nesting, then the mutant genes of that bird will out-reproduce the original genes, because the bird in question will out-reproduce the other individual birds. Over time, the late-March nesting birds will begin to dominate the species and may even drive to extinction the early-April nesting birds. But, the species is no better off.

The example makes the intended point, but it can be used to illustrate more about the theory. There are a variety of significant underlying assumptions that should be made explicit. One assumption is that the act of selecting a nesting place a day (or a week) ahead of everyone else confers an advantage to reproduction. That assumption requires, at a minimum, that there be a shortage of first-rate nesting sites on the island, relative to the number of birds or, at least, that there is a meaningful variation in the quality of the sites that are available. Otherwise, there would be no advantage in being first. (If other factors are limiting population growth, such as availability of food, the presence of predators, *etc.*, then there could be little advantage from a nesting site.) The example also requires that the disadvantages of nesting earlier than the others (such as, for example, greater vulnerability to predators) do not overall outweigh the advantages that might be gained.

Implicitly, the example, much less plausibly, assumes that these relative advantages and disadvantages remain in a similar balance over long periods of time and not be made moot by other changes in the conditions on the island or other mutations in the genes of the birds of the species that overwhelm this process. This assumption is particularly important because the advantage gained is likely to be over just some birds (most likely the ones that would otherwise lose out in the competition for scarce nesting spots anyway) not, at any particular point in time, over most of the other birds of the species. Only, as the early nesters grow in numbers, will they begin to edge out the birds with the original genes.

The speed with which that process could happen will depend upon the extent to which the bird population regularly over-reproduces relative to the available nesting spaces. If, for example, half of the population in any particular year cannot reproduce, or is at a disadvantage in reproducing, because of a shortage of first-rate nesting sites, then the invasion of early nesters could proceed rather rapidly. If the lack of nesting sites affects only a small percentage of bird pairs in any particular year, then the transition will be much slower.

Separately, to the extent that the disadvantages of earlier nesting are not significant, then the transformation of the population into earlier nesting birds may be relatively easy. Of course, the process could continue, as Deutsch suggests, with some birds nesting earlier and earlier, until the disadvantages become material. If they become too material, then the even earlier nesting birds will not flourish. In addition, it is possible that, over time, there will appear a reproductive advantage in nesting later, when the weather may be more favorable, assuming that nesting sites will then be available (for example, if the early nesting birds have vacated their nests already). If so, then the process could effectively reverse itself and the species would move back to April nesting.

This exercise is quite speculative, with great opportunity for creativity and invention. One could construct many alternative scenarios that have the appeal of plausibility. Each theory is not subject to testing as such, only to the search for consistent or inconsistent examples. Moreover, comparisons among species for "similar" behavior may not be very informative, since similarly appearing behavior could occur for quite different reasons. In any event, the point of this illustration is that the jockeying for nesting sites reflects a purely intra-species rivalry, with no benefits accruing to the species itself (relative to other species).

Maladaptive traits: The peacock's tail

There are many examples of the evolution of maladaptive traits. A vivid one is the male peacock's extravagant tail. Possibly, a colorful tail reflects genetic strengths (such as a strong immunological system) and females that are attracted by the display are reproductively more successful as a result of the "good" genes of the male. (This process assumes female sexual selection, as opposed to species in which the males fight one another to become the dominant male and obtain thereby the most breeding opportunities.) Thus, presumably, the genes of the female that caused her to be attracted by the tail display are reproductively successful.

If so, we would suppose that competition among males to attract females would favor more and more dramatic tail displays, at least to the point where the disadvantages of a bigger, brighter tail (such as vulnerability to predators or the weakening of other systems) outweighed, in terms of effect on reproductive success, the benefits of access to more receptive females. *See also*, Dawkins, *The Blind Watchmaker*, pp.199–214.

Studies have documented traits in humans in many animals that have been demonstrably useful in attracting females for purposes of mating but seem not otherwise useful to the males and cannot even be said to be indicators of genetic strengths. *See, e.g.*, Sylvia R. Karasu and T. Byram Karasu, *The Art of Marriage Maintenance* (2005), pp.21-2. Some of these traits in human males (and, presumably, primates generally) actually reflect increased vulnerability to disease, since they are characteristics of increased testosterone which is known to diminish the effectiveness of the immune system. *Id.* In addition, there have been experiments where the length and color of tail feathers have been artificially enhanced, and the females have responded enthusiastically. *Id.*, p.154; *see also*, Dawkins, *The Blind Watchmaker*, pp.213-4 (describing such an experiment by Malte Andersson).

So, one cannot conclude that the intra-species reproductive success is necessarily related to characteristics that are favorable for any reasons other than reproductive success in intra-species competition. At least in this context, natural selection would not necessarily tend to promote the evolution of more "fit" life forms. (*But see*, the discussion below of Beauty.)

THE CENTRALITY OF SCARCITY

The contribution of Malthus

Thomas Malthus generated dire predictions about the prospects of mankind based upon a simple model. He hypothesized that human populations (and other living forms, as well) would increase until limited by the supply of food. In other words, animals, including people tend to produce more offspring than can survive. Thus, if there is excess food, then the number of living people will increase until that surplus disappears, absent intervening events like disease or natural disasters. The only constraint on population size is the available quantity of food and other necessaries to support life.

Given this basic assumption about human beings, there are some powerful conclusions that follow. Perhaps the most poignant is that humanity will never long exist above a subsistence level, absent some external force (governmental, societal, moral) to curtail the impact of human instinct. See Thomas Robert Malthus, *An Essay on the Principle of Population*, 1798 (there were six editions in all, published between 1798 and 1826). Malthus used simple mathematics in his analysis. Population, he assumed, would inevitably grow exponentially while food supplies, he assumed, would only grow arithmetically. It was easy to see the implications, given the assumptions. It is a good example of the power of mathematics to illuminate a proposition. Of course, both of his assumptions turned out to be incorrect. See, e.g., Deutsch, *The Beginning of Infinity*, p.201.

Both Charles Darwin and Alfred Russell Wallace, independently, were greatly influenced by the writings of Malthus, and both recognized that if life regularly produced more individual offsprings than could survive given the existing finite resources available, then there would necessarily be a "competition" among those individuals to survive. One would expect the most advantaged of the individuals to be the survivors. The necessity of competition provides the basis for the Darwin/Wallace theory of evolution, grounded in the concept of the "survival of the fittest" (a phrase introduced, however, by Herbert Spencer in 1864). See, e.g., Nowak, *SuperCooperators*, pp.xi–xiii. See also, David Sloan Wilson, *This View of Life* (2019), pp.16-22 .

One might initially think that this paradigm looks somewhat like Newtonian mechanics, which is based upon the hypotheses that things moving in a particular direction and at a particular speed will continue to do so unless acted upon by an external force, and things standing still will remain so unless acted upon by an external force. However, the Malthusian paradigm is clearly more anthropomorphic. It "understands" certain phenomena as resulting from inherent urges or motivations. Indeed, compared to Newtonian mechanics (as discussed later), Malthusian theory is much more "explanatory" in that it purports to identify the motivating factors, whereas Newtonian mechanics actually offers no explanation of how or why—just assertions as to what happens. See Richard Feynman, *The Character of Physical Law*, pp.4-9, 33.

"Dog eat dog" and "survival of the fittest"

Originally, much emphasis was placed upon conflict arising from the inevitable scarcity of resources, such that survival of living organisms necessarily was a matter of what we would call competition—and fierce or cut-throat competition at that. See, e.g., Shermer, *Why Darwin Matters*, p.9; Nowak, *SuperCooperators*, pp.xi-xii, 13-14; Deutsch, *The Beginning of Infinity*, pp.48-49. In fact, the full title of Darwin's Origin is *On the Origin of Species by Means of Natural Selection, or the Preservation of Favored Races in the Struggle for Life*.

Darwin himself described in his autobiography how he received the inspiration for the theory of natural selection in 1838 while reading Thomas Malthus *On Population*. See, e.g., Dawkins, *The Greatest Show on Earth*, p.17. And, in the Introduction to his book, Darwin described his chapter "Struggle for Existence" as being "the doctrine of Malthus, applied to the whole animal and vegetable kingdom." *The Origin of Species* (1962) (from the sixth edition of 1872, the last revised by Darwin), p.21.

As noted above, the Malthusian world is one of inevitable scarcity and the culling of existing populations through starvation and other deprivation, in significant part as a result of the geometrical rate of reproduction of living organisms. It was perceived inevitably to be a veritable "dog-eat-dog" world. Of course, we also have the specter of Thomas Hobbes lurking in the background: in the state of nature, human life is "nasty, brutish and short." The biological model assumes an abundance of reproduction—anthropomorphized as a response to the enormous dangers that threaten life—creating almost a guarantee that some essential resources will be scarce (and, therefore, that many of the new individual organisms created by the reproduction will be unable to survive). The result is competition for survival.

There was certainly competition for resources among species, but the important point for Darwinism was the prevalence of competition within species. Not only did individuals strive against their relatives for food, shelter and other necessities, but they strived to be the ones that avoided being the victims of predators. As the old joke says, you need not be able to outrun the bear; you need only to outrun your buddies (and only some of your buddies, at that).

This characterization of the environment in which evolution would occur clearly underscored the concept of survival of the fittest. As Darwin put it at the end of Chapter 7 of the initial (1859) edition of *The Origin*: we see "one general law, leading to the advancement of all organic beings, namely, multiply, vary, let the strongest live and the weakest die." Quoted in Dawkins, *The Greatest Show on Earth*, p.400. (Although, as already noted, Darwin did not originate the phrase "survival of the fittest"; he did embrace it in later editions of his book, including the final, sixth edition of 1872. (*See* Chapter 4: "Natural Selection; or the Survival of the Fittest").

This competition must be contrasted with chance and accident. Obviously, some, perhaps a large, number of variations with great potential are prematurely terminated by events as to which no biological defense is conceivable. (At the same time, there are undoubtedly variations that could conceivably have reduced the likelihood of survival that are terminated by accidents.) Other variations are probably also frequently terminated by predictable threats as to which the variation in question provided no benefit, but as to which other potential variations could have been effective. However, the concept that matters here is that the variation makes one relevant unit (for example, a gene) more likely to reproduce successfully than other very similar, but not identical, units. And, it is not really a matter of survival, except that survival enables reproduction. The resulting evolution at the level of the gene will cause changes in the individual organisms making up the group. If the variation is successful, and nothing else changes, the group will ultimately be made up of organisms that reflect the consequences of the successful variation.

However, as suggested above, if we think about the competition paradigm, it should become increasingly apparent that we are dealing with a mere tautology. If circumstances are such that not every individual can survive (because of limited resources), then it is logical to conclude that some will die. The survivors are the ones that were the most fit to survive, by definition. To say something more, we need a more detailed model of the process of "competition," that is, of the process by which selection takes place.

Some Implications and Issues

Gradualism

A commitment to gradualism is a consequence of the traditional neo-Darwinian theory. If change is introduced as a result of the mutations of genes and changes in the genetic makeup of populations occurs through natural selection, then changes in the individuals making up the populations will occur gradually, in very small steps. In part, this conclusion is compelled by the empirical observation that large mutations are almost certain to be fatal to the organism or, if not, would likely prevent it from being able to reproduce with the available mates, so the changes that do occur and which survive must be small. Similarly, it is hard to imagine that large, substantial changes could occur with sufficient frequency to allow a process like natural selection to function.

Thus, natural selection would appear to require extremely large numbers of changes, most of which would be detrimental, in order to generate the series of beneficial changes that appear to have occurred. Finally, very small changes that happen to be beneficial are unlikely, but not inconceivable; so given long periods of time (by human standards) the probabilities of the mutations leading to evolution actually occurring are not disqualifying. In other words, the changes underlying evolution must be in increments sufficiently small that they could reasonably occur by chance. See, e.g., Dawkins, *The Blind Watchmaker*, pp.43, 73.

However,

> "The history of most fossil species includes two features particularly inconsistent with gradualism: (1) Stasis. Most species exhibit no directional change during their tenure on earth. They appear in the fossil record looking much the same as when they disappear; morphological change is usually limited and directionless. (2) Sudden appearance. In any local area, a species does not arise gradually by the steady transformation of its ancestors; it appears all at once and 'fully formed'."

Stephen Jay Gould, quoted in John Hands, *Cosmosapiens: Human Evolution from the Origin of the Universe* (2016), p.296.

Darwin's natural selection seems to provide a persuasive explanation of incremental changes tp color, size, height, speed, etc.; additional challenges arise with respect to phenomena that are not clearly incremental.

A good example is that of the organ or faculty that performs a useful function because of several different elements or characteristics that themselves do not seem to be particularly useful. If numerous random mutations must take place in order to create the beneficial faculty, the theory would seem to require that each of them have survival power in and of themselves so as to have reproduced a sufficient number to have been available when the next mutation occurred, and so on. If the assumption is correct that the individual mutations did not enhance survival or reproduction, then the theory that a series of such changes accumulate over enormous stretches of time to create a new faculty, let alone a new species, seems beyond plausibility, except perhaps for an the isolated freak occurrence.

An example: The eye

One of the most common, and long debated, examples of such an organ proffered by the critics is the eye.[9] The argument is that the high level of complexity and the unique specialization of the task being performed simply cannot plausibly be accomplished by a series of small, incremental changes of the type that mutations could explain, especially where the incremental changes do not seem to provide any particular advantages in survival or reproduction. In such circumstances, it is argued that the end result could only be the product of design and purposeful or directed development.

The Darwinist response generally is to attempt to construct scenarios by which at least some of the essential incremental developments could have had value to the organism, even if it is of a type unrelated to the ultimate function of the eye. See, e.g., Nick Lane, *Life Ascending: The Ten Great Inventions of Evolution* (2009), pp.172–204.[10] The argument is that if some such progress can be made toward posing hypothetical scenarios that would result in the emergence of an eye, then the burden shifts to the critic to demonstrate that the evolution of the eye could not have occurred through random mutations.

Darwin himself focused on the issue raised by the eye. Moreover, he identified the problem in the same terms in which it is discussed today: Could one identify a series of small, incremental steps, each one of which would be advantageous to the success of the organism, that in sequence would lead to the sophisticated eye and the resulting benefit of sight? *The Origin of Species*, pp.178–9. Again, it does seem to be a rather weak form of proof simply to identify, hypothetically, at least one alternative route through which evolution by natural selection could have created the eye.

On the other hand, if one is responding to an argument that because such a development is not conceivable, the eye is evidence of a Divine Creator; then, the construction of such hypotheticals is certainly both fair and responsive.

Indeed, Darwin himself acknowledged that if one could demonstrate that a complex organ could not possibly have arisen through successive small changes, then his "theory would absolutely break down" and, one hundred years later, Dawkins agreed that he would come to the same conclusion if presented with that evidence. *The Blind Watchmaker*, p.91.

It is important to keep focused on the issue that matters. The argument based on the question "what good is 5% of an eye" is silly if interpreted to mean having vision 5% of what humans have today. Of course, 5% is better than nothing and every increment thereafter, no matter how small, is likely to be potentially beneficial to survival and reproduction. See *id.*, pp.81–90. The question, instead, is whether individual parts of the organ that has become an eye were each beneficial—say 5% of the component pieces of the eye, not 5% of the visual capacity. Dawkins asserts that "it is clear that part of an eye is better than no eye at all." *Id*, p.85. But, it is not so clear. It really depends upon what part one is talking about. The question is how the complex mechanism could have come to be constructed out of small, incremental changes where each change presumably must have improved the reproductive success of the animal involved.[11]

Many scenarios for the evolution of the eye have been put forward. See, e.g., Lane, *Life Ascending*, pp.182–3; Dawkins, *The Blind Watchmaker*, pp.77–91. Curiously, there is also evidence that eyesight has appeared independently on multiple occasions over the history of life on Earth. Wright, *Nonzero*, p.275. Indeed, some current biologists believe that sight has originated some 60 or 70 different times on Earth. Nick Lane, Lecture delivered at Cambridge University, 2 March 2014. Such a fact, if true, would support the argument in favor of the evolution of the eye. However, there is also recent evidence of master genes that relate to eyesight, genes that are found across species. See Fodor and Piattelli-Palmarini, *What Darwin Got Wrong*, p.23. That evidence suggests that organisms with eyes share some common genetic features. It is not likely that they are all descendant from some common ancestor, but there are other ways of gene transmission that have been recently discovered and explored, as discussed below.

Irreducible complexity

There are other examples of features that seem at least as hard to explain by gradual change and natural selection.

These examples include behavioral characteristics. Take the technique of parasitical wasps that inject cockroaches in one spot to put them in a kind of stupor that allows the wasp, which would not be capable of dragging the cockroach, to lead it to the nest, followed by another injection in another spot that paralyzes the cockroach so that it can be used as a living food source for the wasp larvae. *Id.*, pp.89–91. It is difficult to conceive of how this technique could have been developed in gradual steps, since the benefit (the food source in which the larvae grow) does not automatically flow from the earlier steps, but it cannot be achieved without them.[12] There are other examples of organisms being genetically "programmed" to alter the behavior of another species. See Christakis, *Blueprints*, pp.332-52. (Of course, there are many animals that are genetically programmed to alter their physical environment—beaver dams, spider webs, pollinated flowers. Christakis refers to all of these genetically driven external effects as "exophenotypic". *Id.*)

This type of problem is often referred to as "irreducible complexity." If there is a system or combination of parts that performs a function and if that function could not be performed if any one of the parts were missing, then such a system would be considered to be irreducibly complex. *See, e.g.*, Michael J. Behe, *Darwin's Black Box: The Biochemical Challenge to Evolution* (1996), p.39. The common mechanical example is the simple mousetrap. Every element, from the platform to the trigger to the spring to the hammer, is necessary for the gadget to perform the function of trapping a mouse. Remove any one element and the gadget is worthless (for the specified purpose).

The argument is that an irreducibly complex system could not evolve incrementally, one part at a time, because none of the parts are useful until they are all assembled. Now, that statement claims too much. Each element could have been useful, serving some other purpose, standing alone. And some combination of the parts could also have been useful, again serving some other purpose. A final minor modification could conceivably bring several otherwise independent pieces together with the result of achieving some new and different function (such as catching a mouse) that provided immediate benefits to the organism. It is possible, but remember that, in the theory, each modification has to have provided an immediate survival/reproduction benefit. Moving closer to a desirable outcome is not adequate—no foresight or planning is permitted. Getting closer to some goal or objective does not count, because there are no goals or objectives, and the modification has to generate a benefit that can be enjoyed immediately.

Behe refines the argument of irreducible complexity. He observes that the last few decades have brought us new fields such as biochemistry and molecular biology with dramatic new insights into how cells function. For natural selection to explain the origin of species or, even, the emergence of complex organs and capabilities, it must be able to do so on the molecular and chemical level. That is where the gradual changes must occur, at the level of the formation and functioning of proteins, in order to give rise to new complex systems. The question is how these complicated systems of closely interrelated parts could possibly have developed as a result of gradual modifications. See, e.g., id., pp.74–97. Can one pose a credible hypothetical scenario for such results? Behe stresses that such a hypothetical demonstration must be based upon physical precursors to the end product, not on conceptual ones. Id., p.43. One must postulate changes in the pre-existing genes that provide direct, immediate benefits and that also lead toward the end result to provide a possible path of evolution through mutations.

We are not talking just about the color of one's feathers or the length of a beak or a tail or even a neck (or arm). One might be able to fashion an explanation of gradual modifications at the molecular level leading to such results, without providing any answer to the important questions. We seek to explain the emergence of the eye, of the clotting function of blood, of the immune system, of the traits of intelligence, and so on. These systems are incredibly complex on a molecular/chemical level.

CONVERGENCE

Proponents find support for the theory in the apparent independent emergence of the same or very similar solutions to environmental challenges. Thus, where the range of types of improvements is more limited, natural selection operating on random mutations will tend to find the same solution repeatedly.[13] I note, however, that continuing research has called into question some of the supposed examples of independent origination. It is possible that some examples of convergence in evolution may reflect the effects of patterns that are essentially built into genes as a consequence of their physical structure, so-called genetic "invariables," rather than natural selection. Fodor and Piattelli-Palmarini, *What Darwin Got Wrong*, pp.22, 31.

An example: *Candida auris*

An example of the combination of convergence, natural selection and competition is the recent rise of a lethal fungus, *Candida auris*. It looks like convergent evolution resulted in the appearance of the fungus in several places, but competition with other fungi kept its presence essentially unnoticeable. Then, other man-made factors (fungicides) reduced the competition, creating an environmental opportunity which *Candida auris* has aggressively exploited (to our concern). Matt Richtel and Andrew Jacobs, "A Mysterious Infection, Spanning the Globe in a Climate of Secrecy," *The New York Times*, April 6, 2019.

FREE-RIDER TRAITS

In addition, we have what has been called the "free-rider" problem. *See, e.g., id.*, pp.xx–xxi, 44, 96-104. There are many examples of phenotype traits that for genetic reasons are connected, that is, that appear in pairs (or triplets, *etc.*). Assume that one of those traits has survival value so that it would be favored by natural selection. In such a case, the other trait(s) will be carried along in the process. They may become dominant in a population even though they provide directly no reproductive, adaptability or survival benefits to the organism. Of course, they have an indirect connection—the fact that they appear whenever another trait that does have such benefits appears. While one can say that the dominance of the free-riding trait is a result of natural selection, one cannot (under this hypothetical situation) properly claim that there was selection for the trait. It is a coincidence or fluke caused by the otherwise irrelevant fact that the trait is paired with another trait. Thus, the trait's presence is not adequately explained by natural selection, at least not by natural selection alone.

SUCCESSIVE MUTATIONS

There are other types of examples of characteristics that play important roles in more sophisticated or complex organisms that seem not to have arisen initially for survival or adaptation. Apparently, they somehow persist until a subsequent variation makes them useful.

A study was reported in *Nature*, dealing with the examination of proteins with structural flaws called dehydrons. Jason Palmer, "Protein flaws responsible for complex life," BBC News, 19 May 2011. The defect makes the protein less stable in water and more likely to stick together. These traits would seem to be to the disadvantage of the protein. However, the researchers discovered that the resulting multiple-protein structures frequently appear in more complex organisms. The combination of proteins enables them collectively to mask or compensate for the defects that occur in each of them. Moreover, the more complex proteins seem to have facilitated or participated in the development of more complex organisms. Id.

These proteins may be an example of an adverse change that was subsequently utilized in a manner beneficial from the standpoint of natural selection, in contrast to the neo-Darwinian theory of adaptive changes being selected. It has been suggested that mechanisms to ameliorate the negative consequences of the defects developed first, making the appearance of the defects effectively beneficial changes favored by natural selection. But, even if true, in that scenario, the original appearance of a mechanism that would subsequently solve a problem that arises from a mutation that should be adverse itself seems to be a fluke or accident, not explainable by natural selection.

Professor Michael Lynch, one of the researchers, was quoted as saying: "'We've opened up the idea that the roots of complexity don't have to reside in purely adaptational arguments. It's opening up a new evolutionary pathway that didn't exist before.'" Id.

THE ORIGIN OF SPECIES

Now we come to the question of the emergence of new species. Despite the title of his famous book, one thing that Charles Darwin did not even purport to explain was the origin of species—not the origin of species in general and certainly not the origin of any species in particular. See, e.g., Bill Bryson, A *Short History of Nearly Everything* (2003), p.473. (As Fodor and Piattelli-Palmarini put it: With respect to a causal theory of the origin of species, "that's exactly what Darwin *did not do*... ." *What Darwin Got Wrong*, p.2.) So, we ask, how well does natural selection explain the origin of species?

SPECIES: MULES, LIGERS AND A VARIETY OF HUMANS

This question raises the same problems already addressed, as well as some new ones. A first observation should be that the concept of a species is not as clear-cut as we generally tend to assume. Examples of dramatically different looking and acting species are abundant; but, there are ample areas where the dividing line is not so sharp. The common lay distinction is understood to be the ability to reproduce offspring that are capable of reproducing themselves. Animals that can successfully mate would generally be considered to be of the same species, despite differences in size and appearance (for example, dogs).[14]

Even if differences in size make it unlikely that the female could carry the offspring to term and deliver a live birth, we consider large and small dogs to be of the same species, just of different breeds. Horses and donkeys, however, can produce mules, but the mules are sterile.[15] There are other examples that also challenge the easy definition: various large cats are able to reproduce with varying degrees of success (such as the male lion and female tiger, producing a very large "liger");[16] although, because of life-style differences (habits and habitats), they are quite unlikely to try to do so in a natural environment (as opposed to captivity).

What about *Homo sapiens*?

There has been considerable publicity about the purported discovery of direct DNA evidence that there was interbreeding between the Neanderthal and *Homo sapiens*. It appears that a few "genes" of the Neanderthal appear in the DNA of modern man. See, e.g., *National Geographic*, May 6, 2010. Researchers began collecting pieces of the DNA of Neanderthals in the 1990s. By 2010, most of the Neanderthal genome was reconstructed. When that genome is compared to the genomes of a handful of modern European and Asian humans, there are some very similar pieces or segments identified. Carl Zimmer, "Neanderthals Leave Their Mark on Us," *The New York Times*, January 29, 2014.

There is some controversy about whether the bits could be remnants from a common ancestor, but the evidence suggests that these "genes" appeared in *Homo sapiens* less than 60,000 years ago (and more than 30,000 years ago, when the Neanderthal disappeared), long after the Neanderthal and *Homo sapiens* lines had diverged.[17]

Also, it has been thought that the Neanderthal "genes" do not appear in the DNA of people of purely African descent (there seem to have been no Neanderthals in Africa). So, the introduction of the "genes" presumably occurred in Europe after the early *Homo sapiens* migrated there. However, a recent new analysis suggests that Neanderthal DNA does appear in persons of African decent:

> "[R]esearchers from Princeton University now believe, based on a new computational method, that Africans do in fact have Neanderthal DNA and that very early human history was more complex than many might think. ...their data indicated that a wave of modern humans left Africa approximately 200,000 years ago and this group interbred with Neanderthals. This ancient group of Europeans then migrated back into Africa, introducing Neanderthal ancestry to African populations."

Katie Hunt, "All modern humans have Neanderthal DNA, new research finds," CNN.com, January 30, 2020 ("Previous studies had relied on reference populations, or panels, that were [erroneously] assumed to have no Neanderthal DNA").

Neanderthal genes also appear in the DNA of people of Asian descent, presumably brought from Europe. Michael Marshall, "Human and Neanderthal interbreeding questioned," *New Scientist*, 13 August 2012. Sir Paul A. Mellars, an archaeologist at the University of Cambridge, observed that once *Homo sapiens* had left Africa, there would have been ample opportunity for mating with Neanderthals. Carl Zimmer, "Neanderthals Leave Their Mark on Us," *The New York Times*, January 29, 2014.[18]

If there had been significant, widespread interbreeding, however, one would expect to find more examples of Neanderthal genes. Of course, it may be that most Neanderthal genes were not useful to survival or reproduction, resulting in their disappearance as individuals with those genes became more and more scarce. See id.

A more complete sequencing of the Neanderthal DNA was recently achieved using a new sample from a toe bone. A study published in *Nature* found small segments of Neanderthal DNA in the genomes of some 1000 living people used in the study. These Neanderthal genes appear to be very common, indicating that they were useful to survival as the species developed. Id. Another study, published in *Science*, found apparently mutated Neanderthal genes related to skin and hair in living Europeans and Asians, which may have been beneficial in adapting to the colder environments outside of Africa. See id. There is also evidence that the presence of Neanderthal DNA has actually provided *Homo sapiens* with resistance to various viruses causing diseases like influenza and flue. See, e.g., Josh Gabbatiss, "Interbreeding with Neanderthals gave humans ability to fight off dangerous diseases," *The Independent*, 4 October 2018.

Other recent discoveries indicate that there was interbreeding with another type of early human coexisting with *Homo sapiens* and Neanderthals, called the Denisovans, found in Asia. Robert Lee Hotz, "Early Human Interbreeding More Widespread Than Thought, Study Suggests," *The Wall Street Journal*, December 4, 2013 (describing research using mitochondria from fossils found preserved a cave in Spain that date to about 40,000 years ago); Sarah Pruitt, "Early Humans Slept Around with More than Just Neanderthals," *History.com*, March 16, 2018. There has been recent suggestions that the results of some genetic inheritance from the Denisovans the include the unusual ability of the Inuits in northern Canada to withstand the cold and the Sherpas' abilities to live at high altitudes. See Steph Yin, "Cold Tolerance Among Inuit May Come From Extinct Human Relatives," *The New York Times*, December 23, 2016.

> "A human jawbone found in a cave on the Tibetan plateau has revealed new details about the appearance and lifestyle of a mysterious ancient species called Denisovans. The 160,000-year-old fossil ... shows that Denisovans lived at extremely high altitude and, through interbreeding, may have passed on gene adaptations for this lifestyle to modern-day Sherpas in the region."

Hannah Devlin, "'Spectacular' jawbone discovery sheds light on ancient Denisovans," The Guardian, 1 May 2019.

> "Mitochondrial DNA (mtDNA) analysis indicated [that the group] Denisovan ... diverged from the common lineage leading to mod-

ern humans and Neanderthals about one million years ago, that is, about twice as far back in time as the divergence between Neanderthal and modern humans, and so was an earlier species. ...However, subsequent analyses of nuclear DNA, based on several assumptions including the date of the human-chimpanzee divergence, indicate that the individual shared a common origin with Neanderthals about 640,000 years ago, and that the Denisovan genome differed from that of modern humans by about the same as did the Neanderthal genome. ...Denisovans made no genetic contribution to Eurasian humans as did the Neanderthals, but did contribute 4-6 per cent of their genome to that of present-day Melanesians, although not to other East Asian populations... ."

Hands, *Cosmosapiens*, p.442.

Part of the mystery is that "DNA data that genetically tracks human evolution only covers the last 400,000 years." University of Cambridge, "'Game-changing' research could solve evolution mysteries," *Research*, 11 September 2019. "Scientists ... don't know what the genetic links are between us and extinct species such as Homo erectus – the oldest known species of human to have had modern human-like body proportions – because everything that is currently known is almost exclusively based on anatomical information, not genetic information." Id.

How does a new species arise?

Obviously, postulating that several mutations occurred simultaneously in one organism sufficient to create a new species is inadequate, as well as impossibly unlikely, because one needs two of the species to reproduce. So, instead, the concept is that some subgroup slowly and continuously differentiates itself from other groups in the species. Presumably, the one group becomes more divergent (physically or, perhaps, geographically) from, but still capable of reproducing with, the rest of the species. Ultimately, however, inter-breeding must become increasingly uncommon and, finally, more or less impossible.

We have not been able to observe this phenomenon (the emergence of a new species); but that fact can be dismissed with the assertion that the process takes so much time that human history (or, even, *Homo sapien* existence) would not permit such observation in real time. But, neither can we find clear evidence of the process in the fossil record.[19]

Again, as discussed below, the fossil record is so scant as to be no more than suggestive at best, so no definitive conclusions can be based upon what we have not found. There is evidence consistent with the theory; but, as we have discussed, such consistency is not proof of a scientific theory. And, it appears that there is no experiment we can imagine that would enable us to test the theory. (There is hypothetical evidence that could be viewed as contradicting the theory, if it were unearthed; but, that approach—identifying potential evidence that would contradict the theory and then looking to see if it can be found—is only a limited "test.")

So, with what are we left? Take it on faith, because it seems to make sense? But, does it really make sense?

In any event, natural selection does not directly address the emergence of new species. We would not generally say that the newer species are more fit to survive than the older species. Thus, the claim seems to be, at best, simply that the emergence of new species is merely an incidental side effect of the process of natural selection.

Bacteria and bugs

Clearly, bacteria have demonstrated strong survival value. They are ubiquitous. They exist in the soil, the water and the air.[20] They exist on, within and even as integral parts of animals and plants.

Much credit has been given to the ability of bacteria to adapt quickly through mutations to changing environments or specific challenges (like antibiotics). There are examples of bacteria (in places like the deep ocean floor) that appear to have survived unchanged for billions of years,.That stability presumably occurs because the particular environment in which they exist has not changed and random mutations have not improved upon their suitability for that environment. See, e.g., Douglas Quenqua, "Unchanged for More Than Two Billion Years," *The New York Times*, February 9, 2015.

A similar survival success can be claimed for insects, moving up in complexity. Indeed, there are some examples of incredibly successful life forms. Ants have prospered in almost every dry land ecosystem in the world. There are almost 14,000 known species, and it is estimated that the total number of species of ants number at least 20,000. Living in societies that have survived since the days of the dinosaurs, ants have a global mass that is frequently said to be roughly equal to the global mass of mankind. See, e.g., Nowak, *SuperCooperators*, pp.155–56. Indeed, one can even find estimates made of the number of ants in the world based upon the assumption that the total weight of the ants equals that of the people. See, e.g., "How many ants are there alive on the planet?" *AntBlog*, April 8, 2013.

But, is the biomass of ants equal to the biomass of mankind? Actually, no one knows how many ants there are in the world or what the average mass of an ant is. Different species range from one or two up to 60 milligrams per ant. Estimates of the total number of ants on Earth range from 100 trillion to 10,000 trillion.

The oft-asserted claim that ants weigh as much as mankind comes from a proposition set forth by Edward O. Wilson and Bert Hoelldobler in their 1994 book entitled *Journey to the Ants*. Hannah Moore, "Are all the ants as heavy as all the humans?" *BBC News Magazine*, 22 September 2014. Given Wilson's eminence in the field of biology and in the study of ants, it seems that people just accepted the claim as factual. Apparently, Wilson and Hoelldobler derived their assertion based upon an earlier "estimate" by C.B. Williams of the total number of insects alive on Earth as one million trillion. They then "conservatively" estimated the percentage of insects that are ants as 1%, or 10,000 trillion, and that the average person weighs one million times as much as the average ant.

These estimates and the asserted conclusion are clearly suspect. If the average ant weighed one millionth of the weight of the average person, it would weigh 50–60 milligrams if the average person weighed 50–60 kilograms—way too big for the average ant. If the average ant weighed 5 milligrams and there were 10,000 trillion ants, then the total mass would be 50 billion kilograms, while the 7 billion people on Earth weighing an average of 50–60 kilograms each would have total mass of 350–420 billion kilograms. The total of ants could have weighed as much as the humans only when there were many fewer and smaller people. See Id.[21]

Of course, there are also many species that have not survived. And, in some cases, it can be hypothesized that the disappearance of certain species was attributable to competition from new species that proved more robust under the particular circumstances confronting them all. But, one would not say generally that new species survive because they are more "fit" than those that preceded them, only that the new species were sufficiently fit to survive. We know that they were sufficiently fit, because they did survive. (I am aware that these comments are directed more to the slogan than to the underlying theory, but I think that they are worth making because of the illusion that the slogan provides the explanatory model that gives the theory some of its attractiveness.)

Randomness (again)

Interestingly, efforts can and have been made to test whether the assumption of randomness (as the operative source of change in natural selection) is consistent with the actual evidence of the appearance of new species. For example, if Darwinism is correct, would one expect to see the number of species increase or decrease or remain relatively constant over time?

One could expect a decrease if a particular species appeared that was so superior from a survival perspective that it monopolized sufficient resources to cause the extinction of many other species. That phenomenon seems to have begun to occur with the appearance of *Homo sapiens*. Absent such a dominant species, however, it would seem likely that environmental niches arising from the huge numbers of minor differences in localized conditions would result in opportunities for an expansion in the number of species, at least up to a point. Curiously, the fossil record evidence seems to suggest that the total number of species has remained relatively constant or has decreased somewhat (and is continuing to do so), independent of the impact of man. See, e.g., David Deutsch, *The Beginning of Infinity*, p.48.

Evolutionary biologist Mark Pagel has reasoned that if natural selection operating on a large number of small changes caused the appearance of new species, then there would be certain statistical characteristics that one would find in the resulting historical record. He postulated that the time periods between the appearances of new species would follow a bell-shaped curve, presenting a normal distribution. That characteristic would be expected because, by hypothesis, we are dealing with a very large number of small changes generated by random events. If one could examine a sufficient number of sufficiently accurate evolutionary "trees" and calculate the time periods involved in the appearance of new branches, then one could test whether the appearance of the "new" follow a normal distribution or some other form of distribution or showed no pattern at all.

With modern advances in DNA sequencing, making the process much faster and less expensive, such evidence arguably could be found. Pagel and his team collected and analyzed evidence of more than 130 DNA-based evolutionary trees. The result was the apparent lack of a normal distribution and a reasonable fit to an exponential distribution. Pagel concluded that such a pattern suggested infrequent, unpredictable events, rather than a relentless progression of new species through natural selection. In other words, new species appear because of rare and dramatic events. This evidence seems to present problems for the advocates of gradualism.

Indeed,

> "The normal pattern of fossil evidence for animals is one of morphological stasis with minor and often oscillating changes punctuated by the geologically sudden—tens of thousands of years—appearance of new species that then remain substantially unchanged for some tens or even hundreds of millions of years until they disappear from the fossil record or else continue to the present day."

Hands, *Cosmosapiens*, p.344.

So, it appears that efforts to "test" the theory by making "predictions" of what would have occurred and then examining the available evidence to see if the facts are consistent with what one expected do not consistently support neo-Darwinism. The problem, however, is that we have no theory to explain alternatives like, for example, "punctuated equilibrium".

Islands

But, back to the basics. Presumably, many mutations can occur without affecting the ability of the organisms to mate. Therefore, truly beneficial mutations would be expected gradually to take over the population of the species, to the extent that the population existed in a single area or in overlapping environmental regions. For further mutations, the same thing would be true. As a result, in order for many, many mutations over time to create a new species, the affected group must be prevented from mating with the rest of the population. Eventually, it is plausible that one subgroup would then become sufficiently different from another subgroup that the two could not interbreed.

Thus, the emergence of a new species could most likely occur only where some groups of an existing species are isolated from one another for long periods of time. In other words (in the words of Richard Dawkins), evolution of new species requires "islands"—not necessarily pieces of land surrounded by water, but habitable environments that are isolated by uninhabitable barriers, separating them from other habitable environments. Such an "island" could be a land mass, a pond or lake, an area of shallow water surrounded by deep water, a valley surrounded by sufficiently high elevations or elevated terrain surrounded by areas too low for survival of the species of interest. See Dawkins, *The Greatest Show on Earth*, pp.253-73.

The requirement is that subgroups of a species be reproductively separated from the rest of the species, so that successful changes that do occur by mutation do not spread to the rest of the population.

There is experimental evidence that adjacent groups of plants allowed to cross-pollinate can develop differently if they face different environmental conditions. See Jonathan B. Losos, *Improbable Destinies: Fate, Chance, and the Future of Evolution* (2017), pp.27-79. Of course, each group of plants is stuck in its particular environmental niche and cannot move next door. We can also find examples in two adjacent, but significantly different, environments of organisms that appear to evolve differently, despite the likelihood of crossbreeding. *Id.*, pp.81-107. However, one must be sure that genetic changes are underlying what is being observed. For example, where an area of dark rock lies next to an area of light colored soil, one may find dark moths on the rocks and light moths of the same species on the soil. That circumstance may not represent evolutionary differences. If the moths normally display a range of color differentiation, then what we observe may simply reflect that the light moths on dark rocks and the dark moths on light soil get eaten more frequently, resulting in the apparent differences in the two populations. That is not what we mean here by natural selection.

Without "islands," it is still conceivable that a species could change sufficiently over thousands or millions of years so as to no longer be capable of breeding with their ancestor species, if any examples of the ancestral organism continued to exist. (But, without "islands," no such examples are likely to continue to exist.) In such a case, we might want to say that it appears that one species became extinct and another emerged, but we are very unlikely to have the information sufficient to make the necessary observations and draw such a conclusion.

Consequently, it seems that the emergence of the species through evolution by natural selection must have required many thousands or millions of "islands". Now, that requirement does not seem so onerous in and of itself. One could probably find many thousands of "islands," in the relevant sense, in most geographically diverse countries today. The more challenging aspect of the theory is that thousands of these islands must have remained islands for thousands, if not millions, of years, for the thousands of new species to have evolved.

Somehow, the realization of these requirements for the appearance of new species makes the "tree of life" considerably more incredible than it may at first have appeared. Indeed, the "tree of life" is now viewed as a misleading metaphor. Modern research using DNA has indicated that very many current species seem to have descended from a relative handful of ancestral forms, while some branches have changed little or produced only a few new species.[22]

Dramatic changes and missing links

It may be that the significant variations, including the origin of many species, arose from dramatic, infrequent changes. We will talk below about internal constraints that may significantly limit the possibilities of different phenotypes emerging. Perhaps, occasional changes in such internal constraints gave rise to the explosive emergence of new phenotypes, creating the huge diversity of life forms that we see today. Id., p.50.

There is significant evidence of the occurrence of these brief periods of hyper-diversification. Probably the best known is the Cambrian explosion some 545 million years ago. The fossil record indicates that there were dramatic periods of change in even earlier times. Id., pp.50-1. It may be that natural selection, with gradual modification, worked to make adjustments and alter the balance of life forms following events of great change and upheaval. The question then is "whether the mutations that result in real novelty are the same mutations that happen day to day or are the ones that occur only rarely, on a geological time scale." Id., p.30.

There have been many assertions that the fossil record disproves Darwinian theories because it contradicts (or fails to support) the conclusion of gradualism. The various "missing links" in the fossil record between apparently related organisms are legion. Of course, one cannot establish that a link is "missing," only that it has not (yet) been found.[23] So, the search for examples of intermediate links in the fossil record goes on. And, even if such links are never found, that does not establish that the links never existed, only that fossil records of the links either were never created or have not survived.

Interestingly, this issue arises not only with respect to the animal kingdom and the fossil record. It is also relevant with respect to molecular building blocks of life, such as proteins. Researchers recently reported in the journal *Structure* on their efforts to reconstruct ancient proteins that were apparently instrumental in the early development of life on Earth. See Simon Redfern, "Resurrected protein's clue to origins of life," BBC

News, 8 August 2013. Their efforts to trace the evolution of this protein suggest that the length of the protein changed "in fits and starts, with its helix structure suddenly lengthening at the point that cells started to develop a nucleus (the transition from prokaryote to eukaryote), paving the way for higher life." Id. This evidence indicates that even at the molecular level evolution may have occurred "in discrete jumps rather than along continuous pathways." Id.

AND, SO...

In the end, the search for fossils will never definitively resolve the debate over gradualism. It is certainly not possible that a complete fossil record could ever be compiled. ("Absence of evidence is not evidence of absence.") Additional finds, like remaining examples of missing links, simply will not establish that gradualism is an accurate description of evolution or that it is not. Indeed, there is no reason why there should be an either/or answer. Most likely, there were many examples of gradual change. Also, most likely there were discontinuous jumps. The issue for scientific theory is which type of change is the most prevalent and most significant and what theories best capture and explain the phenomena that matter.

For purposes of this section, the proposition I leave with the reader is that, despite the heady and sometimes foolishly aggressive proclamations with which the recent anniversary of Darwin's theory has been greeted, it is a very charitable understatement to say that we do not really have (yet) a full or satisfactory account for the origin of species.

A brief digression on the genus Homo

The simple version of the story is that Homo first appeared about 2.5 million years ago in eastern Africa. Some groups (particularly, Homo Erectus) migrated out of Africa 2 million years ago, where they continued to evolve in various regions stretching from Western Europe to Indonesia. Homo Erectus appears to have returned to Africa, possibly because of an "ice age." Homo sapiens appeared about 250,000 years ago in eastern Africa. About 70,000 years ago, they migrated to the Arabian peninsula, from where they expanded, eventually, throughout the world: "Australia 47,000–65,000 years ago, and Europe nearly 45,000 years ago. They settled Beringia approximately 25,000 years ago, crossing the land bridge over the Bering Strait during several periods of the Pleistocene Ice Age, and entered deeper into the Americas 14,000–23,000 years ago." Oded Galor, *The Journey of Humanity: The Origins of Wealth and Inequality* (2022), p.19. There was some sporadic (but, apparently not wholesale) interbreeding with the other groups of humans they encountered. *See* Harari, *Sapiens*, pp.4-19.

In Africa, "our biological ancestors were distributed everywhere from Morocco to the Cape. Some of those populations remained isolated from each another for tens or even hundreds of thousands of years, cut off from their nearest relatives by deserts and rainforests. Strong regional traits developed." David Graeber and David Wengrow, *The Dawn of Everything: A New History of Humanity* (2021). p.81. " It seems reasonable to assume that behaviours like mating and child-rearing practices, the presence or absence of dominance hierarchies or forms of language and proto-language must have varied at least as much as physical types, and probably far more." *Id.*

As of 2015, it appeared that there had been at least six types of humans (genus *Homo*) existing in the world 100,000 years ago. *Id.*, pp.4-19. But, there were probably more.

In addition to *Homo sapiens*, two others in Africa, the Neanderthals and the Denisovans, there is evidence of an early human, the *Homo floreseinsis*–the"hobbit"–in Indonesia. Recent work done on skull fragments and some teeth found in China in 1976 indicates the possibility of yet another type of human living contemporaneous with the others. *See* Melissa Hogenboom, "Is this a new species of human?" *BBC Discovery*, 25 January 2015.

A new study based upon the examination of fossils from a discovery in cave system in South Africa appears to find evidence of yet another new species of *Homo*, named *Homo naledi*. *See, e.g.*, Pallab Ghosh, "New human-like species discovered in S. Africa," *The BBC News*, 10 September 2015; Robert Lee Hotz, "Remains of Humanlike Ancestors Found in South Africa,"*The Wall Street Journal*, September 10, 2015; John Noble Wilford, "New Species of Human Ancestor Is Found in a South Africa Cave," *The New York Times*, September 10, 2015. The bones of some 15 individuals were found in a chamber deep underground. The initial speculation was that these individuals lived more than 1.9 million years ago, before *Homo erectus* and long, long before the Neanderthals (which appeared about 200,000 years ago). *Id.*

However, intensive work to attempt to date these bones has led researchers recently to conclude that the bones may be only between 226,000 and 335,000 years old. Michael Greshko, "Did This Mysterious Ape-Human Once Live Alongside Our Ancestors?" *National Geographic*,

May 9, 2017. That dating would make *Homo naledi* almost a contemporary of the Neanderthals. That dating also makes the location of the find particularly surprising (sub-Saharan Africa).

In any event, this discovery further complicates the evolutionary story of the *hominim* line, since the physical characteristics revealed combine old and modern features in a novel way, inconsistent with existing concepts of the chronology of descent. Id.

To add further to the confusion, a paper published in *Nature* at the beginning of June 2017 reported on the results of a redating of fossils discovered in Morocco in 1961, and fossils recently recovered from the same site, which determined a date of about 300,000 years ago, rather than the much earlier date attributed. The fossils are believed by many (but not all) to belong to early *Homo sapiens*. That would make this find the oldest known example of our species, pushing back significantly the date that we appear to have diverged from our predecessor species. sPreviously, the oldest example of *Homo sapiens* was from finds in Ethiopia estimated to be between 160,000 and 195,000 years old. So, it is not only the date, it is also the location of these fossils that challenges the established understanding that *Homo sapiens* originated in eastern Africa. See Kate Wong, "Ancient Fossils from Morocco Mess Up Modern Human Origins," *Scientific American*, June 8, 2017. The results of this paper were widely reported in the press.

During 2018, scientists found a piece of a bone apparently from a girl whose "mother was a Neanderthal and ... father was a Denisovan. She is the only first-generation hominin hybrid ever found." Michael Marshall, "The Four Biggest New Things We Learned About Human Evolution In 2018," *Forbes*, December 20, 2018. Moreover, "[b]y combining deep learning algorithms and statistical methods, investigators have identified, in the genome of Asian individuals, the footprint of a new hominid who cross bred with its ancestors tens of thousands of years ago." Center for Genomic Regulation, "Artificial intelligence applied to the genome identifies an unknown human ancestor," *Science News*, January 16, 2019. The hypothesis is that crossbreeding between Neanderthals and Denisovans led to yet another new human relative.

> *"For those who split them into separate species, any one or none of* **Homo habilis** *,* **Homo rudolfensis** *, or* **Homo ergaster** *may have been the ancestor of* **Homo erectus** *, and any one or none of these may have been the ancestor of* **Homo heidelbergensis, Homo neanderthalensis,** *and* **Homo sapiens,** *while* Homo heidelbergensis

may or may not have been the ancestor of either one or both of Homo neanderthalensis and Homo sapiens."

Hands, *Cosmosapiens*, p.439 (emphasis added).

Given the large differences that are supported today across the modern human population and the evidence of ancient interbreeding (discussed above), it may be that all or most of these human families represent a single species with significant physical differences across a single species. Harari, *Sapiens*, p.19.

By 13,000 years ago, all but *Homo sapiens* had disappeared. *Id.* "Only after those other populations became extinct can we really begin talking about a single, human 'us' inhabiting the planet." Graeber and Wengrow, *The Dawn of Everything.* p.82.

The Role of Altruistic Behavior

Examples of apparent altruism bothered Darwin, and many of his followers have tried to find explanations for the phenomenon of self-sacrifice that occurs among many different species. Beyond the narrow issue of apparently altruistic acts, there is the related challenge of explaining the whole phenomenon of social interdependence and community living. The issue here is not simply working together in the pursuit of a common goal, such as hunting prey in groups. It is the concept of individual sacrifice for the benefit of the group. *See* Nowak, *SuperCooperators: Altruism, Evolution, and Why We Need Each other to Succeed* (2011), p.xiii. Of course, we are discussing here behavior that appears altruistic, not actual altruism.

Kin selection

An answer arose from the consideration of the mathematical or probabilistic frequencies of common genetics among individuals of various degrees of familial relationship. That analysis gave rise to the theory of "kin selection," in recognition of the fact that particular genes can propagate through the survival of an individual with the gene or through the survival of his close relatives, who are likely to have the same gene, with the degree of relatedness directly determining the probabilities. *Id.*, pp.95–98.

Kin selection provides some biological basis for such altruistic behavior consistent with "the selfish gene," in the sense that the sacrifice of an individual for the survival of a larger number of individuals with many of the same genes could clearly promote the survival and reproduction of those particular genes.

The key event in this story was the publication of William D. Hamilton's doctoral work at the University of Cambridge. As reported in his obituary in The New York Times:

> "Dr. Hamilton, a professor at Oxford University since 1984, burst into the field of evolution while still a graduate student at Cambridge University. In 1963 and 1964, he published two papers based on his doctoral work that have proved so seminal to evolutionary biology that it is virtually impossible to read a contemporary study in the discipline without encountering his name and the term he coined, inclusive fitness, also known as kin selection. Through the model of inclusive fitness, Dr. Hamilton proposed **an elegant and mathematically sophisticated way of understanding altruistic behavior,** a problem that had baffled naturalists from Darwin onward."

Natalie Angier, "William Hamilton, 63, Dies; An Evolutionary Biologist," The New York Times, March 10, 2000 (emphasis added).

Hamilton focused upon the fitness of the group, which he called "inclusive fitness." The theory has also been called "kin selection," the label that stuck. Among other things, his theory was able to generate interesting implications for evolution based upon differences in the degree of relatedness among members of a family or community (meaning the extent of common genes).

For example, in bees, the females have two copies of each chromosome (like humans and most animals, a characteristic referred to as diploid), while the males have just one set (referred to as haploid). Unfertilized eggs from the queen become males (there is no father), while most fertilized eggs become female. The male offspring have one set of chromosomes, inherited entirely from the female. The female offspring have two sets, half from the mother and half from the father. Thus, the females are 50% related to the mother and 50% related to the father (the males are 100% related to the mother, and there is no father). The sister bees, however, will be at least 75% related to each other. They all have 100% of the father's genes and 50% of the mother's genes (but not necessarily the same genes from the mother). If the females have female offspring, the relatedness will again be 50%

Thus, in terms of promoting the reproduction of the genes, actions of the females (who are the workers maintaining the hive and caring for the larvae) that benefit their sisters will be more productive than actions that benefit their own offspring. Female offspring will have 50% of the same genes and female offspring of those daughters will have only 25% of the grandmother's genes. In contrast, additional sisters will have 75% of the genes. Dave Goulson, *A Sting in the Tail: My adventures with Bumblebees* (2014), pp.110-3. This unusual genetic characteristic is a plausible reason why significant social behavior can be found among bees and other insects of the Hymenoptera group (which includes wasps and ants) but not often among other organisms. See id., p.111.

GROUP SELECTION

In the 1960s, there was considerable debate around the theory of "group selection," a theory based o the observation that groups more successful at cooperative behavior prosper and tend to win in the competition with other groups, leading to the more successful reproduction of one group and its emergence as more significant presence or to the less successful reproduction of another group and its diminution or disappearance.

Participants in this debate included British zoologist Vero Wynne-Edwards (proponent), evolutionary biologist George C. Williams (critic) and naturalist Edward O. Wilson (defender). Nowak, *SuperCooperators*, pp.84-7. This analytical framework suggested that group competition could have given rise to higher levels of cooperation among individuals within groups, because of the greater success of groups with individual having a genetic inclination toward cooperative behavior.

(This theory has been claimed to suggest that natural selection can contribute to the evolution of cultures, as well as of biological species. *Id.* pp.90-91. But, in my view, any type of cultural selection fundamentally rest on the competition model, in contrast to the biological evolution model.)

MULTI-LEVEL SELECTION

Biologists have looked for social activity or society elsewhere in the animal kingdom.

It appears that most complex social organizations have arisen through "eusociality—roughly 'true' social conditions." Edward O. Wilson, "The Riddle of the Human Species," *The New York Times*, February 24, 2013. Such eusocial groups raise their offspring with group participation, often across generations, and display some division of labor. As Wilson observes, groups achieving such advanced social behavior have been exceptionally successful. Ants and termites, representing only a couple percent of the existing insect species, supposedly constitute about half of the world's insects when measured by body weight.

Wilson, who died at the very end of 2021, noted that such eusociality is "extremely" rare, appearing only about two dozen times among the hundreds of thousands of "evolving lines of animals on land during the past 400 million years" and did so quite late in the "history of life" with termites and ants, appearing only 150 to 200 million years ago. Id. "Evolutionary biologists have searched for the grandmaster of advanced social evolution, the combination of forces and environmental circumstances that bestowed greater longevity and more successful reproduction on the possession of high social intelligence." Id. He described two competing theories: kin selection, enabling altruism among related individuals, and "multilevel selection," a theory that Wilson pioneered in which natural selection is postulated to occur on two levels, individual selection within a group and group selection arising out of competition among groups. Id.

Wilson observed that part of the increased interest in multilevel selection is the result of mathematical exercises that suggest that successful evolution of complex societies through kin selection alone would require the existence of rather restrictive conditions likely not present in, at least, all of the known examples of such evolution. Id.

QUORUM SENSING

A discovery made relatively recently is that some bacteria engage in a type of behavior that could almost be called collective or group activity. It is a phenomenon referred to as "quorum sensing." Certain things happen only when there is a sufficient density or quantity of individual bacteria present. It will not occur with lesser numbers, that is, with less than a "quorum" present.

This phenomenon has been discovered in some bacteria that have a relationship with the cells of particular plants or animals. The relationships can be either symbiotic or pathogenic. In fact, the phenomenon was first observed in a bacterium species that lives in a symbiotic relationship with a small, thumb-sized squid (the Hawaiian bobtail squid). The bacteria live within the mantel of the squid. The squid feeds the bacteria through its secretions; while the bacteria are bioluminescent, rendering the squid invisible to predators swimming below it. However, the bacteria are not bioluminescent if less than the quorum is present. This phenomenon seems to mean that the bacteria have some means of sensing the number of other bacteria in the vicinity and of reacting to that information. In other words, there seems to be a primitive form of limited communication among these bacteria and what almost resembles collective action.

To be more specific, for certain bacteria, the expression of a particular phenotype (physical characteristic) by a gene is triggered only by a sufficient aggregation of bacteria around a host or victim cell. Only when there is a quorum will the gene expression occur. The first obvious curiosity is about how the bacteria "sense" the numbers around them—that is, what causes them to act as if there is a quorum present. The sensing appears to arise through chemical signaling. The individual bacteria emit a chemical, and when the density of that chemical is high enough; the receptors in each bacterium trigger the creation of the new protein that effects the appearance of the phenotype. Thus, the reaction itself would seem not really to be collective, but individual. Nonetheless, the behavior evidenced by the group is not what one would normally expect from lowly bacteria.

This phenomenon has obvious interest for medical science. Where the gene expression that is induced when a quorum is present is pathogen-causing disease, the suppression of the system by which the presence of a quorum is detected would likely prevent the appearance of the pathogen. See Dr. James Hodge Kinston, "Learning the chemical language of bacteria," Master's Fellows' Research Talks, Trinity College (Cambridge), 4 February 2014. For our discussion here, the interesting issue concerns how such a characteristic could have arisen through evolution or adaptation. As noted, it seems that examples of this feature can be found in connection with both symbiotic and pathogenic relationships between the bacteria and the host. Thus, the explanation must work for both relationships that benefit the host and relationships that are detrimental to the host.

One could speculate that for a pathogenic phenotype, there would be benefits in not having it appear until the bacterial load was adequate to resist the victim's immune response. Thus, bacteria that happened to require a certain quorum to trigger the expression would have a survival advantage over the bacteria that incorporated a lower or no quorum. For symbiotic phenotypes, it is possible that there would be little or no adaptational value to the expression of the phenotype until there is some threshold level of bacterial load. But, it is harder to see the advantages to the bacterium's survival unless the expression in lower concentrations gave rise to some adverse consequence.

With respect to the example of the bioluminescent squids, discussed above, the expression of the bioluminescence would have an energy cost that could be quite substantial to the bacteria. That cost could be well overcome by the food supply provided by the squid. Somehow, the package came together. See, e.g., Stephen P. Diggle, et al., "Evolutionary theory of bacterial quorum sensing; when is a signal not a signal?," *Philosophical Transactions of the Royal Society B*, 2007.

FINAL NOTES

An important point to recognize is that inclusive fitness, kin selection and group selection are not based upon the intentions of or decisions by the individuals. There is no strategic behavior involved. These theories do not even necessarily require that one organism "recognize" the degree of relatedness with another. If the related individuals are in close enough proximity so that most interactions are with related individuals, then the process could work. Kinship groups could prosper relative to other groups. In any event, we are far from the concept of cooperation. That subject is examined in a later chapter.

I also discuss in a later chapter recent research into the biological (neurological) roots of altruistic behavior in humans. It appears that charitable acts give rise to stimulation of specific parts of the brain that are associated with the pleasure derived from satisfying cravings like those for food and sex. See Elizabeth Svoboda, "Hard-Wired for Giving," *The Wall Street Journal*, August 31, 2013. If certain altruistic conduct gives rise to pleasurable feelings, is the reason cultural or biological?

THE ROLE OF BEAUTY

In 1871, Darwin published *The Descent of Man, and Selection in Relation to Sex*. In this, his second major work, he set forth explicitly his conclusion that mankind had descended from a "hairy quadruped". He also expounded in detail his theory of sexual selection, which he had referenced in *The Origin of Species*. The book was heavily criticized with respect to the robust theory of sexual selection, which argued that subjective assessments of attractiveness by the ones doing the choosing played an important role in the selection of mates and that such sexual selection was an evolutionary force separate from natural selection.

The immediate criticisms of Darwin's theory of sexual selection reflected three strong prejudices of scientists (and much of the educated public) of the time: first, that natural selection was the all-inclusive and sufficient explanation of evolution (completely displacing religion); second, that animals lacked the sensory and cognitive capabilities for making aesthetic judgments and, third, that female preferences or likes and dislikes were fickle and capricious, insufficiently systematic to have an important impact on the nature of the world (or, that the idea of female autonomy, especially in sexual preferences, was inherently unacceptable). Richard O. Prum, *The Evolution of Beauty: How Darwin's Forgotten Theory of Mate Choice Shapes the Animal World – and Us* (2017), pp.27-35.

By the later part of the twentieth century, the second and third prejudices had largely been rejected by the scientific community based upon new research into the cognitive capabilities of animals and the feminist movement. *Id.*, pp.35-41. The first prejudice, however, arguably had strengthened, for the reason I have already discussed (*i.e.*, scientific animosity to religion). As a result, the scientific community was, and still is, not prepared to entertain the idea that "natural selection is not the only source of organic design in nature." *Id.*,p.134.

SELECTION BY MATE CHOICE

Darwin believed that natural selection and sexual selection, operating independently but in combination, explaining the incredible diversity observed in the natural world. Today, neo-Darwinism asserts that sexual selection is simply an aspect or part of natural selection. The neo-Darwinist argument is that differences in ornamentation, display and courting behavior provide relevant information about the suitability of prospects as mates (that is, whether a prospect is vigorous, healthy, free of parasites and/or has good quality genes). This characteristic is often referred to as "honest advertisement". Thus, sexual selection is adaptive.

But, for the neo-Darwinists to be correct, however, there must be an adaptive function for every detail of ornamentation and display. That seems hardly plausible to me. To put this issue in perspective, Darwin also identified the role of coercion in the mating of many species. Thus, the development of weapons, like tusks or antlers, and of greater physical strength played an important role in the mating "choices" for such species. In such cases, the female has no choice, her only alternative will be determined by the outcome of competition among the males. We find this reproductive model to be common among mammals.

In contrast, the reproductive strategies for most birds and many reptiles and fish reflect active mate selection. The partner making the choice could, in theory, be either male or female; but, in the vast majority of instances, it is the female that selects. So, I shall discuss this issue hereafter in terms of female mate selection. In simple terms, such selection supposedly occurs as follows. Various males presented themselves to the female, who makes a choice based upon some trait or characteristic or conduct that she finds most attractive (or preferable for some reason). The mating couple are likely then to have offspring among which the males have the trait for which their father was selected and the females have the same preference for or attraction to that trait as did the mother (which assumes that the aesthetic preference of the female is heritable). Thereby, both the preference and the preferred trait will appear in subsequent generations. This model of evolution will be most effective where the aesthetic preferences are common among multiple females.

If these circumstances exist, then it is likely that the female preference and the male trait will co-evolve, to change or develop in the same direction. In fact, the evolving female preferences will tend to "direct" the evolution of the male traits (variations in which appear randomly, but the variations that succeed will be determined by the females' aesthetic preferences evidenced by their mating choices). Over time, one would expect changes, refinements or elaborations to occur in the preferred trait, whether it be in the ornamentation, the nature of the male display, the courting performance of the male or all three. Thus arises the great diversity we find even among closely related species.

There is another type of relevant relationship. The male trait that the female finds attractive is likely to impose some costs on the male, whether in terms of its development, the burdens of maintaining it or the possible consequences of having the trait, for example, being an attraction to predators. Therefore, as the preferences and traits coevolve, there is likely to be an equilibrium reached at which the costs of the trait are balanced against the strength of its attractiveness to potential mates (and its consequent usefulness in reproductive success).

Independent but interacting

Non-adaptive mate selection was Darwin's explanation for the incredible diversity found among birds—coloration, plumage, songs (and other forms of making sounds), display behavior and other courting performances. As Prun says: "Throughout the living world whenever the opportunity has arisen, the subjective experiences and cognitive choices of animals have aesthetically shaped the evolution of biodiversity." *The Evolution of Beauty*, p.120.

An illustration of the possible interaction of selection by mate choice and natural selection can be found in the flight of birds. One might assume that fully developed feathers evolved to assist in flight (adaptive evolution). However, the fossil record demonstrates that feathers organized with flat surfaces arose long before flight did (with the dinosaurs). The planar, two dimensional surface of feathers is ideal for decoration. And, there appears to be no adaptive purpose served by such a development from down (or fur) to feathers. So, we can assume that feathers were the result of non-adaptive sexual selection. Thereafter, some feather-equipped creatures began to fly, assisted by the existence of feathers. In the words of Prun: "Feathers did not evolve for flight; rather, flight evolved from feathers. And among the best hypothesis for the key innovation [feathers] that allowed birds to launch into the air is the desire for beauty." *Id.*, p.147.

Another example can be found in experiments with small fish (guppies). See Losos, *Improbable Destinies*, pp.123-53. Researchers discovered in Trinidad that there were examples of colorful male guppies in some natural pools while in other pools the males were drab like the females. Further investigation determined that the pools with the drab males contained predator fish while the pools with the colorful males did not. Based upon this observation, researchers conducted an experiment. In some of the pools with the colorful males, they introduced predator fish. Over time, the male guppies observed in those pools were drab. Moreover, sometime after the predator fish were removed from those pools, the males found in these pools again were colorful. *Id.* Assuming that the changes in the male populations of fish were genetically based, we appear to have examples of both non-adaptive sexual selection (favoring the bright colors) and natural selection (favoring the drab colors).

THE "DEFAULT POSITION"

Now, the issue is how does one determine whether Darwin or the neo-Darwinist is correct? Undoubtedly, some aspects of sexual selection are adaptive. But, is that true for everything that influences mate choice in those species where choice is active? How does one tell if a particular feature is adaptive?

In large part, we are faced with a question of the "burden of proof". What is the normal or natural expectation when confronting the trait in question, that is, what does one assume as the "default position"? Does one assume that something is adaptive, the result of natural selection, and ask for proof that it is not? Or, does one assume that it just is (what Prun calls "Beauty Happens"), and ask for proof that it exists for a reason (that it is adaptive)? (Prun refers to this assumption as the "null model" or "null hypothesis".) Id., p.72.

For reasons that should be obvious, the "default position" chosen makes a big difference. For example, one could never prove that a trait is not adaptive. (Proof of a negative is virtually impossible.) At the same time, one could never really prove that a particular trait is actually adaptive—one can only present a plausible scenario in which the trait has survival or reproductive benefits, but that only shows that the trait could be adaptive. In some circumstances, other historical or experimental evidence might indicate that the trait is likely to be adaptive. But, nothing can prove that it, in fact, is. Similarly, if one cannot conceive of a scenario in which the trait provides survival or reproductive benefits, then one should conclude that the trait is not adaptive. However, there is something uncomfortable about making the scientific conclusion depend on the imagination of the brightest people one can find.

In short, the outcome is largely determined by one's choice in the adoption of the "default position." As Prun observed, "...the ultimate research goal for every young scientist or graduate student is to demonstrate what everyone already knows to be true in some delightfully unexpected, new way that no one has ever imagined before." Id. p.72.

A TENDENCY TOWARD BEAUTY?

I have referred to the female choice as an aesthetic judgment. Is that fair? Females clearly make choices. We assume that they are not all adaptive. Are the non-adaptive choices just mistakes or misjudgments? That conclusion must assume some intention (that is not realized) on the part of the females, for which we have no evidence. Assuming that one finds the argument for non-adaptive mate choice to be persuasive, then we come to the question of whether evolution has a built-in bias in favor of beauty? Certainly, by most human standards, it appears that the result of evolution has been a cornucopia of beauty, but our subjective view is inevitably subjective and may be biased (*e.g.*, we may be programmed to find the natural world to be beautiful or our concepts of beauty may be culturally determined}. The question is whether preferences in mate selection are merely arbitrary flukes or represent some form of innate animal aesthetics, even one similar to ours.

The "New" Biology

I have discussed some of the questions that can be posed to the theory of natural selection as, essentially, a logical matter. In a sense, the issues are whether natural selection was a plausible theory of the origin of species and how much of evolution did it appear to explain. Now, I briefly consider here some recent scientific discoveries about factors other than genes that influence how life evolves. These discoveries clearly are consistent with the concept of evolution; however, they can be seen as undermining the theory of natural selection, not as a matter of the logical persuasiveness of the theory, but as a matter of the actual facts of the world. See Fodor and Piattelli-Palmarini, *What Darwin Got Wrong*, p.55. (Many of the developments discussed by the authors arose in a relatively new field called evolutionary developmental biology or "evo-devo".)

These developments raise significant challenges to neo-Darwinism as the definitive theory of human evolution. It is not that these empirical findings are individually directly or necessarily inconsistent with neo-Darwinian theory; many of them could be incorporated into and accommodated by that theory. The problem is that as the theory acquires more and more caveats and addendums and exceptions, one should certainly ask whether the central theory is really central, that is, does it explain the more important things that are going on? See *id.*, p.21.

Their central point concerning the new discoveries of biology is that are "strong, often decisive, endogenous constraints and hosts of regulations on the phenotypic options that exogenous selection operates on." *Id.*, p.21. In other words, in contrast to the model of natural selection in which random mutations in genes create a multiplicity of new phenotypic traits among which natural selection makes selections, there are a large number of other factors significantly shaping the traits that can emerge. In addition, the substantial interrelationships among genes and the built-in repair and regulatory mechanisms undermine the paradigm of "gene equals trait".

Interactions of genes

We understand that phenotypes arise not just from traditional genes, but also from the interactions of genes. In addition, while only 1.8% of the human genome consists of coding DNA for our genes, there are other portions of the DNA that produce RNA, some of which is useful or, even, essential. Some of the RNA creates proteins that are key to regulating growth and cell differentiation. Chris Ponting of the University of Oxford has reportedly estimated that as much as 8% of the DNA is functional. Carl Zimmer, "Is Most of Our DNA Garbage?" *The New York Times*, March 5, 2015.

The internal controls and regulatory processes are very complex and can operate across multiple genes or gene complexes. There is even evidence of joint regulation across different chromosomes. Fodor and Piattelli-Palmarini, *What Darwin Got Wrong*,, p.26. Proteins that perform DNA repair functions, other units of regulation and internal developmental filters may moderate or eliminate the effect of mutations or rescue the defect genes, and they all affect the phenotypic effects, if any, that might appear as a result of genetic mutation. *Id.*, pp.26-7, 32-5. Consequently, "even if they were actually random, mutations would not always produce random novel phenotypes." *Id.*, p.33. These features of the organism mean that there is a type of internal selection process that takes place before any external (natural) selection can occur. *Id.*, p.39.

Fodor and Piattelli-Palmarini point to the apparent robustness of certain traits relative to genetic variation, that is, the tendency of the trait to persist despite genetic mutations. *Id.*, pp.42-3. They also describe the role of so-called "master genes" that determine the affect of many different traits. *Id.*, pp.44-6. Some of these master genes appear to have survived through hundreds of millions of years of evolution and to exist across multiple and distant species. *Id.*, pp.24, 29. There are strong similarities of genes found in distant species. *Id.*, p.29.

Horizontal transfers of genes

Fodor and Piattelli-Palmarini also report evidence indicating that "horizontal" gene transfer, that is, the transfer of genetic material from one organism to another that is not a direct or immediate descendant has been commonplace. *Id.*, pp.67-8. Perhaps, 45% of even the human genes have derived from horizontal transfers. *Id.*, pp.69, 68. It is the rule in microorganisms, and it seems likely that horizontal transfers were a common occurrence among microorganisms in the early stages of the development of life. *See, e.g.*, Collins, *The Language of God*, pp.89-9.

A recent study by a group from the Department of Chemical Engineering and Biotechnology at the University of Cambridge suggests that it has been widespread. The lead author of the paper, Alastair Crisp, has been quoted as saying: "This is the first study to show how widely horizontal gene transfer occurs in animals, including humans, giving rise to tens or hundreds of active 'foreign' genes. Surprisingly, far from being a rare occurrence, it appears that this has contributed to the evolution of many, perhaps all, animals and that the process is ongoing." University of Cambridge, *Research Bulletin*, "Human genome includes 'foreign' genes not from our ancestors," 12 March 2015.[24] They confirmed that 17 human genes previously reported as foreign were so, and identified another 128 human genes that were of foreign origin, mainly related to enzymes involved in metabolism. They also confirmed that the ABO gene that determines a person's blood type was acquired by vertebrates through horizontal transfer. *Id*. The transfers appeared all to be quite ancient and the sources appear to include bacteria and other microorganisms as well as fungi.

There is speculation that viruses have played a role in the horizontal transmission of genes. Frank Ryan, *Virolution*, p.23. It is possible that much of the supposed "junk" DNA in the human genome is the result of invasions by viruses over the millennia. See *id*., pp.102, 128–9. Ryan argues that viruses have played an important role in evolution, with natural selection operating on the symbiotic combination of the virus and host, favoring various mutations in both the virus and host. *Id*., pp.5–60, 67, 134, 150. Viruses, which seem to be ubiquitous, appear to have been able to cause the creation of new genetic structures and the rearrangement of genetic material. Thus, viruses may play a crucial role in speciation. *Id*., pp.117–21. This process may have created new species much more rapidly and in greater numbers and with greater diversity than could natural selection operating on random mutations. *Id*., pp.124, 128.

EPIGENETICS

Importantly, a field of study has arisen concerning the apparent heritability of characteristics that are not contained in the DNA double helix. See Dan Hurley, "Grandma's Experiences Leave a Mark on Your Genes," *Discover*, June 11, 2013. As Hurley explains, since the 1970s, scientists have been aware that there is a need for something extra (beyond the DNA itself) to enable the selection of the particular genes that will be expressed or transcribed for the creation of proteins in particular cells. One of the known extra elements is the methyl group, which is a regular structural component of organic molecules. It is attached to proteins that support the DNA structure but is separate from and outside of the DNA. Thus, it has been called "epigenetic" material.

Originally, it was thought that changes in epigenetic material ("epigenomes") occurred only during fetal development, but many subsequent experiments have demonstrated that changes can occur in the living person, especially during childhood. The changes can arise from diet, exposure to chemicals and, interestingly, from experiences–both traumatic and comforting. Most importantly, the changes in the epigenetic structure will be passed on to offspring. That is, the impact of various experiences can be biologically heritable, resulting in potentially long-term changes in gene function. Significant work has been done on the impact of the mother's experiences. *See* Michael Meaney and Moshe Szyf, "Epigenetic Programming by Maternal Behavior," *Nature Neuroscience*, June 2004.

Epigenetic effects can mimic genetic ones in particular organisms, creating different phenotypes despite identical DNA. Fodor and Piattelli-Palmarini, *What Darwin Got Wrong*, pp.45–6, 65. It is unknown to what extent epigenetic effects can be inherited by the third or fourth generation. *Id.*, pp.65–6. However, there have been studies in fruit flies in which traits that arise from chemical shocks at a particular point in the development process become heritable traits (with no such shocks) after a few generations. *Id.*, pp.57–9. Other recent studies indicate that the effects of stress prior to pregnancy may be inheritable, based again on experiments with rats. *See* Inna Gaisler-Salomon, "Inheriting Stress," *The New York Times*, March 7, 2014; "Study documents paternal transmission of epigenetic memory via sperm," *University of California - Santa Cruz. com*, October 17, 2018 ("Susan Strome's lab ... has been making steady progress in unraveling the mechanisms behind this phenomenon, using a tiny roundworm called Caenorhabditis elegans to show how marks on chromosomes that affect gene expression, called 'epigenetic' marks, can be transmitted from parents to offspring").

Since the early 2000s, there has been an awareness that epigenetic factors are transmitted from the father as well as the mother. In the last few years, researchers have gained further insight into how that phenomenon occurs. "[I][n two complementary studies, scientists tell us...[a]s sperm traverse the male reproductive system, they jettison and acquire non-genetic cargo that fundamentally alters sperm before ejaculation. These modifications not only communicate the father's current state of well-being, but can also have drastic consequences on the viability of future offspring." Katherine J. Wu, "Dads Pass On More Than Genetics in Their Sperm," *Smithsonian.com*, July 26, 2018. However, some recent research at Cambridge University suggests that the impact of epigenetic factors in humans is actually relatively rare. See, "Studies raise questions over how epigenetic information is inherited," University of Cambridge Research Bulletin, Friday, 2 November (2018). See also, Benedict Carey, "Can We Really Inherit Trauma?" NYTimes.com, December 10, 2018.[25]

In any event, epigenetics provides a new and different perspective to at least part of the evolutionary process.

OTHER NON-GENETIC FACTORS

Plasticity

In addition, there are various physical things other than genes that have been identified as affecting the ability of an organism to adapt and survive (and, therefore, to reproduce). For example, there have been discoveries of surprising abilities of the human body to adapt to environmental factors during developmental stages, causing outward characteristics or capabilities that are not based upon genetic differences. The phenomenon is called "plasticity." See, e.g., Fodor and Piattelli-Palmarini, *What Darwin Got Wrong*, p.52. Plasticity with respect to the brain is discussed further in the chapter Consciousness

Culture

There has also been increasing awareness of the impact of human culture on evolutionary, even genetic, development. An example is lactose intolerance, with respect to which it is believed that genetic changes preventing the gene that enables the digestion of lactose from switching off in childhood after weaning have occurred in response to cultural practice. In other words, the selection pressures on genetic characteristics can arise not just from external environmental factors, like climate change or availability of food sources, but also from cultural forces generated within the species. See, e.g., Kevin N. Laland, et al., "How culture shaped the human genome: bringing genetics and the human sciences together," *Nature Reviews Genetics*, 11, February 2010, pp.137-48; Nicholas Wade, "Human Culture, An Evolutionary Force," *The New York Times*, March 1, 2010.

The "microbiome"

There has been much recent attention devoted to what has been called the "microbiome", the collection of microbes (bacteria) that are part of every living person. The quantity of these microbes is astonish: it is estimated that as much as one half of the cells in our body are not human cells (that is, do not contain our DNA) and that the collective genomes of these microbes exceed by 500 times the quantity of genes in the human DNA. See Jonathan Weiner, "Human Cells Make Up Only Have Our Bodies. A New Book Explains Why," *The New York Times*, August 15, 2016 (a review of a book by Ed Yong entitled *I Contain Multitudes: The Microbes Within Us and a Greater View of Life*). These microbes form an integral, necessary part of us.

One example is the apparent importance of the bacterial composition of the intestines to health and fitness, including nutrition and weight control. A recent discovery involved the differences in the intestinal bacteria of twins who differed substantially in the percentage of body fat in a study published online by the journal Science. Gina Kolata, "In Gut Research's Latest Advance, Bacteria From Humans Can Slim Mice Down," *The New York Times*, September 5, 2013. The researchers located twins where one was fat and one was thin (measured by percentage of body fat). They then collected intestinal bacteria from each twin and introduced it into groups of mice. They discovered that on a normal diet, the mice receiving bacteria from the thin twin became thin and mice receiving bacteria from the fat twin became fat. When mice from the two groups were put in a cage together, they shared bacteria (since the mice naturally eat the feces of each other). In that case, the fat mice became thin. Thus, the bacteria from the thin twin dominated. Id. On its face, this result suggests that the thin twin had bacteria that the fat twin did not, so that the bacteria "causing" thinness was extra, in addition to the other bacteria which may have been common to the two twins. If so, then it would have been the introduction of this additional type of bacterium into the guts of the fat mice that caused the loss of body fat. Interestingly, the researchers also found that diet made a difference. Id. When the fat mice were fed a special high fat diet, they did not become thin despite the sharing of bacteria. Only when they were put on a low fat diet (derived from fruit and vegetables) did the new bacteria cause the fat mice to slim down.

The research suggests that some combination of intestinal bacteria and diet leads to low body fat. So, whether a propensity to put on body fat is a good or bad thing would seem to depend upon the environment of the organism. If food is scarce or erratically available, then the thinness-causing bacteria would probably not be desirable, since body fat would be a means of storing energy supplies. Where food is sufficient or abundant, then relatively low body fat could well provide other health benefits. The type of bacterium also is, in a sense, transmittable from one generation to another, at least among animals that eat feces.

GENETIC VARIABILITY

To give a further idea of how complicated the situation has become, we can look at an example of some recent research and publications that have received press attention. This work indicates that it is not uncommon for a person to have genetic variations within his or her own body. See Carl Zimmer, "DNA Double Take," *The New York Times*, September 16, 2013. As the cost of DNA sequencing has fallen dramatically, more of it is being done. In the process, researchers have found examples of genetic variations within the same person in what appear to be unusually high percentages of cases (such as 60% of the individuals examined). *Id.* (The types of anomalies found include a person with two different blood types, women with Y-chromosomes in the cells of brain or breast tissue, women with some eggs with one DNA sequence and other eggs with another sequence.) In such cases, a DNA sample taken from, for example, a mouth swab may not accurately reflect the DNA of a patient's organ that is of interest. The genetic variations can arise from things like pregnancy or transfers between twins in the womb, as well as from mutations that occur within a person's body that survive but affect only some of the body's cells. *Id.* The existence of multiple genomes may have links to various diseases. There is also evidence that genetic variations can have helpful effects.

SYMBIOTIC RELATIONSHIPS

There are many examples of symbiotic relationship among organisms, such as cleaner fish that eat parasites and dead skin off of other fish (even fish that would normally be considered predators of the cleaner fish) and birds like the oxpecker that similarly eat potentially harmful parasites on the hippopotamus. More importantly, there are examples of apparent organisms that, upon close examination, turn out to be two separate organisms living in a close relationship. Swiss botanist Simon Schwendener discovered in 1868 that lichens found growing on rocks, particularly noticeable at higher altitudes, previously believed to be a single, discrete organism, actually consisted of an alga and a fungus living in very close association. Ryan, *Virolution*, pp.60–61. The relationship is not parasitic, as we generally understand the term, but mutually beneficial and supportive and necessary to both organisms.

Similarly, it has been discovered that many plants live in close relationships with particular fungi, in which relationships the plant and the fungus each provide the other with important nutrients in a mutually beneficial exchange. *Id.* p.82. In addition, most plants derived benefits from fungi in the soil.

> "The success of fungi results largely from their unique way of feeding. Rather than absorbing sunlight like plants or devouring other organisms like animals, fungi spew out powerful enzymes. These break down surrounding cells or even rock, which the fungi slurp up... [And, many] fungi are vital partners to animals and plants. Cows grow fungi in their stomachs to help break down the tough grass they eat. Most plants intertwine their roots with networks of underground fungal threads that supply nutrients."

Carl Zimmer, "A Billion-Year-Old Fungus May Hold Clues to Life's Arrival on Land: A cache of microscopic fossils from the Arctic hints that fungi evolved long before plants," The New York Times, May 22, 2019.

Bacteria and plants can also have a symbiotic relationship, For example, the legume family of plants, like peas and beans but including clovers and other flowering plants that can be found in wild meadows, have nodules on their roots that contain the bacterium *Rhizobium* which captures nitrogen from the air and turns it into a form of nitrate that can be used by the plants. *See* Goulson, A *Sting in the Tale*, p.196. Plants require nitrates to grow, so in poor soil, without fertilizers, many plants are severely hampered. Legumes, however, produce sugars through photosynthesis which they feed to the bacteria residing in their root nodules which bacteria, in turn, generate usable nitrates (from the abundant nitrogen in the air) to sustain the plants' growth. *Id.* The plants feed the bacteria and the bacteria feed the plants. Furthermore, the pollen of legumes is particularly high in proteins with amino acids. It seems that bumblebees display a marked preference for the flowers of legumes. So bumblebees are fed by the protein-rich pollen that is produced by the legumes because of the bacteria and, in the process, pollinate the legumes enabling them to reproduce. *Id.*, pp.196–7.

There are also some examples where the symbiotic relationship appears to have a genetic element. Frank Ryan describes *Elysia chlorotica*, which is a sea slug that can be found on the Eastern seaboard of the United States. *Virolution*, at pp.9–12. Starting perhaps 3 billion years ago, certain bacteria became photosynthesizing microbes. Those microbes were then incorporated into cells that were to evolve into algae and plants. Now, some such chloroplasts are a crucial part of this sea slug. It is green and lives through photosynthesis, just like a plant. However, when the eggs of this slug hatch, the larvae are not able to live on sunlight. Instead, they began to eat the contents of the cells of a particular kind of green alga, whose contents include chloroplasts. The chloroplasts are absorbed by the slug, forming a layer just under the skin. There, under the transparent skin of the slug, the chloroplasts continue to perform the process of photosynthesis, now providing energy to the slug rather than to the alga from which they came. At some point, certain genes from the nucleus of the alga must have been transferred to the nucleus of the slug, enabling the slug to incorporate the chloroplasts into itself, so as to function with the chloroplasts as a single organism. Id.

ENDOSYMBIOSIS

More directly relevant to mainstream evolutionary theory, there appear to be some very significant symbiotic relationships in which the two organisms eventually have truly become one. The best-known example is mitochondria, essential elements of the animal cell that are passed solely from mother to offspring, and chloroplasts, integral to plant cells. Both mitochondria and chloroplasts are believed to have once been independent life forms—primitive bacteria.[26]

The evidence often cited for this theory is that they each have their own DNA (different from the DNA of the host organism), both have double membranes (which may be a remnant of their prior independent existence) and both reproduce like bacteria do (replication of the DNA, followed by cell division). See, "Endosymbiosis," learn.genetics.utah.edu. Indeed, the theory proposes that early eukaryotic cells (the type of cells found in all animals and most plants, but not in bacteria, viruses and blue-green algae) arose from a symbiotic relationship between two separate organisms that, perhaps following an infection of one by the other, came to live together permanently as a single organism. See Roger Penrose, Shadows of the Mind (1994), p.361, citing Lynn Sagan (Margulis), "On the origin of the mitosing cells," Journal of Theoretical Biology, 14 (1967) pp.225-74.

Because of the commonality of characteristics, it is hypothesized that all eukaryotic organisms descended from a common ancestor. The important feature of that common ancestor was that it consisted of a cell or cells that contained within themselves what were previously independent microbes with their own DNA, what we now call mitochondria or chloroplasts. Thus, it may be that endosymbiosis occurred just once, about 1.5 billion years ago (bacteria and archaea having been present for some 2 billion years prior to that). The microbes that became part of the eukaryotic cell took on the function of generating the energy for the cell and gradually lost up to 99% of their genes, retaining only the genes necessary to perform the specialized function that the mitochondria had. It appears that a result of this symbiosis was a dramatic increase in the ability of the cells to produce energy relative to the quantity of genes they contained. In other words, the division of labor and associated specialization appear to have permitted the evolution of mitochondria that were not burdened with additional genetic apparatus (which was no longer needed, or even useful), leading to a huge increase in efficiency.[27]

> "While most evolutionary biologists now accept that symbiogenesis provides the most plausible explanation for the origin of mitochondria and chloroplasts in eukaryotic cells, they do not accept Margulis's hypothesis for the origin of eukaryotic cells and the evolution of the taxonomic kingdoms. ...According to Margulis, although symbiogenesis is not the only cause of evolutionary innovation, it is the major one. When two or more different kinds of organism merge their identities, the process generates novel behaviours and morphologies: new tissues, new organs or metabolic pathways, and new groups of organisms, including new species."

Hands, *Cosmosapiens*, p.388.

So, endosymbiosis may be the key to the evolution of more complex life forms. The large genomic asymmetry within the eukaryotic cell (massive genome of the nucleus surrounded by thousands of tiny mitochondrial genomes) generates a huge energetic payback, which enables the experimentation with other, more complex structures. It is possible that bacteria and archaea have successfully survived and evolved for possibly 4 billion years without evolving into more complex organisms because of energetic constraints simply inherent in their prokaryotic nature. If this theory is correct, however, it suggests that some highly improbable dramatic change in the structure of cells was necessary for the development of the diverse plant and animal life on Earth. Yet, we do not know how it actually happened.

So...?

Now, what does all of this evidence have to do with evolution? Certainly, it indicates that there are physical factors other than genes that affect fitness and adaptability to environmental change. It demonstrates that genetic changes can occur in ways other than reproduction. It also indicates that these other things may be very important, indeed, maybe dominant. In short, the new biology suggests that there may be much more to evolution than "natural selection" of the reproductive outcomes of random variations. These new discoveries open exciting avenues for further research and theory construction, and they do seem to pose some serious problems for neo-Darwinism as formulated in the twentieth century.

SOME OTHER MATTERS

THE THREE DOMAINS OF LIFE

Current thinking describes the domains of life as being three: Bacteria, Archaea (both being prokaryotes) and Eurkarya (eukaryotes). The third domain contains all complex life forms, including all animals, plants and fungi.

Archaea (generally found in very extreme environmental conditions, such as very hot or very cold water) were not recognized as a domain separate from bacteria until 1977, based upon an analysis of their nucleic acid. Prior categorizations were based largely upon morphology and metabolic functions. Under such criteria, bacteria and archaea appeared to be parts of the same family. The archaea were called archaebacteria. The study of the genetic material demonstrated an independent evolutionary history, as well as various differences in their biochemistry. In addition, there is evidence that the archaea and viruses had an evolutionary relationship as early as 2 billion years ago. C. Michael Hogan, "Archaea," *Encyclopedia of Life*, 2012. The bacteria and archaea look quite similar morphologically, but they do have relatively complex and diverse genetic structures and different biochemistry (in particular the membranes). They are both based upon a single genome, which is loose within the cell (not enclosed in a nucleus with its own membrane).

On Earth, the eukaryotes are greatly outnumbered by the prokaryotes (microbes) but the aggregate biomass of the two may be about the same. The eukaryotes display huge complexity and diversity morphologically, which extends all the way down to the cellular level. However, the cells of the eukaryotes all have certain very important characteristics in common. They all have a nucleus enclosed by a membrane and organelles including mitochondria (the organelles are also enclosed by a membrane, all inside the cell which has its own membrane). Such cells all reproduce through mitosis (where the cell divides into two identical cells). The multi-celled eukaryotic organisms largely have gender differentiation and reproduce through sex.

Despite the huge differences between prokaryotes and eukaryotes, they have sufficient similarities to suggest that they may have a common ancestor. They all use RNA and DNA, they all use proteins for the performance of virtually all functions within the cell, they all use lipids and have membranes. The eukaryotes would have appeared after the prokaryotes, since they seem to be based upon a combination of prokaryotes. So, the prokaryotes are likely to have emerged from the common ancestor (which may have appeared about 3.4 to 3.5 billion years ago), followed by the eukaryotes emerging from the prokaryotes (less than 2.0 billion years ago).

There is new evidence that eukaryotes may actually be a combination of bacteria and archaea: the presence in eukaryotes of double membranes with the separate biochemical characteristics of bacterial membranes and archaea membranes. Ed Turner, "Biological Sciences: the story of four billion years of evolution in seven hours," *Institute of Continuing Education*, University of Cambridge, 8 March 2015.

VIRUSES

What about viruses? Viruses may or may not be considered to be alive; but, unlike the very much larger bacteria, they certainly are not stand-alone living organisms. They are essentially bits of genetic material that become alive only within a host organism. Viruses may consist of pieces of DNA or they may consist of pieces of RNA, in which case they are referred to as retroviruses. A retrovirus, like HIV, is able to insert its genome into the nucleus of a host cell, creating DNA based upon the sequence of the RNA of the virus. The resulting DNA then resides in the host cell. The process is in sharp contrast to the normal transcription of RNA from the DNA of the cell, where RNA then leaves the nucleus to perform various functions (like the creation of proteins). Ryan, *Virolution*, pp.20, 74-6. Significantly, viruses mutate very, very rapidly. Id., pp.18, 23-7.

> "[S]ome giant viruses have been found. The hunt for giant viruses has been particularly productive in the Siberian permafrost. Researchers ... have reported reviving a 30,000-year-old virus found frozen in Siberia. The virus is about 1000 times bigger than the typical virus and has more than 2500 genes. It infects (or lives within) amoebae and is said to pose no danger to humans. Dr. Abergel reported that "[s]ixty percent of its gene content don't resemble anything on earth."

Carl Zimmer, "Out of Siberian Ice, a Virus Revived," *The New York Times*, March 3, 2014.

The benefits potentially provided by the virus to the host could include the culling weak members of the group, assisting in the competition with members or groups of the same species or of other species, and assisting in the subjugation of other organisms for the host's own use. For example, apparently the highly publicized success of the non-native grey squirrel over the native red squirrel in the United Kingdom is, at least in part, caused by a virus carried by the grey squirrel without ill effect that is damaging to the red squirrel. Id., p.96. Another example is a parasitic wasp which carries a virus that weakens caterpillars on which the wasp and its offspring will feed, even to the point of preventing the metamorphosis of the caterpillar into a butterfly, with the effect thereby of prolonging the availability of the food source. Id., pp.94-5.

Is a billion years enough?

The rough estimate is that the Solar System appeared about 4.57 billion years ago; a collision gave the Earth its moon, orbit and wobble about 4.0 billion years ago. It is now thought that life first appeared about 3.5 billion years ago. The first cells may have appeared about 3.4-3.5 billion years ago (with evidence of possible fossils dated to about 3.5 billion years ago and evidence of structure and fossils appearing from about 3.0 billion years ago); eukaryotes less than 2 billion years ago and multi-celled organisms in the last billion years. (The "oxidization" of the Earth is thought to have happened about 2.4 billion years ago, followed by a billion years of low oxygen.)

The Cambrian Explosion was about 530 million years ago (we have switched to millions); the first fish, about 500 million; the first mammals, about 300 million; the first primates, perhaps 100 million; the genus Homo, about 2.4 million (more than 2 million, less than 3 million), and Homo sapiens about 250,000 years ago. See Luisa, The Emergence of Life, p.12. So, it seems to have taken about 2 billion years for single cell organisms to evolve to multi-cell organisms, another billion years to evolve diverse complex organisms, another 500 million years for the first primates to appear, but perhaps just 2 million years for Homo sapiens to evolve from the earliest examples of the genus Homo.

At the same time, it is estimated that, from various causes, there have been five major mass extinctions that have occurred on Earth over the last 500 million years, the most recent one some 66 million years ago (and possibly an earlier one some 2 billion years ago). As a result of these events and the normal extinctions that occur over time, it is thought that over 90% (probably more like 99%) of all organisms that ever existed on Earth are now extinct. See, e.g., Michael Greshko, "What are mass extinctions, and what causes them?", *nationalgeographic.com*, September 26, 2019; Connie Reichert, "Mass extinction 2 billion years ago reportedly killed 99% of life on Earth," *Cnet.com*, September 3, 2019.

Yet, these extinctions have provided the opportunities for a vast multitude of new life forms to emerge, including us. See, e.g., Brad Plumer, "There have been five mass extinctions in Earth's history. Now we're facing a sixth," *The Washington Post*, February 11, 2014 (an interview with Elizabeth Kolbat, author of a new book titled *The Sixth Extinction: An Unnatural History*). In fact, environmental upheavals may have led to the emergence of *Homo sapiens* as the dominant bipedal species. See Maya Wei-Haas, "Surprising leap in ancient human technology tied to environmental upheaval: Sediment core evidence reveals the critical factors that may have given rise to strikingly complex behaviors some 320,000 years ago, around the time the first members of our species appeared," *National Geographic*, October 21, 2020 ("Scientists have long pointed to changes in climate, such as the onset of wet or dry periods, as the key driving force behind the adaptation of our early ancestors").

"Billions of years" seems to me to be beyond anything that I can really even attempt to comprehend. But, so does "millions of years." Nonetheless, serious questions have been raised about whether even several billion years is realistically sufficient time for purely random mutations and natural selection to have created the diversity and sophistication or complexity of life that we observe on Earth today. See, e.g., Nagel, *Mind & Cosmos*, p.6. The cases of optimization that appear to occur in nature seem not to be realizable purely through trial and error, even with hundreds of millions of years. See Fodor and Piattelli-Palmarini, *What Darwin Got Wrong*, p.79. (This issue may be slightly ameliorated by recent developments in evolutionary developmental biology which can be taken as a potentially significant caveat to the assumption of pure randomness. See, e.g., Nagel, *Mind & Cosmos*, p.48, n9.)

The Nature of Neo-Darwinian Explanation

Non-predictive?

Clearly, neo-Darwinism does not fit the prototypical model of a scientific theory (deductive and predictive) and cannot satisfy the broadly recognized criterion for a scientific theory of falsifiability, because it cannot make testable empirical predictions. Its incompatibility with the falsifiability test is not just because of the *ceteris paribus* condition, as is the case generally in economics, and it is not that Neo-Darwinism is probabilistic; it is that the theory simply is not and does not purport to be predictive (although, some probabilistic predictions are possible).

Instead, neo-Darwinism is a framework for the generation of historical explanations or historical narratives: "narratives that purport to articulate the causal chain of events leading to the event that is to be explained." Fodor and Piattelli-Palmarini, *What Darwin Got Wrong*, pp.132-3. Such explanations can certainly be useful and meaningful; they can often be "thoroughly persuasive." *Id.*, p.132. They can also incorporate scientific laws or theories to connect pieces of the story; but, they do not (generally) purport to derive ultimate consequences from initial conditions through Laws. (I say generally because there have been some well-known, but largely unsuccessful, efforts to identify laws of historical or economic development, such as the theories of Hegel and Karl Marx.)

At a minimum, any such biological laws would have to be extraordinarily complicated and detailed. Indeed, relative fitness seems to be entirely context determined. It does not easily transfer from one situation to another. See Fodor and Piattelli-Palmarini, *What Darwin Got Wrong*, pp.52, 131-5, 183-4. While the environmental context is highly important; the fitness-enhancing characteristic of a trait also necessarily depend upon the whole phenotype—it cannot be assessed or categorized in isolation. *Id.*, p.125. Thus, it seems highly unlikely that generalized rules, such as A is better than B, could ever be formulated. At best, some sets of highly qualified statements in the form of "in the following set of circumstances, A will be more fit than B, everything else equal (*ceteris paribus*)" might be imaginable. But, how much of the living world would such statements be likely to explain? Without such Laws, the theory must remain definitional or circular. *Id.*, p.142

Natural selection almost inevitably occurs, effectively by definition, but its ability to explain what we observe resulting is largely *post hoc* and *ad hoc.* Here we have a science that does not even purport to predict, but only to explain. Neo-Darwinism is an example (perhaps, the prime example) of the explanatory paradigm discussed in the opening chapter. It enables us to feel, after the fact, that we probably understand what occurred and why.

Yet, this characterization is not completely fair. I have already noted the evidence of some predictability in the mutations of bacteria. *See above,* n28 to the first chapter. In addition, Jonathan Losos devotes much of his book *Improbable Destinies* (2017) to examples of how evolutionary biology is increasingly becoming an experimental science. He discusses experiments much like those in the field of history where one hypothesizes certain results in certain circumstances and looks to the natural world or world history to see if there are examples consistent with the hypothesis. He also provides examples of experimental work done by constructing similar environmental circumstances to run in parallel, often for years and even decades, with researchers periodically compare the results. Finally, he provides many examples of experiments conducted with organisms having very short life-cycles, so that thousands of generations of development may be observed, such as with bacteria developing antibiotic resistance.

None of these experimental techniques are able to test for the origin of species, of course; but, these experiments are able to generate insights into and a better understanding of how the processes of natural selection operate. They can produce interesting and informative results with respect to traits that are able to evolve relatively quickly, such as coloration, increased tolerance to, *e.g.,* low oxygen levels or the presence of certain toxins, resistance to antibiotics or the ability to utilize new food sources. They also give some insights into the nature of the processes involved.

Non-disconfirmable (non-falsifiable)?

As discussed in the first chapter, the test of a theory according to some philosophers of science is whether it generates predictions of propositions that can be disproved, that is, the concept of falsifiability. If a theory is inherently non-predictive, it obviously cannot satisfy that test. Is there an alternative?

We discussed the matter of the burden of proof or "default position" above in connection with sexual selection. In that discussion, we may have dismissed too readily the proposition that the burden should be on the critic of natural selection to prove that it cannot achieve the results for which it has been given credit. (And, the same for the critic of non-adaptive sexual selection.) Is that not what we tentatively concluded in our discussion of theories and knowledge? So, maybe, an important question is whether the critic can identify some phenomenon that the theory, even with the benefit of hindsight, simply cannot explain. However, it does not seem to be enough to support a theory simply to identify plausible facts that would contradict the theory but that are not found by observation. *Cf.* Dawkins, *The Greatest Show on Earth*, pp.146–7. In contrast, one should clearly be troubled if no set of plausible facts could be hypothesized that would be inconsistent with the theory. In such a case, the theory, in fact, is not contingent, that is, it has no empirical or real world content. it is not science, but only mathematics (or logic) or is simply circular.

I think that we should keep in mind four considerations as a matter of philosophy of science:

- First, proof of a negative (that something could never happen) is unusually difficult, if not impossible, and generally should not be considered an appropriate test.
- Second, one needs to be skeptical of explanations conjured up with hindsight and recognize that such explanations have much less force than would successful predictions of future events.
- There is something intellectually unsatisfying about a methodology that makes acceptance of a theory turn on whether proponents are able to provide such after-the-fact explanations upon demand. ("...[T]he ultimate research goal for every young scientist or graduate student is to demonstrate what everyone already knows to be true in some delightfully unexpected, new way that no one has ever imagined before." Prum, *The Evolution of Beauty*, p.72.)
- Finally, the difficulty of proving a negative applies here as well–the mere failure of the brightest to provide an explanation does not establish that one does not exist.

Circular?

One can readily acknowledge the elegance and purported power of the concept of natural selection; it does seem inevitable: scarcity, competition and survival (followed by reproduction). But, one can also see upon reflection that the generality of the theory, that it appears to be a potential explanation for so much, arises from the very fact that the theory, like the model of market competition discussed above, is essentially tautological or circular.

Here is an interesting illustration of the circularity. It is argued that natural selection enables organisms to adapt to ecological niches and that the emergence of such niches provides opportunities for natural selection to operate. However, as Fodor and Piattelli-Palmarini point out, the identification of the niches occurs with hindsight, reflecting the existence of organisms that are observed to be well-adapted to those niches. Clearly, organisms will live in circumstances for which their phenotypic characteristics make them suitable. In effect, the organisms define the niches by their traits. *What Darwin Got Wrong*,, p.142. So, in a fundamental sense, it is a word game. There is and can be no objective, universal definition of what characteristics or factors enhance survival (or set of Laws or general rules), since survival will depend upon the particular environmental factors that happen to obtain. In addition, for the same reason, it is not realistic to think that one could consistently predict what genetic changes would enhance survival. The supposedly successful adaptations are generally identified with the benefit of hindsight–by the fact that the changes did in fact survive.

It is argued that there is a criterion of fitness for genes, which "is the influence that they have on an organism's longevity and fertility–its ability to find food and sex... ." Kate Distin, *The Selfish Meme*, p.57 The test is whether there is something that has the effect of increasing the amount of reproduction of the organisms in which the gene and its off-spring reside. Does that test really avoid or even ameliorate the circularity problem? Is it really different from saying that some survive (out-reproduce) while others do not? Of course, the theory allows for accidents and chance–some that are more fit to survive do not survive. And, some that are less fit survive because of a serendipitous lack of environmental challenge. But, in the long run, over time, chance will be overcome by the rigors of continuous competition and struggle for survival (and reproduction), and only the more fit (for survival or reproduction) will persistently survive. This conceptualization is "true" by definition: how could those less fit for survival have been the ones that survived?

As discussed in the first chapter, there are meaningful uses of tautologies in science.[28] The mental activity of categorization and creation of definitions may often significantly assist "understanding." One important example is man's drive to categorize and name everything, which gives order "by definition". Much of early biology or botany and of chemistry was of such a nature–organization by definitions without theoretical underpinnings or pretensions.

The "best" explanation?

Is natural selection the best explanation available for the origin of species and the evolution of life? Perhaps. The concept of "survival of the fittest," even if tautological or definitional, has nonetheless enabled man to order observations, construct theories of causality and, in the process, better "understand" the world around him. The concept has had powerful applications in biology and other studies of changing, living processes. The theory has been like the spectacles that suddenly enable one to see more clearly or even to see things that were not previously observable (or like the "map" that enables one to find one's way).

But, as that metaphor suggests, the concept carries with it more than the organization of observations and the ordering of analytical processes; it carries the suggestion of a deeper understanding, or a level of truth that matters, which for humans tends to implicate normative judgments. The fact that it has been useful is relevant, but we must be conscious of the potential for misuse. (As we have seen, this problem is not unknown in other areas of science.)

In all events, we may at least seriously challenge the claim that evolution by natural selection is "the single best idea anyone has ever had... ."

Endnotes

[1] "[T]he lineages that led to modern humans and to the chimp – the living species genetically closest to humans – branched apart around six to seven million years ago which means scientists currently have no genetic information for more than 90 per cent of the evolutionary path that led to modern humans." University of Cambridge, "'Game-changing' research could solve evolution mysteries," *Research*, 11 September 2019. Interestingly, as reported, recent breakthroughs could lead to much greater understanding of evolutionary developments:

> "Researchers have now used ancient protein sequencing – based on ground-breaking technology called mass spectrometry – to retrieve genetic information from the tooth of a 1.77 million year old Stephanorhinus.... Researchers took samples of dental enamel ...and used mass spectrometry to sequence the ancient protein and retrieved genetic information previously unobtainable using DNA testing. ...[T]he set of proteins it contains lasts longer than DNA and is more genetically informative than collagen, the only other protein so far retrieved from fossils older than one million years."

The referenced paper itself reports "that proteomic investigation of ancient dental enamel—which is the hardest tissue in vertebrates[ ], and is highly abundant in the fossil record—can push the reconstruction of molecular evolution further back into the Early Pleistocene epoch..." Enrico Cappellini, Frido Welker, ... Eske Willerslev, "Early Pleistocene enamel proteome from Dmanisi resolves Stephanorhinus phylogeny," *Nature*, 11 September 2019.

[2] More controversial than this phrase, however, was Spencer's related belief that the process of evolution was progressive, biased toward gradual improvement and greater sophistication, making the eventual appearance of intelligent beings likely. It has been suggested that Darwin "flirted" with this conception of evolution in response to public pressure, but rejected it. See Ryan, *Virolution*, pp.29–30.

[3] Macbeth asserted that there is a general human tendency, particularly pronounced among evolutionary theorists, to believe that if a theory is the best available theory known, then that fact is evidence of the theory's correctness. He refers to the inclination as "the best-in-field fallacy," which

he believes he was the first clearly to identify as a logical fallacy. *Id.*, p.78. I think that, with respect to this last claim, Macbeth simply was not sufficiently versed in the history and philosophy of science. As should be clear from the earlier chapters, even the best theories are at most approximations to the truth and it is quite proper methodology to cling to the best available theory until a more promising successor is proposed.

[4] Dawkins ignores the similar types of problems, addressed above, that exist with respect to all scientific theories and explanations. If one requires a "scientific" explanation of the world, then God does not comfortably fit anywhere in the answer. However, if one concludes that reality is, in the end, simply beyond our capability to understand or that a meaningful explanation is one that results in a feeling of insight and understanding, then the religious answer may be sufficient and, indeed, the most simple.

[5] We have already examined at some length why no theory, not even the theory of natural selection, can ever be considered to have been proven, no matter how long and well established it is. As succinctly stated by Frank Ryan: "[N]atural selection will never be capable of absolute proof, since a theory is a construct in logic, not fact." *Virolution* (2009), pp.3-4. In contrast, similar or identical genetic material shared among different species is a matter of fact that can be true (or false). But, if true, how that fact came to be, is a matter of theory. As we discuss below, there are possible sources of genetic material other than common ancestors.

[6] It is likely that a successful mutation typically alters some part of the development process of the organism as the embryo grows and turns into an adult. "Evolution is not a genetically controlled distortion of one adult form into another; it is a genetically controlled alteration in a development process." Dawkins, *The Greatest Show on Earth*, p.314.

[7] *See also* Young, *Darwin's Metaphor* (1985), pp.87-125. Darwin started his analysis with the powerful results demonstrated by artificial selection, *i.e.*, intentional breeding of plants and animals by man. He went on to contemplate how something comparable might be accomplished in nature.

[8] Physicists and philosophers have long debated determinism vs. free will. For a recent review of the debate, *see* Robert O. Doyle, *Free Will: The Scandal in Philosophy* (2011). As discussed in a subsequent chapter, the twentieth century witnessed the emergence in physics of new, prominent roles for uncertainty and chance (or probability). These developments pose serious challenges to the determinist view (the view that if one

is given a full enough description of the current state of the Universe, then one could predict the future, as well as reconstruct the past, using the rules of physics). At the same time, the assumption of randomness in mutations driving evolution is also incompatible with determinism—one cannot predict the future even with complete information about the present. Of course, the presence of such randomness itself does not prove the existence of free will.

[9] Paley's argument in *Natural Theology* included the example of the eye, but the issue here is a bit different. The question is not one of design as such, but whether the process of natural selection could manage to produce something like an eye as a result of small, gradual steps. Clearly, a single one-time mutation (that is, a one-time chance occurrence) that happened to create an eye is inconceivable.

[10] Dawkins goes further, asserting that most, if not all, mutations of light-sensitive organs improved the ability of the organism to respond to its environment. See, e.g., *The Blind Watchmaker*, pp.82–6.

[11] Interestingly, the construction of the eye is one of the examples of bad design that may seem to be explainable only as a result of incremental, cumulative evolution of the organ. The connections of the photocells to the optic nerve run outward, through the retina (creating a blind spot and presumably considerable interference with the light waves coming into the eye), rather than inward away from the light source. See Dawkins, *The Blind Watchmaker*, p.93. An intentional design by a competent Designer would presumably not have included such a mistake. If it is a "mistake". Although, the image is received upside down by the brain; there may be unidentified advantages in such a physical arrangement. And, the extra burden put on the brain is well within its abilities to handle.

[12] The burying-beetles, ubiquitous in the fens of East Anglia, will locate the carcass of a small rodent like a field mouse. The male and female bury the carcass, preparing it over 48 hours. The preparation involves the removal of the fur (shaving it) and continuously rolling it until if resembles a ball of flesh covered by secretions from the beetles. The female then lays her eggs in an indentation created in the ball. The larvae and parents feed on the carcass. After about 5 days, the male departs. Some 5 days later, the female departs. The secretions that cover the carcass contain some antimicrobials, but it turns out that the secretions result in an increase, not a decrease, in the bacterial load of the carcass. However, the make-up of the bacterial population is changed. Of the dominant types of bac-

teria that appear covering the carcass, one seems to come from the beetles themselves and the other from the gut of the mouse, which the beetles eat, consuming and then excreting the bacteria. The benefits (if any) to the beetles and their larvae of the resulting bacterial load is still under investigation. Lecture by Rebecca Kilner, of the Department of Zoology, Cambridge University, entitled "Social evolution in a grave," 4 March 2014, sponsored by the Trinity College Science Society.

[13] At the same time, there are important examples of physical characteristics that may have arisen by enabling one function but then provided a stepping stone to something quite different. See *below*, the discussion in the section entitled "The Inevitability of Intelligence" in the final chapter.

[14] I note, however, that there are several different species of baboons that have significant physical differences and tend to occupy different, but overlapping geographical regions. They are able to and often do interbreed, producing hybrid offspring. But, the characteristics of the individual species and the separate groups have strongly tended to persist (for millions of years). *See* Carl Zimmer, "Christening the Earliest Members of Our Genus," *The New York Times*, October 24, 2013.

[15] Matt Ridley notes that,

> *"[f]or insuperable practical reasons connected with the pairing of chromosomes during meiosis, cross fertilisation cannot happen between different species of animal. (It can, indeed does, happen between species of bacteria, 80 percent of whose genes have been borrowed from other species on average—one reason bacteria are so darned good at evolving resistance to antibiotics... .)"*

The Rational Optimist, p.271.

Dawkins explains that for individuals within the same species, the DNA has the same number of chromosomes and the locations along those chromosomes have the same addresses. *The Blind Watchmaker*, p.118.

[16] *See, e.g.*, lion_roar.tripod.com/liger_tigon.html, which lists the various hybrid combinations of large cats that have been successfully bred. Apparently, the liger is able to reproduce. It is also considerably larger than either the lion own tiger but tends to have health problems and a shorter lifespan.

[17] More recent evidence suggests that the Neanderthal line died out in Europe even earlier than was believed. Kenneth Chang, "Neanderthals in Europe Died Out Thousands of Years Sooner Than Some Had Thought, Study Says," *The New York Times*, August 20, 2014. The Neanderthals are thought to have emerged about 200,000 years ago, living in Europe and Asia. About 60,000 years ago, modern humans appear to have migrated out of Europe. Neanderthal remains found in Spain in 1995 were dated at 30,000 years ago; but, more recent radiocarbon dating using radioactive carbon 14 has indicated that the date of the remains found at various sites in Europe were some 10,000 years older than original dating has indicated. Id. So, the interbreeding may have occurred 50,000 to 60,000 years ago, and modern man may have rather rapidly displaced Neanderthal man (over some 20,000 years or so).

[18] Christopher Stringer of the Natural History Museum in London has reportedly observed that there is no evidence of interbreeding in Europe, so the interbreeding may have occurred in Asia with the genes subsequently brought to Europe by modern humans from Asia. Kenneth Chang, "Neanderthals in Europe Died Out Thousands of Years Sooner Than Some Had Thought, Study Says," *The New York Times*, August 20, 2014.

[19] Of course, the interpretation of the fossil record is always challenging. For example, it was reported in October 2013 that after many years of study of a skull found in the Republic of Georgia, David Lordkipanidze, a paleoanthropologist from Georgia, concluded that there may have been far fewer separate species in the descent of *Homo sapiens* than has been thought. John Noble Wilford, "Skull Fossil Suggests Simpler Human Lineage," *The New York Times*, October 17, 2013 (reporting on a study published in the journal *Science*). The conclusion was achievable because the scientists were able to study five different skulls from some 1.8 million years ago all found at the Dmanisi site. The skulls reflected differences that may well have been thought to represent different *hominid* species had they been found in disparate locations. In fact, the range of variety found was consistent with the differences identified among fossils that researchers had previously decided belonged to different *hominoid* species. *See also*, Carl Zimmer, "Christening the Earliest Members of Our Genus," *The New York Times*, October 24, 2013.

[20] Indeed, bacteria and other micro-organisms are among the tiny particles that seed the formation of crystalline structuring of water that create snowflakes and ice crystals as the first stage in most rainfall, as well as all snow and hail. Bacteria that are able to withstand the intense ultra-

violet light and lack of nutrients in the upper atmosphere could travel throughout the world in airborne colonies. Presumably, these bacteria affect weather patterns, so changing the microbiome of a region (for example, through agriculture) may affect the weather in that region, or elsewhere. Ferris Jabr, "It's Buggy Out There," *The New York Times Magazine*, February 13, 2015.

[21] With a more limited geographical area or relative homogeneity, it may be possible to estimate the mass of ants with some degree of confidence. For example, one recent study in a Brazilian rain forest reportedly estimated that the total mass of the ants living there was four times the mass of all the animals in the same area. See Carl Zimmer, "Key to Ants' Evolution May Have Started With a Wasp," *The New York Times*, October 20, 2013.

[22] The current revised version of the diagram does not resemble a tree or bush. See Dawkins, *The Greatest Show on Earth*, pp.315–30; Collins, *The Language of God*, p.128. This change, however, does not ameliorate the problem discussed in the text–if anything, it makes it even more puzzling.

[23] New, transitional fossils continue to be found and presumably will continue to be. For example, scientists report the discovery of a 2.8 million year old jawbone (and five teeth) of a *Homo habilis*, predating any previously identified specimens by 400,000 years. See, e.g., John Noble Wilford, "Jawbone Fossil Fills a Gap in Early Human Evolution," *The New York Times*, March 4, 2015. The famous Lucy, of the species *Australopithecus afarensis*, found in the 1970s, was the last known apparent link between the apes and the human genus, *Homo*. Lucy lived about 3 million years ago and, like this new specimen, was found in the current Ethiopia. The original specimens of *Homo habilis* were found by the Leakeys in Tanzania. They were dated to 1.8 million years ago. So, we seem to be making some progress in closing the 1 million year gap during which the genus *Homo* appeared. Id.

[24] The study published in *Genome Biology* analyzed the genomes of multiple species of fruit flies, nematode worms and primates, including humans. The researchers compared the genes from one species to the similar genes from other species to estimate the likelihood that the gene was foreign and when it was acquired.

[25] Carey continues:

> "Critics contend that the biology implied by such studies simply is not plausible. Epigenetics researchers counter that their evidence is solid, even if the biology is not worked out. 'These are, in fact, extraordinary claims, and they are being advanced on less than ordinary evidence,' said Kevin Mitchell, ...[of] Trinity College, Dublin. 'This is a malady in modern science: the more extraordinary and sensational and apparently revolutionary the claim, the lower the bar for the evidence on which it is based, when the opposite should be true.'"

[26] The modern version of the theory of endosymbiosis is attributed to Lynn Margulis. Born Lynn Alexander, she married cosmologist Carl Sagan in 1957. That marriage ended by 1965. She later married Thomas Margulis. Her publications appeared as by Lynn Sagan and then Lynn Margulis. Several of her later writings were co-authored with her son Dorion Sagan. "Lynn Margulis obituary," *The Guardian*, 11 December 2011. She argued that much of the evolution of life came not from competition among life forms, but from cooperation.

[27] These two paragraphs are based upon a lecture given at Cambridge University on 2 March, 2014, by Nick Lane, of the Department of Genetics, Evolution and Environment, at University College London, entitled "Energetic Constraints on the Evolution of Life", sponsored by the Cambridge Biological Society.

[28] Norman Macbeth, the other retired lawyer discussed above, put considerable emphasis on his realization that survival of the fittest was a tautology. *Darwin Retried*, pp.62-6. In addition, he expressed surprise that some highly respected biologists had acknowledged the circularity but were not more bothered by it. *Id*. I think that Macbeth did not fully understand the role of tautologies (and deductive reasoning) in the sciences.

CHEMISTRY

"[A]ncient magic was mostly chemistry."

Theodore Gray
Reactions, p.5.

LEAD TO GOLD

The word chemistry derives from alchemy, which today conjures up images of black (or, maybe, white) magic. Indeed, much of the better-known alchemists' endeavors were focused on trying to transform base metals, like lead, or other ordinary and abundant compounds, into gold. They sought to rediscover the lost "Philosopher's Stone", which would enable such transformations, as well as provide immortality. The image of chemistry as science was not improved by the frequent use of chemical reactions in public demonstrations in the eighteenth and nineteenth centuries, phenomena generally perceived as a result of magic. See Theodore Gray, *Reactions: An Illustrated Exploration of Elements, Molecules, and Change in the Universe* (2017), pp.8-19. (See also, the section entitled "The paranormal," in the following chapter on Physics.)

But, alchemists also performed real science. See Richard Conniff, "Alchemy May Not Have Been the Pseudoscience We All Thought It Was: Although scientists never could quite turn lead into gold, they did attempt some noteworthy experiments," *Smithsonian Magazine*, February 2014 (but, "obsessed with secrecy, [they] deliberately described their experiments in metaphorical terms laden with obscure references to mythology and history").[1] Much of seventeenth century chemistry was built on their work. Gray, *Reactions*, pp.5-7, 20-1. Indeed, Newton was a practicing alchemist (as well as a theologian and physicist). And, so was Robert Boyle (in fact, he got his start in chemistry by learning alchemy). Jane Bosveld,

"Isaac Newton, World's Most Famous Alchemist: For centuries some of the world's greatest geniuses struggled in secret to turn base metals into gold. In a sense they succeeded: In their restless quest, they unlocked some of nature's greatest secrets," Discover, December 27, 2010.

Robert Boyle was one of the founders of modern chemistry; but, he was more. He is perhaps best known for Boyle's Law, stating that as pressure increases, the volume of gas decreases proportionately, set forth in 1660 and 1662. He also pioneered experimental science:

> "At th[e] time even the idea of an experiment was controversial. The established method of 'discovering' something was to argue it out, using the established logical rules Aristotle and others had worked out 2,000 years before. Boyle was more interested in observing nature and drawing his conclusions from what actually happened. He was **the first prominent scientist to perform controlled experiments and publish his work with details concerning procedure, apparatus and observations.**"

"History: Robert Boyle (1627-1691)," BBC.co.uk, 2004 (emphasis added).

> "Chemistry ... escaped from its superstitious roots and branched from physics during the last quarter of the eighteenth century due to the work of a group of French scientists, most notably Antoine Lavoisier. ...Recognition as a separate branch of science came with the establishment of its own learned societies, like the Chemical Society of London in 1841."

John Hands, Cosmosapiens: Human Evolution from the Origin of the Universe, p.519.

The challenges for early chemists were enormous, and it is not surprisingly lhat they were largely insurmountable. What we later came to perceive as the basic building block of matter—atoms—were far, far too small to be detectable, even indirectly, until the nineteenth century. Moreover, what we now recognize to be the elements were (and are) rarely found on their own. Indeed, most matter that we encounter and can observe consists of compounds (molecules) of two or more elements (atoms) chemically bound together. See, e.g., Tim James, Elemental: How the Periodic Table Can Now Explain (Nearly) Everything (2019), p.9. Moreover, "[n]o matter how skillfully the two giants of 17th-century science manipulated the red earth and set their sights on the Philosophers' Stone, they would have failed to make gold. We know now that such a transformation requires not a chemical reaction but a nuclear one, far beyond the reach of the technology of the time." Bosveld, "Isaac Newton, World's Most Famous Alchemist," *Discover*, December 27, 2010.

Some of the few elements that do exist naturally in pure form—like gold, silver, and copper (and even some nickel)—were identified, extracted and used some 5,000 thousand years ago. (The extraction of iron requires relatively high temperatures, above 1500° C, so its use came later.) The gases that can be found naturally, like oxygen and nitrogen, generally appear in mixtures. Scientists had no means of separating such gases from one another until the eighteenth century.

Mixtures differ from compounds in that the atoms or molecules of the particular elements (or compounds) exist independently, just mixed together, like the air of the earth's atmosphere (roughly 80% nitrogen and 20% oxygen). In contrast, with compounds the atoms are chemically bound together into molecules that normally have quite different physical characteristics than the elements from which they are made.

Very many of the discoveries in chemistry before the twentieth century (and a fair number thereafter) occurred by accident; another significant number were achieved through pure experimentation, by which I mean plain-old "trial and error". Thus, up until the twentieth century, chemistry was essentially an inductive science, proceeding by observation and the keeping of records of what happened. The use of deductive reasoning awaited the development of quantum mechanics in the early twentieth century.[2]

By the way, the first successful transformation of one element into another was achieved by Irène Joliot-Curie (the daughter of Marie Curie) and her husband in January 1934 by bombarding aluminum film (element 13) with alpha particles, creating (radioactive) phosphorous (element 15) (because the additional protons in the alpha particles are "absorbed" by some of the aluminum atoms). See Sean Kean, *The Bastard Brigade: The True story of the Renegade Scientists and Spies who Sabotaged the Nazi Atomic Bomb* (2019), pp.32-5 ("It was artificial radioactivity—scientific alchemy").

In addition, in late 1938. Otto Hahn, attempting to disprove an experimental result reported by Irène Joliot-Curie (and assisted in interpreting his results by Lise Meitner), discovered that by bombarding uranium (element 92) with neutrons, he could "split" the uranium atom and create two other elements, lanthanum (element 57) and barium (element 56) (a process named nuclear "fission"). This discovery lead to an enormous increase in the interest in uranium and, ultimately, to the atomic bomb. *Id.*, pp.48-55. However, efforts to turn other elements into gold were totally unsuccessful until the twenty-first century, and the recent achievement of the "creation" of gold from something else certainly does not have very promising commercial prospects! See James, *Elemental*, p.94.[3]

ATOMS AND THEIR PARTS

I would guess that most of us have the impression that scientists have long accepted the view that matter is made up of molecules and molecules are made up of atoms, which in turn consist of nuclei and electrons. In fact, however, the emergence of the modern theories of the atom and its structure and behavior are essentially contemporaneous with the development of Einstein's theories of relativity and the appearance of the theories of quantum mechanics in the first half of the twentieth century. Moreover, even today, the components of the atom and its structure are "known" only through theories that appear to make sense of the observed consequences or effects of those components (such as electrons, protons, neutrons). See, e.g., Chris Baraniuk, "How do we know that things are really made of atoms?" BBC *Earth*, 20 November 2015. The detection of evidence of these components required the technological advances of the twentieth century.

THE STUFF OF THE UNIVERSE

The ancient Greeks (specifically, Democritus) hypothesized that matter ("the stuff of the Universe") was made up of indivisible or elemental things called "atoms." Greene, *The Elegant Universe*, p.7. But, there have been many theories over the years as to what those elementary, indivisible entities were. Boyle advocated "corpuscularism," a form of atomism presented in terms of particles and their motion. "Robert Boyle: Known for his law of gases, Boyle was a 17th-century pioneer of modern chemistry," *Science History Institute*, December 1, 2017.

A relatively complete atomic theory was put forward by John Dalton between 1801 and 1805. See James, *Elemental*, p.20. And, during the nineteenth century, chemists became increasingly interested in the theory of atoms. The demonstrated fact that gases could be compressed only so far suggested that there were "particles" making up the gas of some physical size that could not be reduced by additional pressure. So, atoms, if they existed, must have had some physical dimensions. *Id.*, p.21.

In addition, physicists were considering that the movement of atoms perhaps could explain the phenomenon of heat. We have already discussed Maxwell and the theory of electromagnetic fields. The concept of light as waves of radiation fit comfortably with the discovery that atoms were in constant random motion, described by statistical techniques. The motion of the atoms, which would increase when material was heated, seemed to be a credible source for the resulting light waves. See Seife, *Zero*, pp.163–5.

But, the existence of atoms was still not fully accepted. The modern theory of the atom gained broad acceptance only in the early twentieth century, following the confirmation of a hypothesis set forth by Einstein (discussed in the next section).

The evidence of the existence of atoms was all indirect or circumstantial until the 1980s, when the newly developed electron microscope enabled the first direct observation of an atom. To observe a sample, the ordinary microscope uses light. The degree of resolution depends upon the "size" of the imaging medium (this case, light). Because of the wave length of light, anything smaller than a bacteria will be blurred. Atoms are much too small to be seen.

The electron microscope uses electrons; the synchrotron uses high-intensity x-rays produced by the acceleration of electrons to close to the speed of light. Currently, crystallography and x-ray technologies enable scientist to get "10,000 times the resolution of the normal light microscope," according to Dave Stuart, the life sciences director at the UK national synchrotron facility called the Diamond Light Source. Jonathan Amos, "Diamond to shine light on infections," *BBC News*, 17 February 2013.[4]

The even newer atomic force microscope (the "scanning tunneling microscope" or "STM"), in which a tiny probe moving across (or just barely above) the surface of the object being examined records changes in the chemical structure of the surface, enables the "photography" of individual molecules in which the constituent atoms can be "seen." It does so by dropping tiny particles and measuring how long they take to reach the surface at each point of the scan, creating a map of the surface from the indirect evidence of the "soundings" (similar to the technique used to map the ocean floor with sonic readings). See James, *Elemental*, pp.27-8. The atoms so mapped appear as small, solid beads or balls.

BROWNIAN MOTION AND EINSTEIN

What became known as Brownian motion, the movement of grains of pollen—and any other tiny matter suspended in water—that is observable through a microscope, was named after Robert Brown, a contemporary of Charles Darwin, who made extensive observations of the motion of pollen grains in 1827. His curiosity about the source of the constant movement led him to examine whether the same phenomenon occurred with materials that were dead and, even, with materials that were inorganic. He observed that the motion was always present, demonstrating that it could not be the result of the presence of life in the pollen grains.

In 1863, Ludwig Christian Wiener proposed that the constant movement of atoms in water could explain the erratic movement of larger particles that could be observed—those larger particles moved around as they were bombarded by the invisible atoms or molecules of the liquid in which they were suspended. See David Lindley, Uncertainty (2007), pp.9-17.[5]

The firm establishment of the atomic theory came as a result of one of Einstein's four historic papers published in 1905. Based upon complicated mathematics, Einstein explained how one could conduct an experimental test to determine whether the movement of a suspended particle is what one would expect if the movement is attributable to the bombardment of the particle by atoms in constant motion. In 1908, French physicist Jean Perrin provided the empirical demonstration that such particles in fact behaved as predicted by Einstein's theory. Id., pp.27-29.

Of course, the random motion of atoms raises the question of how matter composed of such atoms can still behave according to the laws of classical (Newtonian) physics. The answer is the so-called "law of large numbers".

As Erwin Schrodinger explained in a well-known lecture:

> "...we know all atoms to perform all the time a completely disorderly heat motion ... [that] does not allow ... a small number of atoms to enrol [sic] themselves according to any recognizable laws. Only in the co-operation of an enormously large number of atoms do **statistical laws begin to operate and control the behaviour of these assemblies with an accuracy increasing as the number of atoms involved increases**. It is in that way that the events acquire truly orderly features. All the physical and chemical laws that are known to play an important part in the life of organisms are of this statistical kind... ."

"What is life? The Physical Aspect of the Living Cell" (1944).

THE DIMENSIONS OF THE ISSUE

The mysteries of the atom become even greater than one might initially imagine when one confronts the actual physical dimensions of the atom and its constituent parts. A relatively typical atom has a diameter of 0.00000008 centimeters. Bryson, *A Short History of Nearly Everything*, p.139. ("If you wanted to see the atoms in [a drop of water], you would have to make the drop 24 kilometers across." Or, an atom is to a millimeter, "as the thickness of a sheet of paper is to the height of the Empire State Building." *Id.*, p.177.) Analogies are amusing, but hardly adequate to convey the reality.[6]

And, although, an atom is astonishingly, incomprehensibly small by the standards of our senses; the protons and neutrons making up the nucleus and the electrons orbiting the nucleus are much, much smaller still and the spaces between the nucleus and the electrons are enormous relative to the particles themselves. "If you imagine a nucleus scaled up to the size of a tennis ball, then the tiny electron would be smaller than a mote of dust orbiting at a distance of a kilometer." Cox and Forshaw, *The Quantum Universe*, p.9. It has been said that if one could stand on the nucleus and look outward, the orbiting electrons would not even be visible. *Id.*, p.91.

Thus, the atom itself is almost entirely (99.9999999999999%) empty space. Andy Parker, Professor of Physics, University of Cambridge, presentation to the Trinity College Science Society, 11 March 2014 (if the atom were the size of the Millennium Dome in London, the nucleus would be about 3 millimeters in diameter).

THE PARTS

The discovery of the existence of atoms was only the beginning. The obvious next question is are atoms really "indivisible" as their name suggests? If not, then of what do atoms consist?

The first constituent particle that was identified was the electron. At the end of the nineteenth century, J.J. ("Joseph John") Thomson, head of Cambridge's Cavendish Laboratory, established that vacuum (cathode ray) tubes emitted streams of electrically charged particles, not rays, which particles came to be called electrons. They have a negative electrical charge. Why and how, and what the charge is, we do not know. James, *Elemental*, p.29.

A few years later, Ernest Rutherford, using an experiment with gold foil, found that there were solid "particles" contained within gold. Gold, because of its malleability, could be flattened into a foil only a few atoms thick (think of "gold leaf"). Rutherford and colleagues (including Hans Geiger, of the Geiger counter) fired alpha particles at such a gold foil and used detectors to record where the particles went after penetrating the foil. Most went straight through, a few veered off wildly, and a very few bounced straight back toward the source. The only logical explanation was that there were widely scattered solid spots inside gold atoms that the alpha particles could not penetrate. Rutherford called those spots "nuclei". *Id.*, pp.30-3. (In this chapter and the next two, we intrude a bit on nuclear physics and atomic physics—the physics of the nucleus and the physics of the atom—fields of study beginning with the work of Thomson and Rutherford at the dawn of the twentieth century.)

Subsequently, scientists concluded that the nucleus of the atom consisted of protons and neutrons. The protons have a positive electrical charge, so one would expect them to repel one another. However, it appears that neutrons, with no charge, assist in the bonding together of protons within the nucleus, creating what is called the "strong nuclear force." In addition, electrons have a negative charge. So, for an atom to be electro-statically neutral, it would have to have an equal number of protons and electrons. The number of neutrons is discussed later.

The strength of the bonds differ by many orders of magnitude:

> *"The electrons that whirl around the nucleus are held together in their orbits by electrical forces. It takes on the order of a few electron-volts to dislodge an electron from the outer shell of an atom. The 'binding energy' of a nucleon is on the order of a million times greater. ... The huge differences in binding energy are one measure of the differences in the quantities of energy derived from nuclear compared to chemical reactions."*

"Basics of Nuclear Physics and Fission," *Institute for Energy and Environmental Research*, May 2012.

The nature of these bonds is an important, but separate, subject, discussed briefly in a later chapter (Particle Physics and Quantum Mechanics).

ELECTRONS: "ORBITALS" AND SHELLS

In about 1914, Niels Bohr proposed an atomic theory in which electrons orbited the nucleus but did so only at specific, discrete levels of energy, reflecting discrete possible orbits. The only allowable orbits were those in which the electron had an angular momentum equal to some "exact multiple of Planck's constant" (described in a later chapter). Barbour, *The End of Time*, p.188.

At this stage, in what was called the Rutherford-Bohrs model, it was still presumed that the electrons in orbit behaved according to the rules of classical mechanics, but that for some unknown reason only certain levels of energy or speeds were "permitted." *Id.*, pp.60-61. The orbits of electrons were, in effect, "quantized"—only specific orbits were possible, so electrons jumped from one to another in discrete intervals, rather than spiraling upward or downward. Cox and Forshaw, *The Quantum Universe*, pp.10-12.

The theory also embodied the rule that electrons jump from one level to another without ever being in between. They would disappear from one "place" and instantly reappeared somewhere else, so to speak. *See, e.g.*, Bryson, *A Short History of Nearly Everything*, p.186. (This difficult to grasp concept would continue to reappear, in various forms, as quantum theory evolved, discussed in a later chapter.)

Bohr's theory explained why atoms did not normally collapse as negatively charged electrons spiraled downward toward the positively charged nucleus. If the electrons could only appear in certain "orbits," then the atom would be stable.

> "The hydrogen atom with the electron as close as it can be to the proton is said to be in its 'ground state' and it is as light as it can be. Add just the right amount of energy and the electron will jump up to the next available orbit and the atom will become a bit heavier, simply because a bit of energy has been added. ...the heavier version will eventually shed some of its mass. It does so by emitting a single particle of light, the photon we met earlier. For example, a next-to-lightest hydrogen atom will at some point spontaneously convert into a lightest hydrogen atom as a consequence of a change in the orbit of the electron. The excess energy is carried away by a photon."

Cox and Forshaw, *Why $E=mc^2$*, pp.150, 151.

Spectroscopy

Bohr's theory also explained spectroscopy, which was already well established by the beginning of the twentieth century. Light through a prism creates a rainbow of colors, because the prism separated white light into its constituent frequencies which correspond to various colors (as we perceive them). Chemists had shown that chemical elements absorb, and emit, light only at specific wavelengths, matching the particular transitions of electrons from one orbit to another in those elements. Lindley, Uncertainty, pp.47-53. That phenomenon can be displayed visually by the use of a prism.

> "All of the atoms in nature come in a tower of energies (or masses), depending on where the electrons are, and since there is more than a single electron in every atom except hydrogen, the light emitted from them spans all the colors of the rainbow and beyond, which is ultimately the reason why the world is so colorful."

Cox and Forshaw, Why $E=mc^2$, p.152

The spectrum of a particular element when heated (the light frequencies being emitted) when examined with a spectroscope will consist of strips of various colors of the rainbow in varying intensities with a host of back spaces where particular frequencies are not generated. Each chemical element has its own distinction pattern of color lines and back spaces. Gray, Reactions, pp.124-5, 130-2.

As a result, spectroscopy enables chemists to identify the chemical components of materials that were heated and producing light. Some elements (such as hydrogen, copper and aluminum) emit only a handful of colored lines; others (such as neon, titanium and nickel) emit a large number of stripes across the rainbow. See Gray, Elements, pp.14, 33, 39, 59, 75, 77 (displaying the spectrum of each of the elements). In addition, some elements emit lines primarily within a narrow part of the spectrum, in which case they will be perceived as glowing a certain color corresponding to those frequencies when heated.

Because elements also absorb frequencies of light in unique patterns, spectroscopy enables astronomers to identify the elements contained in the atmospheres of planets. Id., pp.130-1. The patterns of absorption and emission lines identifiable through spectroscopy (and shifts in the patterns) contain an enormous amount of information (discussed in the chapter Cosmology).

THE PAULI EXCLUSION PRINCIPLE

"A hypothesis advanced by Wolfgang Pauli in 1925 not only explained how every atom of the same element is the same but also made predictions subsequently confirmed by experiment that laid the foundations of our current understanding of chemistry." John Hands, *Cosmosapiens: Human Evolution from the Origin of the Universe* (2016), pp.134-5. Yet,

> "....no theory explains why this should be so ...[;] it has nothing to say about the behaviour of an individual electron but only applies to a system of two or more electrons. ...[;] it makes every atom of the same element identical; and it dictates how an atom can bond with other atoms... [,] explains the phase of an element—how it can be gas, liquid or solid, and if the latter, metallic or crystalline—and the periodic table... ."

The Principle states that no two electrons in an atom or molecule can have the same four quantum numbers.[7] This Principle reduces dramatically the number of configurations that are possible.

QUANTUM MECHANICS

Quantum mechanics portrays various groups of orbitals of electrons as having different shapes. The shape of the first level is referred to as "spherical" or "s". It is presumed that the one or two electrons on that level can "exist" any place in a symmetrical sphere surrounding the nucleus. There are seven possible spherical sizes, one for each of the seven shells, each one further from the nucleus than the prior one. The next shape, called "p" orbitals, are portrayed as bows or figure 8s, with the nucleus at the center. The two loops are on opposite sides of the nucleus. There are three versions of p orbitals, one for each axis. Two other orbital shapes are called "d" and "f", with 5 different "d's" and 7 different "f"s", portrayed as bundles of balloons. See Gray, *Elements*, p.12.

FILLING THE SHELLS

Electrons occupy levels or "shells" around the nucleus. These levels are often called "orbitals", but I shall try avoid that term because it reenforces the misconception of the Solar System analogy (discussed below). I shall simply refer to energy levels. The lowest energy level is portrayed as closest to the nucleus, and the highest energy level is the furthest away. Each level can contain only two electrons (one with an "up" spin snd one with a "down" spin), so as the atom becomes bigger (more protons and neutrons), there must be more energy levels to contain the additional electrons.

The lower energy levels are the more stable, so when it becomes possible, an electron from the next higher energy level will "jump" to the lower level. This event is accompanied by the generation of heat and light (the release of energy). (I will avoid the usage of phrases like "the electron wants to be closer to the nucleus", since such anthropomorphizing of the electron is highly misleading.) The positions of the electrons store potential energy. When an electron moves to a lower potential energy orbit, kinetic energy is released. Gray, *Reactions*, p.47.

It appears that there are, at least, 7 shells of electrons, with each shell after the first having several energy levels. The first shell can hold 2 electrons; the second shell, 8 electrons; the third shell, 18; the fourth shell, 32; and, theoretically, the fifth, sixth and seventh shells, 50, 72 and 98. However, the most electrons to have been identified in any shell is 32. See, "Electron shells and orbitals," *TechnologyUK.net*.

For the first 18 elements (up through the first three shells), additional electrons (balancing the additional protons) fill the orbitals (and the shells) strictly in order of increasing energy level. Then, starting with potassium, something strange happens. Before the fourth shell is filled, the additional electrons begin to fill the lowest orbital of the next shell. Above atomic number 18, each time that the outer-most shell reaches 8 electrons, the next shell receives electrons before that shell is filled further. But, the fifth, sixth and seventh shells never exceed 32, 18 and 8 electrons, respectively. Why? Here is the explanation:

> *"The reason for this seeming anomaly is that the energy level of the s orbital in electron shell 4n (designated as the 4s orbital) is actually slightly lower than the energy level of the d orbitals in electron shell 3n, and we have already seen that electrons will (usually) occupy the orbitals with the lowest energy levels first (just to make things more complicated, the energy level of the 4s orbital increases when the 3d orbitals start to fill up, ...)."*

"Electron shells and orbitals," *TechnologyUK.net*.

As just described, the outer-most shell of any atom will contain between 1 and 8 electrons, no more. Elements with the same number of outer electrons generally share certain characteristics, as described below.

Interestingly, as noted above, all electrons are the same. Exactly so. Utterly indistinguishable. The same is true for protons and neutrons. It is even true for atoms of the same element (ignoring isotopes). In other words, unlike most things we could encounter or observe by our senses, atoms of a particular element are all truly and completely indistinguishable. (And, atoms of different elements are necessarily different, but identical to one another.) Supposedly, this sameness is true across the Universe (or, at least, it is assumed to be. so). See, e.g., Dawkins, *The Greatest Show on Earth*, pp.92-3. Moreover, all of these particles are also virtually eternal (with some caveats).

A NOTE ON COLOR

As earlier noted (see n.1, Knowledge and Understanding), when we see something of color that is not itself emitting light, we perceive the color because the object in question absorbs greater percentages of the frequencies of white light that do not correspond to the color we see than it does of the frequency we "see." In effect, something has a particular color because of the relative absence of other colors in the light it reflects. Gray, Reactions, pp.123-4. For such materials, the "primary" colors are red, yellow and blue (actually, the pure primary colors are cyan, magenta and yellow). Those three pigments can be combined to create the other colors we recognize (e.g., yellow and blue makes green), by subtracting light frequencies (not reflecting them) from what we see. All three combined in equal amounts will produce black—no light reflected. However, from the standpoint of additive light, the "primary" colors are blue, green and red. "When he was 23 years old, Isaac Newton made a revolutionary discovery: By using prisms and mirrors, he could combine the red, green and blue (RGB) regions of a reflected rainbow to create white light. Newton deemed those three colors the 'primary' colors since they were the basic ingredients needed to create clear, white light." Michelle Konstantinovsky, "Primary Colors Are Red, Yellow and Blue, Right? Well, Not Exactly," science.howstuffworks.com, July 2, 2019.

The Solar System Analogy

Atoms are so small that it is a little meaningless to ask what they "look like." However, it is tempting to try to conceptualize a model of an atom. The "most famous model was proposed by the great Danish physicist Niels Bohr" and "is a miniature solar system." Dawkins, *The Greatest Show on Earth*, p.92. See also, Greene, *The Elegant Universe*, p.7; Cox and Forshaw, *The Quantum Universe*, pp.8–9. So to the average school-age child, an atom is a system of electrons orbiting around a nucleus. (As we shall see, current models of the atom are quite different and much less easy to visualize.)

However, there are several pretty obvious and difficult questions not answered by the analogy:

- Given that the nucleus has a positive charge and the electrons have negative charges, one would expect a powerful attraction that would tend to lead to the "collapse" of the atom, like the anticipated disintegration of planets as their orbits began to diminish. What prevents that collapse and provides the obvious stability of the atom?
- Given that protons have positive charges and, so, will necessarily repel one another, how can several protons be tied together in the nuclei of more complex atoms?
- We know that matter is not readily compressible. What prevents two atoms from being pressed into the same space, as would presumably happen if two solar systems were to collide?
- Lastly, light passes readily through the Solar System, but much (though not all) of matter is opaque, where light just bounces off the "surface". Why is that?

See Dawkins, *The Greatest Show on Earth*, p.92.

Moreover, the "image" that the analogy invokes is just not accurate. The one real contribution of the Solar System analogy, however, is that it gives some indication of the vast emptiness of the atom. See, *e.g.*, James, *Elemental*, p.35.

CHEMISTRY

THE ELEMENTS

The early Greeks attempted to identify the basic elements from which everything else could be constructed. Each of several philosopher had his own favorite, but ultimately the Greeks agreed on a compromise proposed by Empedocles in the fifth century BCE: "earth, water, air and fire". James, *Elemental*, pp.52-3.

Of course, they were wrong as to each and every one.

> *"Of the 95 elements, 8 make up more than 98 per cent of the mass of the Earth's crust, which consists mainly of oxygen (47 per cent), silicon (28 per cent), aluminium (8 per cent), and iron (5 per cent); the oceans comprise mainly oxygen (86 per cent) and hydrogen (11 per cent). While we consist of 41 elements, 99 per cent of the mass of an average human body comprises just 6 elements: oxygen (65 per cent by mass), carbon (18 per cent by mass), hydrogen (10 per cent by mass), nitrogen (3 per cent by mass), calcium (2 per cent by mass), and phosphorus (1 per cent by mass)."*

Hands, *Cosmosapiens*, p.132.

STRUCTURE OF ELEMENTS

Unlike every chemistry book that you will encounter, I am not going to describe the periodic table (nor reference Avogado's number, except in ftn.6). I will (and did in the prior section), however, discuss some of the principles upon which the table is constructed.

Atoms contain equal numbers of protons and electrons, so the charges are balanced. Protons come in one size only, and there are no half or quarter protons. So, the smallest atom has one proton and one electron. It is the element hydrogen, with an atomic number of one (the atomic weight is the numbers of protons and neutrons.) It is easy to predict the possible elements. Just keep adding one proton and one electron (and some neutrons). The potential atomic numbers are 1, 2, 3, ... 100, etc. No half sizes. The only question is how big can you go. Naturally, along the way, scientists wondered whether every number had a related element that did or could exist. The answer turned out to be yes, at least up to 118.

After hydrogen, all elements also contain electrically neutral particles called neutrons. Generally, there are the same number of protons and neutrons, giving an atomic weight of twice the atomic number. But, some elements can have more neutrons than protons, forming isotopes, some of which can be radioactive. For example, carbon has 6 protons and 6 electrons. The isotopes carbon 12 has 6 protons and six neutrons; carbon 13 has 7 neutrons but is also stable; carbon 14 has 8 neutrons and is radioactive. The atomic weight of such elements is the average of the atomic weights of all isotopes.

Curiously, although I have referred to "smaller" and "larger" atoms, that reference is to the number of protons in the nucleus (the atomic number). But, atoms also take up a certain amount of space (they have particular diameters). And, it happens that bigger atoms (by number) can be smaller (in diameter), and smaller atoms can be larger. For each of the seven shells of electrons, the larger diameter appears first; then as the atom gets "bigger", it gets smaller until the next shell of electrons is utilized. See Gray, *Elements* (listing the diameters of the atoms of each of the elements). (Remember, of course, that although atoms take up space, they are themselves virtually empty.)

The atom with the largest diameter is number 55, cesium (or "caesium"). *Id.*, pp.12, 130-1. It is the most reactive of the metals, explosively so with water, and becomes a liquid at 28.5° C.

There today are 118 known elements, of which only the first 92 (hydrogen through uranium) occur naturally. The remainder are synthetic or man-made and several are very, vey short-lived. In fact, some of these synthetic elements have been produced only in such minuscule quantities that their characteristics have not been ascertained.

CHARACTERISTICS OF ELEMENTS

Not surprisingly, various groupings of elements have relatively common characteristics, such as conductivity (conductors, nonconductors and semiconductors), reactivity and acidity. But, the similar elements are not the elements next to each other in a listing by atomic number. More surprisingly, the "similar"elements can be grouped on the periodic table (vertically, actually). The reason is that the elements display a form of periodicity (hence, the name "periodic table"). As a result, it has been possible to predict certain characteristics of as then undiscovered elements.

The periodicity is a result of the seven shells of electrons described in an earlier section. Certain characteristics of an atom tend to depend upon the state of the outer-most shell of the atom. If that shell is filled or has 8 electrons, the atom will be nonreactive. If it is partially empty (especially with one empty space), the atom will be reactive (receptive to combinations). So, hydrogen (one proton and one election, with one empty space) readily combines with many other elements, while helium (two protons and two electrons, with no empty space) does not combine with anything.

The elements very dramatically in terms of reactivity. The relatively nonreactive elements (like gold and silver) may often be found existing on their own (generally, in mixtures, see below, with other compounds). The more reactive elements will never appear free in nature, only in compounds. Elements also very greatly in their stability. Generally, the smaller atoms, i.e., the ones with fewer protons (and neutrons and electrons) will be quite stable. As the atoms become larger, i.e., with more protons (and neutrons and electrons), they become increasingly unstable in fact, it appears that there is a maximum size of atom that can exit, which is thought to be 128 protons, beyond which no number of neutrons would be able to hold it together. James, *Elemental*, p.100.

Conductivity, in contrast, depends upon how big (in diameter) the atom is relative to the size of its nucleus (or atomic number). Where the electrons are farther from the nucleus, they are less strongly attached and more susceptible to flowing away as electricity. Such atoms are conductive, while the tighter atoms are nonconductive. The "metals" are all conductive. See Id., pp.104, 108-9.

The elements can appear in different phases (solid, liquid or gas), depending upon the temperature and pressure. At normal temperatures and pressures, there is only one liquid metal, mercury (atomic number 80) (although, there are a few metals that melt at temperatures just above room temperature). The only other element that is liquid under normal conditions is bromine (atomic number 35). All of the other elements are either solids or gases. At high temperatures, many solids will turn to liquids (e.g., molten iron). At very low temperatures, many gases turn to liquids (e.g., liquid nitrogen). (The phases of compounds are discussed below.)

SOURCES OF ELEMENTS

The most common element in the Universe is hydrogen (atomic number 1), estimated to constitute 90% of total matter. Gray, *Reactions*, p.31. It appears that hydrogen atoms formed more than 350,000 years after the Big

Bang, as things began to cool down. Essentially everything else was made in the core of stars. Id., pp.32-4. Some current theories about the sources of the elements is discussed in the chapter Cosmology. But, in short, when gravity created large enough masses, the stars appeared. In the "typical" star (like our Sun) nuclear fusion occurs, creating helium from hydrogen with the conversion of protons into neutrons. That process releases lots of energy in the form of heat and light. The heavier elements, like carbon and oxygen, were produced by collapsing stars (supernovae). Even heavier elements like iron probably required successive generations of stars for their production, requiring extraordinary temperatures. The energy required to create the atoms of the heavier elements is "'available" for release if and when such an atom is altered in a chemical or nuclear reaction. See James, Elemental, p.97.

Radioactive Decay

At the end of the nineteenth century, Marie Skłodowska Curie and her husband Pierre were conducting extensive studies of radioactive elements, identifying isotopes and two new elements (polonium and radium).They won the Noble Prize in Physics in 1903 (and she then won a second Noble Prize, this time in Chemistry, in 1911). See *Nobel Lectures, Physics 1901-1921* (1967). (As noted above, their daughter Irène Joliot-Curie and her husband did work involving "artificial [i.e., man-made] radioactivity," for which they won the Noble Prize in Chemistry in 1935.)

In 1902, Ernest Rutherford and Frederick Soddy presented a theory of transmutation by which radioactive elements changed into other radioactive elements in a chain as they decay. Lindley, *Uncertainty*, pp.35–41. Such decay is caused by the ejection of alpha particles, each one consisting of two protons and two neutrons (like the nucleus of a helium atom), altering the atom's nucleus and atomic number. James, *Elemental*, pp.34-5. Radioactive decay can also result in the emission of beta particles, very fast moving electrons or positrons, ejected when a proton turns into a neutron, thereby increasing the atomic number of the element by one. Radioactivity is a nuclear process, because changes occur in the configuration of the nucleus.

"While all elements have some radioactive isotopes, ...37 elements have no stable isotopes. These elements are considered the 'radioactive elements'." Todd Helmenstine, "What Are the Radioactive Elements?" *sciencenotes.org*, September 9, 2019. The radioactive elements are numbers 43, 61 and 84-118. Each such element has its own unique half-life (the period of time during which one half of the existing radioactive element would turn into the next element in the chain). The rates of decay were unique, different and identifiable, and assumed to be invariable. Radioactive decay appears to follow laws of probability. One can not predict which atom will decay at what time, only that a certain percentage will do so in a certain period. In other words, each atom of the element had a probability of decaying corresponding to the rate of decay of the element. See Lindley, *Uncertainty*, pp.40–41.

The introduction of probabilities, the notion that one could not predict what a particle would do but only the pattern that would emerge with a large number of particles, was in conflict with the classical theories of physics, which had led to a concept of determinism. See Cox and Forshaw, *The Quantum Universe*, p.8. The "puzzling thing about radioactive decay...was that nothing seemed to trigger the emission of the rays; they just popped out of substances spontaneously and unpredictably." *Id.*

Nonetheless, the discovery of the uniform half-life of certain elements proved to provide an incredible tool for the dating of fossils, specific strata of the Earth and other things of interest to biologists, paleontologists and geologists. For example, the radioactive carbon isotope carbon-14, from cosmic rays from outer space, is absorbed in the bones of living creatures. Carbon-14 decays, while carbon-12 is stable, so a comparison of the ratio of the two will reveal the age of the bone sample. Carbon-14, however, is useful for dating objects not much more than 20,000 years old because of its relatively short half-life (about 5,740 years).

Because different elements display a very wide range of known half-lives, from seconds to millions of years, there are elements from which to make selections well suited to the task at hand (for example, to date something about 10,000 years old or something about 10 million years old or something in between). See Dawkins, *The Greatest Show on Earth*, pp.91–107.[8] To some extent, other dating methods can be (and have been) used to confirm the results of calculations from the decay of radioactive elements into other elements.

MOLECULES AND COMPOUNDS

Molecules consist of two or more atoms bound together. If the atoms are of different elements, the molecules are "compounds", of which there seem to be an infinite variety. See, *e.g.*, Theodore Gray, *Molecules: The Elements and the Architecture of Everything*, (2014), p.16. But, molecules can also be made of atoms of the same element. For example, hydrogen itself at room temperature always appears as two atoms bound together (H_2), so the element hydrogen is both an atom and a molecule. See id., p.38. Carbon also readily binds together forming chains, but usually with other elements.

Interestingly, based on analyses of meteorites, space is a source of many chemical molecules found on Earth, including organic compounds like amino acids. This evidence suggests that the basic chemistry is the same throughout, at least, the Solar System. See Pier Luigi Luisa, *The Emergence of Life*, pp.32-3, 34. Other organic compounds were created on Earth from the impact of celestial bodies hitting Earth and others over time from the effects of ultra-violet light, hydrothermal vents, lightening, radioactivity and other energy sources. *Id.*

THE NATURE OF BONDING

The bonding occurs through the sharing or "lending" (and "borrowing") of electrons. The electron exchange takes place in the outer-most shell. If that shell is "full" (for shells 1 or 2 and containing 8 electrons for the other shells), no binding will occur. The elements with "full" shells are the "noble gases": helium, neon, argon, krypton, xenon and radon–atomic numbers 2, 10, 18, 36, 54 and 86.

If the bonding is the result of two atoms exchanging an electron (*e.g.*, one atom with one empty space in its outer shell and one with only one electron in its outer shell), the bond is called "ionic" and the resulting molecule will be polar–one atom in the compound will have a positive charge; the other, a negative charge. (Water is polar.) Gray, *Molecules*, p.56. The atom that becomes electron deficient as a result of the bonding (lending the electron to the other atom), it is called an "ion". James, *Elemental*, p.121. If the atoms share electrons, the bond is called "covalent", and the molecule will be electrically neutral.

The idea is that electrons with negative charges find themselves between two nuclei with positive charges (excuse the anthropomorphism). Each nucleus is attracted to the electrons, which bind them together like a chain: nucleus–electron(s)–nucleus. The strength of the electrostatic charge (like gravity) is inverse to the square of the distance ($1/d^2$). So, since the nuclei are closer to the electrons than to each other, the force of the repulsion between the two nuclei is weaker than the force of the attraction between each nucleus and the electrons. See Gray, *Reactions*, p.39.

It is the electrostatic force that holds compounds together and is the basis of all of chemistry. But, we do not know what that force actually is:

> *"We know a lot about how this force works—how strong it is, how quickly it weakens with distance, how fast it can be transmitted through space, and so on. ... But what the electrostatic force actually is remains a complete and utter mystery.* **It's quite marvelous that something so fundamental is fundamentally so unknown***."*

Gray, Molecules, p.12 (emphasis added).

But, note that "[I]f we need to add energy just to break the molecule apart, then it follows that the molecule is less massive than the sum of the original two hydrogen atoms, just as the hydrogen atom is less massive than the sum of the masses of its constituents." Cox and Forshaw, Why $E=mc^2$, p.153.

CHARACTERISTICS OF COMPOUNDS

Take hydrogen (H2). Add an oxygen atom (O), and you have water (H2O). Then, replace one hydrogen atom with a carbon and three hydrogen atoms and you have an alcohol. Add another carbon and two hydrogen atoms, and you have ethanol (drinking alcohol, as in wine or spirits). Gray, Molecules, pp.38-9. Again, borrowing an example (James, Elemental, p.3), take the following molecules: H_2O, $C_2H_2O_2$ and $C_6H_{12}O_6$. For the second molecule, we have added carbon (and increased the oxygen). The third molecule appears to be roughly three times the size of the second one (but, note that the hydrogen is 6 times more). They look similar; yet, the three constitute water, vinegar and simple sugar (like glucose). Normal table sugar, however, is $C_{12}H_{22}O_{11}$ (a combination of glucose and fructose). Anne Marie Helmenstine, "What Is the Chemical Formula of Sugar?" ThoughtCo, June 22, 2018.

Sodium and chlorine are two very dangerous elements on their own (sodium is highly reactive with water, chlorine is a poisonous gas). Bound into sodium chloride, however; they become salt, a necessity of life. Or, hydrogen snd carbon, just two elements which, in different combinations, form all of the one hundred thousand different named hydrocarbons, from grease to oil to petroleum to plastic. Gray, Molecules, p.11. For example, "[p]etrochemicals that are derived from oil and gas feedstocks form the building blocks for products that range from plastic bottles and beauty products to fertilisers and explosives." Ahmad Ghaddar and Ron Bousso, "Rising use of plastics to drive oil demand to 2050: IEA," *Reuters, Environment*, October 4, 2018 ("Plastics and other petrochemical products will drive global oil demand to 2050, offsetting slower consumption of motor fuel, the International Energy Agency ... said").

Pretty different substances from the same elements. How does that happen? Frankly, we do not really know.

Example: Carbon

Even molecules of the same element can take different forms. Carbon molecules can appear in three different structures (allotropes) creating (i) graphite and graphene (a thin, nonmetallic conductor); (ii) charcoal, bituminous (soft) coal, and anthracite (hard) coal and (iii) diamonds. Again, pretty different substances. Most carbon structures found in nature contain some additional elements as impurities, most often hydrogen but also sulfur, oxygen, and nitrogen. See Gray, *Reactions*, p.54.

Carbon is thought to be the sixth most common element in the Universe and, with 4 empty spaces, is the most prolific of the elements. There are an estimated 10 million different compounds containing carbon. See "Carbon: Chemical Element," *chemistryexplained.com*. With 4 empty spaces, it regularly bonds with 4 other atoms. Gray, *Molecules*, p.18. An example is methane with 1 carbon and 4 hydrogen atoms. Carbon also bonds with itself, usually with some other element, so it can form long chains. Carbon atoms can bond together with one, two or three electrons, giving multiple possibilities. Gray, *Molecules*, p.19. The carbon bonds typically share electrons, so the resulting molecules are non-polar. *Id.*, p.57. Another simple example is ethylene. It has two carbon and four hydrogen atoms. Each carbon atom bonds with two hydrogen atoms, gaining two electrons but leaving two empty spaces. The two carbon atoms then share their two "free" electrons, creating a double bond between the carbon atoms. See Gray, *Reactions*, p.41. A more complicated example is propane, C_3H_8, with three carbons with single bonds and eight hydrogen atoms. *Id.*, p.42. And, octane has eight carbons with 18 hydrogen atoms filling the gaps. *Id*, p.74.

When there are three carbon atoms connected by single bonds, the middle one will be bound to each of the other two carbon atoms leaving only two spaces left to be filled with hydrogen atoms. The two end carbon atoms will each have three hydrogen atoms (*e.g.*, C_3H_8). That pattern continues as thee chain gets longer and longer—three hydrogen atoms on each end and two on each intermediate carbon atom. See *id.*, pp.72-4 (setting out numerous illustrations of hydrocarbons).

Phases

Above, we discussed phases for elements. They also occur for compounds, but often within much smaller temperature ranges, like water (ice, liquid and gas). Well, perhaps not everywhere in the Universe. *See, e.g.*, Michelle Starr, "Scientists Just Created a Bizarre Form of Ice That's Half as Hot as The Sun," *ScienceAlert*, 9 May 2019. "It has taken one of the most powerful lasers on the planet, but scientists have done it. They've confirmed the existence of 'superionic' hot ice –frozen water that can remain solid at thousands of degrees of heat."

On Earth, with its relatively modest pressure, water is solid below $0°$ C, a gas above $100°C$ and a liquid in between.[9] But, it appears that at very high pressures and very high temperatures, water becomes a solid again. Sort of. Actually, a part solid/part liquid. The oxygen atoms bond into a lattice crystal and the hydrogen ions (H^+) flow through it. This super "ionized" ice is black, very hot and about four times denser than water. *Id.* It then melts into a liquid again at about $8,500°$ F. *Id.*

It seems very strange, but superionized ice may be the most abundant form of water in the Universe, apparently found to constitute much of the mass of the large frozen planets like Neptune and Uranus:

> "'Because water ice at Uranus and Neptune's interior conditions has a crystalline lattice, we argue that superionic ice should not flow like a liquid such as the fluid iron outer core of the Earth. Rather, it's probably better to picture that superionic ice would flow similarly to the Earth's mantle, which is made of solid rock, yet flows and supports large-scale convective motions on the very long geological timescales... .'"

Id.

Evidence of this form of ice had been accumulating before scientists managed actually to create it on Earth. A year earlier, its discovery was reported. "This new form, called superionic water, consists of a rigid lattice of oxygen atoms through which positively charged hydrogen nuclei move. It is not known to exist naturally anywhere on Earth, but it may be bountiful farther out in the solar system, including in the mantles of Uranus and

Neptune." Kenneth Chang, "New Form of Water, Both Liquid and Solid, Is 'Really Strange,'" *New York Times*, February 5, 2018.[10]

One may wonder what other bizarre chemical phenomena may exist beyond the limiting conditions of Earth.

MIXTURES

We have talked about compounds and their formation through chemical reactions. Now, I want to say a few things about "mixtures" (solutions, composites, emulsions, *etc.*).

I have mentioned that atmospheric air is a mixture of about 80% nitrogen and 20% oxygen. There are also trace contaminants, such as the current 415 parts per million ("ppm") of carbon dioxide, CO_2.[11] There will also be some varying amounts of water vapor, H_2O. All of these molecules are "floating" independent of one another and susceptible to being separated out. Mixtures are all around us, in the soil, in our waste and even in solid materials where foreign molecules are "trapped" within some other compound (*e.g.*, concrete is a composite of sand, gravel and cement).

Let us look at another type of example. Take a glass of water and pour into it salt crystals. The salt crystals will dissolve and visually appear to disappear in the water. However, when the water evaporates, the salt crystals will be left behind. So saltwater is just another mixture, right? Well, it is not quite that simple. *See* Gray, *Reactions*, pp.101. Both salt and water are polar, meaning that the molecule has "poles", in effect, the atom at one end of the molecule will be negative and the atom at the other end of the molecule will be positive. The salt (sodium chloride) is also polar, with a positive sodium end and a negative chlorine end, again because of the lending of electrons. When salt is added to the water, the first things that happens are that the salt crystals "dissolve" into individual atoms of salt (crystals are alternating ions, electrically-charged particles), and molecules of H_2O separate into positive hydrogen ions (a nucleus with no electrons, because of the "lending" of electrons in the bond that had formed he molecule) and negative HO molecules (hydrogen atoms bound to oxygen atoms with the extra "borrowed" electrons, giving them a negative charge). The charged atoms or ions from the water effectively pull apart the sodium chloride molecules (NaCl) into Na^- and Cl^- ions. The water ions form transitory pairings with the salt ions. When the water evaporates, the H_2O molecules recombine and become a vapor and the separated sodium and chlorine atoms recombine as a solid, forming new crystals. See Gray, *Molecules*, pp.58-9; Gray, *Reactions*, pp.99-100.

By the way, saltwater conducts electricity much better than does plain water because of the ions, and saltwater has a lower freezing point than plain water because the extra ions interfere with the formation of water crystals. *Id*. In contrast, pure or distilled water, with no minerals, is non-conductive.

What about sugar dissolved in water? Non-polar sugar, in contrast to salt, simply disperses in water creating a simple mixture, right? It is closer to a mixture, and generally characterized as such. But, even in sugar water there are chemical reactions in the form of weak hydrogen bonds. Gray, *Reactions*, pp.102-3. But, sometimes when one tries to create a mixture of a substance with non-polar molecules and a substance with polar molecules, they may not even "mix". Gray, *Molecules*, pp.56-7. An example is oil (non-polar) and water (polar). As you have undoubtedly observed, you can shake a bottle of oil and water as vigorously as you want, but when you stop shaking, the compounds begin to separate. Also, the oil is less dense than the water and will rise to the top.

A NOTE ON SOAP AND WATER

Now, you will also have noticed that if you add enough soap, oil appears to dissolve in the water. What, in fact, happens is that the soap molecules, being polar with long hydrocarbon "tails" that are negatively charged and repelled by water and positively charged "heads" that are attracted to water, attach by their "tail" ends to the molecules of oil and, in effect, encapsulate the oil molecules, keeping them separate from the molecules of water and from themselves. In that form, the individual molecules of oil will not "stick together", but will disperse among the molecules of water. Id., p.67. The oil can then be washed away. Soap was discovered by accident thousands of years ago. The primary benefit of washing your hands is that bacteria, dirt and other things tend to stick to the oil on your skin. The soap enables the water to rinse away such material by dissolving the oil as described above. In addition, for many viruses, including COVID-19, the capsid of the virus (essentially a film encapsulating the protein and RNA) is made of lipids that can be dissolved by soap, exposing the interior structures to rapid deterioration. For some bacteria, the membrane, even though more complex than a capsid, will also "dissolve" with soap and water. Then, the water will rinse away the remains. See Ferris Jabr, "Why Soap Works: At the molecular level, soap breaks things apart. At the level of society, it helps hold everything together," The New York Times, March 13, 2020.

REACTIONS

When one thinks of chemical reactions, it is explosions that come to mind. But, probably the most familiar chemical reaction is fire, where oxygen bonds with hydrogen or carbon. in all cases, energy is released. See Gray, Reactions, p.57.

"Electrons are the particles the enable chemical reactions; nucleons take part in nuclear reactions." "Basics of Nuclear Physics and Fission," *Institute for Energy and Environmental Research*, May 2012. All chemical reactions are the creation or destruction of or changes in chemical bonds between atoms. Gray, Reactions, p.47. So, they are almost entirely about the movement of electrons, like electricity. See id., p.78. When electrons move to orbits with lower potential energy, the excess potential energy is released as heat and light.

Before and after the reaction, there are the same number of atoms reflecting the same elements but bound into different compounds. It often will not appear to be so. We will think that something has disappeared., e.g., the fuel burned in a fire. Of course, various compounds and molecules may be gone, but the same number and types of atoms will always remain, just in different compounds (which may be gases, like CO_2). Id., p.43. That is the Law of Conservation of Mass, "discovered" In 1789 by Antoine Lavoisier. It has now been qualified based upon the theories of Albert Einstein (as discussed in the chapter Physics).

The conversion of mass to energy in the typical chemical reaction (like the burning of a carbon fuel) involves about one billionth (.0000000000001) of the mass of the fuel; the conversion in a nuclear reaction may be of as much as 0.7% (or .007) of the mass of the fuel–millions of times more. Rees, *Just Six Numbers*, p.53. At the same time, the conversion of mass into energy through fusion is still slow enough to enable the stars to continue the process and continue to burn for billions of years. Id. Nonetheless, '[s]o ponderous is the conversion of protons into neutrons that, 'kilogram for kilogram,' the sun is several thousand times less efficient than the human body at converting mass to energy." Cox and Forshaw, *Why E=mc^2*, p.172.

Most observable changes in matter are the results of chemical reactions, from the digestion of food and the growth of plants to the propulsion of an automobile or the explosion of nitroglycerin ($C_3H_5N_3O_9$) or TNT ($C_7H_5N_3O_6$). There are, however, also physical changes that we can see, like phase changes, that are not chemical reactions. (In addition, the atomic bomb and the shining of the Sun involve nuclear, not chemical, reactions.) Chemical reactions, and the bonding involved, generate light and heat, which may be almost imperceptible or quite dramatic. The frequencies of the light emitted depend upon the elements involved. The intensity of the reaction depends in part upon the speed of the bonding (or "unbonding").

Rust is the result of the slow oxidization of iron (the combination of the surface atoms of iron with oxygen). Fire is a much more rapid but generally controlled oxidization. Explosions occur when a lot of bonds dissolve or form almost simultaneously. The intensity also depends upon the potential energy present in the atoms prior to the reaction, reflected in the positions (energy levels) of the electrons.

As an example, here is a description of the explosive nature of the reactions of trinitrotoluene ("TNT"):

> *"TNT is explosive for two reasons. First, it contains the elements carbon, oxygen and nitrogen, which means that when the material burns it produces highly stable substances (CO, CO2 and N2) with strong bonds, so releasing a great deal of energy. ... [E]xplosives like TNT, actually have less potential energy than gasoline, but it is the high velocity at which this energy is released that produces the blast pressure. ...The second fact that makes TNT explosive is that it is chemically unstable - the nitro groups are so closely packed that they experience a great deal of strain and hindrance to movement from their neighbouring groups. Thus it doesn't take much of an initiating force to break some of the strained bonds, and the molecule then flies apart.*

Imperial College, London,"TNT", *ch.ic.ac.uk*.

Acids are molecules that release positively charged hydrogen ions (H+) in water. Those ions will bond with other molecules, causing reactions that can be rather dramatic, like the rapid dissolving of various kinds of matter (including flesh). Gray, *Molecules*, p.36.

EMERGENCE

One astonishing thing about chemical reactions is that the creation of new compounds is full of surprises. It seems to be impossible to predict from theory alone what all of the characteristics of any new compound will be. Instead of theoretical predictions, chemists have catalogued the results of innumerable experiments. The other impressive fact about chemical reactions is the utter regularity of the reactions in terms of the creation of compounds. The exact same thing happens each time. Once one knows (from experimentation) what will happen, it does so quite reliably.

(That is not to say that scientists have not been trying to figure how to predict the characteristics of materials from "first principles," like quantum mechanics. *See, e.g.*, website of Chris Pickard, Department of Materials Science & Metallurgy, University of Cambridge. While some impressive results have been achieved in developing models that can predict some characteristics of certain materials in extreme conditions, the results are very modest compared to the broader objective.)

An explanation offered for what underlies these two characteristics is "emergence". While the word may be unfamiliar, the concept of emergence (or emergent phenomena) is neither new nor strange. Its essence is captured in the saying that "the total is greater than the sum of the parts".

The idea of emergent phenomena has a history dating back to the later part of the nineteenth century. It appears in the work of John Stuart Mill, who concluded that there were characteristics in matter at increasing levels of complexity that simply were not present in the same constituent parts of less complex structures. Mill cited chemistry as a particular example of emergence. See Deacon, *Incomplete Nature* (2012), pp.146–52. These ideas were developed by philosophers in the first half of the twentieth century. Deacon, *Incomplete Nature*, pp.154–59. The resulting philosophical developments are quite complex and esoteric. *See, e.g.*, O'Connor and Wong, "Emergent Properties", *The Stanford Encyclopedia of Philosophy* (Summer 2015). Today, the proffered applications of the concept go well beyond chemistry. See Luisa, *The Emergence of Life*, pp.229–30.

Here is the idea:

> "You could spend a lifetime studying an individual water molecule and never deduce the precise hardness or slipperiness of ice. Watch a lone ant under a microscope for as long as you like, and you still couldn't predict that thousands of them might collaboratively build bridges with their bodies to span gaps. Scrutinize the birds in a flock or the fish in a school and you wouldn't find one that's orchestrating the movements of all the others.

John Rennie, "How Complex Wholes Emerge From Simple Parts," *Quanta Magazine*, January 28, 2019.

Laughlin argues that rules governing emergence are independent of the details of the prior circumstances or the individual elements constituting the collective that reflects organization. Thus, for example, multiple variations of the lower level can be consistent with the higher level of organization. *A Different Universe*, p.207. He notes, however, that emergence is not equivalent to complexity. The more complex seeming phenomena could be at the microscopic level (like quantum mechanics) or at the macroscopic level.

A popular illustration of emergence is that of phase transitions (like water changing to ice or to vapor). *See, e.g.*, Robert B. Laughlin, *A Different Universe*, pp.33–5. As noted in the quotation above from John Rennie, a full understanding of one phase is simply not adequate to predict the characteristics of the other phases. A related chemical example is the formation of crystal lattices. *Id.*, p.37. But, probably the best known examples are the results of chemical reactions, such as oxygen and hydrogen producing water or sodium and chloride producing salt (which was the example prominently cited by J.S. Mill in the reference above). A much more recent statement of that second example follows:

> "$NaOH + HCl \rightarrow NaCl + H_2O$(*Sodium hydroxide + hydrochloric acid* **produces** *sodium chloride + water*) *The product of this neutralization reaction, water and a salt,* **is in no sense the sum of the effects of the individual reactants**, *an acid and a base.*"

O'Connor and Wong, "Emergent Properties," *Stanford Encyclopedia of Philosophy* (emphasis added).

Chemistry provides the most vivid and readily recognizable illustrations of emergence. Indeed, chemistry has been described as the study of the emergent properties of physics and biology. See Pier Luige Luisi, "Emergence in Chemistry: Chemistry as the Embodiment of Emergence," *Foundations of Chemistry*, 10-2002, Vol. 4, Issue 3, pp.183–200. Roger Penrose explains that chemical processes involve quantum effects, so that chemistry is not contained in classical physics. *Shadows of the Mind*, p.235. Elements combined into common molecules often have characteristics neither present in nor predictable from the characteristics of the constituent elements themselves.

Most characteristics of what results from chemical reactions are largely predictable now as a result of experimentation. But, many of the chemical characteristics have also become theoretically predictable (or explainable). However, even with an atomic level explanation for how the elements combined and of how they are held together in the new matter formed, the physical characteristics of the new matter (what our senses "perceive") seem simply not generally to be derivable theoretically from the characteristics of the parts. It is as if there are two sets of rules or laws involved—those governing the behavior of the parts and those governing the behavior of the combination. But, it is also the case that the emergent phenomenon can result in changes the characteristics of the constituent parts. See, e.g., Luisa, *The Emergence of Life*, pp.235-6 ("downward causality"). In water, the atoms of hydrogen and oxygen continue to exist, but they appear much different above the atomic level.

Scientists debate whether the appearance of emergence is the result of lack of knowledge and difficulties in application of theory to real world facts, of current (and, therefore, transitory) theoretical inadequacies or of something just not explainable. Some chemists, thus. challenge the assertion that emergence is the heart of chemistry, arguing that with increasing knowledge we will be able to predict and explain that which today appears to be emergent phenomena in the physical world. Predict, perhaps, but the prospect of real explanation seems remote to me.[12]

Of course, the nagging question that arises is that if emergent characteristics of compounds and more complex structures are not "contained" some how in the constituent parts, what is their source? Moreover, as discussed below in the chapter on Consciousness, philosopher Thomas Nagel asserts that reliance on emergence is no explanation unless a case can be made as to why such phenomena may be likely to emerge from the underlying physical elements when such elements are organized in certain ways. *Mind & Cosmos*, pp.55-6, 60.

ENDNOTES

[1] Conniff continues:

> "Historians of science began deciphering alchemical texts—which wasn't easy. ...[There is] growing evidence that the alchemists seem to have performed legitimate experiments, manipulated and analyzed the material world in interesting ways and reported genuine results. And many of the great names in the canon of modern science took note,... ."

[2] I decided to discuss chemistry before quantum mechanics, even though some anticipation of that theoretical work is required. Quantum mechanics raises issues that go beyond chemistry. Of course, biochemistry, for example, is invoked in the matters discussed in the prior chapter on Darwinism and the later chapter on Consciousness. So, there is no strictly logical order.

[3] "Turning lead into gold might not be possible through sacred incantation but, if you take an element like thallium, boil it to a gas, pressurize it, and fire alpha particles through the sample, one in every few thousand thallium atoms would be turned into gold." Id. Since gold is atomic number 79, the transformation requires the liberation of two protons and two neutrons. Note that thallium (atomic number 81) does not occur free in nature, so this process is even more difficult than the quotation suggests.

[4] This technology is now to be used to examine at an atomic level Containment Level 3 pathogens—viruses and bacteria that cause severe to fatal diseases in humans, but are considered treatable (the Containment Levels range from 1 to 4). Id. A new, alternative approach is to expand the thing to be examined. New technology, called "expansion microscopy," allows the enlargement of biological features like neurons through the infusion of cells with a polymer that expands uniformly as it absorbs water. The cells and constituent parts expand with the structure intact. To date, researchers have been able to enlarge cells by four to five times, making the structure (highlighted by fluorescent dye) visible to ordinary optical microscopes. John Markoff, "Expansion Microscopy Stretches Limits of Conventional Microscopes," The New York Times, January 19, 2015. Also, new super-resolution microscopy, based on sequential photos of a sample as the light on different portions of the sample is turned on and off, has permitted the optical resolution of images well beyond (10 to 100 times)

what could be obtained by the normal optical microscope. The selective lighting of various parts of the sample can be achieved by using light-activated fluorophores in the sample or by blocking the external light from illuminating different parts of the sample for each photo. This advance is particularly significant for the biological sciences because it is usable on living cells, unlike the electron microscope. University of Cambridge, "The super-resolution revolution," *Research Bulletin*, 27 February 2015.

[5] During the second half of the nineteenth century, physicists developed the kinetic theory of heat, explaining, for example, why gases expanded when heated (the pressure from more rapidly moving atoms and molecules colliding with each other and with the vessel containing the gas). This theory suggested that the movement of the atoms and of anything with which they collided appeared to be random and unpredictable. Such an inference conflicted with the deterministic view of classical physics. Of course, it was still possible that the movements were predictable in theory (that is, if one could have sufficient information about all of the atoms or molecules involved), but not in practice (too much information to handle). Lindley, *Uncertainty*, pp.20-25.

[6] An idea of the tiny size of an atom can also be suggested by considering the tiny size of much larger molecules made up of atoms. Bill Bryson collected examples of calculations of the enormity of Avogadro's Number, the number of molecules in 2.016 grams of hydrogen gas:

> *"I can report that it is equivalent to the number of popcorn kernels needed to cover the United States to a depth of nine miles, or cupfuls of water in the Pacific Ocean, or soft-drink cans that would, evenly stacked, cover the Earth to a depth of two hundred miles. An equivalent number of American pennies would be enough to make every person on Earth a dollar trillionaire. It is a big number."*

Id., p.139n.

[7] Specifically,

> *"[e]ach orbital is denoted by a principal quantum number n, a function of the distance of the electron from the nucleus. Three other quantum numbers specify the way that an electron interacts with a nucleus: l, the angular quantum number, denotes the shape of the orbital; m_l, the magnetic quantum number, denotes the orientation of the orbital; and m_s, the spin quantum num-*

ber, *denotes the direction of the electron's spin on its axis of orientation."*

Id., p.134.

[8] There are some uncertainties involved in the practice of radioactive dating. It is often the case that the item or material one wishes to date does not itself contain the radioactive element to be used. Therefore, it may be necessary to select material that one can assume with reasonable confidence is from the same period as the sample. In addition, there is the problem of contamination. Such an issue occurred with respect to the dating of the remains of Neanderthals found across Europe. It appears that very small amounts of more recent carbon had contaminated the samples causing the scientists to conclude that they were several thousand years younger than they in fact appear to be. See Kenneth Chang, "Neanderthals in Europe Died Out Thousands of Years Sooner Than Some Thought." *The New York Times*, August 20, 2014.

[9] Kenneth Chang, "New Form of Water, Both Liquid and Solid, Is 'Really Strange,'" *New York Times*, February 5, 2018.

> *"Earlier experiments by other groups had produced conductive water that could have been superionic, but those scientists could not determine if the current were carried by ions and not electrons. Here, Dr. Millot and his colleagues were able to capture the optical appearance of the ice. If electrons were moving around, it would have been reflective. (That is why metals are shiny.) Instead, the sample was opaque. That pointed to the movement of ions instead, indicating a superionic ice."*

[10] In a vacuum (no pressure), water remains a gas until the temperature gets below about 210 K (-63.15° C). Then, it becomes a solid. Outer space not only has a very, very low pressure (essentially zero), it is also very, very cold (approaching absolute zero or 0 K, far below 210 K). So, what happens if water in liquid form is released into space? First, it boils into a vapor; then, when all of the molecules are exposed to the extreme cold, it turns to ice crystals, a solid. (if it were released as ice in space, it would stay as ice). See Ethan Siegel, "Water In Space: Does It Freeze Or Boil?," *Forbes*, December 23, 2016 ("When the astronauts take a leak while on a mission and expel the result into space, it boils violently. The vapor then passes immediately into the solid state ... and you end up with a cloud of very fine crystals of frozen urine").

[11] Peter Dockrill "It's Official: Atmospheric CO2 Just Exceeded 415 ppm For The First Time in Human History," *ScienceAlert*, 13 May 2019 ("a mistaken reading ... suggested the 415 threshold was actually breached on May 3. That false data was subsequently revised, but not before a handful of sites reported the grim accomplishment. This time, unfortunately, there appears to be no doubt").

[12] Philosophers refer to "strong" emergence and "weak" emergence. The former applies when the characteristics of the emergent form are simply not present in or traceable to the component parts from which they arise. Weak emergence exists where the emergent property is theoretically reducible to the components, but practical difficulties prevent accurate predictions of the emergent properties. See, *id.*, pp.233-4.

PHYSICS

> "...[T]he most perfect physical science,
> embodying the ideal toward which
> all other branches of inquiry ought to aspire."
>
> Ernest Nagel
> *The Structure of Science*, p.154.

Classical (Newtonian) mechanics was based upon three laws of motion from which the theory could be deduced, *i.e.*, derived using the laws of deduction or logic. Mechanics began with the study of the mechanical advantages of the lever. Whitehead, *An Introduction to Mathematics*, pp.46-7. "By the middle of the nineteenth century mechanics was widely acknowledged as the most perfect physical science... ." Ernest Nagel, *The Structure of Science*, p.154. In this chapter, however, I will skip over Newton's work establishing classical mechanics, but I do discuss the subsequent modifications to it. My primary focus is on fields, forces, space and time.

ACTION AT A DISTANCE

> "Some [world-models], such as **astrology, telepathy, clairvoyance, and witchcraft, bring in forces that act powerfully across big separations in space and time.** Others, such as **extrasensory perception, telekinesis, prayer-induced miracles, and magical thinking, assign prominent roles in shaping the course of physical events to mind and will.** Most of those ideas are 'reasonable' extensions of the world-models we build up as children, in which our mind is disembodied and our will controls our body. **Historically, most people's world-models have accepted many or all of them.**"

Frank Wilczek, *Fundamentals: Ten Keys to Reality* (2021), p.64 (emphasis added).

A commonly accepted distinction between modern and primitive societies is our "scientific" view of the world rather than the older views in which magic, divine intervention and other "superstitions" play an important role in explaining why and how events occur. Earlier world views tended to conceive of things organically and holistically as or like living organisms and, as a result, often as behaving purposively. Paul Davies, *The Mind of God: The Scientific Basis for a Rational World* (1992), p.74.

The heart of the "scientific view" of the world is the belief that there are always causes of events arising from various initial conditions that operate through or as a result of Laws of Nature. We expect to be able to construct theories of causation making use of those Laws and, thereby, explain the events at issue. The scientific perspective is essentially mechanistic, epitomized by Newton's Laws of Motion. That perspective is also fundamentally reductionist, based upon the assumption that one can understand an object or event by taking it apart and examining the elements, including individual causal relationships that may be combined. Id., pp.77-8. In other words, the total is no more than the sum of the parts. And, a key assumption is "locality", that things are affected only by other local (nearby) things.

One might say that spiritual explanations of events are not scientific because they lack a theory that physically connects cause and effect. In fact, the perceived defect has to do not with the lack, but with the nature, of the connection. For example, we find it hard today to imagine how people could believe that the mere mental activity of one mind could cause illness in another body or could alter physical events at some remote location. In part, our skepticism about voodoo and telepathy is a result of our assumption that separate physical things are generally independent of one another; absent some "physical" connection, one cannot influence the other. I cannot read minds, cause someone pain by sticking a pin into a doll or "will" someone to take a particular action, because of the presumed lack of any kind of connection between my mind and the physical or mental state of the other person. Believers in and practitioners of black magic would contend that there are such connections, but they do not articulate the relationships in terms of the physical theories of the natural sciences.

Of course, the physical world confronts us with apparent examples of the very phenomenon of action at a distance. Certain fundamental parts of science were built on models in which objects acted on other objects without any detectable physical connection or apparent communication. Indeed,"Newton was extremely unhappy with one of his most glorious discoveries. ...Newton realized that his law of gravity is not local ... and he did not like it. This perceived flaw was, for Newton and for several generations of scientists who followed him, purely theoretical [or aesthetic or, even, theological]." Wilczek, *Fundamentals*, p.66.

Gravity (again) and Magnetic Forces

Two well-known and readily recognized examples are gravity and magnetic forces. These "forces" operate over even very great distances[1] and can do so through a total vacuum, as well as through solid materials. One has to wonder how this actually happens. What "reaches out" from one object to exert an influence on another?

We have come to accept that heat radiates through the transfer of energy from one molecule to another, whether the molecules compose gases, like air, or a solid, like a cast iron stove. Similarly, sound travels in the form of waves of vibrations through the air and to some extent through solids. But, neither heat nor sound can be transmitted through a vacuum. (Of course, the Sun heats Earth without heating the empty space in between—another example of action at a distance.) These models of the transfer of energy make sense to us because they seem physical. We visualize one molecule touching another and thereby passing on much of the energy. Of course, we are told that they do not really "touch" or even come close to touching and that most of what we think of as solid consists of empty space. But, I suppose that most of us take these as mere details, in part because the structure and composition of atoms is not very relevant to our day-to-day activities.

In contrast, although, gravity and magnetic forces are also clearly part of our day-to-day experiences; we lay persons do not have models that really explain these phenomena. These forces are apparently neither movement through some medium, like waves through water or sound through air, nor objects with a physical presence (or mass), like neutrons and protons (we will come to electrons and other elementary particles below). The same thing is true of radio waves and, obviously, light.[2]

Fields

The modern explanation for these phenomena "fields." As a result, a theoretical physicist today would presumably object to the characterization of any of these phenomena as "action at a distance." "[I]n the mid-nineteenth century, physicists filled the passive 'vacuum'—a nothingness, or Void—that Newton complained about with force-transmitting materials, which we call fields. Fields, rather than particles, are the fundamental building blocks of matter in modern physics." Wilczek, *Fundamentals*, p.67.

The concept of fields arose in a somewhat surprising way. Work in the second half of the nineteenth century with electricity produced observations that were difficult to reconcile with the long-established views of the forces involved in physics based upon Newtonian mechanics. See Albert Einstein and Leopold Infeld, *The Evolution of Physics* (1938), p.125. Scientists undertook to identify and then describe observable electromagnetic forces.

In the 1830s, Michael Faraday discovered the relationship between electricity and magnetism, demonstrated by the electromagnet (where the passage of electric current through the wires wound around the steel bar turns the bar into a magnet while the current is flowing). The experiments made it clear that there was "some kind of influence [that] must pass through empty space" in order for the current passing through the wire to have an impact on the separate object that is attracted by the "magnetism" created. Cox and Forshaw, *Why $E=mc^2$*, p.19. "Faraday realized that he was observing some kind of deep connection between magnetism and electricity, two phenomena that at first sight seem to be completely unrelated." *Id.*

Faraday identified the influence as a "field." There were magnetic fields and electrical fields. Faraday went on to conduct numerous experiments involving magnetism and electricity and recorded the results. A very substantial body of experimental data was collected concerning electromagnetic "fields".

Experiments demonstrated that the flow of electricity through a wire would have an effect on a magnetic needle. The effect was perpendicular to the flow of electricity in the wire and increased with the velocity of the flow. This phenomenon differed from the gravitational force of classical mechanics which appeared to arise through straight lines between the bodies involved and vary only based upon distance (given the masses of the bodies in question). Einstein and Infeld, *The Evolution of Physics*, p.128. And, oscillating electric fields create magnetic fields and *vice versa*.

> "In 1864, three years before Faraday's death, [Scottish physicist James Clerk] Maxwell succeeded in writing down a set of equations that described all of the electric and magnetic phenomena

> Faraday and many others had meticulously observed and documented during the first half of the eighteenth century."

Cox and Forshaw, Why $E=mc^2$, p.21. See also Einstein and Infeld, The Evolution of Physics, pp.125-37.

> "Maxwell wrote down his equations in the language of fields because he had no choice. It was the only way of bringing together the vast range of electric and magnetic phenomena observed by Faraday and his colleagues into a single unified set of equations. ...[I]t led to a simple and unified picture of all observed electric and magnetic phenomena that worked beautifully in the sense that it allowed for the outcome of any and all of the pioneering benchtop experiments of Faraday and his colleagues to be predicted and understood."

Cox and Forshaw, Why $E=mc^2$, pp.23, 24.

From this perspective, one notes, a gravitational field is merely a mathematical description of the apparent lines of force. And, "all the lines of force, or briefly speaking, the *field*, indicate only how a test body would behave if brought into the vicinity of the sphere for which the field is constructed." Einstein and Infeld, The Evolution of Physics,, pp.126-7 (emphasis in original). "Our entire knowledge of the acting forces can be summarized in the construction of a field." Id., p.131.

(A new apparent discovery indicates that "quantum fields" may even enable the transmission of heat through a vacuum, at least at very, very small distances. Kara Manke, "Heat energy leaps through empty space, thanks to quantum weirdness," Phys.org, 11 December 2019.)

WAVES

As noted, Maxwell managed to generate mathematical equations that depicted the relationships reflected in that data, thereby providing a mathematical representation of Faraday's fields. Mazur, The Motion Paradox, p.166. In the process, Maxwell introduced a factor (known as the "displacement factor") that was neither experimentally observed in nor necessary to the experimental data. He added it to the equations because he believed that it was necessary to make the formal mathematics of the equations internally consistent. Cox and Forshaw, Why $E=mc^2$, p.25-6.

However, the introduction of the displacement factor enabled Maxwell to transform his new field equations into a mathematical form known as wave equations. (As noted in the chapter on Mathematics, one can introduce a new factor on both sides of an equation without disturbing the equality; and subsequent manipulation of the equation can lead to a formulation that is suggestive of new relationships or that provides new insights.)

Maxwell's equations described the behavior of electric and magnetic fields, which fields could "carry energy across otherwise empty space." Penrose, *Shadows of the Mind*, pp.217-8. "Maxwell's wave equations describe how these two fields are linked together, oscillating backward and forward. They also predict that these waves must move forward with a particular speed. Perhaps not surprisingly, this speed is fixed by the quantities Faraday measured." Cox and Forshaw, *Why E=mc^2*, p.26-7. Thus, Maxwell's equations imply that the speed of the waves is not effected by the relative motion of the source of the waves or of the observer (neither of which is part of the equations). "Maxwell's waves travel at 299,792,458 meters per second. Astonishingly, this is the speed of light—Maxwell had stumbled across an explanation of light itself." Id., p.28. Indeed, the waves that are created by oscillating electromagnetic fields are "light". Theodore Gray, *Reactions: An Illustrated Exploration of Elements, Molecules, and Change in the Universe* (2017), p.114.

ARE THEY REAL?

The concept of a "field" is merely descriptive. Fields are the set of initial conditions for some type of parameter existing at a point in time (with each and every point in space having a particular value, some or many of which may be zero) and they then evolve over time pursuant to the applicable equations of motion (Laws). Unfortunately, however, we do not know of what such fields consist or through what mechanisms, if any, the evolution of those fields through time (pursuant to the equations) are implemented.

Thus, while the field representation may provide powerful predictive capabilities, it does not explain how or why the things being predicted are happening. To repeat part of a quotation set forth in the first chapter concerning gravity:

> *"no one ever discovered gravity, for the physical reality of this force has never been demonstrated. ... This basic concept of physical science is a complete mystery, and all we know about it*

*is a mathematical law describing the action of a force **as though it were real**."*

Kline, Mathematics, p.361 (emphasis added).

The same is true of waves and fields. With respect to the electromagnetic field, nonetheless, scientists consider it as something that is "real." Einstein and Infeld, *The Evolution of Physics*, p.145. It can be created experimentally; it can be turned on and off by controlling the flow of electricity through the wire; oscillating electrical charges cause changes in electromagnetic fields that generate electromagnetic waves. Id., pp.148-9.

Although, it will take us two chapters, we will see that, for today's physicists, all that exists anywhere are fields.

A NOTE ON THE PARANORMAL

I want to note the curious level of hostility in the scientific community toward explanations for events or, even, claims of the occurrence of events that are often lumped into the category of the "paranormal." We seem to accept that certain types of explanations (like fields and waves) are scientific, whereas others (say, black magic) are not.[3] To be fair, much of the derision seems to arise from the belief that the paranormal phenomena do not really occur, that they are only the creatures of magicians and con men. Thus, the issue is not initially about the nature of the explanation, but whether there is any evidence of the claimed cause and effect. As to that question, of course, scientific inquiry is entirely appropriate.[4] Yet, there does seem to be some paranoia about letting "the nose under the tent." As an interesting aside, even late in the nineteenth century, much of the public face of science was in the form of entertainment. Lectures and demonstrations attracted substantial crowds. Much of the science offered differed little on the surface from the shows by what we would today call magicians and charlatans. See, e.g., Professor Simon Schaffer, Cambridge University, Department of History and Philosophy of Science, "Why trust public experiments?", presentation to the Trinity Science Society, 28 October 2014 (Trinity College, Cambridge).

Light

Aristotle had assumed that the speed of light was infinite or, more accurately, that light traveled instantaneously.

The fact that the speed was finite and calculable was discovered by the Danish astronomer Ole Roemer in 1676. Joseph Mazur, *The Motion Paradox* (2007), pp.157–9. Roemer noted that there was a difference in the timing of the eclipses of a moon of Jupiter (called Io) depending upon the distance Jupiter was from the Earth (the moon would be expected to have had regular eclipses since it was presumably moving at a constant speed around Jupiter). Roemer concluded that the differences arose because of the time it took light to reach Earth. Comparing the timing of the eclipse when Jupiter was closest to Earth with the timing of the eclipse when Jupiter was furthest from Earth, Roemer calculated that the speed of light was 225,000 kilometers per second (75,000 kilometers per second slower than current calculations). *Id.*, p.159. Apart from what has come to be realized was the impressive accuracy of his calculations, Roemer's work led to a general acceptance of the proposition that the speed of light was finite.

The other key question concerning light was simply of what did it consist? A natural possibility was that it was particles of some type. That was the view of Descartes and, generally, of Newton. *Id.*, p.162. (Newton did entertain the possibility that light was propagated as waves.)

It was observed that light can cast sharp shadows, indicating that it does not normally bend around objects and that it travels in a relatively straight line. It could not be a mist or vapor or else the edges of the shadows would be fuzzy or there would be no real shadow at all. In addition, light is able to travel very great distances with relative ease; yet, it can not pass through many materials, unlike sound. But, it can travel through a vacuum, also unlike sound. In addition, light can pass through itself with no apparent interference.

The English physicist Thomas Young had performed what became known as the two-slit experiment (described below) in 1801, finding clear evidence of wave-like behavior being displayed by light. Mazur, *The Motion Paradox*, p.164. However, the authority of Newton was strong, and Young's experiment did not immediately change the prevailing view. *Id.* Newton's theories were based upon particles and forces acting on particles, like gravity.

Maxwell's equations coupled with the experimental data enabled the calculation of the speed predicted by (or assumed in) the equations. The speed derived is virtually equal to what had been calculated as the speed of light. Cox and Forshaw, *Why E=mc²*, pp.28-29. That seems to be a pretty incredible coincidence. See Mazur, *The Motion Paradox*, p.168. But, it was not viewed as a coincidence. Instead, physicists concluded that this result demonstrated that light is an electromagnetic field or, at least, is caused by or dependent upon electromagnetic fields. Cox and Forshaw assert "we learned that light is none other than a symbiosis of electric and magnetic fields surging forward... ." Cox and Forshaw, *Why E=mc²,* p.37. (It is clearly an overstatement to say that we "learned" such a thing. I would say that we had empirical evidence consistent with the hypothesis that light was such a "symbiosis.")

So, at the end of the nineteenth century, physicists presumed that the physical world consisted of two distinct types of "things": particles of matter that had mass and defined positions, with paths and speeds of movement, and electromagnetic fields that "permeated space and behaved like waves." Julian Barbour, *The End of Time: The Next Revolution in Physics* (1999), p.186. And, the accepted view by then was that light fell in the category of waves. Seife, *Zero*, pp.161-2. Indeed, "the evidence for the wave theory of light was strong." Barbour, *The End of Time*, p.185.

The "two-slit" experiment

Young's famous experiment is regularly performed in secondary school physics classes. See, e.g., Greene, *The Elegant Universe*, p.97. A light illuminates a distant wall. When a board with two slits is interposed, so that the light can reach the wall only through the two slits, one finds that the light on the wall displays a pattern of light and dark lines where the light coming through each of the two slits overlaps. The only explanation that can be derived from our experiences with the physical world is that the light travels in waves. New wave patterns emerge from each slit, fanning out again so that the two sets of waves become intermingled. The bright lines result where the peaks of two waves meet and the dark lines appear where the peak of one wave meets the valley of the other, cancelling each other out. This type of interference is a well-known phenomenon of waves.

The ether hypothesis

The problem confronting the Victorian scientists was the identification of the medium through which the light waves could travel. Waves were assumed to require something in which the wave phenomenon could occur—it was believed that waves could not exist on their own, but were a form of motion that occurred within some medium. Scientists in the nineteenth century concluded that there must be "ether" through which the waves traveled. The ether of the late nineteenth century was thought to permeate all matter as well as all space, certainly everything through which light could travel. Mazur, *The Motion Paradox*, p.164. Indeed, the ether hypothesis offered an explanation for other wave-like forces— including, perhaps, gravity—and could be seen as an answer to the action-at-a-distance problem. *Id.*

Moreover, the speed of light would be "relative" to the ether. The idea was that in some sense the ether was stationary, providing a fixed frame of reference for the measurement of speeds. In addition, the characteristics of the ether would determine the speed at which the waves travel independent of the speed or motion of the source of the waves. *See* Joao Magueijo, *Faster than the Speed of Light*, pp.68-9. That would fit with Maxwell's equations.

Our normal experience is that speed is additive; that is, if a platform is moving at the speed of x and one launches a projectile from that platform at the speed of y, then relative to the "stationary" observer, the projectile will be moving at the speed of x plus y. Similarly, if the object is moving at the speed of x and the observer starts moving toward the object at the speed of y, then the object will appear to the observer to be approaching at the speed of x plus y. If the observer starts moving away from the approaching object at the speed of y, then the object will appear to be approaching the observer at a speed of x minus y.

But, the speed of waves in water is independent of the speed of their source. However, the speed of the observer is not irrelevant. If the observer is moving toward the source, against the waves, the waves will appear to be moving more rapidly than when the observer is stationary.

So, based on the example of waves in water, it was concluded that, with light traveling at a constant speed through the ether, it should be possible to observe some differences in the speed of light as observed from Earth, because Earth is moving in an elliptical orbit through space (that is, changing direction continuously) and, presumably, through the ether (that is, the ether is not moving with the Earth). The situation should be analogous to the observed speed of waves from a boat changes when the boat is moving with the waves and when it is moving against the waves. Therefore, at least in principle, it should be possible to test the ether hypothesis.

Elaborate experiments by Albert Michelson and Edward Morley, beginning in 1881 and continuing throughout the 1880s, failed to detect any observable differences in the speed of light as observed from Earth. Magueijo, *Faster than the Speed of Light*, p.69; Cox and Forshaw, *Why E=mc^2*, p.33. They managed to construct an elaborate (and massive) instrument intended to measure the effects of the ether on light. Mazur, *The Motion Paradox*, pp.173-5. Using a half-silvered mirror to split a light beam into a right angle, they placed mirrors at each end to reflect the beams back to the observer. As a result of the construction, the observer would see two beams from the same source emitted at the same instance that had traveled the same distance from the source to the observer.[5]

However, one beam would be traveling in the direction that Earth was moving through the ether, and back again, while the other would be traveling at a right angle to the motion through the ether. They hypothesized that for the beam moving against the ether, the wavelength of the light should be diminished as a result of "ether wind" (like the boat moving toward the waves). In addition, the speed of light effectively moving against and then with the ether should be affected (again, like a boat in water). The other beam traveling perpendicular to the motion of the earth should not be affected by the motion of the Earth relative to the ether. So, there should be a difference observed between the two beams when they return to the point of origin, both in wavelength and speed (that is, arrival time).

Michelson and Morley found that regardless of the direction of the motion of the Earth (taking measurements six months apart), the light beams returned exactly in phase, with the same wavelengths and at the same instance, unaffected by the presumed direction of the light relative to the ether. *Id.* (Actually, not "exactly in phase". There were slight deviations observed, but they were much less than what was predicted. These deviations were put down to experimental error.) The ether, if it existed, had no detectable effect on the speed or the wavelength of light.

Thus, it would appear that there is no ether, analogous to water, through which the light waves travel. A related conclusion appears to be that the speed of light is simply not affected by the motion of the source or by the motion of the observer, *i.e.*, that it is independent of both. There were alternative explanations offered. For example, it was suggested that Earth pulled ether with it as it moved, so that relative to the ether in the immediate vicinity of Earth, the apparatus was not in motion. (Of course, "the ether was always a rather ugly concept, since it would define a benchmark in the universe against which absolute motion could be defined in conflict with Galileo's principle of relativity." Cox and Forshaw, *Why E=mc²*, pp.31-2.)

A few years later, George Francis FitzGerald suggested that the distances traveled might have been different because the motion of Earth (and the apparatus) might cause a shortening of the arm of the apparatus moving in that direction (parallel to Earth's movement). In other words, the length of an object might be affected by its motion in the direction of motion. Mazur, *The Motion Paradox,*, p.176. FitzGerald also suggested that the amount of "shrinkage" was related to the speed of motion. *Id.*, p.179. Hendrik Lorentz made the same proposal and asserted the related proposition that mass increases with the speed of motion. *Id.*, p.180.

Experimental error?

Physicists around the world were surprised by the negative results and were inclined not to believe the conclusion. Many attributed the failure to detect differences in the speed of light as the result of experimental error or lack of sufficiently refined instrumentation. After all, the differences, if any, would have to be very, very small.

I think that it is useful to examine what we mean here by very, very small. I am interested only in orders of magnitude, so I will round the numbers for ease of exposition. The speed of light is a little less than 300 million meters per second. Thus, the speed of light in miles per second is 186,000 (and in miles per hour, about 670 million). That is the speed through empty space or a vacuum; through other materials, the speed will be reduced. (The amount of reduction is quantified in terms of the refractive index. For air in Earth's atmosphere, that index is 1.0003, so the speed of light is reduced only by about 3 one hundreds of one percent—a very small difference.)

The speed at which the Earth circles the Sun is 67,000 miles per hour. So the "impact" of the movement of the Earth would be 0.01 percent or one hundredth of a percent. The Sun also moves through the Milky Way at 486,000 miles per hour. Assuming that one could time the measurements correctly to get the maximum differences in the speed of Earth (its rotation plus the Sun's movement going toward the light and the same going away from the light), then the total difference in speeds would be a little over 1 million miles per hour (553,000 mph toward the source, then 553,000 mph away from the source). The 1 million miles per hour is less than 16 hundreds of one percent of the speed of light (or .0016). So one must measure the speed of light accurately to better than one tenth of one percent to notice a difference. Now, that degree of accuracy is not exceptional in science in general, but it is clear that we are looking for very small differences indeed.

In all events, the Michelson/Morley experiments, confirmed by many subsequent experiments made with increasingly refined technology and constructed somewhat differently, are seen as having experimentally established the constancy of the speed of light through a vacuum and as "disproving" the ether hypothesis. See Mazur, *The Motion Paradox*, pp.176-80.

As always, it is still possible that there could be an "ether", but no "ether wind" and no impact on the measurable speed of light. However, in that case, we have no idea of how to detect the possible existence of it. Folowing the established scientific preference of ignoring all hypotheses that have no observable impact on the world, we "reject" the "ether" hypothesis as untrue (or, more accurately, as irrelevant). See, e.g., Cox and Forshaw, *Why $E=mc^2$*, pp.12-13.

Of course, we are still left with the question of, in the words of Joseph Mazur, "what would be waving" as light or electromagnetic waves travel. *Id.*, p.176.

TIME

**"I am Time,
from whose teeth no one is safe.
Mortals, do not flee!"**

Claudio Monteverdi
Il Ritorno d'Ulisse
(1643)

In the prologue to Monteverdi's final opera (one of the first modern operas written) based on the second half of Homer's Odyssey, Human Frailty bemoans the fact that his fate is inescapably determined by Time (as well as Fortune and Love): "Time, who created me, also fights against me."

Then, Time appears and sings the lines quoted above.

We are born in time and immediately, with the relentless passage of time, begin the process of death. Time appears to be a fundamental part of how the Universe works—processes happen through time; events occur in time and they appear in chronological sequence; stars (like living things) have births and deaths. In other words, it is not just our mortality that gives time its prominence. Time is fundamental to our perceptions of life and of the world around us. To some extent, one might say that time is the one unavoidable and inescapable fact of life. We cannot stop it; certainly, we cannot reverse it. Time is with us always and its overhanging presence affects much of what we do and how we feel.

However, as Julian Barbour observes: "Nothing is more mysterious and elusive than time. It seems to be the most powerful force in the universe, carrying us inexorably from birth to death. But what exactly is it?" *The End of Time* (1999), p.1.

THE NATURE OF TIME

We can see time as another identifying dimension of where something or someone is to be found. *See, e.g.*, Greene, *The Elegant Universe*, p.49. If you are to arrange a meeting, you will specify a place (identified uniquely in the three dimensions of space, especially in a place like New York City where two spacial dimensions will often be inadequate) and a time: the corner of First Avenue and 70th Street, 3rd floor, at 3:00 pm (EST). Leave out any of this information and the meeting will be jeopardized.

But, time is not like the other three dimensions; time moves in only one direction. One can walk uptown or downtown, across town to the west or across town to the east; but, one can move only one direction through time. And, one has no choice about it—not only is the direction fixed, but the movement is compulsory. If you are late for the meeting, you cannot just go back the way you came. These unique characteristics of time underlie many of our models for understanding the world around us.

Also, our conception of causation is time dependent. A cause always, necessarily, precedes its effect. We do not accept that subsequent events can cause prior events. That is so not only because time moves only in one direction, but also because we believe that communication of information cannot occur backward through time. Thus, a future occurrence can have no impact on a past occurrence. Indeed, if it were otherwise, we face the paradox, for example, of an event observed to occur on day "one" actually not occurring because of something that happens on day "three". The event cannot both occur and not occur.

(A "solution" to the paradox could be an explanation that, in fact, the event does not occur on day "one", because the future event on day "three" is predetermined to happen, causing the earlier circumstances to conform to the subsequent cause. That eliminates the paradox, but would we then really be comfortable saying that the event on day "three" caused the event on day "one"? Of course, to the extent that everything that occurs is actually predetermined, it is not very meaningful to talk about causation in a localized context anyway. The only real question about causation in that case would be what "caused" everything to be predetermined in the manner in which it was.[6])

There is the basic question of what time actually is. This question is much more troubling than one might at first assume. Time is obviously neither matter nor energy. It is presumably not a field, like many of the other mysterious phenomena we have discussed. It is part of any process, an inevitable element of change; but, it would be odd to call time itself a process. It has been asserted by Aristotle, St. Augustine and some contemporary cosmologists that time essentially is change, nothing more. Without change, there is no time.

So, has time always existed or did it have an origin? And, of course, we wonder whether there will be an end of time. (The beginning and end of time are discussed later in the chapter on Cosmology.) There is also the other, perhaps more tractable question of whether time (whatever it is and however long it has been around) is absolute.

THE MEASUREMENT OF TIME

We all know that the human experiences of time can vary greatly. Time can pass almost instantly when we are unconscious; it can pass very slowly when we confront a life-threatening event. It is obvious that the "experience" of time (whatever that means for non-conscious things) will be quite different for a butterfly that lives for a few days and a tortoise that lives for hundreds of years. In addition, different people age at varying rates.

But, the concept of precise timekeeping is a relatively recent phenomenon in human history. For a very long time, people must have been aware of day and night, probably morning and afternoon (and high noon) and of the seasonal differences in periods of these events (for people inhabiting regions away from the equator). See, e.g., Adam Frank, *About Time*, p.xiv.

Moreover, the awareness of the general passage of time arose presumably rather early in the development of human consciousness.

> *"For centuries, we have divided time into days. ... For centuries, we have divided the days into hours. But for most of those centuries, however, hours were longer in the summer and shorter in the winter, because the twelve hours divided the time between dawn and sunset: the first hour was dawn, and the twelfth was sunset, regardless of the season... ."*

Rovelli, The Order of Time, p.58.

At the end of the twelfth century, clocks were unknown (although, in some towns, bells were rung to mark certain periods of the day); but, by the end of the fourteenth century, many cities of Europe were "equipped with that most modern of devices: a mechanical clock." Frank, About Time. p.67. During the first half of the fourteenth century, clocks had spread throughout the major cities of northern Italy and then to other parts of Europe. Id., pp.71-4. The subdivision of time into hours and then minutes coincided with and likely contributed to dramatic changes in the organization of the day and to society and its institutions. Id. (The minute hand became widely used only in the seventeenth century. Id., p.91.) With the ever-increasing fineness of the measurement of time, other issues were to arise.

Apart from the issue of the measurement of the passage of time by ordinary people leading ordinary lives, there was a more fundamental issue for scientists: what to utilize as the standard for keeping track of time. Before worrying about the technology of marking off the passage of increments of time, one needs to identify some clearly defined fixed points that can be utilized by everyone and that we can agree are actually fixed, that is, that do not change from one place to the next or over time: a way to know that an hour today (or any other arbitrary increment of time) will actually be the same tomorrow as it is today and will be the same in China as it is in England.[7]

Obvious standards for human beings would be the Sun and the Moon. In fact, an early clock for ancient astronomers was the movement of stars through the sky, while fixed relative to one another (sidereal time). The movement of the Sun established solar time. The two are not identical, the solar day being on average 4 minutes longer than the sidereal day. Barbour, The End of Time, pp.97-8. Although the solar clock was the most readily identifiable, it was discovered that the motion of the Moon could be accurately predicted by using sidereal time but not solar time. *Id.*, pp.98-9. (Sidereal time is the result of the Earth's rotation.)

In the late nineteenth century, however, measurement techniques became sufficiently precise that deficiencies in using the Earth's rotation (with small deviations being caused by the tides arising from the Moon's gravitational attraction) began to be observed. Astronomers were able to devise a more accurate clock based upon the Solar System. Such time was called Newtonian time and, then, ephemeris time. *Id.*, pp.106-7. More recently, atomic time has been adopted as the official standard. *Id.*, p.107.

In all cases, time involves what we apprehend as change.

Today, following developments in fundamental physics, we can ask whether time is continuous or, as the material world is thought to be, made up of small, discrete irreducible units. *See, e.g.*, Deutsch, *The Beginning of Infinity*, p.298. We generally conceive of and model time as a continuous variable, just as the classical models treated matter and energy. Indeed, some famous paradoxes (like Zeno's) are based upon that conception, one in which it is always possible to split the time (or distance) remaining in half. If there is a smallest unit of time, then at some point, such further division becomes impossible. (That would be the instance at which the hare passes the tortoise.)

The problem posed by Zeno's paradox of the tortoise and the hare arises because of the way in which the question is presented. The paradox focuses on identifying the moment at which the hare will catch the tortoise and presents the analysis in the form of what we recognize today as converging infinite series. *See, e.g.*, Terrence W. Deacon, *Incomplete Nature* (2012) p.11. By continually slicing the distance between the two racers in half, the distance measured will become increasingly (infinitesimally) close to the point at which the hare catches the tortoise, but will never actually completely get there. Likewise, the time that elapses will become ever increasingly (infinitesimally) close to the moment of that event, but will never actually completely get to it. Both infinite series approach the moment and position that the hare catches the tortoise, with that moment and position as limits. In the real world, of course, time does not have a limit. The moment of interest will necessarily come and go, and both racers will pass through the point at which they are "neck and neck." *See also*, Mazur, *The Motion Paradox*, pp.130-3.

As with the material world (or three dimensional space), powerful mathematical techniques enable calculations with great precision and considerable convenience based upon the assumption (a simplifying assumption) that space, matter and time are all continuous. Since the supposed units are extraordinarily small, one can safely say that for practical, every day life and purposes, the assumption of continuity is accurate.

THE ROLE OF TIME IN PHYSICS

Contemporary (and non-traditional) physicist Lee Smolin asserts the major developments on the road to modern physics resulted in "the expulsion of time from physics" or, more precisely, "from the physicists' conception of nature." *Time Reborn*, pp.55, 93.

Smolin points first to the construction of most theoretical physics, starting with Newton, in terms of mathematical models. Mathematics is timeless, that is, it exists outside of time and is, essentially, eternal. One might agree with Plato that mathematics exists in a real but separate (and timeless) world, or one might simply recognize that mathematical objects and operations are independent of time. *Id.*, pp.6-10, 36. Mathematics simply does not include time—time can be a variable (as an input or output) in the mathematical models. It is that the models exist independent of the flow of time.[8] Time, like the observer himself, is outside of the theory.

The related aspect of the independence from time is that almost all of our scientific theories in physics are time reversible. A common illustration of this feature is the fact that a video of a ball in motion can be viewed running forward and in reverse and one, with no other information, cannot determine which is which. *Id.*, pp.51-2. The motion is the same in either direction. "Time-reversal symmetry continued to hold up as the scope of the laws expanded. Maxwell's equations of electromagnetism and Einstein's revised equations of gravity both have it, for example, as do the quantum versions... . ...[I]n 1964, ... James Cronin, Val Fitch, and their collaborators discovered a tiny, obscure effect in the decays of K mesons* that violates [time-reversal symmetry]." Wilczek, *Fundamentals*, pp.189,190.

Yet, while the direction of time is irrelevant to the mathematical models, it is not irrelevant to the "sense" of the theories. Motion occurs in time. Absent the passage of time, there would be no motion. Thus, in a fundamental way, our understanding of these timeless theories depends upon time. On the other hand, the Newtonian theories certainly do not explain or even describe time. Instead, time itself is outside of the explanatory paradigm. It exists independently of what is being observed and predicted in the theory. Moreover, time is not affected or influenced by what takes place within the model (which reflects the subsystem under examination).

Note that in Newtonian mechanics, space is also external. It is not explained by the theory; it simply exists. It is the canvas on which the mechanics of motion are depicted, but it is not affected nor influenced by the motion or by the objects in motion. Newton assumed that space was absolute, that is, that it could be considered as a given, presumably timeless and unchanging. Newton thought that time also was absolute, marching on without regard to anything that occurs in this world or in the Universe. Aristotle, in contrast, thought that time marked or counted change and, thus, was dependent upon what actually happens in the Universe. See, e.g., Nina Emery, Ned Markosian, and Meghan Sullivan, "Time", *The Stanford Encyclopedia of Philosophy*, Winter 2020.

Of course, "[t]he laws of elementary physics do not speak of 'causes' but only of 'regularities,' and these are symmetrical with regard to past and future." Carlo Rovelli, *The Order of Time* (2018), p.169.

THERMODYNAMICS

> "...the device by which an organism
> maintains itself stationary
> at a fairly high level of...orderliness
> ...really consists [of] continually
> sucking orderliness from its environment."
>
> Erwin Schrodinger
> "What is life?
> The Physical Aspect of the Living Cell"
> 1944

Thermodynamics is the study of heat and its relationship to energy and the performance of work. The Laws of Thermodynamics have a relatively exalted status in physics.

Man has been fascinated by heat for thousands of years. Aristotle proclaimed that heat was the source of life. But, what is heat? How does it behave? Can it be measured? It appeared that heat naturally "flows" from hot to cold, and not the other way around. That followed from the assumption that heat was "something" (and that cold was just the absence of heat). It was also recognized that the sensation of heat was caused by a differential in temperatures, that is, that the experience of feeling hot and cold is relative.

From Galileo onward, considerable progress was made in devising techniques to measure temperature. The ability to measure temperature reliably facilitated exploration of the nature of heat. Curiously, it was discovered that two different materials in the same oven could emerge with different temperatures even though they were heated the same. So, temperature was a characteristic of things, not a measure of the quantity of heat. But, what is "temperature"? And, what is the relationship between temperature and heat? (Insight into those questions had to wait until the general acceptance of the theory of the atom at the end of the nineteenth century, already described in the chapter Chemistry).

The investigations that led to what became thermodynamics originally dealt with heat flows and mechanical processes (the steam engine) and were largely the domain of engineers. Frank, *About Time*, p.109. At the beginning of the Industrial Age, engineers realized that heat could produce work or movement (notably, the steam engine). Also, they recognized that movement could produce heat (friction). Most importantly, a scientist named Sardis Carnot demonstrated that the work performed by the steam engine was proportional not to the temperature of the boiler, but to the temperature differential between the boiler and the radiator, *i.e.*, the temperature "loss" occurring during the operation of the engine ("Carnot's Principle"). Michael Guillen, *Five Equations that Changed the World: The Power and Poetry of Mathematics* (1996), pp.172-3, 178-9. So, what is the relationship between heat and work?

In the mid-nineteenth century, a Prussian scientist named Rudolf Clausius declared that heat and work were the same thing just in different forms, they were both "energy", as were numerous other known forces. (We define "energy" today as the capacity to perform work, measured by the amount of the mass and the distance it is lifted.) Observing that a steam engine produced mechanical energy but also radiated heat leaking from the steam system and from friction among the moving parts (which represented the inevitable inefficiencies of the machine), Clausius concluded that the total energy involved remained the same, just changing forms. He thus propounded a Law of Energy Conservation. (The Law of Conservation of Mass was set out In 1789 by Antoine Lavoisier.)

(Julius Robert Mayer, another German scientist, had published an apparently little-read article in the 1840s expounding his theory that the

Universe began with an enormous single force which then began breaking down into many other forces, the sum total of which remained the same. Of course, Mayer's theory of the beginning was heresy, but others similarly argued that the totality of forces in the world was constant. Id., pp.190-4.)

Subsequently, Clausius proposed a new concept more comprehensive than energy. It included energy and temperature. He named it "entropy." After some deliberation, he concluded that entropy did not stay constant over time like total energy did, but always increased–a Law of Entropy Non-Conservation. Id., pp.197-208. All natural processes saw energy move from hot to cold, never from cold to hot. To change the direction of energy transfer required the performance of work.

Clausius had effectively stated the First and Second Laws of Classical Thermodynamics, without knowing what energy or entropy are. See also, Peter Atkins, Four Laws That Drive the Universe (2007), pp.54-5.

> "It is [Rudolf Clausius] who grasps the fundamental issue at stake, formulating a law that was destined to become famous: If nothing else around it changes, heat cannot pass from a cold body to a hot one. ...Clausius introduces a quantity that measures this irreversible progress of heat in only one direction and, since he was a cultivated German, he gives it a name taken from ancient Greek–entropy... ."

Carlo Rovelli, The Order of Time, pp.23-4, 25.

In the later nineteenth century, James Clerk Maxwell and Ludwig Boltzmann developed what was to become modern or Statistical Thermodynamics based upon the then relatively novel theory that matter is made up of atoms and that temperature is a reflection of the motion of atoms or molecules. Smolin, Time Reborn, p.201. They used the statistics of very large numbers to develop a model of the dynamics of atoms and molecules in motion. Id. ("It is not that we can never observe the fate of a single small group of atoms or even of a single atom. We can, occasionally. But whenever we do, we find complete irregularity, co-operating to produce regularity only on the average." Schrodinger, "What is life?".)

Curiously, the Laws of Thermodynamics are general propositions not subject to empirical testing in the sense of being falsifiable. It is claimed that every empirical investigation to date has confirmed the First Law, but that we still do not understand why that Law is true. Gray, Reactions, p.61. So, how do we know that it is and always will be true? It is also said that we know intuitively that the Second Law is true and that it is established by mathematics. *See, e.g.*, id, pp.57, 61. Yes, our experience tells us that broken eggs do not put themselves back together again (and, in fact, practically speaking cannot be put back together again), but does that prove that entropy always increases? Of course not.

THE FIRST LAW

The First Law of Thermodynamics states that in a closed system (a system isolated from other sources and uses of energy), total energy is conserved (that is, the total energy stays the same, even if the forms in which it appears change—e.g., from heat to light or energy to work, or *vice versa*). Of course, we have few examples of truly closed systems in nature, except, perhaps, the Universe itself (assuming that there is nothing outside of it that can influence it). Man can create systems that are closed, to varying degrees. Energy is conserved because it cannot escape a closed system, by definition. It changes form, but it does not disappear.

As we shall see, in modern cosmology, the status of the First Law is somewhat compromised or challenged. It is far from clear how the Big Bang could occur, creating the Universe out of nothing, or how particles can pop in and out of existence consistent with the proposition that energy/mass cannot be created or destroyed. Perhaps our Universe is not in fact a closed system or, perhaps, when we understand enough, we will be able to construct an accounting that demonstrates the conservation (or non-conservation) of energy.

Certainly, we might ask what empirical evidence we actually currently have for the First Law? The answer is particularly elusive, because we still do not know what "energy" actually is. We often talk about energy as if it were a "thing". We purport to measure it; we certainly use it. But, energy is actually more a characteristic of things, than a thing itself. Like momentum or velocity or temperature.

A QUOTATION ABOUT INVARIANCE

> *"The requirement that the laws of nature will not change if we spin around and determine them while facing different directions is called rotational invariance. The requirement that the laws will not change if we move from place to place is called translational invariance. These seemingly trivial requirements turned out to be astonishingly powerful in the hands of Emmy*

> Noether, whom Albert Einstein described as the most important woman in the history of mathematics. In 1918 Noether published a theorem that revealed a deep connection between invariance and the conservation of particular physical quantities. ...[I]f the laws of nature remain unchanged irrespective of the direction in which we are facing, then there exists a quantity that is conserved. In this case, the conserved quantity is called angular momentum. For the case of translational invariance, the quantity is called momentum."

Cox and Forshaw, Why $E=mc^2$, pp.59, 60.

THE SECOND LAW

The Second Law is one of the few laws of classical physics that seem to incorporate or reflect the concept of the one-way passage of time: things break, they do not repair themselves; rocks crumble or erode, they do not grow. See, e.g., Frank, About Time, p.110. In addition, the Second Law reflects our practical awareness that any conversion is not perfect or 100% efficient, so when "work" is performed in a subsystem (some part or area of an otherwise closed system), there is always lost energy, generally in the form of heat (e.g., there can be no perfectly efficient steam engine). The consequence is an increase in entropy for the system as a whole.

Since order (organization) or structure in one area (a subsystem) can be created only through the performance of work, energy in an ordered form must be removed from some other subsystems surrounding the one in question; and, since the transfer inevitably results in some loss or dissipation of energy back into the surrounding environment within the system, there must be an increase in disorder for the system as a whole. In other words, the increase organization within the subsystem necessarily will be less than the decrease in order in the rest of the system. But, the quantity of energy does not change.

For example:

> " ... [I]n the case of higher animals we know the kind of orderliness they feed upon well enough, viz. the extremely well-ordered state of matter in more or less complicated organic compounds, which serve them as foodstuffs. After utilizing it they return it in

> *a very much degraded form—not entirely degraded, however, for plants can still make use of it."*

Schrodinger, "What is life?"

Our Sun is the source of low-entropy energy that enables the creation of structured or ordered forms of life on Earth. We receive low-entropy energy from the sun during the day, mainly in the form of high-frequency, high-energy photons constituting yellow light; the Earth returns high-entropy energy to space at night, mainly in the form of low-frequency, low-energy photons in the infrared range. Penrose, *The Road to Reality*, p.705.

It was initially asserted that entropy for the entirety can never decrease over time but subsequently recognized that entropy may very briefly decrease where highly improbable ordered structures occur by chance as a result of unpredictable fluctuations. See Smolin, *Time Reborn*, pp.201-2. Yet, the trend will be for entropy always to increase (even if some transformations occur that increase order in part of the system, like in the building of a structure), at least until a state of complete uniform disorder (maximum entropy) is reached, when all change will stop. See, e.g., Adam Frank, *About Time*, pp.111-3. Where order arises naturally, like in crystallization, there is generally heat released. So, in one location, entropy may decrease; but, for the whole system, it is said to increase.

The Second Law depends on the definition of the "system," i.e., the relevant space. Indeed, the relevant space for us may be a peculiar and particular subset of the Universe with systems in which the Second Law obtains.

> *"With regard to these systems, entropy is constantly increasing. There, and not elsewhere, there are the typical phenomena associated with the flowing of time: life is possible, together with evolution, thought, and our awareness of time passing. There, the apples grow that produce our cider: time. That sweet juice that contains all the ambrosia and all the gall of life."*

Rovelli, *The Order of Time*, p.151.

A curious law

The Second Law is unusual in several ways.

First, unlike other laws of physics, it is presented effectively as an inequality. Second, it is also probabilistic (any closed system in the next moment very likely will have greater entropy than it did in the prior moment of time). Penrose, *The Road to Reality*, p.690. Indeed, it expressly deals with statistical characteristics of large numbers of particles, with averages representing typical rather than actual individual behavior. Penrose, *Shadow of the Mind*, pp.232, 233. It does not state that it is impossible for entropy to decrease (order to increase) in any closed system, especially in any particular finite time period. It only says that it is highly (maybe very, very highly) likely that entropy will increase over time. *Id.*, p.14; Hawking, *A Brief History of Time*, p.103.[9]

(All Laws of traditional physics may actually be probabilistic: "the laws of physics, as we know them, are statistical laws". Schrodinger, "*What is Life*". But, scientists essentially ignore the random activity of individual atoms when dealing with matter as we know it. Indeed, until quantum mechanics, discussed in the next chapter, physics was deemed to be deterministic, dealing with certainties, not probabilities.[10])

It seems that the Second Law is a supplement to, not a consequence of, Newtonian physics (or relativity)—it cannot be deduced from the other Laws of dynamics, such as the Laws of Motion. Roger Penrose, *Cycles of Time*, p.15. Thus, Penrose, for example, does not consider the Second Law to be a "fundamental" Law but more of a law of convenience given our present state of knowledge. *The Road to Reality*, p.692.

The end (and beginning) of everything?

The Second Law means that for the Universe (which we shall assume is the ultimate closed system) entropy will inevitably increase over time and do so until a state of thermal equilibrium is reached, where no further entropy is possible.

At that stage, it would appear that the Universe will be in a complete state of disorder and that the processes generating life-giving heat and light will be extinguished or "dead." The Universe will be cold, lifeless and "empty." Aggregations of particles that constitute our stars and planets and rocks and even the elements (and, of course, living things like ourselves) will be gone forever. *See id.* No meaningful change will occur thereafter. (Of course, it is possible that the increase in entropy could occur like an infinite series converging on, but never actually reaching, infinity. Disorder would continue to increase in smaller and smaller increments, forever.)

But, if there will be an end of the Universe because of entropy, then there also must have been an origin. Otherwise, the end would have already occurred (if the increase in entropy has been going on for an infinite period of time). See id., p.114.

A NOTE ON TEMPERATURE AND HEAT

We did not yet answer two questions posed at the very beginning: What is temperature? and What is heat? The modern answer is as follows: Atoms and molecules appear at several different discrete energy levels. In an object, most possible energy levels will be represented, with the greatest number at the lowest level. The arrangement is referred to as the "Boltzmann distribution." That array (and the "average" energy level resulting from the number of atoms at each particular energy level) reflects the object's temperature. The more atoms or molecules there are at the higher energy levels, the higher the temperature. If two objects with different temperatures are put in thermal contact, the energy from the warmer object will move into the cooler object. The transfer is "heat"—a creature of the differential. See Atkins, Four Laws, pp.12-9, 28. We feel warmth and chills because of these differentials. By the way, our skin cannot "feel" wetness, only the temperature differences it causes.

DEFINING ENTROPY

Thermodynamics has developed into the "science of systems ... [applicable] to any and all systems: mechanical, biological, celestial. ... [A] universally valid theory of energy and its changes...." Frank, About Time, pp.109-10. It is asserted that for any system, "the first and second laws will always hold true." Id., p.113. So, the concept of entropy deserves further examination.

Entropy is often characterized as referring generally to randomness or lack of order. Typical examples used to illustrate the increase in entropy include the dispersion of gases in the atmosphere, the mixing of cold and hot water to become warm water, the tendency of regions of different temperature that are in thermal contact with one another to move toward a uniform temperature and the dispersion of a colored pigment throughout a liquid within an enclosed vessel. These phenomena actually are examples of the consequences of the Brownian motion of molecules. Take a container with a mixture of oil and water. In contrast to the prior examples, over time that mixture will not become more uniform but separate out into areas of oil and water. That phenomenon is a result of other characteristics of the constituent molecules. Indeed, one could say that the mixture reflected less entropy and the separation reflected an increase in entropy. See Penrose, *Cycles of Time*, p.25. In all cases, "concentrations" at particular locations in the container are viewed as indicators of low entropy—order or organization of the molecules as a result of some work that was done. Over time, as the concentrations tend to disappear, the entropy increases.

"Entropy is not an arbitrary quantity, nor a subjective one. It is a relative one, like speed. The speed of an object is not a property of the object alone: it is a property of the object in relation to another object." Rovelli, *The Order of Time*, p.145. Entropy reflects the relative degree of dispersion of energy (not the total quantity) within a particular "state space", *i.e.*, how spread out it is. The state space can be, but is not the same as, a physical space. Gray, *Reactions*, pp.59-60. Energy exists in the motion of atoms and molecules and in the bonds between atoms (which bonds can be "stressed" so as to hold more energy, *id.*, p.45).

The number of possible arrangements of that energy defines the entropy of the "state space." The more concentrated the energy is within the state space, the lower the entropy. Thus, for a particular compound (*e.g.*, water) or an element (like nitrogen), its solid form has less entropy than the liquid form which has less entropy than the gas form, since in each successive phase, the dispersion of the existing energy increases. The energy is "allowed" to be in more places as the state space expands. (At a molecular level, the question is how probable is it that you can predict certain characteristics of a random molecule. The higher the probability, the lower the entropy. See Atkins, *Four Laws*, pp.66-9.)

Again, the argument is made that there are many more high entropy state spaces than low entropy state spaces; and, once a compound is in a high entropy state space, it is extremely unlikely that it will ever return to the low entropy state space (remember, gravity does not seem to operate at the atomic level). Gray, *Reactions*, p.62. The increase in entropy generally involves the radiation of heat, which cannot be reversed.

Ludwig Boltzmann put forward a formal definition of entropy in 1875. Penrose, *Cycles of Time*, pp.17, 35. The goal was to achieve an objective and, at least theoretically, implementable definition so, for example, one could empirically determine whether one state actually has more entropy than another. The test should envisage the possibility of consensus as to the conclusion reached. Such a definition was to be largely (or completely) generalized, applicable to all situations. Id., p.35.

Boltzmann's definition was based upon descriptions of the physical world on two levels–a macro level and a micro level. The macro level (or macro-state) can be described by a relatively few number of variables or characteristics. The micro level (or microstate) is a detailed, precise description based upon the position and motion of every atom. Smolin, *Time Reborn*, p.196. The measure of entropy then is the number of possible micro-states that can give the same macro-state–that is, the number of different arrangements of, *e.g.*, atoms that would look the same when assessed by the parameters that define the macro-level picture. Id. An increase in entropy means an increase in the uniformity of areas of space as measured by various macro-level characteristics. See Frank, *About Time*, pp.28-30. In other words, under Boltmann's definition, the microscopic states reflecting randomness or disorder are the states that are indistinguishable one from another at the macroscopic level. The identification of the criteria of assessment at the macroscopic level (what one considers to be ordered) determines the relative number of states of order and of disorder. See Penrose, *The Road to Reality*, p.691.

> "Boltzmann has shown that entropy exists because we describe the world in a blurred fashion. He has demonstrated that entropy is precisely the quantity that counts how many are the different configurations that our blurred vision does not distinguish between. Heat, entropy, and the lower entropy of the past are notions that belong to an approximate, statistical description of nature."

Rovelli, *The Order of Time*, p.32.

An argument for the conclusion that entropy increases over time is that there are simply many more states of disorganization than of organization, so that the probabilities strongly favor higher entropy. See, *e.g.*, Hawking, *A Brief History of Time*, p.145; Penrose, *The Road to Reality*, pp.696-9. For example, the whole egg lying on a table can be realized in only a relatively small number of states of arrangement of the relevant molecules (more than one, obviously, because many minor differences can be accommodated by the bigger picture); in contrast, there are many, many more arrangements of molecules that would constitute the broken eggs splattered on the floor.

BUT, ...

There is at least a whiff of circularity here. Everything depends upon the definition of the macro state. For example, for any particular precise arrangement of molecules, whether one in which the egg is broken or one in which the egg is whole, there is only one such arrangement. One cannot say that any one such specific arrangement is more probable than another. (Remember, obtaining any specific series of heads and tails from a coin toss is no more probable than obtaining all heads (or all tails)—in both cases, each toss must produce a specific pre-identified result. In contrast, obtaining particular percentage mixes (or numerical counts) of heads and tails will be much more likely, because there are more tosses that can be allowed to give either result and still achieve the required totals.)

"If we think about it carefully, every configuration is particular, every configuration is singular, if we look at all of its details, since every configuration always has something about it that characterizes it in a unique way." Rovelli, *The Order of Time*, pp.31-2.

The characteristics defining the macro state are those that we find relevant, which depends on our world view, and that we are able to distinguish, which depends on our skills and our technology. Thus, the conclusion concerning increasing entropy comes from the fact that we have defined the states that constitute disorder such that they are more numerous than the states of order. *See, e.g.*, Dawkins, *The Blind Watchmaker*, pp.7-9 (there are many fewer arrangement of parts that are useful or "good for something" than arrangements that are not). Penrose acknowledges that "[t]here is still something very subjective about this definition [of Boltzmann]..., despite its being a distinct improvement on earlier notions of more limited applicability, and an undoubted improvement on the idea of just a measure of the 'randomness' of a system." *The Road to Reality*, p.691.

Indeed, the Second Law, like time, may be seen as depending on humankind for its meaning and, perhaps, for its existence.

THE SECOND LAW AND THE UNIVERSE

According to some theories, the early Universe consisted of matter broadly and rather evenly distributed—consistent with our intuitive picture of high entropy gas. Then, according to those theories, aggregations of matter arose due to gravitational attraction. And, ultimately, some day, all matter will be concentrated in various "black holes" of infinite density and, individually, of very small or no size.[11]

Such models would seem to the layperson to suggest dramatic decreases in entropy—movements toward greater organization of matter and energy. The argument of the astrophysicist in response seems to be that black holes display enormous entropy that greatly outweighs any reduction in entropy created by the formation of the aggregations of matter. See, e.g., Penrose, Cycles of Time, pp.74–5; cf. Hawking, A Brief History of Time, p.103. So, extraordinarily dense and compressed aggregations of matter, in which nothing can really be called dispersed, are examples of very, very high entropy. One might conclude, then, that the introduction of the influences of gravitational forces causes the physical manifestations of entropy (what it looks like) to be reversed. See Penrose, Cycles of Time, pp.74–5; The Road to Reality, pp.706–7, 714–6. See also, Rees, Just Six Numbers, pp.115–7. Or, one might conclude that we are defining entropy so as to get the "right" result.

In all events, if the Second Law applies to the Universe, then the cosmologist must explain how it is that the Universe came to achieve such an extraordinarily low level of entropy in the beginning that entropy could have been increasing for at least 10, and perhaps as many as 16, billion years yet there are still dramatic amounts of highly-ordered (low entropy) energy and matter within local regions of space (e.g., the galaxies, the solar systems, the stars, the planets and the living organisms). See, e.g., Penrose, The Road to Reality, p.705.

The original state of the Universe presumably was presumably close to thermal equilibrium, so it "must" have had very high entropy. Penrose, Shadows of the Mind, p.233 (the source of which state Penrose finds very difficult to explain). Then, the galaxies, the stars, the planets and matter itself would be the result of random local fluctuations, decreasing entropy in only a few localized areas (actually, many areas but representing a small part of the whole). Smolin, Time Reborn, p.221. In fact, Boltzmann, believing the Universe to be eternal, tried to explain the fact that the Universe obviously was not (yet) in thermal equilibrium (despite an eternity having passed), as the Second Law would predict, on the basis that local fluctuations, such as that creating our Solar System, were frequent, but temporary, disruptions decreasing entropy and delaying the end. See id., p.210.

However, one puzzling fact is that the Universe seems to be becoming more interesting over time, not less. Id., pp.210–11. Of course, there are vast regions of space that likely have very high entropy. The local abnormalities are complex, self-organizing systems that are not isolated, but have energy flowing through them. Id., p.218. Gravity also plays a highly important role, as a unique force that continuously attracts. Id., pp.223–6. (But, curiously again, it is argued that the low entropy state for bodies affected by gravity is one of wide and uniform dispersion; while the high entropy state is the massive concentration of matter, such as in a black hole. Penrose, *The Road to Reality*, pp.706–7.)

Penrose claims that the really interesting question is not why entropy increases, but why (and how) the origins of the Universe were of such extremely low entropy, as a result of which we have so much low entropy phenomena now or, ultimately, why the Universe is not already "dead." See *Cycles of Time*, pp.52–4. But, then, he makes the odd assertion that "[t]he fact that [the Big Bang] must have had an absurdly low entropy is already evident from the mere existence of the Second Law of Thermodynamics." Id., p.726. Surely, this statement has the matter of proof backwards.

SPECIAL RELATIVITY

So, we come to Albert Einstein and the theory he called the Special Theory of Relativity ("Special Relativity," as a shorthand).

He began with two axioms, assumed to be true:

- The observed speed of light is constant.
- There is no absolute space.

On the basis of these two axioms, Einstein developed his Special Theory of Relativity. It was called the "Special Theory", because it applied only in the "special" cases of uniform motion and recognized only relative motion. Cox and Forshaw, *Why $E=mc^2$*, pp.41-2.

EINSTEIN'S FIRST AXIOM: A CONSTANT SPEED OF LIGHT

Maxwell's work led Einstein to consider trying to formulate a theory based upon the assumption that the speed of light in a vacuum is a constant. Actually, the proposition needs to be stated somewhat differently. The speed of light in a vacuum will be perceived by all observers in uniform (non-accelerating) motion to be the same. Many of his predecessors had struggled to explain how the speed of light could be a constant for all observers. Einstein changed the nature of the problem. See Adam Frank,

About Time, p.132. He started from the premise that the speed of light was constant and undertook to derive the theoretical implications of that assumption. In effect, Einstein decided simply to disregard all of the truths of classical physics and embark on a mental adventure of discovery. So, the first axiom was not an arbitrary assumption—it appeared to be an empirically established, although puzzling, fact. But, it carries with it the implications that both time and distance are also not absolute, since variations in one or the other or both are necessary to maintain the condition that the observed speed of light is the same for all observers in uniform motion. *See, e.g., id.*, pp.42-44.

EINSTEIN'S SECOND AXIOM: ABSOLUTE SPACE

If Earth, or the Sun, was at the center of the Universe, then one could specify positions in space and measure motion relative to the Earth (or Sun). But, astronomers concluded that Earth moved around the Sun, the Sun moved within the galaxy and the galaxy moved within the Universe. What about the Universe? Does it move relative to something else?

The concept of relative motion is not strange to us. We readily recognize that on an airplane in flight, one can walk up and down the aisle or sit in one's seat. Certainly, we say we are moving when we are walking, but we are obviously also moving even when sitting. Of course, we also know that Earth is in motion, so that even standing still on the ground, we are moving. More generally, we can easily see that for two objects in isolation, it is an arbitrary decision to say that one is moving and one at rest, or *vice versa*, or that both are moving, but differently (i.e., at different speeds or in different directions or both). The important insight is credited to Galileo, refined by Newton. *See, e.g.*, Hawking, A *Brief History of Time*, p.17. Galileo abandoned the concept of absolute space and assumed that all we can do is talk about motion of objects relative to each other (and they, then, relative to some other object and so on).

> *"Galileo's great insight was to draw a profound conclusion from this seeming paradox. It feels like we are standing still, even though we know we are moving in orbit around the sun, because there is no way, not even in principle, of deciding what is standing still and what is moving. In other words, it only ever makes sense to speak of motion relative to something else."*

Cox and Forshaw, $E=mc^2$, p.11.

Did Galileo's analysis and Newton's mathematics establish that there is no such thing as absolute space—that is, a grid that is stationary and relative to which the location of everything else can be specified and its movement calculated? Not exactly. Aristotle believed that a body could be "at rest" and that that was its natural state, which would be altered only when being acted upon by a force. Id., p.17. The Newtonian concept of inertia included uniform motion as a natural state, which also can be altered by the application of force, and there is not any necessary implication that there is no absolute standard of rest. The conclusion is just that we have no way of ascertaining what it is or of observing it.

I can assert that there is such a thing, but I must acknowledge that I do not know where or what it is and can identify no evidence of its existence. You can assert that there is no such thing, but you also can point to no evidence. So, we again follow one of the methodological guidelines of modern science: If the existence of something makes no detectable difference to the facts we can identify, we assume that it does not exist. That makes things easier, at least.

(A more interesting question is whether time is absolute. That is, is there a single "clock" by which the passage of all events in the Universe can be measured? This question was not answered, or really even asked, by Galileo. Everyone knew that time was immutable. "Both Aristotle and Newton believed in absolute time. That is, they believed that one could unambiguously measure the interval of time between two events, and that this time would be the same whoever measured it, provided they used a good clock." Hawking, A Brief History of Time, p.18.)

So, the second axiom was derived by logic applied to experience. For a person moving at a constant speed in a straight line (experiencing no acceleration), there is no test or experiment one could imagine that would enable that person to determine whether he was moving or standing still. And, the situation does not change if one introduces a second observer (or object). Thus, Einstein concluded, all non-accelerated motion (i.e., where neither the speed nor direction is changing) is necessarily relative and relative not to some construct of absolute space but to other—all other—objects. There is no absolute space. (Although, it would be possible to determine when there is acceleration, deceleration or change in direction—it is uniform movement that is not detectable). See, e.g., J. Richard Gott III, Time Travel in Einstein's Universe: The Physical Possibilities of Travel through Time (2001), p.41.

THE SPECIAL THEORY

There is no absolute time or absolute distances (or lengths).

Clocks, mirrors and trains

Cox and Forshaw present the issue by hypothesizing, following Einstein, a "clock" that measures time by the reflection of light between two mirrors. Why $E=mc^2$, pp.42-45; see also, e.g., Greene, The Elegant Universe, pp.37-41. One tick is the round trip passage of the light to the opposing mirror and back.

Now, this special clock is put on a fast moving train. For an observer on the train, nothing has changed. However, for the observer on the platform as the train goes by, the light from one mirror back to the first has traveled farther than the distance between the two mirrors, because of the motion of the clock. It not only goes between the two mirrors, but it has to move in the direction that the train is traveling in order to catch up with the opposing mirror that is moving. The resulting path of the light, when observed from the platform, is a triangle. (This phenomenon was recognized in the experiments of Michelson and Morley, discussed above.)

Next, since the distance traveled is different from the perspective of the two observers, if one assumes that the speed of light is constant relative to both observers—one on the platform as well as one on the train—then, the time that passes must be different for the two observers. This conclusion is simple "math" or logic: distance equals speed multiplied by time. If the speeds are the same in two equations and the distances are different, then the times must be different as well. More importantly, this conclusion is a necessary consequence of the assumption that the speed of light is constant—the first axiom. So, if the speed of light is always constant, then time cannot be an absolute—it must vary. See, e.g., Greene, The Elegant Universe, p.36.

(Such a conclusion also would arise from the somewhat different assumption that the speed of light is the universal speed limit or even from the assumption that there is a universal speed limit. See Frank, About Time, pp.133-4.)

Of course, if the speed of light varied depending upon the circumstances, then distance and time could both be constant. In addition, if "distance" varied depending upon circumstances, then the speed of light and time could both be constant.

And, aging bodies

One might (too) quickly conclude that this odd result is a function or characteristic of the strange form of clock being used. Since light is assumed to behave differently than other things with which we are familiar, a light clock may also behave differently.

However, here is where the second axiom comes in. If one assumes that the light clock can give a different time than would a more traditional clock (put one on the train and one on the platform and see if they keep the same time), then it would be possible to decide who is moving and who is not, in contradiction of the second axiom. The reason is that if the clocks can give different times, then one could put a light clock and a traditional clock on the train and one of each on the platform and the pair that fell out of synchronization over time would be the pair that was moving.

But, such an experiment would require that there is a standard or circumstance of no motion in which place the pair of clocks would stay in complete agreement. If that were the case, then we would have a concept of absolute space. See Cox and Forshaw, *Why $E=mc^2$*, pp.44-45; Greene, *The Elegant Universe*, p.40. Note that this same reasoning applies to the body clock, that is, to the aging of people and animals. In other words, if the light clock slows down relative to some other clock when it is moving quickly relative to that other clock, then so must the aging of the body or else one could make observations that would reflect absolute space, that is, observations leading to conclusions as to who is moving and who is standing still.

So, we conclude that there is no absolute time, because we have assumed that there is no absolute space. Again, it is useful to remember that these results are derived logically from the assumptions that are made—in particular, the assumption that a person in constant (non-accelerated) motion cannot possibly determine whether he is at rest or in motion. If that assumption were withdrawn or relaxed, then the glue that ties his argument together, as a matter of irrefutable deductive logic, dissolves.

The differences in the passage of time for bodies in relative motion, however, are almost unobservable until the difference in relative motion reaches 90% or more of the speed of light. *See, e.g.*, Cox and Forshaw, *Why $E=mc^2$*, p.49. To create aging differences that would seem to matter to us as people (not scientists) would require speeds that are unimaginable for anything larger than an elemental particle. *Cf.* Gott, *Time Travel in Ein-*

stein's Universe, p.33. Cox and Forshaw give as an example the calculation of the difference in time on the train compared to the time on the platform that would accumulate over 100 years if the train is moving at 150 kilometers per hour: one-tenth of a millisecond. As a result, one might ask how significant it is for our daily lives that time is not absolute

And, so ...?

In sum, as we noted above in the section on Light, in the late nineteenth century, Fitzgerald hypothesized that length is relative to motion, and Lorentz hypothesized that mass is relative to motion. Then, in the early twentieth century, Einstein hypothesized that time is relative to motion. And, since Special Relativity denies (hypothetically) the existence of absolute space, motion itself is only relative. Relativity.

The construct is certainly impressive, but have we proved anything? We have shown that as a matter of logic (a deductive theory), the necessary consequence of Einstein's two axioms is that time is relative and that relative time depends on the relative speeds of motion. This reasoning is a big step beyond the observation that the motion we observe is always relative to the observer, which simple proposition is not very troubling. Yet, one might think that we are engaged in a lot of maneuvering in order to explain this apparent curious characteristic of light.

There is, perhaps, an easier way to express the conclusion. If the speed of light is constant for all observers and all sources, then the only way that the standard formula for speed can remain true is to have time and/or distance change in different circumstances. Such adjustments are mathematically necessary in order to maintain the equality. Indeed, for the persons or things traveling at very fast speeds, the distances traveled have to shrink, since for them time barely passes.

Einstein's Special Theory of Relativity provides that different entities record the passage of time at different rates based upon their relative motion; so, there is no single clock in the Universe. Each person (and other thing) has his or its own "clock" that will "tick" more slowly or faster than someone else's "clock" whenever there is a meaningful difference in their relative speeds of movement.[12] The differences are measured in relation to the speed of light, so the "clocks" will be practically the same for everyone and everything we see and experience on Earth and in most of the observable Universe. (I note that, as discussed below, the General Theory of Relativity has given rise to experimental evidence that the passage of time is also affected by the presence of objects of large mass.)

By the way, it is not clear whether these theories reflect features of the Universe or only of our abilities (and inabilities) to perceive it. Maybe from the perspective of a super-being, things look different. Of course, if absolute space exists but is something we cannot perceive or detect, then should it not be proper for science to ignore it? What cannot be part of our mathematics or a basis for our predictions is irrelevant to science (or, at least, to technology). But, is it irrelevant to our efforts actually to understand the Universe?

SOME IMPLICATIONS?

TIME TRAVEL

Special Relativity indicates that as a body (say a space traveler) approaches the speed of light relative to another body (say, Earth), then the experience of time by that first body will slow dramatically relative to the experience of time by (or on) the second body. "... [A] fast-moving clock runs slow. A fast clock can indeed travel five light-years for every year that it records if it moves at ninety-eight percent of the speed of light." Rees, *Just Six Numbers*, p.69. This implication means that it could be conceivable for a person to travel to distant galaxies, even though the distances would be in the millions of light years. Traveling at almost the speed of light, the experience of time passing for the traveler over a trip of, say, 3 million light years might only be about 50 years. Of course, on Earth, 3 million years would have passed. *Id.*, p.55.

THE TWINS' PARADOX

Such reasoning creates the so-called Twins' Paradox. One twin stays on Earth; the other travels far into space at close to the speed of light. One has aged 60 years and the other only about 3 years. And, the aging of each is proper, given each one's own particular point of reference. *See, e.g.,* Frank, *About Time*, pp.134-5.

But, here is a curiosity. From the standpoint of our traveler, who—once she reached the speed at which she would then travel—would feel as if she were standing still, Earth could be perceived to be receding at a speed approaching the speed of light. So, relative to the traveler, does time on Earth slow down to a virtual crawl, so that the traveler would experience the passage of time at a much faster pace than would the inhabitants on Earth? Remember that we start this discussion based upon the assumption that there is no absolute standard of space or place, so that all that can be said about two objects is that they are in positions that can be described relative to one another and that they are in motion relative to one another. Either one can be taken as "standing still." So, our conclusions about the passage of time for either are not absolute, only relative to each other. From that perspective then, object A can be said to be moving close to the speed of light, relative to object B, so time passes much more slowly for A than for B. Or, one can say that object B is moving close to the speed of light, relative to object A, so time passes much more slowly for B than for A.

One might say that these statements are paradoxical from the standpoint of an observer C, who can watch A and B and "see" the passage of time and the aging process for both. But, because there is no point of absolute motionlessness (no absolute space); such an observer would also necessarily be moving relative to A and B. Thus, he cannot observe the process untainted by the changing nature of time.[13] Now, what about the scenario where one of the twins reverses direction and ultimately returns to meet the other twin. At that rendezvous, will one twin have aged more than the other? If they have both gotten "younger" (or "older") relative to the other, then perhaps we find that they have each aged the same. Remember, we cannot talk about whether their aging was "normal," because there is no standard for "normal." Similarly, it makes no sense to ask how much either has "really" aged. There is no relevant standard.

This so-called Twins' Paradox is resolved because of the necessary acceleration and deceleration of the space traveler. See, e.g., Cox and Forshaw, Why $E=mc^2$, pp.96-101; Gott, *Time Travel in Einstein's Universe*, pp.66-69. When an object is accelerating (or decelerating), then it is necessarily changing its motion. If there are two objects and one is accelerating, then one cannot claim that there is no difference between the assumptions that one object is standing still and the other is moving or *vice versa*. The object undergoing a change in its speed or direction is clearly not "standing still." Thus, the process by which the space ship stops and turns around to come back affects the relative aging of the two twins. See Cox and Forshaw, Why $E=mc^2$, pp.96-101. The Earth-bound twin appears to age rapidly during the traveler's acceleration and deceleration.

In fact, the process of acceleration and deceleration theoretically causes the differences in the ages. For this point, we have to skip ahead to Einstein's General Theory of Relativity. In brief, acceleration (or deceleration) is equivalent to gravity. And, time slows down in the presence of gravity. So, during the periods of acceleration and deceleration, both outbound and on the return, the traveler would age more slowly than while the ship is traveling at a constant speed. Thus, after a 40-year journey consisting of an acceleration equal to "one g" outbound and a deceleration equal to "one g" inbound to the distant point, and the same in reverse to return to Earth, the traveler would return to an Earth that had aged 59,000 years but he would have aged only 40 years. Id., p.100. In such a case, one twin will have become much older but the other will have been dead for over 50,000 years.

Gott presents a slightly different argument. One of his twins stays on Earth; the other travels at 80% of the speed of light to a star 4 light years away (Alpha Centauri) arriving after 5 years of travel. Because both twins are traveling at a constant speed in a straight line (ignoring the relatively small movement of the Earth), each twin will perceive the other as aging only 3 years (since the other twin's clock will appear to be ticking at only 60% of the speed of the observing twin). So each twin will believe that she has aged more than the other.

To be more specific, the traveling twin believes that only 3 years have passed since her departure, because on her slowed clock, only three years have passed. Since she perceives that the other twin still on Earth has been moving close to the speed of light relative to her, she concludes that the Earth-bound twin has aged only 1.8 years (60% of 3). However, when the traveling twin turns around, decelerating to a stop and accelerating again, things change. *Time Travel*, pp.67–69. At that time, the Earth bound twin will have experienced the passage of 5 years and perceived her traveling sister to have aged only 3 years (60% of 5). At the moment of the reversal of direction by the traveling twin, she suddenly perceives that her Earth-bound sister has actually aged 8.2 years. The return trip takes 5 years Earth-time; it causes only 3 years of aging on the part of the traveler; and the traveler (now again in constant movement, without change of speed or direction) perceives that her sister on Earth has aged 1.8 years.

As a result, at the homecoming, one sister has aged 10 years and the other has aged only 6 years. No paradox, Gott proclaims. Id., p.69.

Travel to the Future?

Gott asserts that this illustration demonstrates the feasibility of traveling into the future. Id. His example again is an astronaut who travels in a rocket that accelerates at a rate of 1g. The astronaut would experience gravity equal to that of the Earth. By the time the astronaut had reached a speed of 99.9992% of the speed of light, she would have traveled 250 light years but have aged only 6 years and 3 weeks. If she slowed down at the same rate, she would travel another 250 light years and have aged another 6 years 3 weeks. At the end, she would be at a star 500 light years away while having aged only 12 years and 6 weeks. Then she repeats the process in reverse, arriving back at an Earth that is now 1000 years older while she is less than 25 years older. Id., pp.34–35. Gott calls this example "travel to the future".

I think that a moment's thought will reveal that the example is not what most of us would consider to be time travel. The astronaut arrives on Earth at a date far in the distant future, an accomplishment that would not be achievable through normal living. However, she cannot go back. In addition, she cannot communicate to her original colleagues any information about that future period. The experience is not unlike suspended animation or some modified form of cryonics (a frozen state): the aging process is suspended, time passes without any awareness or appreciation and one is revived in a different world (in terms of history). The main difference is that the space travel would be much more boring, because the traveler would be sentient throughout. Also, the space traveler would age some. Either way, I wonder who would voluntarily elect to undertake such an experience, especially with no knowledge of the state of the world to be found in that future time (potentially devastated, perhaps even made uninhabitable, by man-made or natural disasters or by disease).

Gott also suggests that we can go back in time by examining light beams that originated millions or billions of years ago. Id., pp.76–83. Again, this argument seems far-fetched, or at least over-sensationalized. Light from stars that originated 100 million years ago is part of our current world—not of any distant past one. Of course, it can convey information about the past (and from the past), but so do ancient documents and fossils.

SPACETIME

As noted above, we often use time as another dimension to mark an event. Three dimensions to identify where, one for when. So, space and time can precisely identify an event.

The geometric union of space and time was first proposed by the mathematician Hermann Minkowski in 1908 as a way to reformulate the Special Theory of Relativity set out in Einstein's 1905 papers. Frank, *About Time*, pp.137–8. Minkowski's work incorporated what has become called the Lorentz transformations. Mazur, *The Motion Paradox*, pp.186–8. It is a non-Euclidean "space" where Pythagoras' theorem has a minus sign instead of a plus sign. Cox and Forshaw, *Why $E=mc^2$*, pp.82–7. "[O]ur immediate challenge is to search for a new measure of distance in spacetime that does not change depending on how we move around relative to each other." Id., p.65.

Minkowski spacetime can be represented graphically using a vertical axis for time and a horizontal axis for space. Any movement in spacetime can be depicted as an arrow that graph, with a direction and a length. To generate a value for the length of the arrow, we multiply the time distance by a speed (distance/time) to create the vertical axis (to get units that can be combined). For convenience, we use the speed of light, c. "We can ... convert distances into times and vice versa if we use the equation we met earlier, $v = x/t$. With a miniscule bit of algebra we can write time $t = x/v$, or distance $x = vt$. In other words, distance and time can be interchanged using something that has the currency of a speed." Id., p.74.

So, the vertical axis measures "ct", the multiplying of time and speed to get a distance. The horizontal axis measures "d", distance in space. Since Minkowski spacetime is curved, non-Euclidean space ("hyperbolic space"), we use the Pythagorean Theorem with a minus sign. Assuming all of spacetime is curved the same, the length of vectors in spacetime will be the square root of the square of change in ct minus the square of the change in d. "The length of the arrow must be determined by Minkowski's formula for the distance in spacetime, and it is therefore specified by $(\Delta s)^2 = (c\Delta t)^2 - (\Delta x)^2$." Id., p.124.

We have made the assumption (hypothesize) that the resulting length of the arrow in spacetime is the same for all observers (of course, the apparent movement in space or in time separately will continue to differ). That invariant length as seen by all possible observers will appear on the time/distance graph, where the values on each axis reflect particular observers' viewpoints. And, finally, we introduce the assumption of a universal speed limit to preserve our sense of causality in the spacetime world (causes always occur before effects). Id., pp.87-9. Thus, we get (or create) another universal invariant value like the speed of light—movement in spacetime.

If we identify all the points in spacetime that have a spacetime vector of a certain length, we have a parabola. (Actually, four parabolas, but in order to preserve the time sequence of causality, only the one centered on the "ct" axis above zero can be valid.) "The velocity of an object moving through spacetime always has length c and it points in the direction in spacetime in which the object travels." Id., p.126. "To finish our construction of the new spacetime momentum arrow, all we need to do is multiply the spacetime velocity vector by the mass m. It follows that our proposed momentum arrow always has **a length equal to mc** and points in the direction of travel of the object in spacetime." Id. (Emphasis added.)

THE UNIVERSAL SPEED LIMIT?

We assert that the speed of light is a constant relative to all observers; but, as we all have been told, it is also a cosmic speed limit–nothing can go faster. How do we derive that conclusion? There actually are several arguments that give us that result.[14]

Proofs

Assume that the light clock made with the two mirrors moves faster and faster. As its speed becomes a significant percentage of the speed of light, the passage of time must slow noticeably, because the light beam is traveling farther and farther to bounce off the two mirrors. Suppose that the speed of the clock equals the speed of light. Then, the bouncing light beam will never be able to catch the opposing mirror, since the angled path between the mirrors is necessarily longer than the path along which the clock is moving. By our reasoning so far, time for the clock must actually stand still when the speed of the clock reaches the speed of light. If the speed of the clock were to exceed the speed of light, time would have to travel backwards, which we assume to be impossible. Magueijo, *Faster than the Speed of Light*, pp.32-3, 35-6.

In the late nineteenth century, Dutch physicist Hendrik Lorentz developed a formula relating the mass of a charged particle at rest to its mass when in motion. According to that formula, mass increases as its speed is increased, reaching infinity at the speed of light. See Mazur, *The Motion Paradox*, pp.180-1. So, the particle could never go faster than the speed of light.

And, since Newton, we have recognized that the faster an object moves, the more energy it has (by definition, kinetic energy=$MV^2/2$). If, as a particle approaches the speed of light, its mass continues to increase; then the force required to push it faster will increase. At the speed of light, the energy would have to be infinite. Thus, for a particle to travel at or in excess of the speed of light would clearly be impossible. *See, e.g.*, Greene, *The Elegant Universe*, p.52; Hawking, *A Brief History of Time*, pp.20-21; Magueijo, *Faster than the Speed of Light*, pp.35-6. So, actually, "T[t]e c in E = mc^2 should ... be seen more correctly as the speed of massless particles, which are absolutely forced to fly around the universe at this speed." Cox and Forshaw, Why $E=mc^2$, p.139.

Einstein theorized that all things travel through four dimensions, space plus time, and that the speed of things through the four dimensions always equals the speed of light. For the type of phenomena that we experience with our senses, most of the movement is through time and only a very small portion of the total speed is the result of motion through space. However, if something were to move through space at the speed of light, there would be no motion through time at all (all of the motion having been used up in the motion through space). So, the speed of light must be the limit. Incidentally, this reasoning implies that light does not age or get old—it is timeless. Greene, The Elegant Universe, pp.50-51.

It has been observed that an absolute speed limit marked by the speed of light is not, in fact, crucial to Einstein's Special Theory of Relativity: The key elements are that the speed of light is constant or the same for all observers, regardless of their relative positions or motions, and that all positions of observation are equal, that is, there is no absolute space or point of reference. See, e.g., Brian Greene, The Fabric of the Cosmos: Space, Time, and the Texture of Reality (2004), p.118. However, the theory does depend upon there being some universal speed limit, even if it is not light. See Frank, About Time, pp.133-34; Cox and Forshaw, Why $E=mc^2$, pp.87-9. The speed limit is the speed of a mass-less particle or an electro-magnetic wave. If it should turn out that photons do have mass (contrary to current belief), then the speed of light is very slightly slower than the limit.

The cone of light

Physicists are fond of using the image of the "cone of light." See, e.g., Hawking, A Brief History of Time, pp.24-9; Lee Smolin, Three Roads to Quantum Gravity (2001), pp.57-8. It reflects our understanding of causality—the cause must happen before the effect.

Assume a universal speed limit (e.g., the speed of light). On a diagram with time as the vertical axis, an event can be "seen" to have an inverted cone (or triangle in two dimensional representations) reflecting its relevant past, with the point appearing at the event. The further back in time one looks (lower on the vertical axis), the wider the area of phenomena that can have affected the event, reflecting the finite speed limit. The event will also have a cone for its future, depicting the range of future phenomena that could be affected by the event. That cone will have its point at the event and then get wider the further up the time scale one goes (the further into the future).

The central point is that anything outside of the two cones must be irrelevant and unrelated to the event. That is so because, given a finite speed limit, things that happened too far away from the event in space, for each point in time in the past, cannot have reached the point of the event before the event occurred. Similarly, parts of space too far away from the event, given the amount of time that will have passed at each point in the future, cannot be affected by the event because light will not have had enough time to get there. Thus, for example, the light cone for the disappearance of the Sun will not include Earth for the first 8 minutes. See Hawking, *A Brief History of Time*, p.27 (figure 2.6).

The heart of this concept is the assertion that information takes time to be transferred, because it cannot travel faster than the speed of light. Therefore, anything that occurred further away from an event, in distance and time, than light could travel before the event occurred could not possibly cause the event–the occurrence is necessarily irrelevant to the event.[15]

$E=mc^2$

The famous equation $E=mc^2$ arose out of Einstein's inference (or hunch) that energy and mass were theoretically interchangeable, providing the final piece of his Special Theory. See, e.g., Guillen, *Five Equations that Changed the World*, p.247.

We were long familiar with equations involving mass that relate to energy. Newton's formula for force is mass times acceleration, where acceleration is measured in d/t (for example, feet per second). If force is a measure of energy, then force equals m(d/t). The formula for kinetic energy is $mv^2/2$, where "v" is velocity. From the spacetime momentum arrow, the assumption that energy is a function of mass (and mass is a function of energy) and the conservation laws, the final formula can be derived, assuming c is the speed limit. See, e.g., Id., pp.247-53; Cox and Forshaw, *Why E=mc2*, pp.130-3.[16]

That is the formula for mass at rest, not moving. The total energy of a moving object is $mc^2 + mv^2/2$. (The impacts of speed are additive here.)

> "Mass contributes to inertia and gravity, but it is not the only factor. In particular, a moving particle has more inertia, and exerts more gravity, than a particle at rest. Indeed, the theory of relativity teaches us that it is energy, not mass, that controls inertia and gravity. **For bodies whose speed is close to the speed of light, however, E = mc2 is way off.** [P]hotons carry energy, and thus that they have inertia and exert gravity, despite having zero mass."

Wilczek, *Fundamentals*, p.232 (emphasis added).

We now recognize that (small) mass and energy transformations occur all the time.

> "A box full of hot gas has more mass than an identical box containing the same gas at a lower temperature. The temperature measures how fast the molecules are whizzing around inside the box—the hotter the gas, the faster the molecules move around. Because they are moving faster, they have more kinetic energy ... and hence the box has more mass. The logic extends to everything that stores energy. A new battery is more massive than a used battery, a hot flask of coffee is more massive than a cold one... ."

Id., pp.144-5.

Of course, the changes in mass are very, very small. A tiny amount of mass equals a lot of energy. (By the way, we have made reference to various Laws of Conservation. This formulation complicates such Laws, but does not overturn them.) We previously raised the question of what energy is: a thing or a characteristic of things. It resembles a characteristic of things, like weight or or color or momentum, rather than a thing itself. So, we have the proposition that two characteristics of all things are interchangeable and are "conserved" in total. But, matter is a thing, right? So, what gives matter mass (or, for that matter, what gives it energy)? (Keep reading.)

EXPERIMENTAL ERROR?

Let us leave the philosophy for a moment. The key question for the theory set out above is whether it predicts something that can be experimentally verified or tested. For example, can we empirically find an example of time passing at different rates?

In fact, scientists claim to have demonstrated the phenomenon of speed affecting the passage of tme. See, e.g., Cox and Forshaw, Why $E=mc^2$, pp.51-2; Greene, The Elegant Universe, p.42. They have done so through the observation of particles called muons, which are produced when cosmic rays hit the Earth's atmosphere. Magueijo, Faster than the Speed of Light, p.29. Muons are identical to electrons but heavier. When they "die," they turn into electrons and other particles called neutrinos. Of relevance here, muons have a very short and regular lifespan—about 2.2 microseconds—and are also very easy to accelerate to high speeds. Muons can be "shot" by a particle accelerator. So, scientists have accelerated muons around a 14-meter-diameter ring to a speed of 99.94% of the speed of light.

Simple calculations indicate that with a lifespan of 2.2 microseconds and a speed close to the speed of light, the muons should be able to make 15 trips around the ring during their life. However, observations reveal that these fast-moving muons make close to 400 trips. Cox and Forshaw, Why $E=mc^2$, p.52. Therefore, it must be the case that time has "slowed down" for the moving muons, so that they can travel 29 times as far as they would have been able to travel in a normal lifespan.

But, a version of the paradox reappears. From the viewpoint of the muon, how can it travel so far in so little time? The "answer" given is that from the standpoint of the muon, the distance (the circumference of the ring) shrinks so that it travels only the distance generated by the formula of 2.2 microseconds times the speed of 99.94% of the speed of light. Id., pp.53-54. Thus, for the scientist observer, the muon lives 29 times longer than it should; from the viewpoint of the muon, it travels 1/29 of the distance that the scientist measured! See also, Greene, The Elegant Universe, p.42

Efforts to measure the speed of light have resulted in different results over time. See Sheldrake, The Science Delusion, pp.92-3. Variations over time in measurements have occurred with respect to most, if not all, of the fundamental constants. Id., p.89. Such results can be attributed to experimental error and improving technology.

Some recent experiments conducted at a particle accelerator at CERN (the European Center for Nuclear Research) appear to have recorded tiny particles called neutrinos traveling a tiny bit faster than the speed of light. The difference was infinitesimal. The announcement of the results generated a deluge of publicity. See, e.g., Dennis Overbye, "Tiny Neutrinos May

Have Broken Cosmic Speed Limit," *The New York Times*, September 22, 2011. The New York Times quoted Alvaro de Rujula, a theorist at CERN, as calling the announcement "flabbergasting. If it is true, then we truly haven't understood anything about anything. It looks too big to be true." Id.

This test was conducted by a research team using the acronym Opera. Within 6 months, another European research team, called Icarus, reported that their efforts to measure the speed of neutrinos fired from the accelerator at CERN reflected that the neutrinos traveled at, not faster than, the speed of light. The prior results from the other team may have been the results of technical "misfunctioning." Both teams plan to conduct further retests. Associated Press. "Neutrinos Not Faster Than Light in Retest, Proving Einstein Right," *International New York Times*, 17 March 2012. (On 21 March 2012, the paper published a correction to the earlier headline, noting that the retest did not prove Einstein was right, only that a challenge to his theory was rebutted.)

In a subsequent "op-ed" piece, professor of theoretical physics at the City University of New York, Michio Kaku, expressed barely disguised pleasure at the results of the retesting. "No, You Still Can't Go Faster Than Light," *The Wall Street Journal*, March 21, 2012. Professor Kaku, explained that as a consequence of the announcement of the original test results, "[t]he world of physics was thrown into turmoil because the bedrock of modern physics would disintegrate if [the results] were true. My research is in string theory, for example, which extends Einstein's theory of relativity. So if his theory is wrong, my own life's work would go out the window as well." (For reasons noted above, he clearly overreacted.) He describes in some detail the supposed mistakes apparently made in the first experiments, referring to them as "spectacular errors."

He notes, then, that "[e]xperiments like this are fiendishly difficult to perform, with scientists pushing the boundaries of particle physics and high-tech measurements." These experiments require the control measurement of phenomena at a scale that is almost impossibly precise. So, the errors were not really "spectacular". Mistakes will continue to happen. The appropriate response is to repeat the tests with great care.

Shortly thereafter, Overbye also reported on the updates. He quoted Sergio Bertoluci, CERN's research director, as saying, "The evidence is beginning to point toward the Opera result being an artifact of the measurement." Overbye notes that there is currently no theoretical basis for faster-than-light neutrinos (a bit of an overstatement), connecting his by-line: "The Trouble with Data That Outpaces a Theory." *The New York Times*, March 27, 2012. Overbye explains that a fact in science must "be understood, fitted into a conceptual framework that we have reason to believe in, or confirmed independently some other way... ." Id. As for the apparent mistake in the first experiment, Overbye is even more reserved. The conclusion is that that experiment yielded no usable results. The experiment needs to be repeated with the errors corrected. He paraphrases Dr. Laura Patrizzi, a member of the Opera team, as saying that "the neutrino affair had produced a wonderful discussion and a great lesson in how science is done. ...[S]cientists—even Einstein—proceed by trial and error." Id.

I add this recent development not because I am predicting that Einstein was wrong, but to illustrate that the real life (or real world) process of science is not always as open-minded or objective as we might pretend. Scientists are people too, and as people they face strong incentives to try to protect their personal investments, their reputations and their achievements.

At the same time, mistakes can tend to generate constructive introspection and appropriate. humility.

Accelerated motion?

This brings me to my last point about Special Relativity. What about the impact of acceleration on a single body?

We know that acceleration creates an experience of weightiness equivalent to the effects of gravity. Thus, the experience of weightlessness can be simulated by downward motion (toward Earth) at a speed that offsets the pull of gravity, and the experience of having weight can be simulated by acceleration in a weightless or (relatively) gravity-free environment like outer space. As is well known, these techniques are used in the training for space travel and in the design of living quarters for space residents. A similar example is that of the spinning bucket of water or the spinning person—the water rises along the edges of the bucket, creating a concave surface, and the person's arms begin to float or fly outward—the consequences of what we call momentum and inertia. These principles are utilized in certain amusement park rides, which most of us have experienced (or observed).

Acceleration presumes some point of reference; although, it need not be an absolute frame of reference—it can be any framework. In the case of two bodies moving at an increasing or decreasing speed relative to one another, one or both could be experiencing the effects of acceleration. We normally think of the effects of acceleration relative to Earth. But, if one could find a place in which there is no observable influence from other bodies (no light, no gravitational pull, nothing), would the same phenomena be experienced? That is, could one ascertain that one is spinning (accelerating)? Could one experience the feeling of weight (or of "gravity")? I believe that the physicists assume that the answer to all is yes. See, e.g., Greene, *The Fabric of the Cosmos*, p.74. However, there have been debates about whether it is yes and, if it is, then why. *Id*., pp.23–38, 72–4.

There have been concepts or theories explored relevant to this question that suggest something like Newton's absolute space. Ernest Mach proposed that acceleration is defined in relation to the combined mass—matter and energy—of space, that there is a theoretically identifiable and measurable concept of the cosmic setting in which one finds oneself and that accelerated motion relative to that setting could occur. *Id*. We might even think of the Mach space as the existing Universe (which suggests that there is something beyond or outside of the Universe—even if that something is nothing.) But, then, in our hypothetical world of complete nothingness (outside of our Universe), the normal consequences of acceleration would not occur or be experienced—indeed, the concept of acceleration itself would be meaningless.

Where would it leave Special Relativity, if the theory's central tenet that there is no absolute or preferred point of reference does not really hold?[17]

General Relativity

The goal of Einstein's General Theory of Relativity (1916) was to incorporate gravity into relativity, broadening the theory from the special case to general application. Einstein's General Theory introduces the concept of the "warping" of space and time as the means by which the supposed force of gravity is communicated between objects with mass. Greene, *The Elegant Universe*, p.53. Thus, the influences of gravity are transmitted at the speed of light. Einstein used a four-dimensional spacetime ("Minkowski spacetime"). Gravity is no longer thought of as a force, as in the Newtonian system, but as a cause of a "warping" of spacetime, an effect described explicitly by a set of equations formulated by Einstein. The result is a "curved" spacetime.

Supposedly, Einstein's work was inspired by a perceived conflict between Newton's Theory of Gravity and the Special Theory of Relativity: Newton's theory assumed that gravity operates instantaneously, while Einstein's Special Theory of Relativity yields the proposition that nothing can travel faster than the speed of light. See Greene, *The Elegant Universe*, pp.5–6, 53–7; Hawking, *A Brief History of Time*, pp.28–9. I must say that I find the claimed "conflict" to be a bit overblown (but the question of the speed at which gravity acts is interesting in itself).

The example offered is that if Earth's Sun were suddenly to disappear, the lack of light would take 8 minutes to be observed on Earth, whereas Newton's theory predicts that the lack of gravitational pull between Earth and the Sun would be felt immediately. *Id.*, p.56. Is this a meaningful illustration? What needs to disappear is the gravitational influence of the Sun. So one must ask, how does the Sun suddenly "disappear"? Even if the Sun were sucked into a black hole (a phenomenon discussed later), its mass and, therefore, its gravitational influence would continue; although, that influence would be minuscule compared to the gravitational pull of the black hole. If the Sun disintegrated, the process would take time and the dissipation of the mass generating the gravitational force would only gradually occur. Thus, it is hard to visualize practical situations in which the sudden, instantaneous change in mass (or the position of mass) could occur to give rise to the "problem."

This point brings us to my second comment. One could argue that Newton's theory of gravitation simply does not address the speed at which gravitational forces are conducted. "The force between two bodies, according to Newton's law, depends only on distance; time does not enter the picture." Einstein and Infeld, *The Evolution of Physics*, p.127. The mathematical formulation assumes that the effect is instantaneous, but unless the assumption matters to some real world observation or measurement, then it can be viewed as an innocent simplifying assumption, of which there are many in science. Newton's theory describes a state of gravitational attraction in which mass and distance determine its strength. Most matter and energy is already in place. Changes in distance will occur at speeds far, far less than the speed of light. The same is true for changes in mass.

So, it makes little practical difference whether changes in the gravitational force travel at the speed of light or instantaneously. One could presumably introduce a time element into Newton's theory without altering the results of the calculations made, only the ease of making them. It would be much more difficult to alter Einstein's Field Equations to change the speed of gravity.)

WHAT IS GRAVITY?

Johann Kepler, some 400 years ago, was able to analyze empirical data that had been collected by a Danish astronomer Tycho Brahe in the late 1500s and formulate three propositions or "laws" that were able quite effectively to forecast the movement of the planets. Then, Galileo "discovered" the principle of inertia—that something moving will continue to move in the exact same direction and speed unless acted upon by something else. Newton added the force of gravity to "explain" why the planets move in curving paths rather than continue in straight lines and the mathematical formula for the gravitational force, a function of the multiplication of the masses of the two bodies divided by the square of the distance between them. See Feynman, *The Character of Physical Law*, pp.4-9.

The strength of the gravitational attraction decreases in proportion to the square of the distance between the two bodies; if the distance doubles, then the pull of gravity is only one-fourth as strong. (If the distance triples, the pull is only one-ninth as strong.) So, the force of gravity diminishes rather quickly as the distance increases. The squared function appears pretty frequently in physics. Perhaps, the reason is that it is the basis of the geometrical relationship between the area of a circle and its radius (Pi times r squared). There is a similar important universality (in our three dimensional world) in the relationship between the length of a side of a cube and the volume or the radius and volume of a sphere—the third power.

The result was a theory or model of gravity that "explained" the movement of the planets and of other objects with mass, including cannon balls. For example, two objects of substantially different masses (a billiards ball and a cannon ball) would fall toward the ground at the same speed, because the greater gravitational attraction of the cannonball would be exactly offset by the greater inertia of the larger mass (as reflected in the mathematical formula). See, e.g., Hawking, *A Brief History of Time*, pp.16-7.

Yet, despite the great precision with which movements could be predicted, the phenomenon of gravity was not "explained" at all. Our predictions now are even more precise, but (as discussed as an example in the first chapter) we still cannot explain gravity. What is it? How does it work? As Richard Feynman observed, "Up to today, from the time of Newton, no one has invented another theoretical description of the mathematical machinery behind this law.... . So there is no model of the theory of gravitation today, other than the mathematical form." *The Character of Physical Law*, p.33. *See also*, Kline, *Mathematics*, p.361; Mazur, *The Motion Paradox*, p.181.

In fact, our lack of understanding is more profound. We do not even have an explanation for the foundational concept of inertia. "[T]he motion to keep the planet going in a straight line has no known reason. The reason why things coast forever has never been found out. The law of inertia has no known origin." Feynman, *The Character of Physical Law*, p.9. (However, I admit that to me, the law of inertia seems intuitively sensible, if not compelling.)

At some level, we are left with a set of laws, expressed mathematically, that seem to reveal important discoveries about the functioning of the physical world, the result of which creates a feeling of "understanding" that is, at best, misleading and, perhaps, just plain false.

Einstein's answer

In Einstein's General Theory of Relativity gravity is not a "force" that acts through or across space; the effects of what we call gravity occur as a result of the warping or distortion of Minkowski spacetime by objects with mass. And, the effects occur at the speed of light, not instantaneously.

(I note, however, that research and theorizing on gravity continue. For example, Claudia de Rham, a theoretical physicist at Imperial College, is a proponent of the theory of "massive gravity", discussed in the chapter on Cosmology. Her "more recent work covers other aspects of gravity. She is interested in the speed of gravity, which has never been directly measured and which theory predicts could in some circumstances be faster than light. She is also investigating whether gravity, like light, moves at different speeds through different materials." Hannah Devlin, "Has physicist's gravity theory solved 'impossible' dark energy riddle?" *The Guardian*, 25 January 2020.)

Einstein again utilized some simple assumptions and applied logic. He approached the analysis with a fresh start, free from biases arising from the accepted wisdom of classical theory, beginning with just basic experience and observations and utilizing the principle of equivalence (where two situations yield identical experimental results, they must in some fundamental way be equivalent). Frank, *About Time*, pp.138–39.

Accelerated motion again

As in the development of the Special Theory of Relativity, Einstein started with the proposition that a person in a sealed capsule could not determine whether he was in a state of standing still or of uniform motion. However, when the capsule was accelerating (changing speed or direction), then the occupant would experience the pull of inertia and that pull would enable him easily to distinguish the situation from that of uniform motion. (Accelerated motion was discussed at the end of the prior section.)

Then, Einstein asserted (or recognized) that the "pull" that would be experienced would indistinguishable from the perceived effects of gravity. If the experiences were equivalent, Einstein concluded; then there could be no fundamental differences between acceleration and gravity. See id. So, Einstein's "solution" of the problem of gravity, therefore, was based upon his recognition that "the force of gravity was inherently fictitious." Robert Laughlin, A *Different Universe (Reinventing Physics from the Bottom Down)* (2005), p.122.

Adding the experimental conclusion that two bodies under the force of gravity fall at the same speed regardless of mass, Einstein recognized that one could eliminate the force of gravity (or "turn it off") by allowing an object and everything immediately around it to fall freely. A contemporary example is the feeling of weightlessness that an astronaut has in a space craft orbiting the Earth—it is not a lack of gravitational force on the astronaut from the Earth (the distance is not sufficient to materially diminish the gravitational pull), but the fact that the astronaut and space craft are both in free fall. The same phenomenon occurs when an airplane is allowed to fall freely, creating an experience of weightlessness for the passengers. *Id.*, pp.121–2.

Thus, Einstein reasoned, gravity was not a force of attraction but the result of the free-fall paths of objects converging (so that, if allowed to continue in a free fall, the objects would collide). *Id.*, p.122.

Curvature of what?

Einstein proposed the theory that space was a medium that was subject to the influence of bodies with mass. Large mass objects cause curvatures in space, which curvatures have the consequence of altering the free-fall paths of other objects. An analogy is the convergence at the poles of the longitudinal lines on the Earth. *Id.* Latitudinal lines remain parallel going around the Earth (like concentric circles, if viewed from above). Longitudinal lines are furthest apart at the equator and then get closer and closer together until they meet at the north and south poles. When one attempts to create a flat map of the Earth, various distortions are necessary to accommodate the curvature. Thus, the shortest flight path on the globe (as a ball) may often be quite different from what it would appear to be on a flat map (such as over the pole, rather than directly west to east).

Of course, as discussed above, Einstein added time as a fourth dimension to create spacetime. Thus, the effects we call gravity would be the results of the curvature or warping of spacetime. General Relativity "links space and time in a four-dimensional mathematical curved universe that geometrically models the mass of an object as something that depends upon its velocity, its rest mass (mass at zero velocity), and the velocity of light." Mazur, *The Motion Paradox*, p.188. (The popularized analogy of a bowling ball on a trampoline was discussed critically at some length in the chapter Knowledge and Understanding.)

EINSTEIN'S "FIELD EQUATIONS"

We have been discussing the General Theory of Relativity in the form of the descriptive analogy that Einstein used. But, he set out the theory in terms of field equations, using the mathematics that had been successfully employed for electromagnetic fields. These equations (actually, one equation with ten sets of parameters called "tensors" corresponding to different coordinate systems) provided a powerful analytical tool. In addition, the mathematical formulation recognized gravity as a phenomenon of fields. At each and every point of space, there is a gravitational field value, and across points, those values will (generally) be different. The pattern of those values will reflect the strength of what we call gravitational attraction. Moreover, those values will change pursuant to predictable equations (relationships), with the changes propagating at the speed of light. And, these are equations "that can be applied by any observer and incorporate the remarkable circumstance that the speed of light, measured in any 'local' experiment, is the same however the observer is moving." Rees, *Just Six Numbers*, p.36.

However, Einstein and Infeld observed that the field representation in the case of gravity is an interesting and sometimes helpful means of displaying the relationship identified by Newton, but if someone used the representation to conclude that there were forces "traveling" through the force lines of the field, then they would be confronted with the absurdity of assuming that "the speed of the actions along the lines of force must be …infinitely great!" *The Evolution of Physics*, p.127. Thus, they treat the field concept as a mere construct. Fields become something real only in the context of Maxwell's theory of the electromagnetic field with its calculation of the speed of the waves. *Id.*, p.145. *Cf.* Rovelli, *The Order of Time* ("there is also a 'gravitational field: it is the origin of the force of gravity, but it is also the texture that forms Newton's space and time, the fabric on which the rest of the world is drawn"). p.74..

EMPIRICAL EVIDENCE

The predictions of General Relativity are very close to those of Newtonian mechanics, but observed minute deviations in the orbits of, first, Mercury and then (with even smaller deviations) other planets were consistent with the equations of General Relativity but not Newtonian mechanics, providing empirical confirmation of Einstein's theory. Stephen Hawking, *A Brief History of Time* (1988), at pp.30–1.

Einstein hypothesized a spacetime framework in which objects traveled in straight lines (unless acted upon by some force). Those straight lines in timespace, however, could appear to be curved lines in three-dimensional space. *See id.* So, a high-mass object like a star will cause a distortion in spacetime such that an orbiting planet will be traveling in a straight-line in spacetime but will appear to be traveling in an elliptical orbit when observed in three-dimensional space. General Relativity predicts that light waves will also follow straight lines in spacetime and, therefore, may appear to be bent by gravitational forces when observed in three dimensional space.

Observations by Sir Arthur Eddington in 1919, during a solar eclipse (so that light from the stars passing close to the Sun could be seen) demonstrated that light was affected by the gravity of the Sun. *Id.*, p.32. It was observed that the position of stars appearing near the edge of the Sun would appear to be in different positions relative to other stars than when those stars were seen to be further away from the Sun. It was concluded that when the light from the stars under examination did not pass close to the Sun, the light came in straight lines to the Earth (relatively, at least), and when the light passed close to the Sun, its path was bent, causing its position relative to other stars to appear to be different.

Note that this evidence does not mean that photons of light are subject to gravity as such (they are supposedly massless); it is argued that their paths follow straight lines in spacetime, which is curved by gravity. However, General Relativity predicts that light traveling away from a large-mass body will lose energy. The loss of energy will cause the frequency to decrease, meaning that the time between the crests of the waves will increase. Id.

GPS

General Theory of Relativity also predicts that time as well as space will be affected for an observer by the presence of bodies of large mass. Frank, About Time, p.141. Specifically, time will pass more slowly near a large mass. The impact of gravity on time is much more relevant to us than the impact of relative speed. One now important consequence of this phenomenon is that time will pass more quickly for an object in orbit around the Earth than for the same object on the surface of the Earth. This prediction was confirmed in 1962, using two very accurate clocks located at the bottom and top of a tall water tower. Hawking, A Brief History of Time, pp.32-3. This discovery has been used to calibrate the GPS navigation systems in wide use today, providing much greater accuracy over time. Id., pp.32-3; Rees, Just Six Numbers, p.38.

Gravitational waves

Although, in Einstein's General Theory of Relativity, gravity is not a separate force, the operation of gravity according to Einstein's field equations generates a flow of energy (in very small amounts) in the form of gravitational waves or ripples moving out from, for example, two orbiting bodies. Penrose, The Road to Reality, pp.464-6. Thus, the gravitational field contains energy, which is part of the total energy (and mass) of a system. Id., p.345.

Using a radio telescope at the South Pole, where the dry cold air generates less distortion and contamination, scientists led by John M. Kovac of the Harvard-Smithsonian Center for Astrophysics conducted what is called the Bicep2 experiment ("Background Imaging of Cosmic Extragalactic Polarization 2"). The researchers spent 6 years collecting data and three years analyzing it. They were looking for evidence of patterns of polarization in the background radiation, which required strenuous efforts to eliminate contamination from terrestrial and extra-terrestrial sources. (The existence of polarization was discovered in 2002. The polarization was detectable by very small temperature differences.)

In March 2014, the researchers announced at a news conference at the Harvard-Smithsonian Center and in two papers submitted to the *Astrophysical Journal* that they had detected swirling patterns in the background radiation itself that they identified as evidence of gravitational waves generated by the Big Bang. Dennis Overbye, "Space Ripples Reveal

Big Bang's Smoking Gun," *The New York Times*, March 17, 2014; Robert Lee Hotz and Gautam Naik, "Discovery Bolsters Big-Bang Theory," *The Wall Street Journal*, March 17, 2014. However, subsequent debate among scientists, some of which was conducted on Facebook, resulted in an acknowledgement by the authors of the study that interstellar dust may have affected the findings, producing much or even all of the signals that they detected. See Dennis Overbye, "Astronomers Hedge on Big Bang Detection Claim," *The New York Times*, June 19, 2014. The research team did further calculations and concluded that further observations were necessary to determine definitively whether the signal detected is of primordial origin. Raphael Faluger, J. Colin Hill, David N. Spergel, "Toward an Understanding of Foreground Emission in the BICEP2 Region," *Cornell University Library*, 28 May 2014, revised 20 June 2014.

So, the search went on.

In the United States, over some 40 years, the National Science Foundation spent about $1.1 billion on the search, funding two duplicate dedicated observatories, one in Louisiana and one in Washington State. These "LIGO" (Laser Interferometer Gravitational-wave Observatory) detectors are operated by Caltech and MIT. After a decade of no success, the detectors were shut down in 2010 for a $200 million upgrade, increasing their sensitivity.

In September 2015, the detectors appeared to observe evidence of gravitational waves from a massive collision of two black holes about 1.3 billion light years from Earth. More than 1000 researchers from some 15 countries spent the next several months checking the data. In February 2016, they announced the results: gravitational waves had been detected. Robert Lee Hotz, "Gravitational Waves Detected, Verifying Part of Albert Einstein's Theory of General Relativity," *The Wall Street Journal*, February 11, 2016; Dennis Overbye, "Gravitational Waves Detected, Confirming Einstein's Theory," *The New York Times*, February 11, 2016. The collision that was detected are believed to have involved two massive black holes that were 36 and 29 times the mass of the Sun. When they collided and merged into one, the resulting black hole had the mass 62 times that of our Sun. So, some three solar masses were lost in the collision, converted to energy presumably in the form of gravitational waves. That energy burst was 50 times the simultaneous energy output of all of the stars in the Universe combined. *Id*. It must have been a truly stupendous event.

The event was detected on Earth by apparatus consisting of two vacuum tubes, each some 2.5 miles long, placed at right angles to each other. Laser beams were being shot down each pair of tubes simultaneously, to be reflected back to the origin by mirrors hung at the ends of the tubes. New, improved instrumentation measuring the times at which each light beam returned to the point of origin was capable of detecting differences as small as one ten thousands of the diameter of a proton–a size too small to be seen even with our most powerful microscopes. This massive event caused the arms to move four one thousandths of the width of a proton, infinitesimal but far more than the minimum that the detector was capable of detecting. *Id.* These numbers give one some idea of the nature and force of the supposed gravitational waves.

Now, efforts are being made to use the detection of gravitational waves as a tool to discover other cosmic phenomena. *See, e.g.*, Davide Castelvecchi, "Gravitational waves hint at detection of black hole eating star: LIGO and Virgo observatories have spotted ripples from what could be the first-ever detection of this long-sought event," *Nature* 569, 15-16, 26 April 2019.

Some issues

Laughlin calls it "ironic" that Einstein started with the denial of absolute space in this Special Theory of Relativity and ended up with a General Theory of Relativity "conceptualizing space as a medium." *A Different Universe*, p.120. Something that can be "warped" would normally be thought to have content that was more than just relative relationships. Does that mean that there is such a thing as absolute space, so that an object can actually be at rest? We leave that question for now.

Similarly, while Einstein's Special Theory of Relativity supposedly eliminated the need for the "ether" in which light waves are propagated; Einstein's General Theory of Relativity (actually, the field equations in which he expressed the theory) incorporated a cosmological constant that could be thought to represent the mass density of the ether. *Id.*, p.123. And, experiments made possible by powerful, modern particle accelerators (and related theories) indicate that the vacuum of space is not actually empty but may be full of matter. *See, e.g., id.*, p.121.

In addition, in the Special Theory of Relativity, "the distance between two points in Minkowski spacetime does always satisfy $s2 = (ct)2 - x2$, and this means that it curves in the same way everywhere." Cox and Forshaw, *Why E=mc2*, p.220. But, in the General Theory, spacetime does not curve uniformly because of the warping caused by objects. with mass.

Laughlin asserts that "Einstein's Theory of Gravity...was an invention, something not on the verge of being discovered accidentally in the laboratory. It is still controversial and largely beyond the reach of experiment. ...The view of space-time as a nonsubstance with substance-like properties is neither logical nor consistent with the facts." Id., pp.119, 123.

TIME REVISITED

Einstein's theories, as noted, altered the conceptions of space and time. Time, like space, becomes both relative to the observer and simply another dimension in the time-space construct. Indeed, "[a]ccording to General Relativity, 'time' is merely a particular choice of coordinate... ." Penrose, *Shadows of the Mind*, p.384. "[O]ne has just a 'static' four-dimensional space-time, with no 'flowing' about it." Id.

This point deserves to be emphasized.

The concept of spacetime eliminates time as we know it. It suggests that any point in time (and space) is simply a location in an already existing structure. See Tegmark, *Our Mathematical Universe: My Quest for the Ultimate Nature of Reality* (2014), p.272. So, Einstein's General Theory of Relativity, unlike Newtonian mechanics, does expressly incorporate time. But, time, like the other dimensions, is reversible. Just as one can move up and down and back and forth, one can move into the past and into the future. ("Many laws of physics are time-reversible. One is Newtonian mechanics, another is General Relativity, still another is quantum mechanics. The Standard Model of Particle Physics is almost time-reversible, but not fully so." Id., p.52. See also, Barbour, *The End of Time*, p.22.)

In sum, as we noted above in the section on Light, in the late nineteenth century, Fitzgerald hypothesized that length is relative to motion, and Lorentz hypothesized that mass is relative to motion. Then, in the earl/y twentieth century, Einstein hypothesized that time is relative to motion. And, since Special Relativity denied (hypothetically) the existence of absolute space, motion itself is only relative. Smolin, *Time Reborn*, p.71.

In his General Theory, Einstein created a closed Universe (in his model), so that everything that exists must be inside the Universe, including time (or, at least, all clocks used to measure time). And, time is also affected by gravity. In sum, the one-way flow of time is not part of modern physics. Perhaps, the passage of time simply is not part of physical reality. Clearly, we experience it profoundly, but that may be a human phenomenon. The passage of time may be an integral part of how we think and perceive the world, but perhaps it is not part of the world outside of us.

So, again, what is time?

An Illusion?

Einstein's General Theory of Relativity, in which time is just another dimension of spacetime, suggests the possibility that everything, the past and the future, exists simultaneously and frozen for all time (please excuse the usage of the word). The only question for us, then, is where in the time and space dimensions do we happen to be focused. So, Smolin concludes: "If General Relativity is a true description of our universe, it's hard to escape the conclusion that time cannot be fundamental." *Time Reborn*, p.71.

Julian Barbour (another contemporary and non-traditional theoretical physicist), who Lee Smolin described as presenting "[t]he best thought-through approach to making sense of quantum cosmology" (*id.*, p.85), has argued that the Universe consists of an infinite series of slices or snapshots or frozen moments, all of which exist now and always. Each slice contains everything that exists at the particular instance that the slice represents, including all memories of past moments. See, *e.g.*, Barbour, *The End of Time*, pp.35–57. We experience individual slices at various moments, but all of the slices exist simultaneously. Thus, there is no real change that ever occurs, only the appearance (or illusion) of change as we perceive different slices.

This theory solves the problem of causality given the ambiguity of simultaneity, because there is no actual causality. As Smolin summarizes the theory: "Nothing can be the cause of anything, because in actuality nothing is happening in the universe." *Time Reborn*, p.86. The slices that correspond to what we would consider to be latter times are inevitably larger or more voluminous (since more things are included), which may explain why we perceive time to be moving in the direction of what we call the future. *See id.*, p.87. (There are, still, some marks of the passage of time in the world and in physics, which must be explained.)

Fundamental?

Of course, physics could be wrong, or misleadingly incomplete.

Lee Smolin argues that time is both real and fundamental. He asserts that the difficulties in modern theoretical physics result from the failure to treat time as an integral part of the theory. Thus, a new theoretical construction of physics is needed, with time as central. Interestingly, we know that the relentless march of time is a reality of our experience of the world. The "fact" that Newtonian physics accurately describes the path of a cannonball from the mouth of the cannon to the target and, as equally, the path that would be taken if traced from the target back to the mouth of the cannon does not mean that we might see the cannon ball fly backwards. There are at least two different things at work here: the causal explanation for the flight of the cannonball and the mathematical description of its path. One is a phenomenon that occurs in time; the other is description of a parabola that is independent of time.

Smolin questions the most basic premises of Einstein's theories of relativity. Perhaps, there is a preferred observer of natural phenomena, an observer for whom there is absolute motion and absolute time and an absolute position of rest. *Time Reborn*, pp.163–4. In such a world, simultaneity would not actually be relative; although, it may appear to be so for localized observers. Thus, locality has important, but potentially misleading, effects. (Typically, however, we are most concerned with things that are local, are close to us, since they are the most likely things to have an impact on our lives. *Id.*, p.174.)

Smolin also speculates that there may be connections among things much more complex than suggested by three-dimensional space—connections that effectively eliminate locality. Thus, quantum entanglement could reflect direct connections that must be measured by standards different from our normal concept of distance. *Id.*, p.183. Such connections would be fundamental. At that level, time is global and absolute. Space, therefore, as we know it, may not be fundamental but an emergent phenomenon, emerging from the physics of atoms. *Id.*, pp.172, 175. Perhaps, space is not infinitely divisible, but also is quantumized in irreducible "atoms" of space. He references "shape dynamics," a theory in which shapes are real but size (and distance) is relative. *Id.*, p.172. In these scenarios, Special Relativity is simply a feature of our experience of locality. *Id.*, p.191. According to Smolin, with the emergence of space, locality arises; with locality, the speed of light becomes a "universal speed limit." Similarly, within at least small regions of that space, the Newtonian paradigm applies. *Id.*, p.180.

Under this approach, as with other theories of emergence, there is the prospect that the laws of nature change. There are certain laws before the emergent phenomena and different laws after; although, perhaps, at a deeper level, the different laws can be found to be consistent or explained.

Alternatively, there may be meta-laws that specify how the laws of nature evolve or change. *See id.*, pp.242–3.

Smolin outlines the type of new theory that he envisages. First, the Laws of Nature are not timeless and unchanging. They change and evolve over time. Second, the process is central. Using a Darwinian model, Smolin suggests that there has been a series of universes created, and they have had different laws governing them. The universes with laws that favor the creation of new universes eventually come to dominate the set of all universes. Our Universe, he explains, has Laws that favor the creation of new universes and, therefore, not surprisingly, is readily available to be the one in which we find ourselves. *Time Reborn*, pp.123–39.

His Darwinian model, however, is lacking the necessary element–the survival pressure–that makes the process work. Also, he purports to base his theory on something other than the ability to sustain intelligent life, but that something happens to correspond to the conditions necessary for life. *See id.*, pp.130–1. He offers potent criticisms of other theories on the grounds that they do not appear likely ever to yield propositions that are falsifiable, but his claim that his approach avoids that fault by predicting a general value of the cosmological constant, which could be true or false, is overstated. *See id.*, pp.131–9. I would suggest, instead, that he has constructed a theory that incorporates the prediction that recent work indicates is the value of the constant.[18]

A HUMAN PHENOMENON?

It is part of our experience that we perceive things simultaneously through our senses that have actually occurred at different points in time. Touch has a delay measured in nanoseconds. Sight, a slightly longer one (the time it takes the light to travel to the eye plus the time it takes the eye to communicate to the brain). Hearing, considerably longer. The varying delays are the results of the differing rates or speeds at which information is transmitted. Moreover, our own internal mental processing takes time, so even the conscious experience of an event is not simultaneous with the occurrence of the physical stimuli impacting our sense organs.

Although we are keenly aware of the passage of time, our observations are all necessarily made in a particular moment, the present. So, we do not, and cannot, actually observe time passing. The "movement" of the hands of the clock is an inference from two or more observations.

As St. Augustine explained at the very beginning of the 5th century:

> "In Book XI of the *Confessions*, Augustine asks himself about the nature of time. ... [H]ow we can be aware of duration – or even be capable of evaluating it – if we are always only in a present that is, by definition, instantaneous. How can we come to know so clearly about the past, about time, if we are always in the present ? Here and now, there is no past and no future. Where are they ? Augustine concludes that they are within us: '**It is within my mind, then, that I measure time. I must not allow my mind to insist that time is something objective. When I measure time, I am measuring something in the present of my mind.**'"

Rovelli, *The Order of Time*, pp.181-2 (emphasis added).

We perceive the passage of time on a clock only because of memory, the memory of the prior observation. Similarly, our awareness of time is based upon memory and upon our ability (or compulsion) to anticipate the future.

> "This space – memory – combined with our continuous process of anticipation, is the source of our sensing time as time, and ourselves as ourselves. ...Time, then, is the form in which we beings, whose brains are made up essentially of memory and foresight, interact with the world: it is the source of our identity."

Id., p.189.

For example, music consists of our instantaneous perception of a note enhanced by memory of those preceding and anticipation of those to come. Thus, the concept of time may just be imbedded in the human mind, more of Kant's *a priori* knowledge discussed in the first chapter.

> "Kant ... interprets both space and time as a priori forms of knowledge—that is to say, things that do not just relate to the objective world but also to the way in which a subject apprehends it. But he also observes that whereas space is shaped by our external sense, that is to say, by our way of ordering things that we see in the world outside of us, time is shaped by our internal sense, that is to say, by our way of ordering internal states within ourselves."

Rovelli, *The Order of Time*, p.184.

We have already noted that entropy and the Second Law might be creatures of the human mind and perception. *See* Rovelli, *The Order of Time*, pp.148-9, 155. (There is evidence that some other animals perceive the passage of time, but not evidence that they are aware a future or a past.) However, note that this proposition, even if true, does not answer the question of the role of time in the physical world.

"Perhaps, ... the flow of time is not a characteristic of the universe: like the rotation of the heavens, it is due to the particular perspective that we have from our corner of it." *Id.* p. 150.

We need to be sensitive to the consequences that result from the choices made in the construction of a theory. If one leaves time outside of the box of things that one is trying to explain, then the resulting theory will not explain time.

To the extent that we are attempting to understand the Universe, leaving out time implicitly assumes that time is not part of the Universe. The formulation will necessarily be consistent with the belief that time is eternal. Alternatively, if time is assumed to be a part of the Universe, then one is implicitly assuming that time began when the Universe began. If it had a beginning, it will have an end.

So, how does one proceed if the questions one wants to answer include whether time is part of the Universe? The traditional approach would seem to be to try to construct theories under both assumptions and look for ways to evaluate the satisfactoriness of the alternative, competing results.

ENDNOTES

[1] We know that gravity operates over enormous distances of empty space, but magnetic influences can also reach substantial distances. For example, NASA's Voyager I, launched in 1977 (as was Voyager II), reached the outer edge of the Solar System in the summer of 2013 (over 35 years later). Instrumentation has recorded that, as expected, the "solar winds" consisting of particles emitted by the Sun had essentially vanished, no longer striking Voyager I in perceptible quantities. In contrast, and contrary to expectations, the magnetic forces generated by the Sun continued to be detectable some 11.5 billion miles from Earth. See Kenneth Chang, "Going, Going, Still Going? Voyager 1 at Solar System's Edge," *The New York Times*, June 27, 2013.

[2] For a long time, scientists thought that light moved through an "ether" that filled the Universe. Such ether would have provided a mechanical explanation for waves (and fields) extending through space; but, following Einstein, the concept of an ether largely disappeared (although, current thinking asserts that empty space is not actually empty after all).

[3] The concept of fields is captured by elegant mathematics. But, absent a more concrete physical explanation, the concept can be used to purport to explain phenomena in a rather unscientific way. For example, Rupert Sheldrake, a trained biochemist who has been pursuing theories in the realm of the para-normal for over two decades, suggests that there may be "fields" related to the mind that can interact with the fields of other minds and with inanimate objects. See *The Science Delusion*, pp.212-29. He speculates that psychic phenomena, like telepathy, may actually exist and be effectuated through such fields. Sheldrake emphasizes efforts to test empirically the fact of telepathic phenomena, but he offers no suggestions as to how to test his hypothesis of "fields."

[4] Sheldrake makes the reasonable argument that since paranormal phenomena would have identifiable physical consequences, science should be willing to undertake experimental studies to test such hypotheses. *The Science Delusion*, pp.214, 253-7. He cites some such studies that have been done, mainly by people we would tend to categorize as social scientists.

> "Extrasensory perception (ESP) is the claimed ability to acquire information without the use of the five senses. It includes telepathy (the information comes from another person), remote view-

> *ing (the information comes from a distant object or event), and precognition or retrocognition (the information comes from the future or the past). There is no reason in principle to assume that such a thing is impossible."*

Hands, *Cosmosapiens*, p.569.

The orthodox view of such efforts is probably accurately reflected in the comment of Richard Dawkins who, when referring to a study that tried to ascertain whether prayers for sick persons had any measurable impact on their recovery, observed that "[t]he very idea of doing such experiments is open to a generous measure of ridicule... ." *The God Delusion*, p.86. Dawkins forgets how many of our current scientific theories would have seemed preposterous to leading scientists of prior generations. By the way, Dawkins reported that that study of the effects of prayer (offered by one person for another) failed to find any. Id., pp.87-8.

[5] Actually, as described, there will be a difference in the distance traveled because the perpendicular beam has to travel at an angle, along the hypotenuse of a triangle, to reach the moving mirror and again to return to the observer. By tilting the splitting mirror slightly from 45°, it is possible to make the distances equal.

[6] With our existing understanding of time, however, we can (and some do) postulate that with given initial conditions and the Laws of Nature, the course of subsequent events is predetermined or given and that the Laws reflect meaningful causal relationships.

[7] This same type of problem arose in other areas of science. The difficulty of identifying and agreeing on a standard or fixed reference for temperature is laid out in detail in Hasok Chang, *Inventing Temperature: Measurement and Scientific Progress* (2004).

[8] The same is true, by the way, of the observer. It is an implicit assumption that neither the observer nor the fact of observation (or measurement) has any impact on the phenomena under examination. The ball moves independently of both. As we shall see, this assumption about the relevance of observation or measurement is altered in quantum mechanics.

[9] In addition, pursuant to the Second Law, if two systems are combined, the entropy of the resulting combined system will be greater than the sum

of the entropy of the two systems alone. Hawking, *A Brief History of Time*, p.102. Thus, with respect to addition as well, entropy is different.

[10] Those models were deductive theories in which the conclusions are logically necessary. Given a set of initial conditions, the rules of motion, for example, determine the outcome. The objective of the model is to predict the consequences of actions. There can be no surprises. (The problem of how the initial conditions are established still exists and is a question that is outside of the theories.) The flow of time plays no role. Smolin, *Time Reborn*, pp.50-1. So, classical mechanics is deterministic and reversible. See Susskind and Hrabovsky, *The Theoretical Minimum*, p.2. Pierre-Simon Laplace stated that "[w]e may regard the present state of the universe as the effect of its past and the cause of its future." *Id.*, p.1. The past determines the future and the present reflects or embodies the past.

[11] There are scientists who dispute that such laws can be applied to the Universe as a whole or can be useful in cosmology. *Id.*, p.115. *See also*, Smolin, *Time Reborn*, pp.76-88. *Cf.* Penrose, *The Road to Reality*, p.700 ("the 'system' under consideration really has to be the universe as a whole").

[12] This line of reasoning seems to suggest that something could move away from (or towards) us at an observed speed (distance divided by time) that is greater than the speed of light if the object is traveling at close to the speed of light, because the clock for the moving object is recording a much slower passage of time. *See* Rees, *Just Six Numbers*, p.69.

[13] Stephen Hawking says simply that there is a paradox only because of our persistent subconscious assumption of absolute time; if we accept that each person has his or her own relative time, then the paradox goes away. *A Brief History of Time*, pp.32-3. Julian Barbour makes the point that much of the perplexity that arises over relativity occurs because people persist in thinking of time as something external to the particular region of the Universe on which they are focused (or external to the Universe itself). Thus, it is difficult to understand how time can be different for different observers. The confusion is reduced if one remembers that we are talking about the different clocks (however constructed) of different observers. *See The End of Time*, p.137.

[14] Since Einstein, the speed of light has been believed to be the maximum speed at which anything, including information, can travel–a universal "speed limit." However, it happens that when traveling through a medium other than a vacuum, the speed at which light travels is reduced.

Interestingly, it appears that in certain circumstances, it is possible that other forms of energy when entering that other medium can continue at speeds greater than the reduced speed of light through that medium. John D. Barrow, *The Constants of Nature* (2003), p.301 (n.3 to Chapter 3). When that happens, radiation is emitted, known as Cerenkov radiation, after the Russian physicist who discovered the phenomenon. *Id.*

[15] The relativity of time still seems to cause certain ambiguities. Most obviously, if for one observer the occurrence happens before the event and for another observer it does not, can the occurrence in reality be a cause of the event? Does not the concept of causeation require a preferred or standard observer?

[16] The equation $E=mc^2$ incorporates the speed of light as the conversion factor (when squared) between energy and mass. See Rees, *Just Six Numbers*, p.52. If the speed of light varies over time or between places, then the relationship between energy and mass would vary as well.

[17] Indeed, some physicists have even concluded that there is such a thing as a cosmic clock, even though "local" time is still relative (to motion). See Greene, *The Fabric of the Cosmos*, pp.234–6.

[18] I am attracted to the assertion that time is fundamental. However, Smolin's concept of a new theory leaves me unimpressed.

Particle Physics and Quantum Mechanics

> "Babe, you're just a wave,
> you're not the water."
>
> Jimmy Dale Gilmore
> "Just A Wave, Not The Water"

Particle physics deals with the elemental or basic particles of matter (and, as we shall see, of energy) and the forces that connect these particles.[1] The simple question being addressed is: Of what does matter consist? If it is some type of particle, then we would want to know how those particles move around and what holds them together or pushes them apart. See, e.g., Brian Cox and Jeff Forshaw, *The Quantum Universe: Everything that Can Happen Does Happen* (2011), p.27. That investigation ultimately led to what is called quantum mechanics.

As we shall see, current theory postulates that these fundamental particles are not really bits of matter, as we would know it, but are more like waves in fields. Or, only waves. See, e.g., Hawking, *A Brief History of Time*, p.173. The developments in quantum mechanics certainly seem to undercut the theme of Jimmy Dale Gilmore's lyrics "Babe, you're just a wave, you're not the water." The assertion in the song is intended to diminish the importance of the singer in the life of his former lover—being "just a wave." Modern science suggests that it is only the waves that matter.

The (Sub) Microscopic World

When the modern, familiar theory of atoms emerged in the early 1930s, the elemental things were thought to be electrons, protons and neutrons. As previously noted, the earliest identified elemental particle was the electron, and it has remained relatively inviolate.

But,

> "[s]ince the advent of particle colliders in the 1930s, physicists [have] been slamming particles together at ever higher energies. ...Altogether, physicists detected hundreds of particles, all of which were unstable and decayed quickly. These particles seemed to have no apparent relation to each other, ...Then came Murray Gell-Mann. ...[T]he observed particles were composed of smaller entities that—for reasons not well understood at the time—had never been detected in isolation. Gell-Mann called the smaller constituents 'quarks.'"

Sabine Hossenfelder, *Lost in Math: How Beauty Leads Physics Astray* (2018), pp.24, 25.

The current theory, informed and tested by the use of enormous particle accelerators, holds that protons and neutrons are not elemental particles. They are comprised of three smaller particles called "quarks," which come in two forms—"up" and "down". (In fact, however, a proton cannot actually be physically split into three quarks: any attempt to split a proton results in six quarks.) Although, quantum theory says that the proton is made up of three quarks, it can be treated as indivisible and elemental. James, *Elemental*, p.169 (Appendix 2). So, now, "[e]verything you see in the terrestrial world and the heavens above appears to be made from combinations of electrons, up-quarks, and down-quarks. No experimental evidence indicates that any of these three particles is built up from something smaller." Cox and Forshaw, *The Quantum Universe: Everything that Can Happen Does Happen* (2011), pp.7–8.

It appears, nonetheless, that there is also prevalent in the Universe an additional particle called the neutrino, which plays an important role in and emerges from the process of fusion within stars like the Sun. Neutrinos are ejected from the Sun (and other bodies) and travel through space, rarely interacting at all with other matter. Thus, there are supposedly four particles that constitute the minimum necessary set of particles from which a Universe like ours can be constructed. Id,, p.198.

Yet, physicists have also concluded that there are or have been a number of additional particle types: the muon, four more quarks, the tau, the muon-neutrino and the tau-neutrino. These particles perform functions other than that of being building blocks of what we call matter. The muon, which was discussed in connection with the experiments concerning the Special Theory of Relativity, was detected in cosmic showers of particles. It is like a very heavy electron. The other additional particles made a very brief appearance in the Universe at the beginning and still can be made to appear (again, only very, very briefly) in connection with the high-powered collisions of particles of matter produced by increasingly advanced particle accelerators. Id., pp.8-10.[2] All of these particles are considered to be fundamental because it appears that none of them can be split into constituent parts.

Furthermore, each of these particles supposedly has an opposite or "anti-particle" (such as the positron, which is the anti-particle of the electron). It would necessarily be rare to find both particles and their antimatter equivalent in proximity to one another, because if a particle comes into contact with its anti-particle, the two destroy one another, generating energy but no other particles. Id., pp.8-9. (In theory, one could have a world made up of anti-particles, which would look pretty much like a world made of particles, and it is possible that one part of the Universe contains particles and a part elsewhere is made up of anti-particles.)

And, now the Higgs boson,

> *"...proposed independently by several researchers in the early 1960s, was the last fundamental particle to be discovered (in 2012), but it was not the last particle to be predicted. Last predicted—in 1973—were the top and bottom quarks, whose existence was experimentally confirmed in 1995 and 1977, respectively. ...But since 1973 there hasn't been any successful new prediction...."*

Hossenfelder, Lost in Math, pp.55-6.

The Standard Model

The so-called Standard Model emerged in the 1970s. See, e.g., Bryson, *A Short History of Nearly Everything*, p.211. See also, Leonard Susskind, *The Cosmic Landscape: String Theory and the Illusion of Intelligent Design* (2006), pp. 33-8, 51-61. It incorporates a large collection of particles and three of the four fundamental forces (excluding gravity). See, e.g., Cox and Forshaw, *The Quantum Universe*, p.196.

The assumption of "symmetry" has been frequently and very productively employed by theoretical physicists. The concept is that one can make a significant change but everything will still look the same. Think of a simple circle. You can flip it any direction and it is still a circle. Or a sphere. Roll it around, and it still looks the same. Theoretical physicists apply the assumption to mathematical formulations and see what happens. Often, it happens that the resulting equations are able accurately to predict experimental results. For particle physics, theorists made the assumption of a particular symmetry, called "gauge symmetry", and discovered an equation that describes the interactions of the elementary particles, by significantly narrowing the possible outcomes.

The result is the Standard Model of Particle Physics.

> "If we express this symmetry requirement in mathematical form, we see that it entirely fixes how the particles must behave. ...Moreover, the symmetry requirement automatically adds the force-carrying gauge bosons that must come with it. ...The standard model is an exquisite construct of abstract math, a quantum field theory with gauge symmetries."

Hossenfelder, *Lost in Math*, pp.52, 53, 54.

The equation is a (possibly incomplete) list of elemental particles and of the forces that connect them with mathematics specifying how these particles interact, based in part on theory, but also on observation. For example, the mass and the charge of electrons have to be determined experimentally and used as inputs to the Model. Indeed, there are 17 numbers that have to be determined by experiment. See, e.g., Alexander Unzicker and Sheilla Jones, *Bankrupting Physics: How Today's Top Scientists Are Gambling Away Their Credibility* (2013), pp.8-9. "[T]he number of elementary particles remains bewildering large, and the equations still involve numbers that ... can't be derived from theory alone." Rees, *Just Six*

Numbers, p.136. And, there is nothing that is easily visualized as a "model" comprehensible in terms of analogies to the world we know and experience. And, it is certainly not elegant or aesthetically pleasing. Leonard Susskind, *The Cosmic Landscape*, pp. 12-3, 111-30.

THE FUNDAMENTAL FORCES

We start with "forces". (Because a force can be attractive or repulsive, the forces are often referred to as interactions, *e.g.*, the "weak interaction".)

Physicists say that there are four:

- the electromagnetic force;
- the strong nuclear force;
- the weak nuclear force; and
- gravity (yet, in General Relativity, gravity is not a "force"!)

We described the electromagnetic force above in the discussion of light. We introduced he strong nuclear force in the chapter on Chemistry. When it was observed that atoms and various particles sometimes decay, the question arose as to why they do not always and rapidly decay, that is, what holds the constituent parts together? The answer is the weak nuclear force. These three forces have observable consequences, from the attraction of two things through magnetism, despite mass and inertia, to the stable structure of the atom (despite the attractive and repulsive forces of the opposite and same electric charges). Then, there is gravity.

It is important to note the enormous disparity in the strengths of the fundamental forces. Compared to gravity, the electrical forces that affect the constituent parts of atoms (the electrons, neutrons and protons) are stronger by a factor of 10 to the 36th power (or 1 followed by 36 zeroes). Rees, *Just Six Numbers*, p.33 (Lord Rees categorizes the relative strength of the electrical and gravitational forces as one of the basic six numbers that constitute a "recipe" for the Universe, *id.*, p.4). That is a huge difference. And, the strong nuclear force is a (mere) million times stronger yet.

Gravity is essentially non-existent at the level of atoms. It is assumed to exist, but to be so relatively weak as to be irrelevant (and unmeasurable) for the chemist concerned with the formation of molecules. Penrose, *Shadows of the Mind*, p.218. (Of course, gravity is always attractive, whereas electrical forces can attract or repel and cancel one another out. When mass becomes great enough, gravity can overpower the electrical forces. Rees, *Just Six Numbers*, p.32.) Gravity, however, is not part of the Standard Model. In fact, the underlying quantum theories have not yet been able to incorporate gravity.

In the late 1960s, Sheldon Glashow, Abdus Salam and Stephen Weinberg proposed a model in which the weak nuclear force and the electromagnetic force were two aspects of a single force called the electroweak force. For that work, they were awarded the Nobel Prize in Physics in 1979. There has also been speculation that all the forces arose out of a single unified force that existed following the Big Bang. The forces may have separated from one another or otherwise emerged, along with the Laws of physics, as the Universe cooled. *See, e.g.*, Dennis Overbye, "Space Ripples Reveal Bid Bang's Smoking Gun," *The New York Times*, March 17, 2014.

We have already discussed at some length that we do not know what gravity actually is. We also have noted that we similarly do not know the cause or nature of the electrical force. Not surprisingly, the same is true of the strong and weak nuclear forces. All we know is what these forces apparently "do".

THE ELEMENTARY PARTICLES

Fermions and quarks

It seems that categories of elemental particles differ because of something called "spin", proposed in 1925 by Samuel Goudsmit and George Uhlenbeck, both graduate students at Leiden University in the Netherlands. Sean Kean, *The Bastard Brigade: The True story of the Renegade Scientists and Spies who Sabotaged the Nazi Atomic Bomb* (2019), pp.61-3.[3]

"The basic idea of spin is that elementary particles are ideal, frictionless gyroscopes, which never run down." Frank Wilczek, *Fundamentals: Ten Keys to Reality* (2022), p.74. Spin is reflected in angular momentum, which can be measured. Particles with "spin" equal to some half-integer value are called "fermions," and they constitute the elemental particles of which things are made. Cox and Forshaw, *The Quantum Universe*, pp.132-3. (Particles with "spin" equal to some whole integer value or zero are called bosons, discussed below. *Id*. The most familiar boson is the photon or light particle.)

All fermions are subject to gravity, the weak nuclear force and the electromagnetic force. However, only certain fermions respond to the strong nuclear force. They are called "quarks," and they are the building blocks of protons and neutrons (with the neutron having a neutral electrical charge). Quarks come in two varieties called "up" and "down" quarks, the representation of which determines whether a combination of three quarks is a proton or a neutron (two up and one down making a proton and two down and one up making a neutron).[4] The strong nuclear force is the force that binds them into nuclei. (As already noted, the strong force is very strong relative to the other forces, but it operates only over very, very short distances.)

Other fermions respond to the weak nuclear force, but not to the strong nuclear force. They are called leptons, the best known of which is the electron. Neutrinos are also leptons. Like quarks, leptons are subject to gravity, that is, they have mass.

> "Neutrinos are ghostly particles that hardly ever interact with anything, and as such, most of them stream out from the sun as soon as they are produced without hindrance. The neutrino flux is so great, in fact, that around 100 billion of them pass through each square centimeter of the earth every second."

Cox and Forshaw, Why $E=mc^2$, p.163.

There are two other families of heavier quarks and leptons believed to have existed in the early days of the Universe, but all of the members appear to have decayed to the lighter families.

Positrons

Electrons with a positive charge (ultimately named "positrons"), which would be effectively the opposite of normal electrons, were "introduced by Dirac in early 1931 to solve a problem with his quantum mechanical equation for the electron—namely that the equation appeared to predict the existence of particles with negative energy." Cox and Forshaw, The Quantum Universe, p.192. Apparently coincidentally, Carl Anderson detected evidence in experiments observing cosmic ray particles of something that appeared to have a positive charge and a mass comparable to that of an electron. Id. "In order to make sense of a piece of mathematics, Dirac introduced the concept of a new particle—the positron—and a few months later it was found, produced in high-energy cosmic ray collisions." Id. (The relationship of the electrical charge of a particle to the strength of the electromagnetic force can be analogized to the relationship between mass and the strength of gravitational attraction. Greene, *The Elegant Universe*, p.10.)

Bosons

Bosons are the supposed "force carriers".

We know that photons are emitted when various changes occur within an atom. When quarks decay, there is an emission of W and Z bosons. Curiously, these W and Z bosons appear to have mass, unlike the more familiar photon. Symmetry, at least, suggests that there is also a "graviton" that "transmits" the force of gravity (if gravity is a "force"). See, e.g., Greene, *The Elegant Universe*, pp.8-13; Cox and Forshaw, *The Quantum Universe*, pp.197-9. (As discussed later, there is also some theorizing that gravitons have mass.) The emission or absorption of bosons accompanies changes in the nature of the elemental particles. For example, a proton is said to consist of three quarks, two down, each with a positive charge, and one up, with a negative charge. Thus, the proton has a net positive charge. When a down is converted into an up, there is a release of a W boson that carries away some positive charge, leaving a neutral neutron. See James, *Elemental*, p.97. And, when an electron absorbs a W boson, it becomes a neutrino.

But, what does it mean to say that these particles transmit forces? We understand that protons repel each other because they have positive charges. This theoretical formulation would say that they are kept apart by electromagnetic force through "photon exchange" (actions of the relevant force particles). Cox and Forshaw, *The Quantum Universe*, p.201. All that the theory actually says is that when particles decay or otherwise change, there is an emission of bosons. None of these observations tell us how any of the forces actually effectuate attraction (or repulsion), what the forces are or what role in that unknown process is played by the suppose carrier bosons.

ANTI-PARTICLES AND VIRTUAL PARTICLES

When Paul Dirac undertook the comprehensive analysis of quantum theories in the 1920s, he found that the theories could almost be connected to Newtonian mechanics but that they were not consistent with Einstein's laws of relativity. Lawrence Krauss, *A Universe from Nothing: Why There is Something Rather Than Nothing* (2012), pp.59-60. Fiddling with the mathematics, he realized that if one incorporated the fact that an electron has "spin," giving it angular momentum and either a clockwise or counterclockwise direction of rotation, then he could generate more complex mathematics that gave results consistent with relativity. *Id.* The problem was that the equations required the existence of new particles that were like electrons but had positive electrical charges. As a result, Dirac's theory was destined to be rejected, even by Dirac.

Yet, within two years, experiments with cosmic rays revealed evidence of such particles, which have come to be called positrons. *Id.*, pp.61-2. So, surprisingly, Dirac's theory, reinvigorated by this empirical evidence, became the basis for a large expansion in the number of particles thought to exist. So, the accepted theory now says that every elementary particle must have an opposite: an anti-particle. Moreover, to make the mathematics work, it seems that they all must also have companion "virtual" particles. See, e.g., *id.*, pp.61, 67.

THE FINAL LIST?

To summarize, the elemental particles consist of the basic four: electrons, neutrinos, up quarks and down quarks. The first two are called "leptons." The building blocks of matter are the quarks and the leptons. Then there are two additional families consisting of four particles, each with properties identical to the basic four but with masses many, many times larger. These families include the muon and tau as the two heavier electrons. Then, each of the 12 particles presumably has an anti-matter opposite. Finally, there are the particles that "transmit" the three forces: the photon for the electromagnetic force, the gluon for the strong nuclear force and the weak gauge boson for the weak nuclear force. These predicted particles have been found. Actually, to be more accurate, evidence has been found of particles with the correct charges and more or less the correct masses (and certain other characteristics) of the predicted particles.

BUT, LIST OF WHAT?

> "So far **we have done little more than trot off a list of which particles** 'live' in the master equation. The twelve matter particles must be added into the theory a priori, and we don't really know why there are twelve of them. [S]ince it seems necessary to have only four (the up and down quarks, the electron, and the electron neutrino) to build a universe, **the existence of the other eight is a bit of a mystery.**",

Cox and Forshaw, Why $E=mc^2$, p.181 (emphasis added).

The list is, in part, the result of the requirements of the mathematics. That is, these particles need to exist in order for the mathematics not to generate results that are contrary to experimental observations or other established theories such as Special Relativity. The central mathematical problem was that the solutions to the field equations resulted in infinities unless the particles involved are assumed to have no mass. Yet, our knowledge of the physics at issue indicates that many of the particles do have mass.

As discussed below, the introduction of a hypothetical Higgs boson, as the source of what we call mass, broke the symmetry of the equations and allowed the generation of sensible answers from the mathematics.

> *"According to our present best understanding, the primary properties of matter, from which all its other properties can be derived, are these three:* **Mass Charge Spin** *That's it. ...[T]he most important and remarkable point about our trinity of properties—mass, charge, and spin—is simply that there are so few of them. ...[O]nce you've specified the magnitude of those three things, together with its position and velocity, you've described it completely. ...**We have, in place of 'atoms and the void,' spacetime and properties.**"*

Wilczek, *Fundamentals*, p.76 (emphasis added).

So, "[t]hese are not particles in the usual sense of the word. They don't go around bouncing off each other like miniature billiard balls. Instead they interact with each other much more like the way surface waves can interact to produce shadows on the bottom of a swimming pool." Cox and Forshaw, *Why $E=mc^2$*, p.182. So, are they "particles"?

Why the question? One might say that the characteristics define the particles so any particle can be nothing more than the combination of these characteristics. If so, then I am making a distinction without a difference. But, I do not fully agree. It may initially be a matter of semantics whether one refers to a package of characteristics or gives that package a name (like positron). But, I think that the process of naming can create a sense of an identity that can be misleading (remember the example of the "gene"). I think we have a sense of a "thing" as being something more than merely the set of characteristics that describe it.

Particle Physics and Quantum Mechanics   343

A Note on Supersymmetry

In order to address certain apparent deficiencies in the Standard Model, including the apparent need for dark matter to add to the mass of the known particles in the Universe (discussed in the next chapter), physicist have proposed something called "supersymmetry" which predicts the existence of much more massive versions of the particles that have been identified. For each particle identified in the Standard Model, supersymmetry proposes that there is yet another, far larger one. These larger particles are created only at very high energy levels generated through very powerful particle accelerators. The results of experiments with the Large Hadron Collider, however, appear to conflict with the predictions of supersymmetry, calling into question the existence of these hypothesized super-particles. See BBC News, 12 November 2012.

A former practicing supersymmetry theorist explains:

"Supersymmetry postulates that the laws of nature remain the same when bosons are exchanged with fermions. This means that every known boson must have a fermionic partner, and every known fermion must have a bosonic partner. ...Since none of the already known particles match as required, we have concluded there are no supersymmetric pairs among them. Instead, new particles must be waiting to be discovered. ...All we have learned so far is that the superpartners, if they exist, are so heavy that the energy of our experiments isn't yet large enough to create them. ...Supersymmetry has much going for it. Besides revealing that bosons and fermions are two sides of the same coin, [it] also aids in the unification of fundamental forces and has the potential to explain several numerical coincidences. Moreover, some of the supersymmetric particles have just the right properties to make up dark matter."

Hossenfelder, <u>Lost in Math,</u> pp.11, 12.

Theorists have revised their predictions with respect to the additional potential particles. Experimental investigations will have to await the achievement of the full operational power of the Large Hadron Collider or, even, the expected construction of even larger colliders capable of causing collisions generating even higher energy levels (with a tunnel of perhaps 80 kilometers in length, more than three times the length of the Large Hadron Collider). Parker, presentation to the Trinity College Science Society, 11 March 2014. [5]

THE HIGGS BOSON

The Standard Model "stops making meaningful predictions at the energy levels involved in the collisions of almost light-speed protons" in the modern particle accelerators. Cox and Forshaw, *The Quantum Universe*, p.196. The problems with the mathematics can be solved by assuming the existence of another particle, which has been called the "Higgs boson," named after Peter Higgs (see below), the man who predicted its existence in 1964. Id., pp.197, 207.

The efforts to find underlying unifying theories had led to a single mathematical framework in which electromagnetism and the weak force could be presented as manifestations of a single force. Krauss, A *Universe from Nothing*, pp.xviii–xix. However, that framework required some explanation of how the particles involved, as well as the photon and the particles that convey the weak force, can have such dramatically different masses (none in the case of the photon). The hypothesis of a Higgs field with respect to which different particles react quite differently, giving some the characteristics of mass to varying degrees, including some that appear to be without mass, provided an answer. Id.

The concept of a field that was responsible for the existence of mass (and different masses) was proposed in three separate papers published within a few weeks in 1964, with five listed authors. A physicist from the United Kingdom, Peter Higgs, whose paper was actually published second, received the most recognition, leading to the identification of the field as a Higgs field. In 2013, Peter Higgs and Francois Englert (from Belgium) were awarded the Nobel Prize for Physics for this theoretical discovery, which had received its first empirical corroboration in 2012 (discussed below). Englert's co-author, Robert Brout, died in 2011, making him ineligible for the Nobel Prize. The three co-authors of the third paper lost out, presumably because their publication had been a few weeks later. See, *e.g.*, Gautum Naik, "Francois Englert and Peter Higgs Win Nobel Prize in Physics," *The Wall Street Journal*, October 8, 2013; Dennis Overbye, "For Nobel, They Can Thank the 'God Particle,'" *The New York Times*, October 8, 2013.

The proposed concept was that this Higgs field filled all space and that particles moving through it experienced different "drag" or resistance from the field, slowing them down and creating what we know as mass.

Certain particles, like protons and neutrons, experience significant drag and, therefore, have substantial mass (relatively speaking), while electrons encounter less resistance and photons none. Absent this field, all particles would travel at the speed of light and have no mass. Under such circumstances, matter as we know it would not (and could not) exist. There would be no planets, stars, rocks or living things. As with all other fields, as we saw above, the Standard Model postulates that there must be associated particles. Thus, we have the Higgs particle or Higgs boson.

The hypothesis is that so-called empty space is actually filled with particles. In addition, these Higgs bosons interact with certain other particles, causing those other particles to "zig-zag" through space in the process of going from one point to another. Which particles bounce off the Higgs bosons? The ones that have mass. The particles with no mass move in straight lines with no interaction. That means that the "cause" of mass in certain particles (even particles that may have no size, as discussed below) is the interaction of those particles with the Higgs bosons, which "selectively retard" the motion of those particles, slowing them and giving them what we identify as mass. Cox and Forshaw, *The Quantum Universe*, pp.203-10. Matter that consists of particles with mass will have the characteristics that we associate with mass, like inertia, momentum and gravitational attraction.

THE HUNT

The mathematics of the Standard Model predicted the existence of the Higgs boson, but no physical evidence of its existence had been found.[6] As is well known, CERN undertook the construction of the Large Hadron Collider (LHC) under the Jura Mountains to search for, among other possible particles, this Higgs boson. The "discovery or exclusion" of the Higgs boson was "one of the primary scientific goals of the Large Hadron Collider" built at the cost of billions of dollars at CERN. "Observations of a new boson at a mass of 125 GeV with the CMS experiment at the LHC," *Physics Letters B*, Vol. 716, Issue 1, 17 September 2012.[7] As the work proceeded over the years, the hunt was often characterized in the media as the ultimate test of the Standard Model. As a result, there was much public attention to and anticipation of the results of the first complete set of experiments to look for the Higgs boson, which were completed and reported in 2012.

THE RESULTS

On September 17, 2012, two papers reporting on the results were published in a peer-reviewed journal. The journal is *Physics Letters B*, the same journal that published the original paper by Peter Higgs in 1964. *New Scientist*, September 15, 2012, p.4. There were two papers, with lists of contributing authors several pages long, because there were two separate experiments.

The first paper (mentioned above) is titled "Observations of a new boson at a mass of 125 GeV with the CMS experiment at the LHC." *Physics Letters B*, Vol. 716, Issue 1, 17 September 2012, pp.30–61. This paper noted that the Standard Model theory did not predict the mass of the Higgs boson, so the researchers searched over a substantial range. However, the Higgs boson is predicted to have zero "spin." Based upon the statistical analysis of the data generated by the CMS experiment, the researchers concluded that the results were "consistent, within uncertainties, with expectations for Higgs." The second paper, entitled "Observation of a new particle in the search for the Standard Model Higgs boson with the ATLAS detector at the LHC," reported on the data collected through the use of ATLAS. Again, the statistical analysis of the results led to the conclusion that the experiment had generated "data consistent with" the existence of the Higgs boson. *Id.*, pp.1–29.

The initial announcement was made on July 4, 2012. A few weeks later, the team released further results that indicated that the observations of the new particle were found at the 5.9 sigma level in the Atlas tests and 4.9 to 5.0 sigma levels in the CMS experiment. A level of 5.0 sigma is a statistical result that is believed to indicate that the probability that the observations made could have occurred by chance, rather than as a result of the existence of a new particle, were one in 3.5 million. That level is deemed to be the "formal threshold for claiming the discovery of a new particle." *BBC News*, 1 August 2012. Both papers concluded with the observation that more data and more analysis were needed. The scientists involved were certainly more circumspect about the meaning of what had been accomplished than were the popular press reports or even other physicists, at least when writing for the general public. *See, e.g.*, Adam Frank, "Cracking the Quantum Safe," *The New York Times*, October 14, 2012 (he expected a Nobel Prize for "the discovery, after a multi-billion dollar effort, of the Higgs boson").

The BBC report of the initial announcement noted that "many questions remain as to whether the particle is indeed the long-sought Higgs boson;

the announcement was carefully worded to describe a 'Higgs-like' particle." BBC News, 1 August 2012. However, almost a year later, CERN announced that the analyses of the full set of data reported in 2012 appeared consistent with the predicted Higgs boson. Dennis Overbye, "CERN Physicists See Higgs Boson in New Particle," The New York Times, March 17, 2013.

YET, ...

A question remains whether the particle discovered was *the* Higgs boson or only one of a family of Higgs bosons. Id.

Part of the uncertainty is caused by the fact that the two different teams generated different estimates of the "weight" of the new particle. The Atlas team used two methods to measure the particle's mass, getting two results—126.6 GeV (billion electron volts) and 123.5 GeV—that vary sufficiently to appear to be statistically different results, rather than numbers within an expected measure of error. Dennis Overbye, "All Signs Point to Higgs, but Scientific Certainty Is a Waiting Game," The New York Times, March 4, 2013. The two results of the CERN team were closer together and consistent with a mass of 125.8 GeV. Id. Additional experimentation, data collection and data analysis may resolve the uncertainty. If so, then perhaps the usefulness of these extraordinarily expensive "collider" facilities may soon have peaked, at least with the energy levels currently attainable. Dennis Overbye, "Particle Physicists in U.S. Worry About Being Left Behind," The New York Times, March 4, 2013.

Interestingly, the basic use of particle accelerators has been to attempt to generate particles that were thought to have existed in the very early nano-seconds of the Universe, immediately following the proposed Big Bang, when the temperatures and levels of energy were vastly higher than what exists in the known Universe today. The accelerator is able to deliver to the particles that collide with one another very high energy levels, enabling the appearance—generally for a very, very brief instance—of these more exotic particles that are thought to have been abundant for a short time at the beginning. These particles are exotic and unnatural, thought to have once occurred naturally and to be part of what made the Universe we know, but not currently part of that Universe. Curiously, the Higgs boson, in contrast, is proposed as something that does exist and is very prevalent in the Universe today. We may wonder is there any way of observing it "in the wild," so to speak?

Quantum Mechanics

The "Discovery" of Quanta

In the second half of the nineteenth century, the wave theory of electromagnetic radiation encountered a theoretical obstacle in connection with the explanation of the heating of ovens, the solution to which proposed by Max Planck in 1900 would contribute to the development of "quantum theory" over the next ten brief years and, then "quantum mechanics" in the mid-1920s. See Julian Barbour, *The End of Time*, p.185.

Black-body radiation

The problem was the relationship between existing theory and actual experience with respect to the process that heats an enclosed cavity, referred to as black-body radiation (the emission of heat from black surfaces—surfaces that are not radiating light).

Prevailing theory derived from Maxwell established that energy from the walls of an oven would heat the interior of the oven by the creation of electromagnetic waves. Each wavelength would have the same energy level. Within a contained box, like an oven, the only waves generated by the energy from the walls of the oven would be ones that had peaks and troughs occurring in whole numbers that fit evenly between the walls of the oven. Nonetheless, even though the wavelengths present would be limited to those with whole numbers that fit, there would still be an infinite number of such wavelengths. If each wavelength had the same energy level, the sum of an infinite number of wavelengths would still be infinite. Thus, it would require an infinite amount of energy to heat the oven to any particular temperature. That was clearly a monumental flaw in the theory, because actual experience was to the contrary. Greene, *The Elegant Universe*, pp.88–90.

Note, however, that this problem was essentially one of the formulation of the mathematical or logical model. The way in which the theory was constructed gave rise to a scenario that did not make sense. But, the problem implicated both electrodynamics and thermodynamics (particularly, the Second Law and the theory of entropy). Still at issue at the time was "the fundamental question of whether all matter is composed of atoms." Helge Kragh, "Max Planck: the reluctant revolutionary," *physicsworld.com*, 1 December 2000.

Particle Physics and Quantum Mechanics

Planck's Constant

Planck found a "solution" by assuming the energy in electromagnetic waves is "emitted in little packets of energy, which he called 'quanta.' The word itself means 'packets' or 'discrete.'" Cox and Forshaw, *The Quantum Universe*, pp.5, 62–5. Planck assumed that the energy in the waves came in such discrete units and that the wavelengths that could occur were only those that contained total energy expressed in whole units—no fractions. Greene, *The Elegant Universe*, pp.91–2. Thus, the word quantum "entered physics in 1900, through the work of Max Planck." Cox and Forshaw, *The Quantum Universe*, p.5.[8]

Furthermore, Planck "posited that the minimum energy a wave could have is proportional to its frequency: larger frequency (shorter wavelength) implies larger minimum energy; smaller frequency (longer wavelength) implies smaller minimum energy." *Id.*, p.93. Thus, longer wavelengths that required minimum energy levels smaller than the minimum size unit in which energy came could not provide any energy inside the box. This limitation solved the problem of infinite energy by limiting the energy carrying wavelengths. In addition, Planck was able to calculate the size of such packets that would give results consistent with experimental observations. Planck computed the factor that established the proportionality between the frequency of a wave and the minimum energy of the wave, which factor came to be called the "Planck constant." *Id.*, p.93.

Planck, thus, introduced a "new constant of nature, the quantum of action, now called Planck's constant... ." Barbour, *The End of Time*, pp.185–6. Planck's constant came to be considered the smallest unit of "action," that is, of anything and everything, including motion. Up to this time, physicists had been assuming, and their mathematical models presuming, that "all physical quantities vary continuously." *Id.*, p.186. However, this smallest unit is very, very "tiny." *Id.* And, because it is so very small, our senses and even our most sensitive instruments at the time could not detect the individual units of matter or motion. But, the substitution of discrete, tiny increments in place of continuous functions (smooth variations in values) invited the use of different types of mathematics. Newton's differential calculus, in which limits are approached through increasingly small increments, no longer appeared literally to apply.

Of course, the idea that light might be a stream of "particles" was not new, going back to Newton. However, it appears that Planck's discovery was not the result of conceptual theorizing about the nature of light or matter or motion. Planck simply discovered that the introduction of quanta enabled the mathematics to produce results that were not nonsensical; indeed, it enabled calculations that corresponded closely to experimental observations. See Greene, The Elegant Universe, pp.92, 94. It was, in effect, a purely mathematical solution. Helge Kragh has referred to him as "the reluctant revolutionary".

> "As he explained in a letter written in 1931, the introduction of energy quanta in 1900 was 'a purely formal assumption and I really did not give it much thought except that no matter what the cost, I must bring about a positive result'. Planck did not emphasize the discrete nature of energy processes and was unconcerned with the detailed behaviour of his abstract oscillators. Far more interesting than the quantum discontinuity (whatever it meant) was the impressive accuracy of the new radiation law and the constants of nature that appeared in it."

"Max Planck: the reluctant revolutionary," physicsworld.com, 1 December 2000.

The question was whether Planck's solution corresponded to anything theoretically observable in the real, physical world.

EINSTEIN'S CONTRIBUTIONS

Planck's solution was supported by one of Einstein's famous four papers of 1905. In that paper (for which he was awarded the Nobel Prize in 1921), Einstein set out a theory of light as packets of energy or light quanta (later called "photons"). He was attempting to explain a phenomenon called the photoelectric effect. It was known that when light shines on certain metals, the metals will give off electrons. The energy from the light causes the electrons to come loose. The curious fact that was observed was that as the intensity of the light increased, a greater number of electrons would be ejected, but the speed of the ejected electrons was a function of the frequency (the "color") of the light and independent of its brightness. Moreover, if the frequency of the light became too low, then no electrons were ejected (the speed falls to zero) regardless of the intensity or brightness. Greene, The Elegant Universe, pp.94–95.

Einstein theorized that the electrons would be knocked off if the energy of the packets of light (the photons) was sufficient, utilizing Planck's theory that the energy of the individual photons was proportional to the frequency (the "color") of the light. Id., pp.95–6. As was typical for Einstein, he set out along with the theory a plausible experiment that could, conceptually, be used to test the predictions of his theory against reality. Lindley, *Uncertainty*, p.66.

The experiment was actually conducted by Arthur Compton, and the results were published in 1923. Compton obtained the predicted results. *Id.*, pp.96-7. So, the work of Maxwell and experimentation by Arthur Compton and his team ("bouncing" light off electrons) demonstrated that light had properties of particles, as well as of waves. *See also*, Cox and Forshaw, *The Quantum Universe*, pp.5-6.

> "'The hypothesis of quanta will never vanish from the world,' [Planck] proudly declared in a lecture of 1911. 'I do not believe I am going too far if I express the opinion that with **this hypothesis the foundation is laid for the construction of a theory which is someday destined to permeate the swift and delicate events of the molecular world with a new light.'** ... The changed status of quantum theory was recognized institutionally with the first Solvay conference of 1911... .. ...[I]t was not believed that quantum theory had anything to do with atomic structure. Two years later [1913], **with the advent of Niels Bohr's atomic theory, quantum theory took a new turn that eventually would lead to quantum mechanics and a new foundation of the physicists' world picture.**"

Kragh, "Max Planck: the reluctant revolutionary," *physicsworld.com*, 1 December 2000 (emphasis added).

THE EMERGENCE OF QUANTUM THEORY

THE UNCERTAINTY PRINCIPLE

We now come to what some now consider to be the "heart" of quantum mechanics: the Heisenberg Uncertainty Principle, set out in 1927. See, e.g., Greene, *The Elegant Universe*, pp.116, 204; Lindley, *Uncertainty*, p.145. The principle states that certain pairs of properties of a particle—most famously, position and momentum—cannot be precisely determined simultaneously.

Suppose one shines a light on a moving electron with the objective of determining the position and momentum of the electron by analyzing the scattering of light that bounces off the electron. Since light consists of photons, the collision of the photons and the electron would be a quantum event. Max Born had demonstrated that such a quantum event would not result in a single predictable outcome. Instead, there would be a series of probable outcomes (many possible outcomes with different probabilities of being realized). If many different paths are possible for the reflected light beam, then one cannot calculate backwards from the observable outcome to a single position and momentum for the electron under observation. Lindley, Uncertainty, pp.146-47. Thus, a single experiment cannot determine both position and momentum of the particle. More specifically, "if you multiply the uncertainty in the position of a particle by the uncertainty in its momentum (its mass times its velocity) the result can never be smaller than a certain fixed quantity, called Planck's constant." Hawking and Mlodinow, The Grand Design, pp.70-2 . This conclusion is known as the Heisenberg Uncertainty Principle.

(From our perspective, Plank's constant is almost vanishingly small, but from the perspective of an atom, it creates a moderate degree of uncertainty.)

Note, however, that this argument itself does not necessarily demonstrate that the electron did not in fact have a single position and specific, discrete momentum at the time of measurement, only that the two characteristics could not both be measured. Niels Bohr interpreted Heisenberg's Uncertainty Principle to declare that the act of measurement could never be passive, but inevitably would interact with the thing being measured. The measurements were disturbances or events in the phenomena under consideration; the "measurement defines what is being measured ... and thus fatally restricts the information that any future measurement might yield." Lindley, Uncertainty, p.155.

Continuing the analogy to waves, it is possible that the act of measurement collapses the wave function for the characteristic being measured first (say, momentum), giving it a specific value. But, the collapse of the wave function also means that the wave function no longer exists so as to enable an attempt to measure another characteristic (say position). See Barbour, The End of Time, pp.200-204. Interesting.

PARTICLES OR WAVES?

The behavior of the particles in the two-slit experiment certainly suggests that each individual particle somehow manages, in effect, to be in at least two places at once ("superposition"). See, e.g., Cox and Forshaw, The Quantum Universe, pp.7-26. "Between one appearance and another, the electron has no precise position, as if it were dispersed in a cloud of probability. In the jargon of physicists, we say that it is in a 'superposition' of positions." Rovelli, *The Order of Time*, p.87.

The patterns arise even when individual electrons are sent one at a time, each striking the screen before the next one is sent. Therefore, it cannot be that the various electrons are somehow simultaneously interfering with one another to create the pattern. Instead, each electron must be interfering with itself (or electrons sent at different times are interfering with one another). Barbour, *The End of Time*, p.197. See also, Cox and Forshaw, *The Quantum Universe*, pp.22–24. We see that "the patterns are always built up by individual 'hits'. This is extraordinarily strong evidence for particles. But if particles are creating the patterns, they must somehow explore all of the slits at once. They must do what the very concept of a particle denies—be everywhere at once." Julian Barbour, *The End of Time*, pp.196-7.

Physicists were confronting experimental evidence that was apparently inconsistent with the then existing theories. See, e.g., Cox and Forshaw, *The Quantum Universe*, pp.4–5. The objective became to imagine and explore physical theories that could be consistent with this evidence and then test the theories with other experiments. Id., pp.7–26. The challenge for the theorists was that light must consist of small, individual photons, but those photons clearly displayed the characteristics of waves. "The great mystery was how light could consist of particles yet exhibit wave behavior." Barbour, *The End of Time*, p.187.

The mystery grew even more when it was discovered that beams of electrons display the same wave characteristics as do beams of light. The French physicist Louis de Broglie in 1924 speculated that electrons might, like light, also "behave both as wave and particle." Barbour, *The End of Time*, p.189. This speculation was confirmed in 1927 by an experiment performed by George Thomson and then an experiment performed by Clinton Davisson and Lester Germer at Bell Laboratories. Id., p.190; Cox and Forshaw, *The Quantum Universe*, p.20. Following the approach of an experiment with x-rays performed earlier in the century, they beamed electrons through crystal formations and studied the diffraction patterns. They found patterns very much like what had been obtained with x-rays. The implications of this result with particles can best be appreciated if the results of the experiment is recast in terms of the simple two-slit light demonstration already mentioned. See id. ("Every account of quantum mechanics includes the famous two-slit experiment... ." Barbour, *The End of Time*, p.194.)

Imagine sending a beam of scattering electrons toward a sheet of recording material. One would expect to get many small hits on the material that would generate a pattern that resembles a circle, with a greater density of hits in the center than around the edges. Then, send the beam through a slit positioned between the source and the recording medium. If the slit is relatively narrow, the resulting pattern should be an oblong shape. Then, position a slit a short distance to one side of the first slit and cover the first slit. As one would expect, a similar oblong shape will appear on the sheet beside the first one. Then, open both slits and send the beam of electrons toward the sheet. One would expect to get two patterns much like the two generated separately, only overlapping. However, that is not the result that one obtains. Instead, the result from the use of the two slits simultaneously will evidence an interference pattern (stripes of hits alternating with stripes of nothing) just like the pattern generated by the light beam (discussed in the preceding chapter). Moreover, the result is the same even when the rate at which the electrons are shot at the target is slowed to the point that only one electron goes through the slit at a time. See, e.g., Cox and Forshaw, The Quantum Universe, pp.20-25; Barbour, The End of Time, pp.194-98.

This result seems inexplicable if the electrons are simply particles. "[S]urely a particle can pass through only one slit, and what it does then will depend upon the properties of that slit. It cannot 'know' whether the other slit is open or closed and change its behaviour accordingly." Barbour, The End of Time, p.196. The existence of the slit through which it does not go should be irrelevant to where it strikes the sheet on which its "hit" is recorded. So how can the particles become grouped into patterns resembling wave interference just because there is more than one slit open? The mystery relates obviously both to the extra "grouping" of "hits" into bands and to the blank stripes of no "hits". The issue is exactly the same with respect to light when one assumes that light consists of a stream of photons One might think that a particle with mass (like an electron) would behave differently than a "particle" without mass (like a photon). Or, said differently, one might believe that it is conceivable that a mass-less "particle" is physically different and potentially more "wave-like" than what we would consider a "real" particle (something with mass). But, we now have the same behavior from electrons, which do have mass, and even when they are "fired" one by one.

It has subsequently been demonstrated that particles of even greater mass, like large atoms, can display the same behavior. "In 1999 a team of physicists in Austria fired a series of soccer-ball-shaped molecules toward a barrier. Those molecules, each made of sixty carbon atoms, are sometimes called buckyballs because the architect Buckminster Fuller built buildings of that shape." Hawking and Mlodinow, *The Grand Design*, p.63. The buckyballs demonstrated the same behavior. See also, e.g., George C. Knee, "Viewpoint: Do Quantum Superpositions Have a Size Limit?", *Physics*, January 20, 2015 ("demonstrat[ing] that a cesium atom travels in a truly nonclassical fashion, moving as a quantum superposition of states and thus occupying more than one distinct location at a time").

Recently, as reported in *Nature*, scientists at the University of Vienna successfully performed an experiment showing that very large, synthetically constructed molecules (consisting of 2,000 atoms) display quantum interference. Jennifer Leman, "The Biggest Quantum Breakthrough Yet—Literally," *Popular Mechanics*, October 2, 2019 ("... the scientists built a special [two meter long] interferometer, called the Long-Baseline Universal Matter-Wave Interferometer, or LUMI. They used a green laser beam to propel the molecules into a tube, which shot them toward a series of slotted barriers to reveal patterns in a screen behind").

The mystery is even greater. Experiments show that if one "observes" the particles in motion before they pass through the slits, then the interference patterns disappear.

> *"[W]hen the experiment is performed, turning on a light changes the results from the interference pattern... . Moreover, we can vary the experiment by employing very faint light so that not all of the particles interact with the light. In that case we are able to obtain which-path information for only some subset of the particles. If we then divide the data on particle arrivals according to whether or not we obtained which-path information, we find that data pertaining to the subset for which we have no which-path information will form an interference pattern, and the subset of data pertaining to the particles for which we do have which-path information will not show interference."*

Hawking and Mlodinow, *The Grand Design*, pp.63-4.

An experiment designed by physicist John Wheeler showed an even more surprising result.

> *"[Y]ou postpone your decision about whether or not to observe the path until just before the particle hits the detection screen. Delayed-choice experiments result in data identical to those we get when we choose to observe (or not observe) the which-path information by watching the slits themselves. But in this case the*

> *path each particle takes—that is, its past—is determined long after it passed through the slits ... ."*

Id.

It appears that the act of observation determines not just the future, but also the past, of the particle.

One implication, building on the Uncertainty Principle, is that a particle can exist in more than one location until (or unless) it is observed. This phenomenon is called "quantum superposition". With respect to this superposition, physicist Adam Frank has summarized the theoretical possibilities as three:

- "That a deeper reality lies hidden beneath all the quantum weirdness" and that once it is understood, the apparent superposition will disappear.
- "That potential realities matter just as much as the single, fully manifested one we experience" until we engage in observation or measurement, at which time the one actuality emerges from the multiple potentialities.
- "That the universe splits itself into parallel realities at the moment of measurement... [resulting in] an infinite number of ever splitting parallel versions of the universe (and us)," only one of which we observe.

Frank also claims that the new discoveries in quantum information theory, the area in which the 2012 Nobel Prize in Physics was awarded, have meant that the first potential explanation he tendered (hidden variables that would explain away the quantum weirdness) has been "completely ruled out." "Cracking the Quantum Safe," *The New York Times*, October 14, 2012.

Entanglement

Emerging out of Pauli's Exclusion Principle of 1925, quantum theory predicts that certain pairs of subatomic particles will continue to behave in conjunction with one another even when separate by large, even very large, distances. For example, at the moment one determines the "spin" of one particle in the pair, the other will be found to be spinning in the other direction. The phenomenon was actually observed in 1997 by physicists at the University of Geneva with photons seven miles apart—interference with one instantly caused a response in the other. Bryson, A *Short History of Nearly Everything*, pp.145-6.

The theory assumes that everything is related to and interacts with everything else in the Universe and that these interactions occur instantaneously, that is, faster than the speed of light. This concept is in dramatic contrast with our normal assumption that two things not in contact and with no identifiable means of communication behave independently—indeed, could not do otherwise. The most sophisticated aspects of the theory of quantum mechanics are based upon the fantastical assumption that each particle is effectively "aware of" every other particle, at least, the mathematics is constructed on that assumption and generates very, very accurate predictions. See, e.g., Laughlin, *A Different Universe*, pp.52-3.

THE THEORIES

It is important to note that much of what we are discussing next appears as a consequence of the mathematics.[9] Both Erwin Schrodinger and Werner Heisenberg developed models that seemed capable of generating the behavior observed if electrons followed rules different from the rules of classical mechanics applied to normal size objects. Krauss, *A Universe from Nothing*, p.59.

Wave or matrix equations

In papers published in 1926, Schrodinger developed a wave equation that appeared to present the behavior of electrons as wave patterns that were continuous functions that generated standing waves, or static field patterns, in only a handful of discrete values, suggestive of the discrete different orbits of the electron. Lindley, *Uncertainty*, pp.120-22. The challenge was to find a mathematical system that would set out the rules for the superimposition of wave functions upon one another. Cox and Forshaw, *The Quantum Universe*, pp.32-43.[12]

During the same time period, Heisenberg explored the approach of describing electrons by two numbers representing frequency and amplitude. The objective was to be able to predict the future wave pattern of a particle based upon the past wave pattern. It would also be necessary to predict how different particles would interact with one another. Max Born found that matrix algebra, developed as a branch of pure mathematics, was a "ready-made" tool for this approach to quantum mechanics. Lindley, *Uncertainty*, p.123. *See also*, Cox and Forshaw, *The Quantum Universe*, pp.41–43.

The theories of Schrodinger and Heisenberg appeared to present two distinct, unrelated representations of quantum phenomena—wave mechanics and matrix mechanics. But, Schrodinger concluded in 1926 that wave mechanics and matrix mechanics were just two systems of mathematics that generated the same results. Lindley, *Uncertainty*, pp.129–30. In other words, two seemingly different mathematical approaches, one based upon continuous wave functions and one based upon the values of discrete quanta or particles, gave identical answers in quantitative exercises. This correspondence suggested that there could be theoretical avenues that would connect the discrete, quantum world with the continuous, Newtonian world. *See id.*, p.143.

When Paul Dirac prepared a comprehensive statement of quantum mechanics, he found that he could not precisely match the calculations of position and momentum in classical and quantum theory. *Id.*, p.144. *Cf.* Cox and Forshaw, *The Quantum Universe*, p.43. Heisenberg's matrix mechanics seemed to be the more complete and comprehensive approach, that is, one that could be generalized to a wide variety of circumstances. Schrodinger's wave mechanics, in contrast, was tied to the concept of waves. *See* Julian Barbour, *The End of Time*, p.189.

The de Broglie-Bohm Theory

In 1927, Prince Louis de Broglie proposed a simple solution to the mysteries of the newly developed quantum theory. *Id.*, pp.157–8. He suggested that the wave/particle duality observed with light was also present with other particles, including electrons. He theorized that the waves propagate just as waves do in water (resulting in interference, etc.) and that the associated particles followed the waves, being pulled toward the crest of the wave by an additional force. The probabilities were that a particle would be at or near the crest of the associated wave, but its actual position would also be unknown because its starting position would be unknown. However, the particle would always have an actual position, contrary to some interpretations of quantum indeterminacy.

Smolin explains that de Broglie's theory was ignored because John von Neumann published a mathematical proof in 1932 supposedly establishing that such "hidden variables" could not exist. *Id.*, p.158. Within a couple of years, a German mathematician, Grete Hermann, discovered a fatal flaw in von Neumann's analysis, but "her paper was ignored." *Id.* Then, in the early 1950s, physicist David Bohm wrote a textbook on quantum physics in which he reinvented de Broglie's "hidden variable" theory. In response to a reviewer's comment citing the von Neumann proof, Bohm rediscovered the error in the mathematics. This theory has become known as the de Broglie–Bohm approach to quantum mechanics. *Id.*, pp.158–9.

One important issue is that the de Broglie–Bohm theory of hidden variables is not compatible with the relativity of simultaneity on which the Special Theory of Relativity is based. The indeterminacy or statistical nature of the traditional quantum physics can co-exist with Special Relativity, but the de Broglie–Bohm approach contemplates detailed and predictable results in quantum experiments that require a preferred observer and a preferred measure of time, contrary to relativity. *Id.*

SOME PHYSICAL INTERPRETATIONS

A "mist"?

One could imagine that particles when emitted carried with them a penumbra of associated waves, what Julian Barbour characterized as a "mist." *The End of Time*, pp.197–8. Perhaps the mist passes through all open slits in varying degrees of presence (whatever that means). Somehow, it must happen that when the screen or sheet is encountered, the particle is reduced to a single point or "hit." We could speculate that the position of that "hit" is influenced by the experience of the "mist" up until the instant that the screen is impacted. For several particles emitted simultaneously, their individual "mists" could interact as waves and create interference patterns in the placement of "hits" on the screen.

Moreover, these sets of possible outcomes only exist while the particles are in mid-stream, so to speak. Once they have "hit" the screen, each particle's singular position is established and its "mist" or waves disappear. *Id.*, pp.199–200. Thus, we have a set of mathematical rules explaining the behavior of the "mist," but those rules and whatever phenomena they represent are gone at the instant of the "hit." *Id.* The "hit," of course, establishes the position of the particle. Thus, it is like a "measurement" or observation of the particle.

Indeed, experiments demonstrated that if equipment is included that would use light to identify the position of the electron just before it goes through the slits, so that one can determine which slit it would be going through, then the resulting pattern on the screen on the other side of the slits no longer displays an interference pattern. Instead, it looks just as one would have expected if the electrons were behaving purely like particles and not waves. Greene, *The Elegant Universe*, pp.109–10. It appears that the act of locating or measuring the electron before the slits causes the wave to collapse or mist to disappear.

Just probabilities?

Max Born proposed in 1926 that the wave pattern (the "mist" or "cloud") that was part of the emitted particle represented probabilities of where the particle would be found, so that the more intense or thicker part of the "mist" was where the probability of finding the particle was greater and in the more dispersed portions of the "mist," the probabilities were less. Julian Barbour, *The End of Time*, p.198. Born interpreted Schrodinger's waves as reflecting the sets of probabilities that a particle would follow various specific paths. Heisenberg and Niels Bohr, at the Bohr Institute in Copenhagen, propounded the "official" interpretation. The wave function of a particle sets out all of the information about that particle. *Id.*, p.202.[10]

Thus, each of the particles is represented by an elaborate pattern of potentialities or probabilities. For all possible positions, there is an associated probability that the particle is there. For very many of the possible positions, the probability will be zero or exceptionally small; for others, it will be significant. However, if the position of the particle is a range of probabilities, then it makes no sense to ask: "What is the position of the particle?" However, once a measurement is taken, then the position of the particle is known (whether it was a highly probable or quite unlikely position).

The predictions derived from the functions are probabilistic, but the probabilities "reflect a fundamental property of nature, not simply our ignorance." *Id.*, p.203. It is not the case that before the measurement of the particle (e.g., before it hits the screen), it has a particular position and "momentum but we just do not know it"; it is that preceding measurement "all momenta...are present as potentialities, and measurement forces one of them to be actualized." *Id.*, p.203.

Similarly, one cannot specify the particular value of one characteristic that is associated with a particular value of another until one of the characteristics has been established by measurement. One simply cannot say that in the "cloud," a specific momentum is associated with a particular position of the particle. Any position has associated with it a large number of different momenta with differing probabilities of occurring. Clearly, it is hard to see how in this scenario there could be a specific actual position and momentum existing for a particle before measurement. *See* Craig

Callender, "Nothing to See Here: Demoting the Uncertainty Principle," *The New York Times*, July 21, 2013).

One might say that these probabilities are "real", in the sense that they theoretically could be observed in a large number of events, that is, the outcomes of very many experiments. But, that sense of "real" is only definitional. You do the experiments and calculate the probabilities. The probabilities cannot be calculated theoretically in advance. Generally, when we utilize probabilities, we have an explanatory model of the events at issue from which we can predict the probabilities. Think of the flip of a coin or the roll of dice. Our models enable us to calculate the probabilities of various outcomes. If experimentation generates something else, we would want to examine the coin or the die to see if it conformed to the theoretical ideal on which our calculations were based, expecting that we might find some deviation that would explain the results observed.

It is very difficult to conceptualize any intuitively satisfying model of what is actually happening.

Feynman's "every path"?

Richard Feynman proposed yet a somewhat different interpretation of the phenomena being observed. He suggested that each electron, in the absence of measurement, actually went through both slits at once. In fact, he proposed that each electron followed all possible paths between its emission and the screen. *Id.*, pp.109–11. He showed that he could assign numbers to all of the possible paths and generate predictions identical to the probability wave approach.

Thus, one would not visualize a mist or cloud, but a multi-path trajectory. (Cox and Forshaw utilize a conceptualization based upon the Feynman approach. *The Quantum Universe*, p.27.) According to the Feynman theory, "the electron ...simultaneously 'sniffs' out every possible path connecting its starting point with its final destination" and then follows every such path. Greene, *The Elegant Universe*, p.110.

If anything, this "interpretation" of what is taking place is even more difficult to grasp in any intuitively satisfying way. Feynman readily admitted that it conflicted with common sense. In 1988, he wrote that the theory "describes nature as absurd from the point of view of common sense. And it fully agrees with experiment. So I hope you can accept nature as She is–absurd." From *QED: The Strange Theory of Light and Matter* (1988), quoted in Greene, *The Elegant Universe*, p. 111.

One can not take Feynman's description of the behavior of an electron as a description of what actually occurs. Certainly, electrons do not "sniff." But, more importantly, does an electron actually follows all of the possible paths? The mathematical model reflecting what this word description of what occurs gives very accurate predictions of observed phenomena, but that means only that the model has great predictive power. As a characterization of physical events themselves, I suspect that it is seriously lacking.

Parallel universes?

Several of these interpretations of what we appear to observe could be reconciled with, or reduced to, the "simple" hypothesis that what we really have are many, many (infinitely many?) parallel universes existing simultaneously. In other words, all of the possibilities actually exist, just in different, unconnected universes. The act of observation, then, places the observer in one specific universe among the many, at least for that observation. The hypothesis of multiple universes has captured the popular imagination. (I discuss it further in the last chapter.)

Quantum fields?

"[By] quantum field, we mean that the field really describes the presence of particles, and the particles are—as we saw earlier—quantum things." Hossenfelder, Lost in Math, p.54. We discussed fields at the beginning of the prior chapter, noting that they had become the physicists' favorite explanation for everything.

> "*The work of Faraday, Maxwell, and Hertz spanned most of the nineteenth century. It established* **space-filling fields as a new kind of ingredient in the fundamental description of the world.** *At first, fields were considered an additional ingredient in the recipe for the physical world, supplementing particles.* **Over the twentieth century, fields took over completely.** *We now understand particles as manifestations of a deeper, fuller reality.* **Particles are avatars of fields.** *...Fields,* **being continuously extended through space,** *appear to be very different from particles. ...Quantum fields, as their name suggests, are still fields (that is, space-filling media).*"

Wilczek, *Fundamentals*, pp.98, 99 (emphasis added).

"Fields rule. Quantum fields, that is." *Id.*, p.101.

Right. But, what is a field again?

A Note on QED

The mathematical formulation contained in the Standard Model is called Quantum Electrodynamics ("QED") (supplemented by Quantum Chromodynamics or "QCD"). Susskind, <u>The Cosmic Landscape,</u> p.51. It is a relativist quantum field theory, combining Special Relativity with quantum field theory using a type of mathematics called gauge theory. The weak and strong nuclear forces, just as the electromagnetic force, are described by the mathematical structure of the field theory. The mathematics of QED does not incorporate gravity.

Until the 1920s and 1930s, the source of the energy being continuously generated by the Sun was assumed to be nuclear fusion. But, the question was how fusion could actually occur in the core of the Sun, because the Sun was not hot enough to overcome the repulsion of fast moving protons. Somehow the protons needed to overcome that repulsion and get close enough together for fusion to occur (and for helium to be created). The "answer" was provided by QED theory: particular protons can become converted into neutrons by having one of the up quarks change into a down quark, which transformation generates a W particle (quickly turns into a positron and a neutrino, shooting across the Universe). When the proton and the newly-minted neutron get close together, there is no electrostatic repulsion and they fuse together, making deuteron, which rapidly leads to the formation of helium.

In the 1980s, the scientists at CERN in Europe, using particle accelerators, discovered particles of very short lives that also had the other assumed characteristics of the W particle (and of the related Z particle, that can be ejected by an electron). Id.

From Micro to Macro (and Back)

It is not just measurement that disrupts the quantum state. For every macroscopic thing, it must be the case that someplace, at some point, a transition from the quantum world to the normal macroscopic has occurred. That is why we do not observe the strange quantum effects in our daily lives. However, we do not know how that transition occurs or when it occurs. If it is a matter of size, then what is the exact scale at which one world replaces the other? This question has become more acute in light of the discovery that even very large molecules display the interference phenomenon (as discussed above).[11]

In all events, it certainly is clear that the apparent weirdness of quantum phenomena disappears at the macro level, the level at which we experience the world, whether by reason of the phenomena effectively cancelling each other out when aggregated on a larger scale or because of an unexplained emergence of new laws out of the organization implicit in the larger scale. Indeterminacy only appears to exist in the analysis of the small, isolated subsystems because there are some unknown factors of which we are unaware, relevant factors or variables not observable by the scientist examining the subsystem. See Smolin, *Time Reborn*, p.155.

Curiously, nonetheless, as we shall see, quantum mechanics is still thought to have significant relevance for understanding the origin and structure of the Universe, including some of the stranger (macro) objects found therein. So we come back to the question of the nature of the transition or relationship between the quantum world and the macro world. And, underlying the whole matter is the puzzling question of whether the elegant and apparently very effective mathematics involved in quantum mechanics reflect what we would call reality.[12]

But, there are other possibilities. Physicist Robert B. Laughlin, who has been referred to as the Feynman of his generation, argues that there are rules for emergent phenomena that arise out of self-organization (in larger amalgamations of things). Interestingly, Laughlin declares that "[t]he transition to the Age of Emergence brings to an end the myth of the absolute power of mathematics." *Id.*, p.209.

The higher-level groups can take on collective characteristics that can be quite different from the characteristics of the individuals that constitute the group. *A Different Universe*, pp.ix–xvi. Thus, we can have a world in which Newton's laws govern, a subatomic world that has its own rules and no connection between the two in the sense that the subatomic rules could be different without affecting the macroscopic rules. *Id.*, pp.31-2. (This concept is different from the description we have already discussed of macro states reflecting probabilistic characteristics not seen in the micro world.)

Laughlin is highly critical of the reductionist approach in which it is assumed or believed that the construction of accurately predictive theories at the lower or microscopic level will enable the explanation or, at least, prediction of characteristics that will be found at the higher, macroscopic levels. He asserts that such theories will never be applicable to phenomena that are emergent. One can also think of unstable phenomena, where a very small difference in initial conditions can result in vastly different outcomes, or where repeated experiments give different results. A final example is the fact that certain theories are not invariant as to scale, that is, there can come a scale (very small or very large) at which the rules no longer apply. Id., p.154.

The implications are not that the emergent phenomena cannot be explained, but that the explanations lay in the rules of self-organization that in turn result in the characteristics of the collective, not in the rules governing the phenomena in the lower, less organized environment. As a result, for example, quantum mechanics may be correct at the microscopic level but tell us very little about the world that we experience with our senses. "Newton's legendary laws have turned out to be emergent. They are not fundamental at all but a consequence of the aggregation of quantum matter into macroscopic fluids and solids—a collective organizational phenomenon." Id., p.31.

This assertion certainly does not mean that Newton's laws emerge from quantum mechanics: physicists "routinely speak about Newton's laws being an 'approximation' from quantum mechanics, valid when the system size is large enough—even though no legitimate approximation scheme has ever been found." Id. Moreover, the mathematics of quantum mechanics would require enormous numbers of computations to make any predictions about the world we observe—computations far beyond practical implementation. See id., pp.50-51.

Emergence is a concept that seems again to have captured the imaginations of various scientists is recent years as a means of escaping the apparent inherent flaws of reductionism and as a means of explaining the origin of life and the nature of consciousness. Perhaps, the most interesting and compelling of emergent phenomena occur with the appearance of consciousness. See Deacon, Incomplete Nature, p.182. As discussed in the chapter on Chemistry, there is controversy over whether the characteristics of the emergent matter are theoretically predictable from the constituent parts or whether there is a true discontinuity that arises through emergent phenomena as a result of the combination or the structure. See, Id.,pp.158-9.

Alarming Implications?

The emergence of quantum mechanics and related developments introduced into physics some alarming and threatening new elements.

The end of determinism?

Classical physics assumed that position and momentum, including angular momentum, are real and definitive characteristics of an object. They can be defined unambiguously. The Uncertainty Principle interposes a significant objection to the classical view of the world. See, e.g., Barbour, The End of Time, p.204. For example: "The language of classical physics is the differential calculus devised by Newton and independently by Leibniz to deal with continuous variation and incremental change. But in trying to understand the workings of atoms, physicists came up against phenomena that were abrupt, spontaneous, and discontinuous. The atom was in one state, and then it was in another." Lindley, Uncertainty, p.107.

While mechanical determinism was perceived to be the logical conclusion of classical physics, the new physics seems necessarily to incorporate random events "governed" by statistical laws and described in terms of probabilities. See id., p.136. The introduction of probabilities and alternative outcomes from the same starting conditions is, at least in principle, inconsistent with the idea of determinism, that everything can be predicted given a knowledge of the initial conditions and the governing laws of change.

The end of the particle?

Quantum mechanics and the accompanying view that everything ultimately consists of fields, embody a big potential change in world view: the elimination of the point particle itself. Rather than matter and energy and empty space, at each and every position or spot (I am avoiding the word "point," since it suggests a particle), there are values that exist for each of the four elemental particles and for each of the four forces we have already discussed. What we have known as a particle is only a specific field configuration (i.e., set of values). All of those values will vary from spot to spot, creating the effects that we know as particles of matter and forces such as gravity and electromagnetism. In addition, those values will evolve over time pursuant to rules or laws that we call the equations of motion.

Particle Physics and Quantum Mechanics 367

> "though it's convenient to call them 'elementary particles,' they aren't really particles. (That is, they have little in common with what the word 'particle' suggests.) Our modern fundamental ingredients have no intrinsic size or shape. If we insist on visualizing them, we should think of structureless points where concentrations of mass, charge, and spin reside."

Wilczek, *Fundamentals*, p.77.

Indeed,

> "the central lesson of quantum mechanics: there aren't really waves and there aren't really particles. Instead, everything in the universe (including, for all we known, the universe itself) is described by a wave function that has properties of both particles and waves. Sometimes this wave function appears more like a wave, sometimes more like a particle. But fundamentally it is neither—it's a new category in its own right. ...whenever physicists refer to particles, they actually mean a mathematical object called the wave function, which is neither a particle nor a wave but has properties of both. ...The wave function itself does not correspond to an observable quantity... ."

Hossenfelder, *Lost in Math*, pp.50, 51.

This conceptualization of the world would seem to make the question "what is the position and velocity" seem rather meaningless.

Spontaneous Appearance?

We talked above about anti-particles and virtual particles resulting from the theoretical work of Paul Dirac; however, there is another, related derivation from Dirac's theory. It has to do not with the number of particles, but with from where they come. The formulation is the work of Richard Feynman. Krauss, *A Universe from Nothing*, p.62. Feynman observed that if particles moved faster than the speed of light, then they would appear to be going backward in time. An implication of the Heisenberg Uncertainty Principle is that for very short periods of time, quantum mechanics would permit particles to so behave. Yet, "a negative charge moving backward in time is mathematically equivalent to a positive charge moving forward in time! Thus, relativity would require the existence of positively charged particles with the same mass and other properties as electrons." *Id.*, p.63. Feynman then postulated that one could have an electron moving through space and then find a positron/electron pair suddenly appears out of nothing, annihilating each other briefly thereafter, leaving an electron moving through space but in a different position than it had been before. Id., pp.63-4.

INSTANTANEOUS MOVEMENT?

As already mentioned, the mathematics conceptually assigns values (probabilities) of a particle appearing next to every existing, conceivable point in the Universe; and, naturally, many of the assigned probabilities are exceedingly small. A supposed "implication" of this mathematical structure is that an existing particle can appear any place in the Universe in the future, even in the next instance of time. In other words, "the electron wave spreads out to fill the Universe in an instant." Cox and Forshaw, *The Quantum Universe*, p.47.[15] Similarly, a particle that is in one position could appear in a position millions of light years away in the next instance.

What about the speed of light as a universal speed limit?

The answer given is rather perplexing:

> "[T]he idea that a particle can be here and, an instant later, somewhere else very far away is not in itself in contradiction with Einstein's theories, because the real statement is that information cannot travel faster than the speed of light.... As we shall learn, the dynamics corresponding to a particle leaping across the Universe are the very opposite of information transfer, because we cannot know where the particle will leap to beforehand."

Id., p.47.

Now, this explanation is not very satisfying to me. I think that most of us understood the proposition that nothing can travel faster than light to mean exactly that a particle cannot move from here to there in less time than it would take light to travel that distance, irrespective of the concept of information.

SIMULTANEOUS EXISTENCE?

We also have discussed the hypothesis that things (very small things, anyway) will exist in two or more places at once. (Or, perhaps, things jump huge distances instantaneously or exert influence on similar things at great distances, also instantaneously.) Or, in the words of the subtitle of the book by Cox and Forshaw, *The Quantum Universe*: "Everything that Can Happen Does Happen." Contrary to all experience and expectations.

VIRTUAL MATTER?

Again, as noted, the mathematics manage to generate predictions that are exceptionally accurate. *Id.*, p.65. Yet, when a much more sensitive measurement device was developed, there was found to be a small discrepancy between the experimental results and the mathematical calculations. Ultimately, a solution was found in the incorporation into Dirac's equation of "the effects of virtual particles." Krauss, A *Universe from Nothing*, pp.66-7. Thus, for the close analysis of the most detailed structure of the hydrogen atom when it is being heated, physicists utilize the assumed existence of an electron/positron pair that spontaneously appear out of nowhere then almost instantly annihilated one another and of a virtual electron. *Id.*, pp.67-8. Krauss asserts that the use of this mathematical model enables "the *best, most accurate prediction in all of science...*"; so, "[v]irtual particles therefore exist!" *Id.*, pp.68, 69 (emphasis in original). One might conclude that the concept of "existence" in this statement is a bit different from how we normally use the word. But, Krauss further asserts that most of the mass observed in the Universe is contained in the energy fields of these virtual particles, not in the normal particles that we have identified as the building blocks of atoms, molecules and objects detectable by our sense. *See id.*, p.70.

A NOTE ON WEIRDNESS

Some theorists argue that the more bizarre phenomena that appear to emerge from quantum theory arise not within the confines of the quantum world as such, but only when the quantum world interacts with the macroscopic world of our existence. Left on their own, quantum particles arguably behave in a completely deterministic manner. It is only when measurement is introduced that indeterminacy and probabilities appear. See Penrose, Shadows of the Mind, pp.264, 349. It is in the measurement process that the "jumping" from one state to another is "observed."

Theories of Everything

We referred above to the efforts to achieve a unification of quantum theory and General Relativity or the incorporation of gravity into quantum mechanics creating a "theory of everything".[14] What is necessary is to quantize gravity, to find gravity consists of quanta rather than being continuous. Here is a very brief summary of those efforts.

Strings and Superstrings

Substantial and sustained efforts to find a means of integrating gravity into quantum mechanics (or of reconciling General Relativity with quantum mechanics) resulted in what is called "string theory" or "superstring theory." The essence of the creative idea was to hypothesize that particles are not points but, instead, are vibrating loops or strings. The introduction of such loops or strings enables the derivation of mathematical results that cannot be obtained when working with points (partly the problems of zeros and infinity to which I have previously made reference). However, the mathematics is very complicated and is still in the process of being worked out. See, e.g., Greene, *The Elegant Universe*, pp.3–6, 15–20. There are different views among physicists as to whether "superstring theory" will achieve the unification of General Relativity and quantum mechanics or whether some further modification is needed, for example, like "M-theory." *Id.*, pp.373–87.

M-theory

There appear to be many, many superstring theories. So, perhaps they are all subsumed under one master theory.

Stephen Hawking provided a strong endorsement of M-theory as the answer in 2010. But, he acknowledged that the did not know whether it is a single theory or an interconnected network of theories. He did not know what the "M" stands for. But, he did "know" that:

> "...M-theory has eleven space-time dimensions, not ten. ....Also, M-theory can contain not just vibrating strings but also point particles, two-dimensional membranes, three-dimensional blobs, and other objects that are more difficult to picture and occupy even more dimensions of space, up to nine. These objects are called p-branes (where p runs from zero to nine). The laws

of M-theory therefore allow for different universes with different apparent laws, depending on how the internal space is curled. M-theory has solutions that allow for ...10500 different universes, each with its own laws."

Hawking and Mlodinow, *The Grand Design*, pp.117-9.

Moreover, "[I]n the early universe—when the universe was small enough to be governed by both general relativity and quantum theory—there were effectively four dimensions of space and none of time. ...[In] the very early universe, time as we know it does not exist!" Id. The authors seem delighted that this M-theory revives the strong anthropological principle (discussed in the last chapter) and also makes "in the beginning" irrelevant: "It removes the age-old objection to the universe having a beginning, but also means that the beginning of the universe was governed by the laws of science and doesn't need to be set in motion by some god." Id.

Hawking and Mlodinow do not make a very convincing case, however. Several times they assert that just because the theory predicts things that we do not like or are contrary to our beliefs does not make it false. Correct. But also, that fact does not make it true—nor does the fact that one likes the theory's implications. We need evidence.

MANY UNIVERSES

Leonard Susskind argues that string theory tells us that there are very, very many "bubble" universes scattered throughout an immense Universe, each so distant from the others as to make it separate from and inaccessible to the others. *The Cosmic Landscape*. (Susskind prefers the term "megaverse" to the more widely used "multiverse". Id., p. 20 n4.) See, generally, Dawkins, *The God Delusion*, pp.169-80.

ADDITIONAL DIMENSIONS

One important aspect of string theory and the newer alternatives is the introduction of additional "dimensions" beyond the three (or four, if one includes time) with which we are familiar. Indeed, the number of such additional dimensions seem to increase as the theory develops, since it seems that successive problems that arise in the mathematics can be "solved" by the assumption of yet another dimension.

This approach is not new; versions of it were used in the development of quantum mechanics—when the mathematics did not work, introduce another hypothetical variable; if that solves the problem (makes the math work), then one assumes that that variable in some sense "exists". Not the most compelling evidence.

In discussing the extra dimensions introduced by string theory, Charles Seife observed:

> "What do these six extra dimensions mean? Nothing, really. They don't measure anything that we are accustomed to, like length, breadth, width, or time. They are simply mathematical constructs that make the mathematical operations in string theory work in the manner they have to. ...Though it is a strange concept physically, it is the predictive power of the equations that interests scientists, rather than their comprehensibility....."

Zero, p.197.

Science-writer Seife is predictably colorful in his description, but physicist Robert Laughlin is even sharper in his criticism.

He writes:

> "String theory is immensely fun to think about because so many of its internal relationships are unexpectedly simple and beautiful. It has no practical utility, however, other than to sustain the myth of the ultimate theory. There is no experimental evidence for the existence of strings in nature, nor does the special mathematics of string theory enable known experimental behavior to be calculated or predicted more easily."

A Different Universe, pp.211-2.

FINAL OBSERVATIONS

The importance of quantum mechanics is hard to exaggerate. It is claimed that Special Relativity explains the nature of space and time, General Relativity explains gravity and quantum mechanics "deals with everything else.... . [I]t's simply a physical theory that describes the ways things behave. Measured by this pragmatic yardstick, it is quite dazzling in its precision and explanatory power." Cox and Forshaw, The Quantum Universe, p.1. The ability of quantum mechanics to generate predictions of astonishing and unprecedented exactitude seems to be widely established (of course, it deals with things of incredibly small magnitudes). But, I think that Cox and Forshaw get a bit carried away when using the phrase "explanatory power." They presumably do not really mean explanatory power in the sense that we have been using explanation in our discussions above.

THE QUESTION OF "MEANING"

What a theory "means" implicates two somewhat different concepts. The first is the identification of a physical analogy from our world of experience (or from our imagination as to what could be conceivable in the world we know, that is, something that we may not experience but could recognize, if it were actually to exist). The second concept is that of meaning or explanation, generally in the form of causality. There are analogous phenomena that we recognize but do not really understand. Indeed, much of this book deals with examples of such things.

So, there is the nagging question of what the theories of quantum mechanics could mean. Some scientists (like the positivist Ernst Mach) were completely content with just mathematical systems that enabled accurate predictions (and believed that nothing more was possible in any event); others (including Rutherford and Einstein) expected more. These others wanted physics, not just mathematics, and they wanted the physics to make sense. When one asks "where is the physics?," one generally means how does the mathematical system relate to or correspond with concepts that we have grown accustom to recognizing as part of physics or of the physical description of the world that we have accepted. Indeed, in describing Heisenberg's efforts to devise a new theory of matter and forces in 1925, these same authors observed that Heisenberg seemed to embrace a new methodological approach, seeking only: "to predict directly observable things.... It should not be expected to provide some kind of satisfying mental picture for the internal workings of the atom, because this is not necessary and it may not even be possible. In one fell swoop, Heisenberg removed the conceit that the workings of Nature should necessarily accord with common sense." Id., p.13. Of course, we have already seen that Newton had employed that same methodology over 200 years earlier.

Nonetheless, many physicists continue to engage in speculation that goes beyond the mere mathematics and attempt to offer interpretations from which they derive implications of a physical nature about our world, implications that are not compelled by the mathematics (as a merely logical structure). An example is the subtitle of the Cox and Forshaw book: "Everything that can happen does happen" (or in the chapter headings, *e.g.*, "Being in Two Places at Once" or "Everything That Can Happen Does

Happen" or "The Universe in a Pin-Head"). *Id.*, pp. 6-26, 45-74, 116-35. (Another example we shall discuss in the next chapter is the hypothesis of infinite universes and the creation of our Universe out of nothing. The creatively imagined implications of quantum theory have produced a revolutionary expansion of cosmological theorizing.)

To some extent, the problem is that we are dealing with different things. Quantum mechanics does not attempt to explain how the particle moves or the path that it takes in doing so. Motion as such is simply not part of the picture. The theory just addresses where a particle may be found in a future instant of time, not how it gets there. See *id.*, p.46. So, the speed at which it travels is also not part of the picture. In addition, quantum mechanics by definition deals with phenomena at a subatomic level, far smaller than anything with which we interact. And, its "rules" seem to have no application—and, perhaps, no direct relationship—to the macroscopic world of objects and sentient beings.

JUST "WAVES OF NOTHING"?

Cox and Forshaw utilize the analogy to waves as we experience them. Thus, a "standing wave" is what occurs in a contained space, like a swimming pool. The peaks of the waves may be of different heights, but they always appear in the same place. *The Quantum Universe*, pp.93, 95-6. The same phenomenon occurs with a plucked string on a musical instrument. The standing wave with the longest wave length produces the note that is heard; the other standing waves are the higher harmonics. *Id.*, pp.94-5. These standing waves can be characterized as "quantized," because they occur at specific and discrete wavelengths, with no wavelength in between being capable of being generated. *Id.*, p.96.

Therefore, it becomes easy to use the standing wave as an analogy for a particle that is confined to an orbit around a nucleus. Consistent with that analogy, the particle's momentum and its energy level will also be "quantized" by standing waves. *Id.*, pp.100-101. "The unchanging nature of standing waves therefore makes them a clear candidate to describe an electron of definite energy." *Id.*, p.105. This analogy gives us a picture of how electrons stay within a particle energy levels of orbit and how, if an electron in one level is disturbed, the electron will end up only in another specified energy level, not in between levels. See *id.*, pp.108-9. Cox and Forshaw note that they have discussed at length how an electron located at a position in space may suddenly appear far away, essentially anywhere else, in the next instant. But, the electron that is orbiting a nucleus in an atom will forever retain its precise energy level until something happens to change it. *Id.*, p.105. So, it appears that the real theoretical challenge is to explain stability or "conservation of energy." In other words, what we really do not understand is the normal state of affairs, the bases of which we thought science had so successfully been revealing to us over the last 300 years.

However, "[o]ur initial wild proposal that the particle can be anywhere and everywhere in the Universe has been tamed by the orgy of quantum interference, and the Uncertainty Principle is in a sense all the remains of the original anarchy." *The Quantum Universe*, p.73. When many particles are involved, "[t]he upshot is that the sand grain stays where it is and there is almost no probability that it will jump a discernible distance, even though we really have to consider the possibility that it secretly hopped everywhere in the Universe in order to reach that conclusion." *Id.*, p.68. The net effect is that generally a particle "is approximately stationary, although the quantum rules imply that it jiggles around a little." *Id.*, p.75. Similarly, on the phenomena we are able to observe directly with our sense, "quantum entanglement" has no impact. "The quantum entanglement in it has not disappeared but has simply ceased to have experimental consequences that matter." Robert Laughlin, *A Different Universe*, p.53.

Does this perspective resolve the tensions? Not for me.

The mathematics is based upon particles as infinitesimal points, like the familiar point in geometry. The point is just a place. In the mathematics, a point has no dimensions. Of course, in geometry, no one thinks that the "point" corresponds to some actual "thing". Yet, Cox and Forshaw explain that "as far as we can tell, the fundamental particles of Nature are of no size at all." *Id.*, p.116. In other words, they take up no space, in and of themselves. They have no size, no color, no taste, no smell. Yet, they make up everything that does. The very notion of a particle is that it is something; and, by something, we normally mean an object that literally takes up space and has a presence. Yet, we find ourselves with a theory that says that quantum-mechanical matter consists just of waves and that those waves are "waves of nothing." Laughlin, *A Different Universe*, p.55.

So, again, what is mass? And, what is energy?

JUST MATHEMATICS (AGAIN)?

The theory is, in its mathematical form, a technique for generating results that we can in fact "observe," in the special sense that word has in reference to the quantum world. But, in fact, in the calculations, what is called "quantum interference" dramatically limits what in fact will be observed to occur. So, the thesis I am setting forth in this chapter is that much of what evolved as twentieth and, now, twenty-first, century particle physics reflects intensive and imaginative efforts to solve fundamental technical problems that have arisen in the mathematical models being employed, rather than problems of physics, as such. The fact that the elaborate mathematical structure based upon such assumptions actually works in making precise predictions does not mean that the assumptions embody characteristics of the real world.

There are, of course, observable physical phenomena that were posing difficult questions for the prevailing theories, results in experimental work that appeared not only not to be explainable by existing theories, but actually to be incompatible or at odds with the prevailing paradigms. Undoubtedly, the adventures in mathematics that dominated the last 100 years were invited, if not encouraged, by these questions arising from the physicists' experiments. As Cox and Forshaw assert, "[q]uantum theory was precipitated, as is often the case in science, by the discovery of natural phenomena that could not be explained by the scientific paradigms of the time." *The Quantum Universe*, pp.4–5. Physicists explored mathematical models that might be able to accommodate these unusual empirical observations. Some of that mathematics was able to provide accurate predictions of subsequent experimental results. This process can be contrasted with the methodology employed by Einstein in developing his theories of relativity. He proceeded largely through hypothetical, mental exercises to construct theoretical models. Then came the mathematics. In turn, the models were used to predict experimental results that were subject to testing.

Brian Greene has commented:

> "By 1928 or so, many of the mathematical formulas and rules of quantum mechanics had been put in place and, ever since, it has been used to make the most precise and successful numerical predictions in the history of science. ...[However,] few if any people ever grasp quantum mechanics at a 'soulful' level. ...[At] a microscopic level the universe operates in ways so obscure and unfamiliar that the human mind, evolved over eons to cope with phenomena on familiar everyday scales, is unable to fully grasp 'what really goes on.'"

The Elegant Universe, p.87.

Greene concludes this line of thought with the recogntion that "beyond the fact that it is a mathematically coherent theory, **the only reason** we believe in quantum mechanics is because it yields predictions that have been verified to astounding accuracy." Id., p.88 (emphasis added).

However, many years earlier, Alfred North Whitehead warned:

"[T]here is no more common error than to assume that, because prolonged and accurate mathematical calculations have been made, the application of the result to some fact of nature is absolutely certain. The conclusion of no argument can be more certain than the assumptions from which it starts."

An Introduction to Mathematics (1911), p.27.

Has the elaboration of the mathematics now taken on a life of its own and come to dominate theoretical science? Is it possible that we are no longer investigating the real world, but only creating an ever more extravagant mathematical model?

Endnotes

[1] The theories of the very small scale are addressed in this chapter; the theories of the very large events are discussed in the next. These two subjects have so captured the public's imagination that there are dozens of books presenting sophisticated, but popularized treatments of the series of latest developments in the fields, a number of which books have been "best sellers."

[2] Particle accelerators have become increasingly large and staggeringly expensive. Current examples cost billions of dollars to construct and hundreds of millions of dollars a year to operate. The Large Hadron Collider at CERN (which organization employs over 3000 people) consists of a tunnel creating a circular path some 26 kilometers long (that is, in circumference), around which particles are accelerated using huge quantities of energy. When they reach the desired speed, they collide with some other particle. The scientists watch to see if they can detect what results. The results are exceedingly short-lived. *See, e.g.,* Bill Bryson, *A Short History of Nearly Everything,* pp.208-9. The European Organization for Nuclear Research is located near Geneva on the France/Switzerland border. The original name was the European Council for Nuclear Research ("CERN"). The name changed but the acronym did not. CERN has 20 member countries and a research staff numbering in the thousands.

[3] "Physicists had been struggling to make sense of anomalies that appeared in spectra when atoms were immersed in a magnetic field: some spectral levels mysteriously split into two or more." Davide Castelvecchi, "The forgotten quantum pioneer who turned wartime spy: Davide Castelvecchi enjoys the extraordinary life of Sam Goudsmit, the atomic sleuth nominated for a Nobel prize 48 times," *Nature* 563, 320-321, 2 November 2018 (reviewing *Sam Goudsmit and the Hunt for Hitler's Atom Bomb* by Martijn Van Calmthout, 2018). Goudsmit and Uhlenbeck theorized that elemental particles had spin, some clockwise and some counterclockwise. That differing characteristic explained the spectrographic anomalies. Goudsmit was nominated 48 times for the Noble Prize, but never was awarded it. *Id.*

[4] Quarks are truly tiny, essentially like points of light. These point-like particles make up the much, much larger protons and neutrons that compose the nuclei of atoms, meaning that the nuclei are almost completely empty. Parker, presentation to the Trinity College Science Society, Cam-

bridge, 11 March 2014. We have already discussed above how truly empty the atom is. So we have almost empty nuclei and tiny electrons composing largely empty atoms that combine to create matter with which we are familiar, floating through a vast Universe, which we typically think of as otherwise empty.

[5] Supersymmetry predicts "a whole new population of elementary particles—called superpartners to the particles physicists already know about" and would explain why the Higgs boson may have a relatively low mass, but it is simply an elaborate theory. Dennis Overbye, "All Signs Point to Higgs, but Scientific Certainty Is a Waiting Game," *The New York Times*, March 4, 2013. The much discussed superstring theory is a version of supersymmetry theory.

[6] The predicted Higgs boson is far too small to be observed directly. So an enormous particle accelerator is used to create extremely high-energy collisions, the results of which are then observed by the scientists. Actually, the word "observed" must be qualified also because one cannot directly see the results of the collisions. For one thing, apart from size, the results last for only very, very small fractions of a second before disappearing. The new particle promptly decays into other particles and energy in the form of flashes of light. Sophisticated equipment detects and "counts" those products of the short-lived particle.

[7] The Large Hadron Collider is much bigger and more powerful than was required to look for the Higgs particle. At full operating strength, it will be used to search for other more massive particles that have been predicted by theory. Parker, presentation to the Trinity College Science Society, 11 March 2014.

> "The LHC is the highest-energy machine ever built, and it really should settle the question of the Higgs's existence once and for all because it has enough energy to reach well beyond the upper limits set by the Standard Model. In other words, the LHC will either confirm or break the Standard Model. We'll return shortly to explain why we are so sure that the LHC will do the job the earlier machines have failed to do... ."

Cox and Forshaw, *Why E=mc^2*, p.213.

[8] "Only in about 1908, to a large extent influenced by the penetrating analysis of the Dutch physicist Hendrik Lorentz, did Planck convert to the

view that the quantum of action represents an irreducible phenomenon beyond the understanding of classical physics. Over the next three years Planck became convinced that quantum theory marked the beginning of a new chapter in the history of physics and, in this sense, was of a revolutionary nature." Id.

[9] The mathematics used to describe the waves is complex. Using a formula relating wavelength of a particle to its mass and velocity, the mathematics demonstrates that restricting a particle's position requires a broadening of its possible momentum, and *vice versa*. However, the minimum degree of uncertainty is based upon Planck's constant, which itself is a very, very tiny magnitude. Thus, anything of a size that we could actually see can have both a position and a velocity that are definite and ascertainable for all practical purposes. See Barbour, *The End of Time*, p.205; Cox and Forshaw, *The Quantum Universe*, pp.54-6, 67-73.

[10] Heisenberg and Niels Bohr, at the Bohr Institute in Copenhagen, propounded the "official" interpretation. Id., p.202. The wave function of a particle sets out all of the information about that particle.

[11] Quantum effects have been observed at extremely low temperature. Superconductivity and superfluidity are attributed to quantum behavior, a phenomenon called quantum coherence where a collection of particles function in a single quantum state effectively unentangled from its environment. Penrose, *Shadows of the Mind*, p.351. This phenomenon is normally found at temperature just above absolute zero—far, far below anything that would occur naturally on earth. But, there is also some experimental evidence of superconductivity at higher temperatures, even up to –10 degrees Fahrenheit. Id., pp.351-2.

[12] Remember, the Standard Model with which we started does not incorporate gravity and does not include dark matter or dark energy, currently thought to constitute combined some 97% of the matter in the Universe (as discussed in the next chapter).

Cosmology

> "In the beginning
> God created the heavensand the Earth.
> The Earth was without form
> and void... ."
>
> *The King James Bible*
> (1611)

> "You know people've been talkin'
> 'bout the end of time
> ever since time began."
>
> Dolly Parton
> "In the Meantime"

In contemplating the Universe, various questions arise naturally: How did it start or, first, did it even have an origin? Is it doing anything (growing, shrinking, moving)? How big is it? And, in addition, humankind seems to be born with an inclination to anticipate the "end of time."

One cannot help feeling that for many scientists the interest in, and positions taken about, whether the Universe had a beginning has been influenced by the implications for religious doctrine. An eternal Universe would eliminate the question of how and why it began, since there would be no "beginning." (But, that would seem neither to answer nor mute the questions of why and how the Universe exists.) In contrast, the assumption of a beginning does seem to demand answers as to how and why. Although, either scenario poses the question of why there is something rather than nothing; the hypothesis of a beginning, even one that happened some 14 billion years ago, makes the questions appear to be more relevant and accessible.

Setting the Stage

Historical Views of the Cosmos

Aristotle conceived of the Cosmos as consisting of separate domains—sublunar and superlunar. See, e.g., Frank, About Time: Cosmology and Culture at the Twilight of the Big Bang (2011), p.47.

In the second century, Claudius Ptolemy presented an Earth-centered (geocentric) geometric model of the Solar System in which all planetary motion occurred in uniform circles. However, the structure of the circles became quite complex in order to comport with observational data, particularly the apparent retrograde motion of certain planets relative to the Earth. Ptolemy's solution involved having bodies move around smaller circles, called epicycles, which themselves moved on larger circles around the Earth. Id., pp.48-9. Although, Ptolemy's model provided a highly accurate mathematical representation of celestial movement; it did not purport to describe what the Solar System was like.

Some thirteen hundred additional years later, Nicholas Copernicus formulated a Sun-centered (heliocentric) model that was much simpler and more elegant than the Ptolemy model. In addition to centering the system on the Sun, Copernicus introduced the regular rotation of the Earth. Id., pp.75-6. The retrograde movements of the planets were explained by the fact that the planets all orbited the Sun with different periods of rotation (the planets further from the Sun taking longer to complete a rotation) against a fixed background of stars. As the Earth caught up with and passed more distant planets in the rotation around the Sun, those planets appeared to move backwards or in the opposite direction relative to the Earth. Id. (In Ptolemy's system, different combinations of circle sizes and orbit speeds could generate the observed results.)

Interestingly, in the final versions of the Copernican system, the data on the planetary retrograde movements implied an approximate size for the Solar System. That estimate was very much larger than previously suspected. Id., p.77. But, still small, by our current standards.

Copernicus had, like Ptolemy, utilized circles in his model. Johannes Kepler achieved the almost perfectly accurate model with the introduction of ellipses in place of circles. The planets revolved around the Sun in elliptical orbits, pursuant to a few simple rules. However, the planets also moved at varying speeds throughout their orbits. What could explain that? Remarkably, Kepler, from examining the data that had been compiled by

Tycho Brahe, realized that if one looked at the line connecting the planet to the Sun, it would sweep equal areas of two-dimensional space in equal intervals of time.

An important question again arises, as it does with all acts of creative discovery: "How does one stumble on the idea of governing speed by the areas swept? ...[I]s it a stroke of genius or an accident of groping?" Mazur, *The Motion Paradox*, p.90. Regardless, "[t]he descriptive economy of [the resulting] relationships, which came to be called Keplers [sic] Laws, became a model for scientific descriptions of nature." Frank, *About Time*, p.79.[1] For Kepler, however, the assumption of elliptical orbits was merely an *ad hoc* device—and one which he reportedly found to be "repugnant" (ellipses being less perfect than circles)—that happened to fit the data.

Only in 1687 did Sir Isaac Newton provide the mathematical analysis that explained why the orbits were elliptical. *See* Stephen Hawking, *A Brief History of Time*, p.4.

THE FIRST TELESCOPES

The early efforts to examine the skies with telescopes gave rise to questions obviously reflecting the limited understanding of the Universe then existing. For example, the observable great differences in the number of visible stars in different directions from Earth suggested the possibility that the material Universe was finite (consisting of what we now know as the Milky Way galaxy) and, perhaps, of some special oblong shape. *See* Frank, *About Time*, pp.150–1.

In the subsequent years, substantial progress was made in observing the phenomena of the skies, assisted by ever-more capable telescopes and astonishing feats of patience and persistence. A young deaf Scottish astronomer John Goodricke, born in 1764, carefully measured variations in the light of certain stars compiling data that would subsequently be of use in establishing the standard measures by which distances to far off galaxies would come to be estimated in the twentieth century, noted below. Linda French, "Disability history month: John Goodricke the deaf astronomer," *BBC News Magazine*, 18 December 2012. He also speculated that the regularity of the observed dimming of certain stars light could be caused by the passage of orbiting planets between the star and the Earth. *Id*. He was awarded the Copley Medal by the Royal Society of London at age 19, becoming, and still remaining, the youngest recipient of that award. *Id*.[2]

Ultimately, of course, astronomers began to detect galaxies beyond our Solar System. With each discovery and revision of the prevailing model of the Universe, the estimated size of the Universe expanded greatly.

NEBULAE: A LARGER UNIVERSE

Initially, when astronomers detected strangely shaped phenomena known as nebulae (apparently massive structures of gaseous matter), they concluded that these "clouds" of matter were located within the known Universe (consisting then only of our galaxy). The swirling, disk shaped nebulae (spiral nebulae) were of particular interest. Into the 1920s, astronomers debated whether these observable nebulae were part of the Milky Way or, possibly, far distant systems of stars similar to the Milky Way. Frank, *About Time*, pp.150-3. An important intellectual obstacle to the hypothesis that these nebulae constituted additional galaxies was the staggering increase in the size of the Universe that such a hypothesis entailed. *Id.*, p.153.

The completion of the huge 2.5 meter Hooker telescope on Mount Wilson, controlled by motors powered by the recently introduced electricity, enabled the resolution of the controversy about the spiral nebulae and demonstrated that the Universe was, indeed, very, very much larger than anyone had dared to imagine. *Id.*; Krauss, *A Universe from Nothing*, p.11.

In passing, it was recognized that the observable Milky Way in the night sky and the great differences in the apparent densities of the stars in various directions from Earth was the result of the spiral shape of our galaxy and the position of the Earth in it.

STATIC AND ETERNAL?

It seems likely that for our earlier ancestors the common assumption would have been that the Universe was eternal, with no origin, and that it was static or unchanging, as a corollary. See, e.g., Jim Holt, *Why Does the World Exist?* (2012), p.25. In the ancient world, presumably there was little awareness of change taking place in the natural world, so early man would have had little awareness of or concern with history or with time, apart from the very regular passage of day to night and night to day. See Stephen Toulmin and June Goodfield, *The Discovery of Time* (1965, 1982), pp.17-32. There is evidence that the issue of the origin of the cosmos was sometimes addressed, but the question was largely focused on the nature of man's role or position in the world. *Id.* We know that Aristotle believed that the Universe was eternal, and his influence on science was long lasting, even increasing in the thirteenth and fourteenth centuries as translations of his works became available. See Frank, About Time, pp.64-5.

Judeo-Christian beliefs, however, conflicted with the concept of an eternal and unchanging Universe, because Genesis clearly speaks (twice) of a beginning. Early Christian doctrine also differs from other creation myths in the simple lack of an initial state before the creation of the world. See Frank, About Time, p.53; Holt, Why Does the World Exist?, pp.19-20.

By the beginning of the twentieth century, Einstein, and most scientists of his time, believed in an eternal Universe. Indeed, as of 1917, the scientific community was largely in agreement that "the universe was static and eternal, and consisted of a single galaxy, our Milky Way, surrounded by a vast, infinite, dark, and empty space." Krauss, A Universe from Nothing, p.2. The proposition that the Universe was expanding (or contracting) had simply not been seriously considered; most scientists conceived of the controversy as being whether the Universe had always been as it is or had had an origin and, thereafter, had remained essentially the same. Hawking, A Brief History of Time, p.6.

However, if the Universe is in a steady state, what about gravity? Since Newton, classical physics has understood gravity to be a purely attractive force and one that is universal, although its strength diminishes rapidly with distance. If gravity exists throughout the Universe and everything stays in the same positions relative to everything else, then there must be something that counteracts and balances the attractive forces of gravity and does so with remarkable finesse. Otherwise, the Universe would ultimately start collapsing because of gravitational attraction.

This consideration also suggests that the Universe must be infinite or, at least, have no edges, since the matter on any edge that did exist would tend to be pulled back into the Universe by gravity, creating a shrinking of the Universe.[3]

The problem did not go away, of course, with Einstein's theory of gravity contained in his General Theory of Relativity. Einstein himself, however, did not see and was reluctant to accept the implications of his theory. As a "solution", Einstein proposed a cosmological constant that would be a kind of anti-gravity, counteracting the force of gravity of existing matter. Einstein later characterized the cosmological constant as his "biggest mistake".

EXPANDING OR SHRINKING OR BOTH?

As a merely logical matter, one can see that there are alternatives to a static Universe.

An obvious possibility (and the one that appears to be true for our Universe) is that the Universe is expanding. But, if the Universe both is expanding and is eternal, then how can we still see so many other stars (unless new stars are constantly being created)? An expanding Universe also suggests an event that propelled matter outward with a force that is, at least so far, overcoming gravitational attraction.

Alternatively, the Universe could in theory be contracting, suggesting an end point at which gravity would so overwhelm matter that there would be a veritable collapse.

Presumably, it is also possible that the Universe has been expanding, due to an outward propulsion, and will soon be contracting due to gravity, but we happen to have appeared at a moment of equipoise (which could be rather long by human standards), like a ball at the top of its arch, readying itself to begin its descent but appearing to be momentarily suspended.

An expanding Universe in which we can still see other stars is in principle consistent with a model of an eternal, but oscillating Universe. Such a model, of course, raises questions about what happens at the point of transition (including, for example, does any 'information" carry from one phase to the next) and what causes each new expansion. Clearly, there could be various combinations of these scenarios.

FLAT, OPEN OR CLOSED?

The modern analysis of the potential end of the Universe arose out of theoretical calculations based upon Einstein's General Theory of Relativity. These theoretical explorations by Einstein and a handful of others were all premised upon certain simplifying assumptions as a group called the "cosmological principle". This principle assumes that:

- the Universe is homogeneous (on a large scale);
- it is isotropic (that is, that it will look the same to every observer regardless of his position in the Universe (which requires that the Universe be infinite, because it otherwise would look different near the edges); and

- the laws of physics are the same (or apply equally) in all parts of the Universe, that is, everywhere.

These are necessary assumptions to make the mathematics of Einstein's field equations manageable.[4]

Georges Lemaitre, a Belgian physicist, showed, in 1927, that the field equations predicted that the Universe was not static and may be expanding—and Einstein protested the conclusion. Krauss, A *Universe from Nothing*, p.5. Alexander Friedmann, in the early 1920s, also demonstrated that Einstein's equations predicted that the "fabric of space was stretching" and that the galaxies were moving away from one another. Greene, *The Elegant Universe*, p.82. Similar models were developed by Howard Robertson and Arthur Walker in 1935, leading to refinements of Friedmann's model to encompass additional scenarios. *Id.* p.346; Hawking, *A Brief History of Time*, pp.40, 42-4. Lemaitre, continuing his work, proposed in 1930 that the Universe began as a tiny point, which marked the beginning of time. Krauss, A *Universe from Nothing*, p.5. He did so by working backwards, projecting the current Universe back in time.[5]

The mathematics indicated that the Universe could, consistent with the General Theory, "exist in one of three different geometries, so-called open, closed, or flat." Krauss, A *Universe from Nothing*, p.27. In the "flat" geometry, light will travel in straight lines, so that two parallel light beams will remain parallel forever, absent the effects of gravity. In the other geometries, light beams will be curved, albeit over enormous distances. If the nature of the curvature is what is called closed, the equations of General Relativity say that the Universe will some day collapse; if the nature of the curvature is what is called "open," then the Universe, according to the equations, will forever continue to expand at a finite rate. *Id.*, p.28. In a "flat" geometry, the expansion of the Universe will continue but will do so at an ever diminishing rate. *Id.* If one could reasonably estimate the total amount of mass in the Universe, then a calculation of the gravitational effects of such an amount of mass would enable one to determine whether the Universe is flat, open or closed. But, it is not so simple.

In the end, empirical evidence and observation were to provide an answer. But, is it the correct answer?

The Legacy of Edwin Hubble

Almost everyone is familiar with some of the legacy of Edwin Hubble (or, at least, with the large orbiting telescope that was named after him).

A "Standard Candle"

In 1924, Hubble demonstrated that there were other galaxies in the Universe, beyond ours, and attempted to estimate the distances of some of those galaxies from Earth. *See, e.g.*, Hawking, A *Brief History of Time*, pp.36–9. His indirect procedure was based upon the assumption that stars of similar detectable characteristics would "burn" with similar intensities or luminosities, which appeared to be confirmed by observations of the closer stars in our galaxy. So, the inference that was made was that differences in observable brightness observed for other stars would be indicative of differences in their distances from the observer.

This theory was based upon the discovery by Henrietta Swan Leavitt, between 1908 and 1913, that there was a regular relationship between the brightness of Cepheid variable stars and the period of their variation. (Cepheid stars had been observed first in 1784; they were notable in that their brightness varied over regular periods or phases.) She examined such stars from one cluster and found a correlation between a star's rate of variation and its maximum brightness. So, if you knew the star's pulse rate, you could calculate its absolute brightness, assuming that that characteristic of Cepheid variable stars obtained every where. As a result, if one could estimate the distance to one Cepheid star, then the distances to others could be calculated based upon their perceived brightness (with the apparent or observed brightness of stars of identical actual brightness declining inversely with the square of the distance of the star from the observer). Krauss, A *Universe from Nothing*, pp.7–8.

Hubble was able to use the data he compiled on Cepheids combined with Leavitt's discovery concerning the relationship between luminosity and period of variation to estimate the distances of numerous stars. *Id.* Such calculations demonstrated that the observable stars were not conceivably all within our galaxy—the distances were far too great.

The "red shift"

Hubble's contributions went far beyond estimates of the possible size of the Universe. Other major discoveries were based upon the science of spectroscopy.

The only observable characteristic for most stars is the color of the light, from which spectroscopic analysis (as described in the chapter on Chemistry) enables one to determine the composition of the atmosphere of the star (since each chemical element absorbs, and emits, distinctive specific colors of the spectrum). Hawking, A Brief History of Time, pp.36–9. Remarkably, the spectroscopic analyses indicated that at least the visible parts of the Universe are all made of largely the same things. This fact is called by Lawrence Krauss "[o]ne of the most important discoveries in astronomy." A Universe from Nothing, p.9.

As efforts were made to collect information on stars in other galaxies, it was also noticed that the spectra of the fainter stars showed a "shift" toward the red end of the spectrum. Otherwise, however, the patterns were readily identifiable. The similarity of the spectra of observable stars, resulting presumably from the fact that the stars were made of the same elements, made the use of the "red-shift" technique possible. This observed shift in the spectrum was attributed to the Doppler effect, which is demonstrated with sound waves when a train or car approaches us and then recedes.

One can hear the shift in the frequency of the sound waves, causing the sound to become deeper (lower) when the source of the sound passes and begins to move away. Sound waves travel at a constant speed once emitted, independent of the motion of the source. However, the peaks of the waves (establishing the frequency of the waves) are closer together (there will be more peaks per second) when the source is moving in the direction that the waves are emitted and farther apart when the source is moving away from the direction in which the wave is traveling. In effect, the source is either catching up with or running away from the peak that was last generated, closing or opening the gap. With the train or car, the sound is being emitted in all directions. Thus, the frequency of the sound moving in the direction that the vehicle is traveling will be shorter or higher, and the frequency of the sound moving in the opposite direction will be longer or lower.

The same phenomenon occurs with light. Note that this is due to the wave characteristic of light. If light consisted of ordinary particles without wave characteristics, one would observe differing speeds at which the light was traveling based upon the movement of the source of the light relative to the observer. See, e.g., Krauss, A Universe from Nothing, pp.10–11.

Awareness of the Doppler effect has been of considerable practical significance.

AN EXPANDING UNIVERSE

One objective was to compare the red shifts of the spectrum of stars with the estimated distances of the stars from Earth. When the data was compiled by Hubble, he discovered that most of the galaxies studied (that is, the stars studied in other galaxies) displayed a red-shifted spectra. There was not an even mixture of red shifts and blue shifts, which one might have expected. Thus, it appeared that most of the galaxies observed were moving away from us. Moreover, it further appeared that the galaxies were moving away at speeds that increased the further away the galaxy appeared to be. Hubble's findings were published in 1929. Hawking, A Brief History of Time, pp.36–9.

If we assume that on a sufficiently large scale, the Universe is essentially uniform, then we would conclude that the Universe as a whole is expanding and that all galaxies are moving further and further apart from one another. The assumption, however, is one that cannot be proved scientifically, but scientists accept it "on the grounds of modesty." Id., p.42. Otherwise, there would seem to be something special or, at least, not representative about the position of Earth (or of our galaxy), a view that was once just assumed to be obviously true but that is uniformly denied in modern thinking.

So, the empirical investigations using spectroscopy and the "red shift" in the dark bands in the spectrum when an object and observer are moving away from each other (at sufficient speed for the shift to be detectable) enabled Edwin Hubble to give an answer, published in 1925, to the question above: yes, as a matter of fact, the Universe is expanding, at least at this time.

The Doppler effect enables calculations of the velocities at which galaxies are moving away, but there are no known techniques for establishing the distances among galaxies with anything one might call precision. See id., p.45; Bryson, A Short History of Nearly Everything, p.217 ("when astronomers say that galaxy M87 is 60 million light years away, what they really mean ... is that it is somewhere between 40 million and 90 million light years away").[6]

With a "standard candle" (discussed above), one might be able to make reasonable estimates of the relative distances from Earth by comparing the brightness from Earth of two such stars believed to be burning with comparable intensities (one may be twice as far away as another one); but, without knowing the actual distance of at least one such star, there will be inevitable uncertainty about the actual distances involved.

The calculations based on Hubble's data, however, suggested that the origin was no more than 1.5 billion years ago. Krauss, A Universe from Nothing, p.15. That estimate conflicted with evidence that the Earth itself was at least twice that old. It has since been discovered that Hubble's data contained systematic errors. Other, later methods of estimating the distances between stars and Earth, using a particular type of supernova as the "standard candle", have enabled revised calculations of the age of the Universe, giving us the current age of about 13.5 billion years. Id., pp.21, 78-9.[7]

AND, THE STEADY-STATE THEORY?

In any event, the conclusion was that the Universe is not static, but is expanding, at least in relation to the Earth and at least in that part of the Universe that we can observe. As it turned out, the discovery that the Universe as a whole was apparently expanding did not itself, it was argued, disprove the prevailing belief that the Universe was, nonetheless, eternal and essentially static.

In fact, throughout the 1950s, the leading cosmological theory was the "steady-state" model. Penrose, Cycles of Time, pp.ix, 68-9. The theory proposed that the Universe was maintained in a steady state by the continual creation of new matter (at a very, very slow rate) in the empty spaces being created by the existing galaxies moving farther apart from one another. Id.; Hawking, A Brief History of Time, p.47. There were, of course, the questions of from where the new matter came from and of how equilibrium was maintained, something that was very unlikely absent a mechanism of dynamic adjustment (for which no recognized theories had been proposed). But, even though a steady state was improbable (at least, without such a mechanism), the theory was simple and elegant. It also generated predictions about potential empirical observations.

The next challenge arose as a result of subsequent observations made to test those predictions. The steady-state theory seemed to imply that the density of galaxies throughout the Universe should be essentially the same. Efforts by Martin Ryle and a group of astronomers working with him at Cambridge University to detect and identify other galaxies and, based upon the faintness of the light, to estimate their distance from us indicated that the space around the Milky Way was less populated than more distant areas. Hawking, *A Brief History of Time*, pp.47–8. This evidence suggested either that the density of the Universe in the past (when the light from the far galaxies was emitted) was greater than it is now or that the densities in different parts of the Universe vary.

Either explanation, however, seemed to conflict with the steady-state theory. Id., p.48; Penrose, *Cycles of Time*, p.68.

So,

The discoveries of Hubble—supplemented by the work of Martin Ryle, Dicke and Peebles, and Penzias and Wilson, described below—effectively eliminated many cosmological theories. Hawking describes the discovery that the Universe was expanding as "one of the great intellectual revolutions of the twentieth century." *A Brief History of Time*, p.39.

Origin of the Universe

Since Hubble's calculations evidenced a linear relationship between distance from Earth and velocity of the movement away from the Earth, one could work backward and conclude (like Lemaitrei) that the Universe started in some small area, even in a single point, and has been expanding since its origin.

In fact, there is considerable interest in the origins of the Universe, with a contemporary fascination with the so-called Big Bang theory. A fundamental problem is that the origin will almost certainly always be an unknown—at best, we can make guesses and see if the evidence available is consistent with one or more of such guesses. And, to the extent that the Universe today reflects emergent phenomena, there may be multiple explanations of the origin that are consistent with current conditions. The challenge is particularly great with unstable phenomena (where repeated events result in different outcomes).

As stated by physicist Robert Laughlin, with some obvious hyperbole:

> "*Exploding things, such as dynamite or the big bang, are unstable. Theories of explosions, including the first picoseconds of the big bang, ...are inherently unfalsifiable, notwithstanding widely cited supporting 'evidence' such as isotopic abundances at the surface of stars and the cosmic microwave background anisotropy.*"

A *Different Universe*, p.211.

Even Lawrence Krauss, despite his forceful and self-congratulatory exposition of the "something from nothing" theory of the origin of the Universe, acknowledges that "this is how our universe *could* have arisen. I stress the word *could* here, because we may never have enough empirical information to resolve this question unambiguously. But the fact that a universe from nothing is even plausible is certainly significant, at least to me." A *Universe from Nothing*, p.xxiii (emphasis in original).

THE BIG BANG

Stephen Hawking and Roger Penrose demonstrated in 1970 that Einstein's General Theory of Relativity implied that the Universe must have a beginning. A *Brief History of Time*, p.34. In a joint paper, they showed that the mathematical model required that the Universe began in a Big Bang, arising out of a singularity, if the theory of "General Relativity is correct and that the Universe contains as much matter as we observe." *Id.*, p.50. The analysis has become more complicated, as described in the next section. That debate is whether the Universe, prior to the Big Bang, was a "singularity".

CONSEQUENCES OF THE BANG

The implications of a Big Bang were rapidly worked out. Scientists concluded that the Big Bang would be stunningly hot with the resulting particles moving at very high speeds. See, e.g., Hawking, A *Brief History of Time*, p.116. The production of huge quantities of photons would be expected to leave a detectable radiation, of the type that was found by Penzias and Wilson (discussed below). Neutrinos would have existed (and should still be present today), but at energy levels too low to allow direct detection. The temperature would have dropped very quickly (during the first second following the Bang) as the new Universe expanded and particles would have begun to clump together. *Id.*, p.117.

Interestingly, there would have been no light at the moment following the Big Bang, only electromagnetic waves. Matter was created next and, then, the motion of matter generated light waves. See Smolin, *Time Reborn*, p.206. According to Hawking, following the first second after the Big Bang, the collisions of the particles would have created pairs of electrons and anti-electrons and other particle/anti-particle pairs. The opposites of the pairs would frequently annihilate one another, producing photons. A *Brief History of Time*, p.116.

The evolution of stars

Pursuant to the theoretical model of the Big Bang, which is used to explain the creation of the observable elements in the Universe, the Universe would have still had a temperature of 3000 degrees Kelvin when it was 300,000 years old. Although, that temperature is cooler than the temperatures of the initial Bang, it is still very hot; so hot, in fact, that atoms could not exist, only the protons and electrons of which the atoms would be made. Krauss, A *Universe from Nothing*, p.43. These protons and electrons would have constituted "a dense 'plasma' of charged particles" that would have been "opaque" to radiation. *Id.* Thus, the first time at which the radiation that appears to fill the Universe and to be coming equally from all directions would have originated when the Universe was 300,000 years old, not at its birth. *Id.*, p.44. Since that time, the Universe has expanded about 1000 times and has cooled to roughly 3 degrees Kelvin. *Id.*

Quoting Cox and Forshaw:

> "At a temperature of **10,000 degrees**, the electrons are ripped from their orbits around the nuclei, leaving behind a gas of protons and electrons known as a plasma. ...The plasma is rescued from a seemingly irretrievable fall when the temperature approaches **10 million degrees**... . Individual protons fuse together to make a deuteron, which itself can fuse with another proton to produce helium... . In this way the **new star** slowly converts a small fraction of the original mass into energy, which heats up the core of the star and allows it to halt and resist any further gravitational collapse, at least for a few billion years... .
> ...[W]hen a star runs out of **hydrogen fuel** in its core? ...[T]he star will once again start to collapse, getting hotter... . Eventually, at a temperature of around **100 million degrees**, **helium begins to burn** and once again the star's collapse is arrested. ...Eventually, the star will exhaust its supply of helium and begin to collapse

even further. As the core temperature rises past **500 million degrees**, *it becomes* **possible for the carbon to burn**.... *In a dense star, the electrons exert an outward pressure [as a result of the Pauli exclusion principle] that increases as the star collapses until it is eventually so large that it can prevent any further gravitational collapse. Once that happens, the star is trapped in an enfeebled but incredibly long-lived state. It has no fuel to burn ... and it cannot collapse any further because of the electron pressure. Such a star is called a* **white dwarf**... . ...*[But, for others] the large initial mass means that the electrons eventually start to move around at close to the speed of light as the collapse continues.* ...*For these massive stars, the next stop is a* **neutron star**, *in which nuclear fusion steps in for a final time. The protons and electrons move so fast that they reach a point where they have sufficient energy to initiate proton-electron fusion, producing a neutron...* [A]*ll of the protons and electrons gradually convert into neutrons and the star is nothing but a ball of neutrons. The density of a neutron star is phenomenal....*"

Why E=mc², pp.161, 162, 164, 165, 167-8 (emphasis added).

The "light" elements

Eventually, the production of the skimpiest elements would have started: first deuterium ("heavy hydrogen" with one neutron and one proton), then helium and ordinary hydrogen. The extraordinary heat of the Big Bang would account for the abundance of helium in the Universe, which otherwise is difficult to explain. Id., p.118. In fact, the analyses of the elements that constitute the visible matter in the Universe supports the theory of a Big Bang, because the extreme temperature of that Bang provides a plausible explanation for the abundant presence in the Universe of the light elements (like hydrogen, helium and lithium) that is observed. Krauss, A Universe from Nothing, p.18.

The "heavy metals"

"*In 1953, when the understanding of the nuclear physics of stars was still in its infancy, astronomer Fred Hoyle realized that carbon had to be manufactured inside stars, irrespective of what the nuclear physicists told him, because he strongly believed that there is nowhere else in the universe to make it. Coupled with his astute observation that astronomers exist, he theorized that this could happen only if a slightly heavier type of carbon nucle-*

> us exists such that it can be formed very efficiently as the result of fusion between the short-lived beryllium-8 and a third helium nucleus. ...It was a matter of days after Hoyle made his prediction that nuclear physicists working in the Kellogg Laboratory at Caltech confirmed his prediction without any shadow of doubt."

Cox and Forshaw, Why $E=mc^2$, pp.162-3.

The significant presence of the heavier elements like carbon in the Universe indicates that it must have been relatively common for stars to have exploded, spewing those elements into space; because such elements would have been produced in the furnaces of stars, not in the initial Big Bang. Krauss, A Universe from Nothing, pp.18-9. The heavier metals, like iron, can be created only in even more intense furnaces like those of the supernovae. It is also believed that the current stars like our Sun, which have significant amounts of the metals, must be examples of fifth or sixth generation stars, formed from the debris of a series of pre-existing generations of stars that exploded, sending heavier elements into space where they would form clouds that ultimately were aggregated into new stars.

But, there has been a debate about how and where the even heavier elements, like gold, were produced—for example, in supernovae or by the collisions of heavier stars:

> "One of the issues ... is the origin of the 'r-process' elements (the most enticing of which to most people is gold). These elements are so named because they can be produced only in environments that are so rich in neutrons that the neutrons combine with nuclei more rapidly ... than the nuclei decay into stable isotopes. Early work favoured supernovae as the origin of these elements, but over the past few years, analyses have leaned towards the merger of compact objects, such as neutron stars, as [being] the prime r-process factories ... ."

M. Coleman Miller, "Gravitational waves: A golden binary," *Nature*, vol. 551, pp.36-37, November 2, 2017.

> "Recent findings have suggested that much of the gold and other elements heavier than iron on the periodic table [platinum, uranium, etc.] was born in the catastrophic aftermath of colliding neutron stars, which are the ultradense cores of stars left behind after supernova explosions. ...The researchers found a vast

amount of heavy elements in the solar system likely originated from a single neutron-star collision that occurred about 80 million years before the birth of the solar system [and about 1,000 light-years away]."

Charles Q. Choi, "Ancient Neutron-Star Crash Made Enough Gold and Uranium to Fill Earth's Oceans," *Space.com*, May 8, 2019 (reporting on an article published in *Nature* on May 1, 2019).

"[F]rom the simple nuclei of hydrogen, all these processes generate successively larger and more complex nuclei, leading to about 95 naturally occurring elements found in the universe. Two of these, hydrogen (75 per cent) and helium (23 per cent), account for most of the mass of the universe." Hands, *Cosmosapiens*, p.132.

Ancient stars

We have talked about the efforts to examine stars from the early days of the Universe by looking at the images being received from billions of light years away, which presumably were initially formed billions of years ago. It turns out that there are ancient stars that still exist and that can be found relatively nearby. Curtis Brainard, "The Archaeology of the Stars," *The New York Times*, February 10, 2004. The presence of very old stars in nearby galaxies suggests that galaxies like the Milky Way have grown through the cannibalization of older, smaller galaxies, with some of the older stars so absorbed still remaining. *Id.*

These ancient stars appear to be quite different from the bulk of the stars that make up the Milky Way in that they have very low amounts of the heavier elements like iron. It is believed that these "metal-poor" stars must be earlier generation stars, formed before the process of successive star creation and explosion had generated substantial amounts of debris consisting of the heavier elements. Thus, they may be first or second generation stars formed as long as 13 billion years ago. *Id.* As more examples of very old stars are found and it becomes possible to date more of them, it may be possible to establish a pattern that the older the star, the less metal. Such a pattern would provide evidence supporting the theory of how the heavier elements were created and dispersed throughout the Universe.

COSMIC MICROWAVE RADIATION

In 1965, two physicists with the Bell Lab in New Jersey, Arno Penzias and Robert Wilson, discovered that there was a constant level of microwave radiation, creating background noise on a sensitive microwave detector, coming toward the Earth from all directions (the readings on the detector that they were testing were the same regardless of the direction that the detector was pointing, all day and all night and all year long). Hawking, *A Brief History of Time*, p.41. The conclusion that the evidence compelled was that the radiation came from outside of the galaxy, otherwise some variations in the disturbances would have been observed with direction, time or season.

Coincidentally, in about the same period, Robert Dicke and James Peebles, from Princeton University, had reasoned that if the early Universe had been so "very hot and dense" that it glowed white, we should still be able to see evidence of the early light reaching us only now from far distant parts of the Universe. But, because of the degree of red shift, that light would appear as microwave radiation (having shifted beyond the visible wave frequencies). *Id.*, pp.41–2.

When Dicke and Peebles learned of the observations of Penzias and Wilson, they realized that the evidence predicted by their theorizing had already been found. *Id.*, p.42. Penzias and Wilson were awarded the Nobel Prize in 1978. *Id.* The evidence, as interpreted by Dicke and Peebles, appears to support the hypothesis that the Universe experienced a very significant event at some point in the distant past, perhaps even its origin in a Big Bang. This discovery (the combination of the empirical observations and the theoretical model) of what came to be called the Cosmic Microwave Background Radiation (CMBR) or just the Cosmic Microwave Background (CMB) dramatically altered the cosmological debate. *See* Penrose, *Cycles of Time*, pp.68-70.[8]

SOMETHING FROM NOTHING

If we assume that there was a beginning, then it is hard not to ask what was there before and what caused the appearance of the Universe in the beginning. If, of course, there was nothing in existence prior to the Big Bang, then we are still confronted with the question of why there is now something rather than nothing, followed by the question of how the Universe could have been created out of nothing. These questions seem to have recently captured the public imagination. *See, e.g.,* Krauss, *A Universe from Nothing*; Holt, *Why Does the World Exist?*. (Cosmological issues tend to reflect a certain faddishness.)

Nothingness

Quantum cosmology indicates the possibility of an answer and, even, of a variation on the hypothesis of a cycling Universe. The theoretical/mathematical work that postulates the existence of both particles and anti-particles, as well as both positive and negative forces and positive and negative energy, suggests that it is possible that the net contents of the Universe is effectively zero, that is, no mass, no energy, nothing. Holt, *Why Does the World Exist?*, pp.140–41. Then we add the concept of virtual particles.

A theory has been propounded that nothingness is inherently unstable, so that a small fluctuation in the quantum soup could initiate the creation of pairs of matter and anti-matter that would expand at an enormous rate, creating a universe. The originator of this idea was a physicist named Ed Tryon. *Id.*, pp.141–2. One implication of this theory is that there was not really nothing before the creation of the Universe. The quantum vacuum itself would be something—an empty space—and it would not really be empty— it would have to be "saturated with energy fields and seeth[ing] with virtual-particle activity." *Id.*, p.142.

Thus, while our Universe may have arisen out of a chance or random fluctuation, we may find ourselves back to the question of why quantum space exists. Indeed, we might even re-ask the question of whether space is not actually filled with an "ether" through which light waves and other electromagnetic waves occur.[9]

A tiny bit of vacuum

Another "answer" was proposed by Alex Vilenkin, a physicist at Tufts University. He used the analogy of a deflating balloon. If space-time were to shrink to a point and then to a singularity, then there would be true nothingness—"mathematically devoid not only of stuff but also of location and duration." *Id.*, p.143. But, Vilenkin's theory needs a tiny bit of quantum vacuum remaining. It could be very, very tiny. That infinitesimal bit of vacuum could experience "inflation" (just as previously discussed), reaching cosmic size in a very small fraction of a second. That tiny bit of vacuum or of space-time, he asserts, could spontaneously "tunnel" into existence. *Id.*

Well, I guess we have made progress. The "something" that exists rather than nothing is, at least, an incredibly small something and it is something that apparently appears out of nothing for no reason other than chance.

Of course, given infinity, if something is possible, then it will happen, an infinite number of times. Right?

An interesting addendum to this theorizing is that the vacuum instability on which it is based raises the obvious possibility that another fluctuation could cause the almost instantaneous creation of another universe that would perhaps simply replace ours, like a "renewal" of space-time. See, e.g., Jonathan Amos, "Cosmos may be 'inherently unstable,'" BBC News, 18 February 2013. This speculation most recently arose out of a conference session discussing the implications of the Higgs boson. The hypothesis is different from the original cycling Universe, where the expansion and contraction would follow rules; here the random fluctuations would presumably be unpredictable. Thus, the replacement of this Universe with another could happen at any moment and for no discernible reason. With this theory, we avoid the problem of a beginning and the problem of nothingness.

Something?

Related to the idea of an initial period of inflation and the concepts of quantum theory, another theory about the origin of all of the matter and energy in the Universe arose. It has been proposed that when the incredibly brief period of inflation was coming to an end, different microscopic regions of the resulting Universe experienced the slowing down of inflation at different rates because of quantum fluctuations (the fluctuations that assertedly occur spontaneously in the mixture of virtual particles and fields that we traditionally have thought of as empty space). Krauss, A Universe from Nothing, pp.97–8. The resulting variations in density (on a very small scale) then would have given rise to the atoms, molecules and, subsequently, objects that make up the visible matter in the Universe today. Id. The force of gravity can be characterized as negative energy. So, as the Universe expands, positive energy may be created to balance the negative energy of gravity. Id., pp.150–51. In that sense, nothing can be said to produce something. Id., p.153.

There is another angle on the question of something or nothing, which arose out of efforts to address the other question about the Universe—its fate. Efforts to determine what is likely to become of the Universe, discussed in the last section of this chapter, have led to theories postulating enormous amounts of "somethings" that exist in apparently empty space. These somethings have been called "dark matter" and "dark energy". They are "dark" because we cannot see them. More importantly, we do not really have any idea what they are. But, the hypotheses of dark matter and dark energy give a new meaning to the concept of "nothingness."

Apparently empty space may be a rather special kind of nothing, being filled with matter and energy of types that we cannot find on Earth and can only indirectly or inferentially detect in the Universe. See Krauss, A Universe from Nothing, pp.xv–xvii, 56.

So...

The widely held conclusions by the later part of the twentieth century were that:

- the Universe must have had an origin,
- that origin occurred with a Big Bang that propelled the resulting material (particles, energy, whatever) outward, and
- that process of expansion is still continuing.

We should note, again, that the mere fact of the current expansion does not necessarily mean that there was an origin and certainly not an origin in the form of a Big Bang. Of course, some cause for the observed current expansion is required. So, we would need hypotheses, and then we would want to look for evidence in support of or inconsistent with those hypotheses. It seems that nothing that has been offered (at least, to date) has gained any traction other than the hypothesis of an initial "explosion".

And now, in the beginning of 2022, we await the unfurling of the James Webb Space Telescope, launched on Christmas day:

> *"The world's largest and most complex space science observatory will now begin six months of commissioning in space. At the end of commissioning, Webb will deliver its first images. Webb carries four state-of-the-art science instruments with highly sensitive infrared detectors of unprecedented resolution. Webb will study infrared light from celestial objects with much greater clarity than ever before. ... The telescope's revolutionary technology will explore every phase of cosmic history – from within our solar system to the most distant observable galaxies in the early universe, to everything in between. Webb will reveal new and*

*unexpected discoveries and **help humanity understand the origins of the universe and our place in it**."*

NASA, "NASA's Webb Telescope Launches to See First Galaxies, Distant Worlds," *Release* 21-175, December 25, 2021 (emphasis added).

Black Holes

We have briefly mentioned the concept of black holes already. It plays a crucial role in modern cosmology. Although the term "black hole" was introduced by physicist John Wheeler in 1969, the concept has a relatively long history. Hawking, A *Brief History of Time*, p.81.

With the assumption that light consisted of particles, a Cambridge fellow, John Mitchell, presented a paper to the Royal Society of London in 1783 arguing that the light particles were presumably subject to gravity and, if so, then for sufficiently massive stars, it was possible that the strength of the star's gravitational field would be such as to prevent the light particles from escaping the vicinity of the star. In such a case, the star would not be visible. *Id.*, pp.81–2.

Following the introduction of Einstein's theories appearing to vindicate the hypothesis that the speed of light is the same for all observers, it was apparent that the situation had to be more complicated. Gravity seemed to have an effect on light, as demonstrated by various empirical tests. A strong gravitational field can cause light "beams" to bend. Such a field can also alter the frequency of light (increasing the distance between the crests of light waves). But, gravity, consistent with Einstein's theories, cannot cause light to slow down or to stop.

How to get a "black hole"

The theoretical answer is that a strong enough gravitational source (such as a sufficiently dense star) will cause light "beams" to bend. As a star collapses, the observable light would display a pronounced red shift. Ultimately, gravity would cause the light beams to bend sufficiently that they would be trapped within a certain distance above the surface of the star, known as the event horizon. *Id.*, pp.82–7. None of the light could be seen beyond that horizon, making the star dark or black.

Another formulation is in terms of escape velocity, the speed at which an object must be traveling in order to escape the gravitational attraction of another, generally much larger, object (and continue off into space). The escape velocity is a function of the ratio of mass to radius of the object from which one wishes to escape. As mass increases and radius of the star diminishes, a point is reached where the escape velocity exceeds the speed of light, creating a black hole from which nothing can escape, including light. Penrose, *The Road to Reality*, pp.705–6.

To the layperson, this explanation raises the question about what happens to the light particles that are trapped. Do they slow down and eventually stop moving, contrary to the rule that light always travels at the same speed? Or do they continue traveling at the speed of light until reabsorbed into the matter that constitutes the black hole?

CHANDRASEKHAR, WHITE DWARFS AND NEUTRON STARS

An Indian graduate student, Subrahmanyan Chandrasekhar, came to Cambridge University to study with Sir Arthur Eddington in 1928. Hawking, *A Brief History of Time*, p.83. While in transit to the United Kingdom, he calculated the maximum size that a star could be and still resist collapsing under the force of its own gravity after its fuel had been all consumed (so that heat and nuclear reactions no longer counteracted the gravitational pull). The struggle at that stage would be between the force of gravity and the repulsion among elemental particles, initially among electrons. *Id.*

Chandrasekhar calculated that a cold star of more than about one and a half times the mass of our Sun would collapse. That mass has been called the Chandrasekhar limit. *Id.*, p.84. For stars of lesser mass, it is possible that they can become what are called "white dwarfs" as they die and remain in that state indefinitely (with densities of hundreds of tons of mass per cubic inch). Russian scientist Lev Davidovich Landau also calculated the number that became known as the Chandrasekhar limit and reasoned that, in addition to "white dwarfs", another form of stable dead star could exist at a density where the repulsion among protons and neutrons (rather than electrons) could counterbalance gravity. These objects were called neutron stars, and they were predicted to have densities a million times greater than the "white dwarfs". *Id.*

For dying stars with masses above the limit, the question was whether something would cause the star to lose mass as it cooled, so that it would become a white dwarf or a neutron star. If not, then what would happen to it? *Id.* The implication seemed to be that such a star would shrink until its density was infinite and its size was a disappearing point. Several leading scientists, including Einstein and Eddington, pronounced such a result as simply not possible. Id., pp.84-5.

THE IMPACT OF GENERAL RELATIVITY

According to General Relativity, heavy objects in motion generate gravitational waves affecting the curvature of spacetime. Those waves travel at the speed of light and carry energy away from the moving object. Hawking, A Brief History of Time, pp.89-90. As a star collapses, the amount of energy being ejected in the form of gravitational waves increases dramatically. Robert Oppenheimer, in 1939, worked out what would happen in such a situation under the mathematical theory of General Relativity, but the issue was not fully explored until the 1960s. Id., p.85.

In 1965, Roger Penrose demonstrated that, pursuant to General Relativity, the surface area of a star collapsing under its own gravity must ultimately shrink to nothing. Thus, its volume must also shrink to nothing, creating a "singularity" or black hole. Id., p.49. As Hawking explains:

> "The work that Roger Penrose and I did between 1965 and 1970 showed that, according to General Relativity, there must be a singularity of infinite density and space-time curvature within a black hole. ... At this singularity the laws of science and our ability to predict the future would break down. However, any observer who remained outside the black hole would not be affected by this failure of predictability, because neither light nor any other signal could reach him from the singularity."

Id., p.88.

The theoretical calculations predicted that the resulting black hole would be a spherical body distinguishable only by its mass and rate of rotation, independent of whatever materials had been involved in the state of the body prior to the collapse (as long as the mass was great enough to create the black hole in the first place). Id., pp.91-2. The particles that do exist are astonishingly small, indeed, may have no volume at all. So, if the forces holding the particles apart are overcome by gravity, then the amount of the volume that could disappear is virtually the entirety of the object. In fact, it may be that the volume of all black holes is zero.

THE INTRODUCTION OF QUANTUM MECHANICS

Hawking asserts that "one cannot really argue with a mathematical theorem" but notes that "having changed my mind, I am now trying to convince other physicists that there was in fact no singularity," because when the object becomes sufficiently small, the principles of quantum mechanics would apply, negating the conclusions of General Relativity. Hawking, A Brief History of Time, pp.50–51.[9] The issue here is not whether black holes exist, but whether they actually constitutes "singularities," that is, points of infinite density and zero size. In short, General Relativity, the mathematical theory that predicts the existence of black holes, breaks down and renders meaningless results when applied to the conditions inside the black hole. It does not purport to tell us what is happening inside. So, Hawking and others conclude that we must turn to quantum mechanics to understand what is going on in black holes. Quantum theory does not lead to singularities, and the same laws of nature can apply inside a black hole and to our more normal world, *albeit*, only observable in phenomena at a very, very small scale. Id., p.133. In this context, the fact that General Relativity and quantum mechanics cannot be reconciled or fit together in one overarching theory becomes particularly vexing.[10]

DETECTING BLACK HOLES

Hawking has described, in simplified version, the nature of the reasoning that astronomers have used to detect the existence of black holes. In 1963, Maarten Schmidt measured the red shift of a faint body. Id., pp.92–3. If the amount of red shift had been caused by a gravitational field, then the body would be at a distance and of a size that it would have affected the orbits of the planets in our Solar System, which it apparently did not do. Thus, the conclusion was that the body must be very far away and the red shift was being caused by its position in an expanding Universe. In that case, however, the object had to be of very unusual brightness to be seen at all. The amount of energy evidently being expended to burn so brightly suggested that the object was not a single star but a sizable portion of a galaxy that was collapsing to form a black hole. This type of object was called a "quasi-stellar object" or quasar. The quasars that have been detected to date, however, are all too far away for astronomers to detect the actual presence of black holes.

Another example is of an observable star like Cygnus X-1 that appears to be orbiting around another star but the other star cannot be seen. The companion star must either be too faint to be seen or be a black hole. Analysis of radio waves being emitted allowed certain conclusions about the mass and size of the companion star, which conclusions appear to be inconsistent with it being a normal star or even a neutron star. Hawking, A Brief History of Time, pp.94-5.

Interestingly, the prediction of the existence of black holes is often viewed as one of the principle successes of twentieth century cosmology. The phenomenon was predicted by theory, and then "confirmed" by empirical observations. Actually, what has been confirmed is that there are places of very high mass that have detectable gravitational influences but do not emit light of any visible frequency. Thus, they cannot be seen. They also appear to emit radiation, which is thought to be the result of phenomena occurring just beyond the event horizon of the black holes.

What we cannot measure is the precise size of these objects, so we rely on the mathematics to conclude that they are of zero size, constituting so-called singularities. In addition, the theory tells us that we do not and cannot directly determine what is happening inside the event horizon, since no information can escape. See, e.g., Penrose, The Road to Reality, p.711. However, these objects do have an impact upon other objects (including light) within a sufficient distance of them to be notice (the affects of the gravitational attraction).

SOME MISPERCEPTIONS

There is a misperception about black holes (evidenced, for example, in fears about potential disasters that might occur when the Hadron Collider was first operated at design speeds). The fear is that if a black hole suddenly appeared, it would "suck in" surrounding matter like a whirlpool. This issue is a bit like the question discussed above about how long it would take after the disappearance of the Sun for the lack of gravitational attraction to be experienced here on Earth.

Black holes do not appear out of nothing, just like the Sun cannot disappear into nothing. The mass that creates the black hole will have existed (in expanded form) in advance of its collapse into a black hole, and that mass will have been generating gravitational effects for distant objects of the same magnitude as the black hole will generate after it appears. Objects that will be attracted to and pulled in by those gravitational effects will be so attracted whether the object is a large star or a black hole.

The black hole phenomenon occurs within the large star, where the gravitational effects because of the large mass become sufficient to overcome the nuclear, electronic or magnetic forces holding the particles apart. The strength of the gravitational pull between the center of the star and the surface will increase as the diameter of the star diminishes and do so by the inverse of the square of the radius. Thus, while the gravitational attraction of the star of a given mass will not change relative to an object at a distance, the forces collapsing the star will increase dramatically as the star shrinks in size (due, for example, to the consumption of the fuel for the nuclear reactions keeping it burning brightly).

It is the general consensus that "black holes" do exist and that about a hundred of them have now been discovered, scattered among some 900 galaxies that have been surveyed as possible hosts. But, again, we should remember that the evidence of their existence, apart from the mathematics, consists of the apparent influence of their huge gravitational attraction on the observable motions of stars in distant galaxies. By their very nature (absorbing all light as well as everything else within their individual event horizons), they cannot be seen or "contacted" in any traditional way.

HOLES AT THE CENTER

Observations indicate that black holes exist in the center of most large galaxies. *See, e.g.*, Jason Palmer, BBC News, 29 November 2012; Penrose, *The Road to Reality*, p.712. It is currently believed that these "black holes" play an integral part in the regulation of the structure of the galaxies, for example, acting "'a bit like a pressure valve that prevents star formation from running away with itself.'" Caleb Scharf, the director of Columbia University's Center for Astrobiology, quoted in Tom Feilden, "Gas guzzler," BBC News, 13 November 2012.

It appears that over about the next ten years, scientists on Earth will be able to observe the "black hole" at the center of the Milky Way (known as Sagittarius A*) devouring a vast cloud of interstellar material (dust and gases) that has gotten too close to the "black hole." We will be able to "observe" the event, because some of the material will be swirled around the "black hole" and ejected out again into space, rather than being sucked within the event horizon, where it would disappear. *Id.* The transformations of the material that is not consumed will generate huge amounts of energy and radiation.

Interestingly, NASA's Nuclear Spectroscopic Telescope Array (NuSTAR) launched in June of 2012 was in place just in time to begin observing the predicted spectacle. *Id.* The approaching event was being observed and monitored by scientists with the Max Planck Institute for Extraterrestrial Physics in Garching, Germany, using the European-funded Very Large

Telescope ("VLT") located in the Atacama Desert of Chile. Gautam Naik, "How a Black Hole Really Works," *The Wall Street Journal*, August 16, 2013. And,

> "[NuSTAR] has captured an extreme and rare event in the regions immediately surrounding a supermassive black hole. A compact source of X-rays that sits near the black hole, called the corona, has moved closer to the black hole over a period of just days. ...As the corona shifted closer to the black hole, the gravity of the black hole exerted a stronger tug on the X-rays emitted by it. The result was an extreme blurring and stretching of the X-ray light. ...Even though some light falls into a supermassive black hole never to be seen again, other high-energy light emanates from both the corona and the surrounding accretion disk of superheated material. [The coronas] contain particles that move close to the speed of light."

NASA, "NASA's NuSTAR Sees Rare Blurring of Black Hole Light," *Release* 14-210, August 12, 2014.

A new technique called the Event Horizon Telescope (EHT) combines (connects) nine radio antennas (telescopes) on four continents in an attempt to "photograph" a blackhole. The data was collected in mid-2017, and then the very extensive computer analyses began. There was considerable excitement in late 2017 anticipating the first photograph in early 2018. It did not appear. *See, e.g.*, Ethan Siegel, "This Is Why The Event Horizon Telescope Still Doesn't Have An Image Of A Black Hole: The data has been taken, collected, and analyzed. So where is the first image of an event horizon, already?" *Medium.com*, June 11, 2018.

In early April 2019, scientists again eagerly awaited the announcement. *See, e.g*, Marlowe Hood, "Scientists set to unveil first picture of a black hole," *Phs.Org.com*, April 6, 2019. This time, a photograph was revealed, one of a black hole in another galaxy called M87. *See, e.g.*, Dennis Overbye, "Black Hole Picture Revealed for the First Time: Astronomers at last have captured an image of the darkest entities in the cosmos," *NYTimes.com*, April 10. 2019. "'We have seen what we thought was unseeable,' said Shep Doeleman, an astronomer at the Harvard-Smithsonian Center for Astrophysics, and director of the effort to capture the image... ." *Id.*

Finally, in May 2022, the composite photograph of Sagittarius A* from the Event Horizon Telescope was released.

> "M87 is 1,500 times bigger than Sagittarius A*; the latter's smaller size is one of the reasons EHT chose to image it second.........The Sagittarius A* image marks only the second direct evidence of the existence of black holes. Before 2019, scientists could only collect indirect evidence of a black hole by measuring the impact of its gravity on nearby objects, or detecting the gravitational waves emanated by it."

Aylin Woodward, "First Image of Black Hole at Center of Milky Way Galaxy Revealed: Sagittarius A* is 4 million times as massive as the sun and some 26,000 light-years from Earth," WSJ.com, May 12, 2022.

I do not mean to suggest that all of the questions are answered. For example, it was reported in November 2012 that the second largest "black hole" ever identified has been found located in the center of a modest-sized galaxy. The galaxy, NGC 1277, is reported to be just a quarter of the size of our own Milky Way but its "black hole" is 4,000 times larger than ours. Jason Palmer, BBC News, 29 November 2012, based upon a recent report in Nature. This observation seems to conflict with the theory that "black holes" get larger as their host galaxies get larger. Of course, it may be that the process reaches a peak and then the galaxy starts to shrink as the "black hole" continues to grow.

THE FUTURE OF BLACK HOLES

There has been considerable public interest in black holes. In the process, I think that many lay persons jumped to the conclusion that the Universe would inevitably end in the collapse of everything into a black hole. Such a conclusion would have been premature.

In 1974, Stephen Hawking, in exploring the implications of quantum mechanics for black holes, derived the conclusion that a black hole would not be completely cold, but would have a small but positive temperature. Hawking, A Brief History of Time, pp.104-9. Hawking's calculations indicated that a black hole would emit particles. Now, since nothing presumably can escape from a black hole, the emission must come from the "empty" space just beyond the black hole's event horizon–the boundary from which nothing can escape. Id., p.105. The quantum theory indicated to Hawking that the quantum fluctuations just beyond the event horizon would generate virtual particles (like the supposed gravitons) that have effects on the orbits of electrons but cannot be detected directly by particle detectors. Because energy cannot be created out of nothing, the virtual particles must occur in pairs of particles and anti-particles, with positive and negative energy respectively, so that they effectively cancel each other out. However, if some of the anti-particles with the negative energy fall into the black hole and some of their corresponding pairs with positive energy escape, the particles that escape will ultimately become real particles (no longer virtual particles) and may be observed by a distant observer as emissions from the black hole. *Id.*, pp.105–6.

The flow of positive energy away from the black hole will be balanced by the flow of negative energy into the black hole. Pursuant to Einstein's equation, $E=mc^2$, the black hole will lose mass as it absorbs negative energy. *Id.*, pp.106–7. As the black hole very slowly shrinks, the gravitational forces that caused the collapse of the old star into a black hole will lessen until, eventually, the black hole would most likely explode. *Id.*, p.107. The resulting explosion may be quite large by Earth standards, but by cosmic standards and the phenomena that led to the creation of the black hole, the demise of black holes may be characterized as mere "pops"! *See* Penrose, *Cycles of Time*, p.117. In addition, if black holes do have a positive temperature, then there should come a time in the very, very distant future where the temperature of the cosmic radiation (discovered by Penzias and Wilson) will fall below the internal temperatures of the black holes and any remaining black holes will start to lose energy to the surrounding space. The energy loss would, pursuant to the formula $E=mc^2$, be reflected in the slow loss of mass. Over eons of time, these black holes, too, would dissolve and disappear. *See, e.g., id.*, pp.116–7.

That would be the end of the Universe for all practical purposes.

INFLATIONARY COSMOLOGY

Most cosmological theories are premised, to one degree or another, upon a fundamental axiom of physics that the speed of light through a vacuum constitutes a universal "speed limit," that is, nothing can travel faster than the speed of light. This axiom was set forth by Albert Einstein in 1905. We shall discuss the theoretical relevance and role of this axiom below, but the proposition is well known to us all. What may be less obvious is that the axiom as utilized by physicists states not just that nothing travels faster than light now, but that nothing ever did and nothing ever will—the presumed limit is a universal and timeless "constant" of nature. *See, e.g.*, Barrow, *Constants of Nature*; Magueijo, *Faster than the Speed of Light*.

Given this assumption, it may seem strange to some to hear discussions of experimental observations today being made of light or other electromagnetic waves generated at the instance of the initial Big Bang (actually shortly—300,000 years—thereafter). If the Universe originated with a massive explosion, with all matter and energy moving rapidly outward from the center and expanding, how could it be that we can observe waves created at that moment just reaching us today? Those waves would have continuously moved outward at the speed of light, while the matter of which we consist clearly would not have done so (and does not now). Why are those waves not far, far beyond our powers of detection?

INFLATION

The "answer" is in an assumption, or gimmick, called "inflation," giving us "inflationary cosmology." See, e.g., Greene, The Fabric of the Cosmos, p.252; Krauss, A Universe from Nothing, p.97. The theoretical concept was the creation of Russian physicist Andrei Linde, who immigrated to the United States in 1990. Holt, Why Does the Universe Exist?, p.13. The proposition is that in an unimaginably small fraction of a second following the Bang, the Universe (with all of its matter and energy) expanded or inflated from a size smaller than a pea to something roughly approximating its current scale. See, e.g., Magueijo, Faster than the Speed of Light, pp.2–4.

As an indication of the proposed magnitudes of the timing and of the expansion (although, none of the numbers are in any sense actually imaginable), it is estimated that the "inflation" occurred sometime between 10 to the minus 36th and 10 to the minus 32nd seconds after the Big Bang (far, far less than one second after, and lasting far, far less than one second in all) and resulted, during that time period, in an expansion of the Universe of between 10 to the 30th power to 10 to the 60th power times. Penrose, Cycles of Time, p.66. The speed of such inflation would seem to have dwarfed the speed of light (an understatement), but physicists say that the speed limit was not broken!

The rough analogy of the process of inflation is that of the blowing up of a balloon—all of the spots on the surface move farther away from one another even though, relative to the surface of the balloon, none of them move at all. See, e.g., Greene, The Fabric of the Cosmos, pp.231–3. The claim is that, although, billions of individual particles of matter or units of energy went from being literally crushed together to being millions or billions of light years apart in a virtual instant; no one of them moved or traveled anywhere—it is simply that the space (or space-time) in which they existed expanded (or, more descriptively, exploded). Therefore, this "inflation" did not break the universal speed limit. Id., pp.237–8. In other words, the speed of light limits travel through space but imposes no limitations on what space itself can do. Krauss, A *Universe from Nothing*, pp.96-7.[11] In the usage currently in vogue, we say that no "information" was transferred from one point to another through space, just that the points became much further apart. See, e.g., Greene, *The Fabric of the Cosmos*, pp.116-7 (suggesting arguments based upon the transfer of information as responses to certain observed phenomena at the quantum level).[12]

I think that many of us would be more than a little surprised by these explanations—we probably did not understand the universal speed limit to have such a significant qualification.

Some laypersons, at least, have observed that this "'story' is simply inconceivable." See, e.g., Jim Holt, *Why Does the World Exist?*, pp.249-51 (quoting novelist John Updike).

A BIG BANG WITHOUT INFLATION?

It might help to visualize what we are describing if we note some of the elementary things that would be different in our world if there had been a Big Bang but no inflation. At the occurrence of the Bang, light, energy and matter would have begun traveling outward in all directions from the point of the Bang. Presumably, light would have traveled at the speed of light, and other things would have traveled at that speed or more slowly. Over time, light would maintain its speed and do so forever; matter would rapidly begin to slow down. By the time man would have appeared on the scene, the light and radiation of the Bang would have long disappeared out into the depths of the Universe, never to be in contact with the Earth again. Of course, scientists believe that they have detected evidence of the Big Bang, including pervasive background radiation. Thus, that simple story must not be true. If there were a Big Bang, then there must have been inflation (or some other explanation that waits to be found).

There is another need for inflation. Modern cosmology is premised on the Cosmological Principle, which assumes that the Universe is homogeneous on a large scale, that is, that the constituents are rather evenly dispersed and the temperatures at various locations appear to be essentially the same uniform when viewed over a sufficient scale. Astronomical observations generally support this assumption if the scale for comparison is at least 100 million light years across. Id., p.18.[13] (That is a pretty large scale. It is usable for theories about the cosmos; but, much of what would be of interest to us, like our own galaxy, the Milky Way, is of a much smaller scale where the assumption does not apply.)

Physicists conclude that because of the size of the Universe and the limit of the speed of light to the communication of information across the Universe, such uniformity could not have come about over time (for example, through the transfer of heat across the Universe). Thus, the uniformity apparently would have had to exist from the beginning. But, that seems to be very, very unlikely.

The inflation theory asserts that the Universe experienced this staggeringly enormous expansion within a very small fraction of a second during its initial existence and that that dramatic expansion smoothed out the differences that existed at the initial instance of creation. See, e.g., Penrose, Cycles of Time, p.122; Hawking, A Brief History of Time, pp.127–9, 131–2.

And, there is yet another issue that the inflation theory helps address. As noted above, scientists are troubled by the apparent fact that certain empirical characteristics about our Universe seem to be very finely "tuned" in such a way as to make life—and, more particularly, intelligent life—possible. Some comfort has been found in the inflation hypothesis, because that theory would allow a disparate range of initial conditions to lead, through the inflation, to a non-inflationary universe similar to the one in which we find that we are living. Hawking, A Brief History of Time, pp.132–3.

EXPLANATION OR PROBLEM?

Some theorists have been more than uncomfortable with the theory of inflation.[14] "[T]he various inflation models introduced in order to explain admitted contradictions of the basic Big Bang model with observational evidence not only lack reliability but also their central claim is untestable." Hands, Cosmosapiens, p.38. We have some evidence consistent with the theory but also with other possibilities. And, the theories were modified in some aspects to fit the evidence better.

Efforts have been made to find alternative solutions for various of the problematic empirical observations that the inflation theory seems to solve. Generally, the alternatives to inflation involve the assumption that the known Laws of Nature are not immutable or timeless. Perhaps, for example, the speed of light was much faster in the beginning of the Universe or was not always a speed limit on the transfer of information. See, e.g.,

Magueijo, *Faster than the Speed of Light*, pp.4-5, 132-3, 139; Smolin, *Time Reborn*, pp.108, 236-9. Or, it may be that the initial conditions of the Universe did not have to be so carefully specified to result in intelligent life. Maybe, the explanation can be found in theories concerning the evolution or development of the Universe after the initial creation, a subject which could be an interesting area for scientific inquiry (although, I am not impressed by Lee Smolin's effort in this regard, described in the chapter on Physics)

Of course, these questions put the Cosmological Principle itself in doubt.

"Dark" Things

Dark matter

The best estimates of the amount of visible matter in the Universe appear to be far less than would be required to stop the rate of expansion that is observed from the "red shift." However, as noted above, physicists have speculated that the Universe is likely to contain substantial "dark matter." See, e.g., Hawking, *A Brief History of Time*, p.45. This other form of matter could be distributed throughout the Universe so as to result in a total density sfficient to slow or even stop the expansion. Id., p.46.

Moreover, computer simulations indicate that "dark matter" was very important in the creation of our world. See Wilczek, *Fundamentals*, p.200.

> "since [dark matter] doesn't interact with light, [it] cools faster than normal matter. In the early universe, therefore, dark matter is the first to start clumping under its own gravitational pull. Indeed, without the dark matter's early clumping, galaxies wouldn't form in the way that we observe, for the gravitational pull of the already clumpy dark matter is needed to speed up the clumping of normal matter. And it's only when enough normal matter has come together that the formation of large atomic nuclei in stellar cores can begin."

Sabine Hossenfelder, *Lost in Math*, p.61.

The rotation of galaxies

The first empirical evidence of the existence of additional matter came from observations of the motion of stars within galaxies and of particular galaxies within clusters of galaxies. Id. In our Solar System, where the gravitational focus is the Sun, we observe that the more distant planets travel more slowly in orbit around the Sun (and the closer in planets travel more quickly). The same phenomenon does not appear to occur with galaxies, even though the stars are revolving around a center, like a bicycle wheel. Observations of the movements of galaxies indicate that the gravitational forces holding galaxies together are not focused at the center of the galaxies. Moreover, the outer reaches of our galaxy was observed to be rotating faster than would have been possible if the total gravitational mass involved were limited to the visible matter in the galaxy. If the total mass were only that that could be seen, then the speed of rotation would result in the galaxies flying apart.

The study of the rotational rate of our galaxy was done in the 1970s by Vera Rubin and her colleagues. Krauss, A *Universe from Nothing*, p.24. Rubin, an American astronomer, was awarded the Gold Medal of the Royal Astronomical Society. Id. (She passed away in early 2017. See Dennis Overbye, "Vera Rubin, 88, Dies; Opened Doors in Astronomy, and for Women," *The New York Times*, December 27, 2016. She now has a new observatory named after her: "The Vera C. Rubin Observatory, formerly the Large Synoptic Survey Telescope, under construction in Cerro Pachon, Chile." Dennis Overbye, "Vera Rubin Gets a Telescope of Her Own: The astronomer missed her Nobel Prize. But she now has a whole new observatory to her name," *The New York Times*, January 11, 2020.) From these studies, astrophysicists concluded that there must be other areas of mass exerting a gravitational pull on the revolving stars.

Since no such source of additional mass could be seen, the additional matter assumed to exist was referred to as "dark matter." Indeed, the theoretical derivations from the empirical observations suggest that "dark matter" in the Universe may have a collective mass as much as 10 times the mass of the visible particles that make up the stars, galaxies and gases. Krauss, A *Universe from Nothing*,, p.25; Hawking, A *Brief History of Time*, p.45. Interestingly, such a magnitude of dark matter indicates that the dark matter cannot be made up simply of the protons and neutrons that would have been created out of a big bang; there would not be enough of those particles. Thus, there appears to be in space some other type of elementary particle that, at least today, does not exist on Earth. Krauss, A *Universe from Nothing*, pp.24-5.[15]

Gravitational "lensing"

There is other indirect evidence of dark matter. Astro-physicists have done extensive (and on-going) studies of what has been called gravitational "lensing." The idea is that there appear to be large clumps of mass between us and various distant stars. Theory tells us that such clumps with large mass would result in gravitational forces that would bend light from the distant stars, much like an optical lens can concentrate or change the direction of light beams.

This phenomenon of light bending when passing a large mass was predicted by Einstein and, as noted, demonstrated by Sir Arthur Eddington, in the first empirical verification of a prediction of General Relativity in 1919. See, e.g., Greene, *The Elegant Universe*, p.77. Moreover, Einstein published a note in 1936 observing that gravity could bend light acting like a lens to magnify the light perceived by an observer; although, he concluded that it was unlikely that we would ever be able to observe that phenomenon, because the effect would be so small. Krauss, A *Universe from Nothing*, pp.29–30.

Fritz Zwicky submitted a paper in 1937 that proposed methods for using gravitational "lensing" for testing General Relativity and for attempting to measure the mass of distant nebulae. *Id.*, p.31. Fifty years later, in 1987, gravitational lensing was first observed, in which galaxies between Earth and distant quasars affected the light received on Earth from those quasars; in 1998, the first estimate of the total mass of a large cluster was made using gravitational "lensing." *Id.*, p.32. The calculations performed by Tony Tyson and colleagues at the Bell Lab using observations from the Hubble Space Telescope indicated that dark matter existed not only in galaxies but also between galaxies that were part of large clusters and in quantities far greater than expected. *Id.*, pp.32–4.

This "lensing" effect appears to have made visible to us galaxies much more distant than could be observed with only the use of our own telescopes. Moreover, astro-physicists have been able to observe what appear to be multiple images of the same distant star or distorted images, such as crescents or rings. The identification of the observed images as being from the same star is based upon spectroscopy, with the assumption that no two stars could have identical spectroscopic signatures (or, at least, that it would be highly unlikely for two such stars to be in the same relatively small region of the Universe). This work, like much else in modern cosmology, requires sophisticated computer simulations that utilize enormous computer resources. Priyamvada Natarajan (Professor of Astronomy

& Physics, Yale University), Lecture, March 9, 2013, New York City (Cambridge in America Day).

"WIMPs"

Dark matter is invisible and appears to consist of something that does not exist on Earth today, so physicists speculate as to what it might be (assuming that it actually exists). One proposal is that it is a "weakly interacting massive particle" or WIMP. Gautum Naik, "Hint of Dark Matter Found," *The Wall Street Journal*, April 3, 2013. As the name indicates, such particles would have to have little or no interaction with normal matter yet have high mass. It has been postulated that when two WIMPs collide, they may annihilate themselves and produce other particles, like electrons and positrons, which can be detected. *Id*.

Such WIMPs would presumably have been left over from the Big Bang. The theory proposes that they are massive, hundreds of times the mass of a proton, but able to pass freely through other matter, such as the Earth. Thus, it is hoped that some evidence of them could be detected by instrumentation located in chambers deep underground, where the equipment would be sheltered from cosmic rays and other particles that bombard the surface of the Earth. Dennis Overbye, "Dark Matter Experiment Has Detected Nothing, Researchers Say Proudly," *The New York Times*, October 30, 2013. (These experiments were described in the previous chapter.)

The continuing search

Scientists are still looking for some direct evidence of particles that could constitute dark matter. Some suggestive data has recently been collected from a particle detector mounted on the exterior of the international space station. See Gautum Naik, "Hint of Dark Matter Found," *The Wall Street Journal*, April 3, 2013.

The Sanford Underground Research Facility is a mile beneath the surface in a former Homestake Gold Mine in Lead, South Dakota. The mine is the location of the experiment by Raymond Davis Jr. in the 1960s that first detected solar neutrinos (in a "giant tank of dry-cleaning fluid"), for which he was awarded a Nobel Prize in Physics in 2002. Caroline Porter, "Physicists Mine Cosmic Answers Deep Underground," *The Wall Street Journal*, July 30, 2012. Many of the particles that constantly bombard the Earth from outer space are filtered out at this depth. Scientists hope to find evidence of the interaction of dark matter particles with xenon atoms. The experiment is known as LUX (Large Underground Xenon). It is "the biggest, most sensitive dark matter detector yet—a vat of 368 kilograms of liquid xenon cooled to minus 150 degrees Fahrenheit... ." Dennis Overbye, "Dark Matter Experiment Has Detected Nothing, Researchers Say Proudly," *The New York Times*, October 30, 2013.

Now, another very large particle detector is located in the South Pole. It is called IceCube and became operational in 2010, after 7 years of construction. It is "a cubic-kilometer ... [and currently has 86 cables] extending to a depth of about 2,500 meters. [It has a] surface array, IceTop, and a denser inner subdetector, DeepCore... ." IceCube Collaboration, "Ice Cube - South Pole Neutrino Detector," *icecube.wisc.edu.*, 16 July 2019. It has been a success:

> "On June 25, 2019, the National Science Foundation (NSF) approved full funding to upgrade the IceCube detector, extending its scientific capabilities to lower energies and thus enabling IceCube to reach neutrino energies that overlap with the energy ranges of smaller existing neutrino detectors worldwide. The IceCube Upgrade project will introduce seven strings of optical modules at the bottom center of the 86 existing strings, adding more than 700 new and enhanced optical modules to the 5,160 sensors already embedded in the ice beneath the geographic South Pole."

Id.

The search also extends to outer space. The Alpha Magnetic Spectrometer ("AMS") is mounted on the outside of the international space station. Recent experiments sponsored by the U.S. Department of Energy have produced data that could be explained by the annihilation of particles of dark matter. Gautam Naik, "Hint of Dark Matter," *The Wall Street Journal*, April 3, 2013.

DARK ENERGY

Contemporary cosmologists are also experimenting with the reintroduction of a version of Einstein's concept of a cosmological constant—in the form of so-called "dark energy." This constant could be thought of as a general, inherent tendency of the Universe to expand, counteracting gravity. See Hawking, *A Brief History of Time*, p.40. Einstein had introduced his cosmological constant to the left-hand side of the equations of his General Theory of Relativity (describing the curvature of the Universe) as a very small extra repulsive force that would be undetectable on the scale of normal observation, such as on Earth or in the Solar System, but that would be sufficient to keep the Universe from collapsing due to gravitational attraction. Krauss, A *Universe from Nothing*, pp.56-7.

The right-hand side of the equations contain the total amounts of matter and energy in the Universe. With the addition of "dark energy", a new value is introduced to the right-hand side. *Id.*, pp.57-8. However, given the proposition that the Universe has dark matter with an aggregate mass several times the mass of the visible particles, the "quantity" of the dark energy must be very substantial to overcome a gravitational attraction that would be much stronger than originally contemplated (before the introduction of dark matter). Complex mathematical calculations lead to the conclusion that the mass of the Universe is some 73% "dark energy", 23% "dark matter" and 4% regular, visible matter! Priyamvada Natarajan, Lecture, March 9, 2013.[16]

However, "**[w]ithin the solar system, or even the galaxy, the mass contributed by space is utterly negligible compared with the mass contributed by ordinary matter** (or dark matter). But such is the vast emptiness of intergalactic space that this small density, present everywhere, comes to dominate the total mass of the universe. " Wilczek, *Fundamental*, p.198 (emphasis added).

So, what is dark energy?

> "*Dark energy has been described as everything from a fifth force to a new form of matter, but so far, no direct evidence has been found of its existence. Noble Prize winning physicist Adam Riess says: 'I have absolutely no clue what dark energy is. Dark energy appears strong enough to push the entire universe – yet its source is unknown, its location is unknown and its physics are highly speculative.'*"

Max Goldberg, "'The Massive Elephant in the Room' –New Theory Says 'No Need' for Dark Energy," *The Daily Galaxy*, January 28, 2020.

Moreover, if the Universe has experienced inflation, then either the dark energy is now less dense than it was before the inflation or the quantity of it somehow increased during inflation. Curious. Furthermore, there is evidence suggesting that the expansion of the Universe was slowing at one time, due to gravitational attraction but that the rate of expansion then began to accelerate. There is no explanation even proffered as to how or why dark energy could have apparently begun to dominate events in the Universe some 5 billion years ago. Krauss, A *Universe from Nothing*, pp.88-9. *See also,* Jake Burba, "'Unknown Dark Energy'–A Fifth Force or New Form of Matter?" *The Daily Galaxy,* August 19, 2019 ("in a March 2, 2019 post, ...*The Galaxy* described a new, controversial theory suggesting that this dark energy might be getting stronger and denser, leading to a future in which atoms are torn asunder and time ends").

I note that as a matter of logic, there are possible solutions other than dark energy to the apparent riddle. Remember, the problem is that our theory of gravity cannot be reconciled with the observable quantity of matter in the Universe and the apparent rate of expansion of the Universe. But, that calculation could work out if the theory of gravity were modified.

Some possibilities are:

- The reach of gravity is not infinite;
- Gravity losses strength over great distances;
- The speed of gravity is (or was) different from the speed of light; or
- Gravity has mass.

"Massive gravity"

One theory is that the supposed graviton that transmits the gravitational force itself has mass. If so, then the force of gravity may decrease with distance, perhaps allowing the observed acceleration without the assumption of "dark energy."

General Relativity assumed a massless spin-2 particle as the graviton. The idea of introducing a mass for that particle has a rather long history (back to at least 1939 with Swiss physicist Markus Fierz and Austrian theoretical physicist Wolfgang Pauli, but there were persistent technical problems—the mathematics yielded absurd results. Many of these problems were solved by Claudia de Rham and her two co-authors in a 2011 paper in the *Physical Review Letters* (de Rham, Gabadadze and Tolley, "Resummation of Massive Gravity"). See Jake Burba, "Unknown Dark Energy", *The Daily Galaxy,* August 19, 2019. Prof. De Rham herself has written:

> *"For almost a century, the theory of general relativity (GR) has been known to describe the force of gravity with impeccable agreement with observations. Despite all the successes of GR the search for alternatives has been an ongoing challenge since its formulation. Far from a purely academic exercise, the existence of consistent alternatives to describe the theory of gravitation is actually essential to test the theory of GR. Furthermore, the*

> open questions that remain behind the puzzles at the interface between gravity/cosmology and particle physics ... have pushed the search for alternatives to GR."

"Massive Gravity," Living Reviews in Relativity, 25 August 2014.

Of course, problems with the math do not prove that gravity has no mass, and a solution to the math problems does not prove that gravity does have mass. The challenge now is finding empirical evidence for or against the theory.

> "As a new generation of gravitational wave observatories, such as Esa's space-based observatory, Lisa, start gathering faint signals from the cosmos in the next decade, De Rham hopes to close in on some answers. 'It's going to be such a high level of science that we can do,' says De Rham. 'It's beautiful.'"

Hannah Devlin, "Has physicist's gravity theory solved 'impossible' dark energy riddle?" The Guardian, 25 January 2020.

TIME (AGAIN)

> "We've been livin' in the last days
> ever since the first day,
> ever since the dawn of Man."
>
> Dolly Parton
> "In the Meantime"

We started a discussion of time in the chapter Physics, but, there is more to be said, especially about the beginning and the end.

THE BEGINNING OF TIME?

One conclusion that is asserted to emerge from the hypothesis that there is an origin of the Universe, whether as a Big Bang or something else, is that that instance of origin must also be the origin of time. See, e.g., Hawking, A Brief History of Time, p.8. This conclusion is contained in Einstein's General Theory of Relativity with its concept of space-time. Since space would be created at the origin of the Universe, particularly if the origin were a Big Bang; time must be assumed to have also been created at that instance. See, e.g., Paul Davies, *The Mind of God: The Scientific Basis for a Rational World* (1992), pp.48, 49. More generally, if one concludes that time is a part of the Universe; then it would seem to follow that somehow time was created when the Universe was created. (Obviously, these questions do not arise if the Universe is assumed to be eternal.)

Conveniently, this conclusion eliminates the question of what was going on before the origin of the Universe, since our concept of "going on" is a process that occurs in time (which would not exist), just as the question of what existed before the origin is resolved by the assertion that the origin created space (our concept of things is something that happens in space).

However, if one speculates that time is somehow external to or independent of the Universe, then one can envisage time passing before the Universe was created and continuing after it ends (if it does end). Whether time is part of or independent of the Universe will depend upon the theories one has of the creation and evolution of the Universe (otherwise, we would just be arguing over definitions: the Universe means everything that is, including time, or the Universe is some subset of everything that there is). Interestingly, however, according to Hawking, St. Augustine used the same argument about the origin of time in response to the question of what God was doing before he created the world. *A Brief History of Time*, p.8.

There is also the attractive parallel of time moving forward only in one direction and the Universe expanding outward forever. These matters do become a bit awkward if one were to conclude that the Universe was oscillating—a bang followed ultimately by a collapse, followed by another bang, etc. However, it seems clear that even in that scenario, events of a prior universe could have no impact on the current Universe, so that whether there was a prior universe is essentially irrelevant to scientific theories and explanations today, even if not to our intellectual curiosity. See *id.*, p.9.

The Judeo-Christian tradition incorporates a Creation, as do many religions. Moreover, the notion of time is integral to the Judeo-Christian tradition with its fundamental belief in an historical progression from the Creation and subsequent Fall of Man to the Resurrection and Second Coming of Christ. See Davies, *The Mind of God*, p.41. In addition, the implication of a Creation is that God existed independently of and outside of the Universe and that the Creation was an act of will or a decision and not inevitable. See Davies, *The Mind of God*, pp.43, 44–5.

Unlike many other religions, Church dogma also came to embrace the concept of creation *ex nihilo*–creation from nothing. Adam Frank, *About Time*, pp.53, 65; see also Davies, *The Mind of God*, p.44. In the Abrahamic tradition, reflected in Genesis, there is at least some ambiguity about what existed before the creation (both "without form" and "void");[17] but, by the third century after Christ, the prevailing theology asserted creation out of nothingness, which idea later appeared in Islamic theology and in medieval Jewish thought. Jim Holt, *Why Does the World Exist?* (2012), pp.19–20. Thus, the concept of nothingness was established. *Id.* God was the Creator not just of the Universe as we see it, but also of all of the matter and energy of which it is composed.

Although, it is not a necessary inference from such a Creation, theologians starting with St. Augustine further concluded that before the Creation, there was not only nothing, there was also no time. Although He is eternal, God is timeless or outside of time. So, time was created by God simultaneously with His creation of the Universe. Davies, *The Mind of God*, p.42.

Since at least the time of Aristotle up until Newton, the passage of time was generally marked by and may even have been defined by changes in matter, like movement of bodies or, even, the aging of a living creature, just as space existed as the positional relationships among things, among matter. Holt, *Why Does the World Exist?*, at p.93. Thus, time may have had no meaning apart from changes in things. Consistent with prevailing Church doctrine, Newton accepted a Universe infinite in size, but finite in duration–that is, one with an origin . See, *e.g., id.*, p.157.

Einstein, of course, in his theory of General Relativity, again recombined space and time into an inseparable unity defining the arena of our Universe. Not surprisingly, Einstein also then concluded that the Universe was eternal and essentially static. *Id.*

THE END OF TIME?

Even if the Universe is and has been expanding since a very dramatic event some 14 billion years ago, one need not necessarily conclude that the Universe had an origin or beginning. It is conceivable that the Universe has experienced expansions followed by contractions followed by expansions, in cycles, for all of eternity. However, each explosion would presumably create an insurmountable barrier between the new universe and the old, so it would still be fair to say that time began (again). Yet, the theory of a continually recycling Universe would seem to side-step the issue of a Creator and to present a rather fundamental conflict with the Judeo-Christian belief about Creation. Thus, it seems to be necessary to consider the potential future of our existing Universe in order to assess our conclusions about whether there was a beginning.

Assuming that the Universe is indeed expanding now, and regardless of whether the expansion is a result of an initial Big Bang, we might ask what can be said about the future of the Universe. The logical possibilities are that the expansion we now observe will continue forever, with the galaxies becoming more and more isolated; that the expansion will slow and slow, approaching, but never quite reaching a static state; or that the expansion will end, to be followed by a contraction and collapse. Stephen Hawking, A Brief History of Time, p.44.

The choice among these alternatives would depend upon the rate of expansion versus the strength of the gravitational attraction. One or the other wins out in the end, or they almost exactly balance. Of course, if we could determine whether the speed of expansion is slowing and, if so, at what rate, we might be able to narrow the possibilities. But, we do not have comparative data that would answer that question; and we will likely not have it for thousands of years, when the time between observations might be great enough to make a deceleration detectable.

Evidence of a "flat" Universe

The efforts to assess the total mass of the Universe have resulted in the conclusion that while dark matter is an enormous portion of the actual mass that exists, the total is still insufficient to result in a "flat" Universe, so the likely fate of the Universe would be continual expansion. Krauss, A Universe from Nothing, p.37. Yet, there is evidence that the Universe is "flat".

Since gravity is presumed to travel no faster than the speed of light, gravitational attraction arising immediately after the inflation following the Big Bang could not have reached throughout the resulting inflated Universe. By the time of the CMBR that we can observe, some 300,000 years after the Big Bang, the force of gravity could only be affecting particles that were within 300,000 light years of one another (because there was not time for the force to reach farther). Krauss, A Universe from Nothing, pp.44-5. Consequently, one would expect to find some tendency toward clumping together of particles within areas of 300,000 light years across, but not between particles farther away from one another. Ultimately, those clumps would begin to attract one another when the Universe became old enough for gravity to reach from one clump to another. Id. So, if it were possible to obtain a sufficiently detailed map of the CMB (cosmic microwave background), one would find "hot spots" of greater density.

Such a capability has now been achieved and the variations in the CMB were observed. Id., pp.46-7. The first such observations were accomplished in 1997 by a high-altitude balloon carrying a microwave radiometer traveling in a circle around Antarctica (called BOOMERANG), which was able to get a picture of the microwave radiation at close to 3 degrees Kelvin, untainted by the much warmer temperatures on Earth. Id., pp.47-9.

Given the theory described above about the clumping of matter into regions of no more than 300,000 light years across and assumptions about the age of the Universe to determine how far away from Earth the observed surface of radiation was, geometry (constructing a triangle with one side 300,000 light years across and the other two sides being in light years the assumed age of the Universe, with the Earth at that apex) enables one to calculate the expected sizes of the hot or dense spots in the CMB that one would observe on the assumption that the Universe is "flat," that is, the assumption that light travels in straight lines.

The angle at the point of observation on Earth should, given our understanding about the age of the Universe, be about 1 degree. In a "closed" Universe, the angle would be greater and the size of the hot spots would appear larger; in an "open" Universe, the angle would be less and the spots would appear smaller.

Analysis of the data generated by BOOMERANG gave surprising strong support for the "flat" Universe hypothesis. Id., pp.50-2. Subsequently, even more sensitive measurements have been possible based upon maps generated by satellites like the Wilkinson Microwave Anisotropy Probe (WMAP) launched by NASA in 2001. These measurements confirmed the conclusion that the Universe appears to be "flat." Thus, there must be even more invisible mass than was estimated from the study of gravitational effects.

The European Space Agency recently released a new map, more detailed, of the cosmic microwave background (CMB) that can be "seen" from Earth, generated from massive data compiled over 15 months through the use of the Planck space telescope. Jonathan Amos, "Plank satellite:

Maps detail Universe's ancient light," BBC News, 21 March 2012. Reportedly, analysis of the new map results in minor modifications of the results of prior calculations based on the earlier collected data. In addition, the map reveals various deviations from prior conclusions about the magnitudes and distribution of temperature variations across the sky. Id.

This data indicates that the Universe in 13.82 billion years old (instead of 13.72 billion years) and that the contents of the Universe consist of 68.3% "dark matter", meaning that there is slightly more matter than previously estimated (most of which is still "dark"). Id.

(The test for "flatness" depends upon an assumed age of the Universe. Errors in that number will alter the strength of the conclusions derived from the analysis. There is clearly a temptation to find confirmation when the use of certain assumptions results in a conclusion that one anticipates or hopes to find. It is also natural to see the attractiveness of the whole package as evidence of the correctness of the individual pieces, but there is clearly the potential for being misled by what may be effectively coincidence.)

Accelerating expansion?

Nevertheless, there is yet another possibility that has gained some currency in the last 20 years. Rather than a flat, open or closed Universe, we may live in one for which the rate of expansion may actually be increasing.

There is, in fact, recent evidence that the expansion of the Universe is accelerating. Further studies of the Type Ia supernova (a particular type of exploding star that was adopted in the 1990s as the "standard candle" for estimating distances) were undertaken in an attempt to see how the velocities at which such supernova were moving away varied based upon their estimated distances from the Earth. A plot of velocity and distance would be expected to curve upward if the expansion of the Universe were accelerating and downward if it were slowing (more distant, therefore older, supernovae moving more slowly or more rapidly). In 1998, one research team published a paper announcing that the expansion of the Universe seemed to be accelerating. Krauss, A *Universe from Nothing*, pp.79–82. Subsequent work has provided further details and confirmation of the phenomenon. Id., pp.83–7.

If the rate is accelerating, astro-physicists conclude that there then must be some force at work that is more than compensating for the gravitational force that should be causing the expansion (if caused only by an initial "bang") to slow down. Otherwise, gravity would necessarily be tending to slow the expansion that emerged from the Big Bang. Thus, we seem to be back to Einstein's original cosmological constant, not operating to maintain a static, steady state Universe as he conceived it, but actually causing the expansion to continue to accelerate (the "dark energy" discussed above). See, e.g., Krauss, A Universe from Nothing, pp.56-8, 219.

A dark and lonely place?

With respect to the fate of the Universe, the current theoretical constructs suggest that the expansion will continue forever—perhaps, at an accelerating rate, until the Universe becomes a cold, dark and lonely place. The distances between stars and galaxies will increase. If there were anyone left to watch, the stars in our sky will gradually fade from view. See, e.g., Krauss, A Universe from Nothing, pp.105-19.

At that stage, if there were any astronomers or astro-physicists attempting to understand the nature of the Universe, the information we seem now to have about what happened and how the Universe began would simply be unavailable to them. Id., p.116. Indeed, in language reminiscent of Carl Sagan's misplaced excitement about living at the crucial point in biological sciences (see the Introduction above), Lawrence Krause speculated that we may be in the very unique and fortunate position of being able to predict the future, because of the current relationship between dark energy and the density of matter in the Universe, and to understand the past in ways that would not have been possible a few billion years ago and will again not be possible a few billion years hence. Id. pp.117-9, 121-6.

Of course, this particular unique window of opportunity (this "'special' time") lasts for a couple of billion years. Id., pp.122-3.[18]

Endnotes

[1] It took another ten years for Kepler to discover his third law, that the square of the period of the planet (the time of one orbit) is proportional to the cube of its mean distance from the Sun. *Id.*, p.91.

[2] The techniques have changed, while the objectives remain the same. It was reported in December 2012 that scientists at the University of Hertfordshire had detected five planets orbiting Tau Ceti, the nearest Sun-like star to the Earth. *BBC News*, 19 December 2012. The technique used did not rely on observing evidence of the dimming of the light of planets through telescopes, but upon detecting very small shifts in the color of the light from Tau Ceti thought to arise as a result of the gravitational force of the planets moving the star slightly.

[3] Stephen Hawking asserted that Newton's conclusion that, in an infinite Universe of stars distributed relatively evenly, there would be no collapse because there would be no "center" toward which the collapse could occur was wrong. If you start with a finite system, in which there would a tendency to collapse, then continually add additional space and stars to that hypothetical universe, no addition of stars could make a difference. *A Brief History of Time*, p.4. In part, this is another example of the problem with the concept of infinity. If the Universe really is infinite, there would be no center. So, if Hawking is right about the analysis, where does the collapsed Universe end up being? The apparent answer would never have occurred to Newton. If the Universe were to collapse, so would space. Thus, the place or point to which the collapse would converge would be everyplace and no place, since when it occurred, all space would be contained therein. *See, e.g.*, Greene, *The Elegant Universe*, p.346 (with respect to the similar question of where the Big Bang "occurred").

[4] On the cosmic scale, Einstein utilized the "cosmological principle" as a simplifying mechanism to enable the analysis of the equations of his theory of General Relativity. Greene, *The Hidden Reality*, pp.17-20. In the General Theory of Relativity, the description of how density of matter changes over time consists of ten equations, computations with which would be very difficult. With the inclusion of the cosmological principle, the number of equations reduces to one, making analysis and calculation much more feasible. *Id.*, p.19.

[5] "[T]he Belgian Jesuit and scientist Georges Lemaître who incorporated Hubble's data into his own 1927 ideas and ran the expansion of the universe backwards to produce his hypothesis of the primeval atom." John Hands, *Cosmosapiens: Human Evolution from the Origin of the Universe* (2016), p.19.

[6] Bryson provides a concise description of the difficulties in determining, and the widely varying calculations of, the age of the Universe utilizing Hubble's Law ($H_0 = v/d$). Because the velocity of retreat reflected in the red shift appears to be nearly proportional to the apparent distance of the observed galaxies from the Earth, Hubble stated that there should be a constant value ("H_0") that is equal to the velocity divided by the distance. From that constant, if it were accurately determined, one could calculate the age of the Universe. Since the late 1920s, the ages calculated have ranged from Hubble's 2 billion years to 20 billion years, with a 2003 pronouncement that the answer is 13.7 billion years. These wide differences reflect greatly varying estimates of the Hubble Constant (H_0), resulting from significantly different estimates of the distances involved (and a relative paucity of data because of the limited telescopic resources). *Id.*, pp.215-17.

[7] These supernovae are very bright so that they can be detected at great distances. They flare and fade at apparently regular intervals. However, the ones that show the greatest red shifts, indicating that they are the furthest away from us, appear to cycle in brightness more slowly than those that are believed to be closer to us. This result is what one would predict for an object moving away from the observer more rapidly. Each subsequent incidence of a flare up or fading stage will be further away and, therefore, take longer to reach us. Rees, *Just Six Numbers*, pp.104-5.

[8] But see, Hands, *Cosmosapiens*, p.21 (emphasis added):

> "Geoffrey Burbidge, astrophysics professor at the University of California, [has] pointed out that Andrew McKellar had discovered the cosmic microwave background radiation, estimating its temperature as between 1.8 and 3.4K, and had published his findings in 1941; he [has] alleged that Gamow, at least, knew of these results and so did not **predict** the cosmic microwave background radiation that subsequent observation confirmed."

[9] Quantum field theory declares that an apparent vacuum is not empty or without activity. Gravitationally-related processes on a very small scale

(the Planck scale) are constantly going on, generating virtual particles from "vacuum fluctuations." Penrose, *Cycles of Time*, p.200.

[10] One attempt to develop a quantum theory of gravity involves the use of "imaginary time," like imaginary numbers. See Hawking, *A Brief History of Time*, pp.139-43. What is the physical meaning of imaginary time? The answer that would be that we do not need a physical interpretation of the concept as long as the mathematics gives accurate predictions.

[11] Now the meter, the measure of distance, has been defined in terms of the speed of light. Sheldrake, *The Science Delusion*,, pp.92-3. So, different results are no longer possible!

[12] Physicist Max Tegmark comments that such "inflation" violates the Special Theory of Relativity, but not the General Theory of Relativity. *Our Mathematical Universe*, pp.47-8.

[13] Hands, *Cosmosapiens*, pp.30-1:

> "In 1989 Geller and Huchra identified a nearly two-dimensional structure approximately 650 million light-years long that they dubbed the Great Wall. In 2005 Gott and colleagues detected the Sloan Great Wall at more than twice that length, 1.3 billion light-years, located approximately a billion light-years away. In 2013 Roger Clowes and colleagues identified a high-membership quasar group that is 4 billion light-years long at a distance of 8 to 9 billion light-years. In 2014 István Horváth and colleagues reported their 2013 discovery of an object more than six times the size of the Sloan Great Wall, around 7 to 10 billion light-years long, at a distance of approximately 10 billion light-years. The sizes of these objects contradict the assumptions of isotropy and homogeneity."

[14] Roger Penrose is not very fond of the inflation theory. His principle objection is that it does not in fact solve what he thinks is the real problem—the very low entropy that has to have then existed for the Second Law of Thermodynamics to be valid (that is, for entropy to have been consistently increasing ever since). The low entropy must have existed before the inflation (if such inflation did in fact occur). *Cycles of Time*, pp.122-35. John Hands quotes other scientists that criticize the theory. *Cosmosapiens*, pp.32-7.

[15] Throughout this book, we have had occasions to note the importance of the rapid advances in the technologies and related instrumentation for the observation, measurement and analysis of data about the physical world. The examples range from the microscope (increasing resolution of optical images followed by electron imaging and then even finer tools like crystallography and x-ray) to medical imaging (from x-rays to fMRIs) to telescopes (from Galileo's handheld telescope to the Hubble space telescope). At the same time, the enormous increase in computer capabilities has been essential to the effective utilization of these other tools. An investigation of dark energy is being pursued through the Dark Energy Survey utilizing the most powerful digital camera in the world (570 megapixels) to photograph large portions of the southern sky. The camera is able to see galaxies up to 8 billion light years from Earth. Astronomers recently announced the discovery of nine new dwarf satellites orbiting the Milky Way and several new dwarf galaxies. These objects are potentially significant sources of dark energy. "Welcome to the neighbourhood: new dwarf galaxies discovered in orbit around the Milky Way," *University of Cambridge Research Bulletin*, 10 March 2015.

[16] It seems that:

> "an unknown type of energy—dubbed 'dark energy'— ...constitutes a whopping 68.3 percent of the total matter-energy budget. We don't know if dark energy has a microscopic substructure; we only know the effect it has. Dark energy speeds up the expansion of the universe. The remaining 31.7 percent of what fills the universe is matter, but—more surprise—most of it isn't the kind of matter that we're familiar with. Instead, 85 percent of matter (26.8 percent of the total energy-matter budget) is what's called 'dark matter.' The only thing we know about dark matter is that it interacts rarely, both with itself and with other matter. The remaining 15 percent of the matter in the universe (4.9 percent of the total energy-matter budget) is stable standard-model particles—stuff similar to what we are made of... ."

Hossenfelder, *Lost in Math*, pp.59-60.

[17] The language of the standard King James translation reads: "In the beginning God created the heavens and the Earth. The Earth was without form and void... ." The New English Bible (1970) begins: "In the beginning of creation, when God made heaven and Earth, the Earth was without form and void... ." The language of both is subject to two interpretations. It may be that God's action "in the beginning" resulted in the appearance of the

formless Earth that was, thereafter, formed into the world we know or that some type of formlessness preceded God's act of creation. See also, Davies, *The Mind of God*, p.42.

[18] Stephen Hawking asserted that there are at least three relevant "arrows" of time: the thermodynamic arrow pursuant to which entropy increases; the psychological arrow that we experience; and the cosmological arrow pursuant to which the Universe expands. *A Brief History of Time*, p.145. The first two arrows always, necessarily point in the same direction (because we perceive time as an increase in entropy). The third arrow can point backwards if and when the Universe contracts. Id., pp.145–7. However, even if the Universe began to contract, reversing the cosmological arrow of time, the other two arrows would continue to point to the future. So entropy would continue to increase. Thus, the contraction of the Universe would not be the exact reverse of the expansion. Id., pp.150–1. If the Universe does contract, then the third arrow of time will eventually reach the beginning and could reverse again. But, because entropy would have continued to increase, the Universe will have reached such a strong state of disorder that the thermodynamic (first) arrow would become essentially non-existent (entropy could increase no more). Id., p.151. Thus, the first arrow of time would end. Finally, of course, the psychological arrow of time would cease to have any meaning, since there would be no "minds" left to experience it.

CONSCIOUSNESS

"...[W]hile the mind remains central
to 21st-century Western thought,
...prominent neuroscientists and philosophers
inform us that it surely does not exist."

George Makari,
Soul Machine, p. xi.

"...[A mind] register[s] the effects of events
that are in the past (but are remembered)
...that are in the future (but are anticipated)
...that are merely possible
(but are contemplated)... ."

Fodor and Piattelli-Palmarini
What Darwin Got Wrong, p.11.

A Focus on Consciousness

There is an array of things that seem indisputably to be real, that is, a part of the world we experience, but that are also indisputably beyond the reach of current scientific theories. I would include in this category self-awareness, creative intelligence, imagination, aesthetic judgment (or appreciation), laughter and a variety of emotional reactions (enjoyment, sorrow, awe, etc.).

There is another array of things that are similarly beyond the reach of current science but about which there may be a debate as to whether they are features of the external world or merely illusions that arose as human cultural constructs. In this category, I would include moral values, meaning (or purpose), choice (as in the apparent exercise of free will) and, perhaps, certain human feelings like love and empathy.[1]

People probably differ as to each list, but that is not important here. As to both categories, we are talking in part about (human) awareness of the feeling or experience of these things which depend upon (or are the result of) consciousness and self-reflection.

Roger Penrose has written: "Consciousness is part of our universe ...yet no physical, biological, or computational theory ... comes very close to explaining our consciousness and consequent intelligence ... ." *Shadows of the Mind: A Search for the Missing Science of Consciousness* (1994), p.8. A fundamental question is whether the things involved, including the awareness of feelings, are beyond scientific understanding or simply beyond the capabilities (or scope) of current scientific knowledge, practice and methods.

Consciousness, according to Penrose, has a passive aspect and an active aspect, but there is not always a clear distinction between the two. The passive aspect consists of the experience of awareness—awareness of sensory perceptions, of emotions and simply of being. The active aspect involves the exercise of free will or choice, including the simple decisions to do something (like stand up). *Shadows of the Mind*, p.39. Penrose goes on to refine his usage of terms by specifying that intelligence requires understanding and that understanding requires awareness. Id., p.37. So consciousness, not surprisingly, underlies or is a requirement for intelligence, as the concept is used in connection with the human mind.

It is common to speculate about what precisely it is that makes humans different from other animals. Some combination of self-awareness and rational thinking must be high on the list. Feelings, introspection and spirituality are perhaps some of the results of these capabilities. The resulting qualities are not just what makes us different; they are what we think makes us special. And, there lurks the persistent, nagging feeling that just perhaps our specialness has some significance or relevance to the rest of the Universe.

Rebecca Neuberger Goldstein (philosopher and writer) writes of an epiphany she had, as a college student, while riding a New York City subway:

> "[S]uddenly my faith in hard and dry reductive materialism was shattered, never to be reassembled. ...The fact that some hunks of

matter have an inner life—sometimes too much inner life—is unlike any other properties of matter that we have yet encountered, much less accounted for."

"The Hard Problem of Consciousness and the Solitude of the Poet," Tin House 13(3), Issue 51 (2012), p.46.

Philosopher Thomas Nagel wrote a notable essay in 1974 entitled "What it is like to be a bat," in which he argued that even a complete physical description of the state and processes of a particular mind would be inadequate to enable the recreation of the mental experiences of that mind or to explain that mind's state of consciousness; thus, for example, no one can know what it is like to be a bat. His recent book, *Mind & Cosmos: Why the Materialist Neo-Darwinian Conception of Nature is Almost Certainly False* (2012), builds upon the argument that science based upon particle physics and laws of motion can never explain the human mind and presents some speculative thoughts about what types of explanations might be possible.

In *Mind & Cosmos*, Nagel stresses that the distinctive characteristic of human intelligence—rationality or cognition—is an even more significant challenge to scientific explanation than is consciousness. It is the ability to reason, to search for the real truth despite appearances and to change one's mind when presented with persuasive evidence or arguments for another position or belief ("thought, reasoning, and evaluation"), a collection of characteristics that appear to be found only in human beings, "though their beginnings may be found in a few other species." *Mind & Cosmos*, p.71; see, id., pp.71-95.[2] "It is not merely the subjectivity of thought but its capacity to transcend subjectivity and to discover what is objectively the case... ." Id., p.72.

Similarly, "the essence of politics...[is] the ability to reflect consciously on different directions one's society could take, and to make explicit arguments why it should take one path rather than another. In this sense, one could say Aristotle was right when he described human beings 'political animals' —since this is precisely what other primates never do, at least not to our knowledge." Graeber and Wengrow, *The Dawn of Everything* (2021), p,86. This subject brings us to matters involving not just the limits of (current) human knowledge, but potentially to the limits of science itself. "While our own psyches seem abundantly clear to us, attempts to objectively establish their existence have been mired in seemingly insoluble problems." George Makari, *Soul Machine: The Invention of the Modern Mind* (2015), p. xi.[3]

The questions for this chapter are whether "science" can tell us what consciousness is and how it arose. I am not going to rehash the philosophical mind/body dualism debate (whether the mind consists of something other than the matter that makes up the body). Instead, I shall try to approach the question from a scientific perspective: what do we currently know about the brain and the mind, what are we likely to find out through further research and experimentation and what would a scientific explanation of consciousness look like?

AT THE LIMITS OF SCIENCE?

BEYOND SCIENCE?

Many leading scientists have for decades contended that the human mind and its experiences will eventually be explained in physical terms.

In the early 1960s, Francis Crick (the co-discoverer of the structure of DNA) and Sydney Brenner reportedly proclaimed to a group of Cambridge graduate students that the two remaining important problems in biology were the process of physical development and the nature of consciousness. Sheldrake, *The Science Delusion: Freeing the Spirit of Inquiry* (2012), p.9. "They had not been solved [according to Crick and Brenner] because the people working on them were not molecular biologists—or very bright. Crick and Brenner were going to find the answers within ten years, or maybe twenty. Brenner would take developmental biology, and Crick would take consciousness." Id. In other words, the subjects were amenable to scientific explanation, if not in terms of existing biology, then in terms of the underlying physics and chemistry.

Similarly, in the 1970s, Carl Sagan confidently declared: "My fundamental premise about the brain is that its workings—what we sometimes call 'mind'—are a consequence of its anatomy and physiology, and nothing more." *The Dragons of Eden* (1977), p.7.

The same types of views have been recently expressed by Sebastian Seung, Professor of Computational Neuroscience at MIT. *Connectome: How the Brain's Wiring Makes Us Who We Are* (2012), pp.61, 262-3. Other outspoken proponents of this view include Stephen Hawking and his occasional co-author and Caltech physicist Leonard Mlodinow. See, *e.g.*, Hawking and Mlodinow, *The Grand Design* (2010), pp.32, 34.

Crick died in 2004, and in the forty years that had passed, he failed to achieve his objective of explaining consciousness in purely materialistic terms. (Brenner also did not achieve his goal of explaining the process of development.) Sheldrake, *The Science Delusion*, pp.9-10.

Indeed, the question of consciousness is one that is very much still subject to debate and awaits some credible empirical evidence. Even Sam Harris in his best-selling attack on religion acknowledges that we do not know what consciousness is and the belief that it is the brain that produces it is "little more than an article of faith." *The End of Faith: Religion, Terror, and the Future of Reason* (2004, 2005), p.208. Of course, we have already quoted several noted scientists, like Lord Martin Rees, expressing doubts as to whether science will ever be able to explain human consciousness.

ALTERNATIVES TO SCIENCE?

Rebecca Neuberger Goldstein, following Galileo, explains that the language of science is mathematics and that there are things that simply cannot be expressed in that language—for Galileo, things like our sensory perceptions (such as taste and color). She notes that it may be that mathematics is not the language of the physical world as such (as some scientists seem to suggest), but only of our knowledge of the physical world, that is, that it is the language of science, not of nature. "The Hard Problem of Consciousness and the Solitude of the Poet," *Tin House*, p.48.

Goldstein concludes that matters of consciousness, feeling, emotion and meaning may be beyond science because the language of science is inadequate for and unsuited to those aspects of reality. She suggests that we look instead to "the language of poetry". *Id.*, p.50. Similarly, essayist Curtis White urges that something other than science is needed to understand important aspects of the world, like another worldview. He suggests the philosophy of nineteenth century Romanticism (Wordsworth, Keats, Friedrich Schiller, Byron) as an alternative to scientism and the language of mathematics. *The Science Delusion: Asking the Big Questions in a Culture of Easy Answers* (2013), pp.57-74, 151-60 (this is the second book I reference with the title *The Science Delusion*, the other being by Rupert Sheldrake).

The analogy to language is interesting. Certainly, mathematics has a vocabulary (actually, many) and a grammatical structure (many). The types of phenomena or relationships that it can describe are necessarily circumscribed by that vocabulary and grammar. However, as we discussed in an earlier chapter, mathematics also carries within it necessary logical implications and conclusions in a way that language as such (as I understand it) does not. So, I think that the analogy is not very useful in the end (and that Galileo probably meant something more than Goldstein does—that is, that it is nature itself that is reflected in or captured by mathematics). So, I do not find the language of the poet or the philosophy of the Romantics to be the answer to the questions being posed here.

If, as some scientists and philosophers suggest, we may conclude that it is questionable whether we can achieve an explanation of consciousness through the theories of physics and chemistry, even aided by natural selection; then the hopes noted at the beginning of this book to achieve a theory of everything or the reduction of all knowledge of the world to a fundamental and comprehensive theory seem doomed. Moreover, the implications could go beyond the conclusion that there is no theory of everything. They can raise questions about the validity of the scientific theories that we do have. As Nagel observed: "If one doubts the reducibility of the mental to the physical, and likewise of all those other things that go with the mental, such as value and meaning, then there is some reason to doubt a reductive materialism can apply even in biology, and therefore reason to doubt that materialism can give an adequate account even of the physical world." *Mind & Cosmos*, p.14.

DOES IT MATTER?

Everyone seems to agree with Roger Penrose that: "A scientific worldview which does not profoundly come to terms with the problem of conscious minds can have no serious pretensions of completeness." *Shadows of the Mind*, p.8.

Well, perhaps not everyone.

Steven Pinker (spouse of Rebecca Neuberger Goldstein, whose is quoted above) acknowledges that we do not understand the origin or nature of consciousness and speculates that we may never be able to do so. Yet, he claims not to be troubled by that prospect:

> *"[It is] to be expected, even welcomed. ...We don't poke fun at the eagle for its clumsiness on the ground or fret that the eye is not very good at hearing, because we know that a design can excel at one challenge only by compromising at others. Our bafflement at the mysteries of the ages* **may have been the price we paid** *for a combinatorial mind that opened up a world of words and sen-*

tences, of theories and equations, or poems and melodies, of jokes and stories, the very things that make a mind worth having."

How the Mind Works (1997), p.565 (emphasis added).

THE MAKEUP OF THE BRAIN

...[T]he adult male human brain contains approximately 86 billion neurons ...passing signals to each other via some 500 trillion ... synaptic connections, making it the most complex thing in the known universe.

John Hands
Cosmosapiens, p. 403.

This section probably should be headed "What we currently know about the makeup of the brain." There is not much challenge in identifying important things we do not know about this subject.

In fact, it may be that 2013 will become known as the year in which the study of the brain took center stage. Following a reference to the potential returns on government investment in science and technology in President Barack Obama's State of the Union address in January 2013, it was revealed that the Obama administration planned to promote a decade-long, multi-billion dollar project to study and map the human brain, roughly modeled on the successful quest to map the human genome that began in 1990 and achieved its objective in April 2003, at an estimated cost of $3.8 billion. John Markoff, "Obama Seeking to Boost Study of Human Brain," *The New York Times*, February 17, 2012. The initial news report noted that many scientists believe that the project to map the brain will be a "drastically more significant challenge" than mapping the human genome. Id. See also Philip M. Boffey, "The Next Frontier Is Inside Your Brain," *The New York Times*, February 23, 2013. The Brain Initiative or Human Connectome Project, as it is now called, is run by the National Institutes of Health and has engaged several consortia of universities and research centers.[4]

One project being undertaken at Washington University is the attempt to create "an interactive wiring diagram of the living, working human brain" based upon data being collected through MRI scans of 1200 volunteers. James Gorman, "The Brain, in Exquisite Detail," *The New York Times*, January 6, 2014. The new focus on the study of the brain is also stimulating the development of new technologies to obtain and analyze data about the brain. *Id.* By June 2014, the estimated price tag had risen to $4.5 billion, to be spread over 12 years. *See* James Gorman, "N.H.I. Seeks $4.5 Billion to Try to Crack the Code of How Brains Function," *The New York Times*, June 5, 2014. The Institutes already spend about $4.5 billion a year on neuroscience research. *Id.*

Other research looks at what we do know about the structure of the brain and attempts to utilize the insights to improve artificial intelligence and computer functionality. For example, IBM has disclosed the development of a new computer chip that tries to imitate the manner in which the brain achieves pattern recognition, utilizing huge numbers of circuits working in parallel. The design supposedly utilizes dramatically less electricity to operate and achieves new levels of miniaturization. The new chip contains 5.4 billion transistors but draws only 70 milliwatts of power. In comparison, modern Intel chips may have 1.4 billion transistors and utilize 35 to 140 watts of power. John Markoff, "IBM Develops a New Chip That Functions Like a Brain," *The New York Times*, August 7, 2014. The new chip, very much still in the development stage, is named TrueNorth.

Although the state of our knowledge of the brain should change rapidly over the decade, it is still valuable to outline what scientists think that they do know now or, more often, what theories and hypotheses they are currently pursuing as promising. This overview of the science will put the philosophical discussions in a more constructive context.

A review of the current state of the science certainly will not answer the question of whether consciousness, intelligence and other distinctive aspects of the mind are susceptible to scientific explanation. It is clear that the current state of our knowledge does not demonstrate that a scientific explanation is possible; but, it is also true that the limited progress that has been made does not prove that we will never be able to understand the human mind.

THE INVESTIGATION OF THE BRAIN

Discussions of the subjects of human consciousness and intelligence have a very long history. The topics were essentially the domain of philosophers and theologians for many hundreds of years.

The importance of the subject to human beings, especially those of a contemplative nature, has been undeniable. The scientific exploration of the subject, however, has been hindered by the practical and ethical difficulties of scientific investigation and experimentation. The main problem was that there was no way directly to observe the functioning of the brain or even to experiment directly on the living brain.

The initial serious scientific work on brain function, beginning in the second half of the nineteenth century, was undertaken through the study of persons with severe brain injuries. See, e.g., Carl Sagan, Broca's Brain (1979), pp.6-9 (briefly discussing the work of French neurologist/anthropologist Paul Broca in the mid-nineteenth century); Seung, Connectome, pp.10-13; Michael S. Gazzaniga, Who's In Charge: Free Will and the Science of the Brain (2011), pp.44-51.

Scientists were able to study the impact on brain function, as evidenced through bodily functioning, of injury to specific, identifiable parts of the brain. These studies strongly indicated that certain brain functions controlling various bodily functions and actions like speech were at least significantly located in specific local regions of the cerebral cortex. Sagan, The Dragons of Eden, pp.30-32. For example, "[s]pecific brain sites below the cerebral cortex have been found to be concerned with appetite, balance, thermal regulation, the circulation of blood, precision movements and breathing." Id., pp.30-31.

Other investigations used the brains of deceased—neuroanatomy or neuropathology. Seung, Connectome, p.xvii. Autopsies enabled studies of the sizes of brains and comparisons of the sizes of various parts of the brain among different individuals. Also, autopsies of the brains of people who had been subject to study while alive and observed to display various types of abnormal behavior or functioning enabled some conclusions to be drawn about the roles of different parts of the brain by relating defects discovered through the autopsies to the behavior previously seen.

Much of the research efforts of neuroscientists has involved the study of disorders such as autism and schizophrenia. See, e.g., Sagan, The Dragons of Eden , pp.17-21, 281-2. That research served multiple purposes. Such efforts helped gain insight into the disorders themselves, which may have enabled more definitive or earlier diagnosis or, hopefully, treatment. The research also aided the efforts to understand better the workings of the "normal" brain, based upon the inferences that can be drawn from the location or nature of the disorder and the impact upon the patient's functioning. (Sometimes, we can gain information on the way a process or mechanism works by looking at examples that are broken.)

In the 1940s, neurosurgeons introduced a treatment for debilitating seizures suffered by persons with mal epilepsy that consisted of severing the corpus callosum that constituted the primary nerve fiber connection between the two hemispheres of the brain. These individuals with "split brains" offered scientists the opportunity to test the functioning of each hemisphere independently and of the apparent consequences of lack of communication between the hemispheres. See, e.g., Gazzaniga, Who's in Charge?, pp.51-55; Sagan, The Dragons of Eden, pp.158-69.

More recently, direct medical experiments involving the electrical stimulation of specific parts of the brain, often in efforts to treat diseases or injuries, have become possible. Such experiments have revealed further evidence of localization of certain functions and, even, of specific memories. See, e.g., Sagan, The Dragons of Eden, p.31. Since the 1980s, advances in medical imaging, like PET and functional MRI, have certainly made the examination of brains much richer.[5] Examination of living brains has become possible, even enabling the documentation of changes over time. The imaging technology enables the observation of certain brain activity, for example, in connection with particular activities or in response to specific stimuli. It also has enabled the identification of various pathologies within the brain while the subject of observation is still alive. In addition, other technological improvements have greatly facilitated the ability of scientists to study brain matter. For example, a new imaging technique called diffusion tensor imaging allows scientists actually to trace nerve fibers. Gazzaniga, Who's In Charge?, p.36.

Yet, even today, the technology available for examining the activities of a living, active brain has limitations. The insertion of an electrode into the brain provides the most detailed and locationally specific information, but it has limited applicability because it is intrusive and dangerous and because it is not viable for assessing multiple areas of activity simultaneously. See Penelope Lewis, The Secret World of Sleep: The Surprising Science of the Mind at Rest (2013), p.65. The attachment of external electrodes to the skull (electroencephalography or EEG) is non-invasive and suitable for taking multiple measurements at one time, but it lacks in spatial resolution since it is measuring internal activity from the surface. Id., pp.65-6. The development of another non-invasive technique called functional magnetic resonance imaging "fMRI"), which responds to changes in blood flow that accompany neural activity, has enabled greater spatial precision but does not permit accurate temporal identification, because the blood flow response to neural activity can take several seconds to appear. In addition, it is useful only with respect to relatively large changes in activity. Id., p.66.

Such analyses establish that there are differences among people's brains, and some at least weak statistical correlations between sizes of brains or parts of brains and characteristics such as IQ and musical ability. Sagan, *The Dragons of Eden*, pp.4-6, 14-5. Of course, such observed relationships tell little about any possible causal relationships, such as whether the brain size is the reason for any of the noticed abilities or whether the activities of the possessor of the brain contributed to the difference in size. Id., pp.24-5.[6]

With the introduction of MRI imaging of functioning brains, scientists have been able to ascertain much more about the physical differences in human brains. Such studies attempt to construct images of the brain based upon data assembled from numerous different persons.

More invasive (or fatal) studies have been and are being made of the brains of animals, such as rats and birds, which provide useful information about relationships between environment factors and brain changes. See, e.g., id., pp.23-4, 189-95. Of course, one must be careful in trying to extrapolate results obtained in the experimentation on animals to humans. The brains and nervous systems are clearly different, so one needs to ask what, if any, implications arise from the differences. For example, some early research on salamanders gave misleading evidence concerning the operations of the human brain because the lower vertebrates, unlike mammals, have regenerative nervous systems. See Gazzaniga, *Who's In Charge?*, p.15. These potential problems do not make animal research useless, obviously, simply not conclusive.

THE STRUCTURE OF THE BRAIN

In vertebrates, the stalk on which the brain sits is the spinal cord, at the top of which is the brain stem. The other major parts of the brain are the cerebrum, consisting primarily of two symmetrical hemispheres, and the cerebellum, which is a smaller part at the lower back of the brain. The parts closest to the brain stem are the oldest in evolutionary terms, with the newer parts arranged as additions moving outward to the skull. Generally, the oldest parts control the basic processes of animal life, like heartbeat and body temperature. The newest parts control what we consider to be thinking and consciousness. See Lewis, *The Secret World of Sleep*, p.59.

Immediately above the brain stem is the midbrain, on top of which sit the two symmetrical hemispheres connected by a thick cable, called the corpus callosum. Between the hemispheres and the midbrain is the thalamus, half in each hemisphere. The thalamus acts as a sort of communications center or relay station among the various regions of the brain. It is surrounded by the basal ganglia, next to which projects the hippocampus. *Id.* p.60. All of these parts constitute the cerebrum and are present in the brains of all vertebrates. However, the relative sizes of the regions vary with the hippocampus, for example, being dramatically larger in mammals than in other vertebrates. *Id.*, p.61.

The cerebrum "is widely regarded as the most important of the three parts for human intelligence; it is critical for virtually all of our mental abilities. It is also the largest of the three parts, occupying about 85% of the total brain volume." Seung, *Connectome*, pp.7-8. In mammals, there is a cerebral cortex, a thin membrane that is densely folded so it appears to be wrinkled. Most of it consists of six layers and is often referred to as the "neo-cortex", with other smaller portions with different numbers of layers having various other designations. See *id.*, p.301 (note to p.174).

Paul MacLean, chief of the Laboratory of Brain Evolution and Behavior of the National Institute of Mental Health from 1971 to 1985, argued that the human brain is comprised of three separate but interconnected systems mounted on the spinal cord. The systems are of very different ages, representing evolutionary stages, and are effectively built upon one another. *See* Sagan, *The Dragons of Eden*, pp.51-5. MacLean developed this theory while a professor at the Yale Medical School, before joining the Laboratory of Brain Evolution and Behavior 1971, and set it forth in detail in the book *The Triune Brain*, published in 1990. *See, e.g.*, Gazzaniga, *Who's In Charge?*, p.30n. The oldest system is shared with other mammals and reptiles. MacLean called it the reptilian-complex. Next in time is the limbic system, shared with other mammals. Finally, there is the neocortex, found in the higher mammals. The neocortex becomes increasingly large and complex in the more advanced mammals, such as primates and, most markedly, humans. Sagan, *The Dragons of Eden* pp.55-6.

Interestingly, the limbic system appears to be involved in the generation of strong emotions. *Id.*, p.62. However, the neocortex is recognized to be responsible for most human cognitive functions. *Id.*, p.69. The three systems are interconnected, and various brain functions have been shown to involve more than one system. For example, olfactory processing involves the neocortex as well as the limbic system, where it primarily occurs. *Id.*, p.74.

The neo-cortex

The neo-cortex covers the cerebrum and, in humans, effectively fills all of the extra spaces within the skull. The folds or wrinkles enable a considerably greater surface area of neo-cortex to fit within the volume available.

> "The neo-cortex is the seat of higher thought—executive control, attention, and the ability to reason. It is likely that it also contains the key to consciousness, but this remains to be established. Along the edges of the neo-cortex, several structures that were present before the expansion of this huge organ swamped the rest of the brain have also developed and taken on new forms. These include the hippocampus and the amygdale."

Lewis, *The Secret World of Sleep*, pp.61–2.

> "[It] develops from the cortex as a thin—about 2mm in depth—but increasingly convoluted new outer layer. ...[T]he striking evolutionary change ... is the growth of the neocortex. In humans it is only twice as thick as in a mouse but has about a thousand times more surface area; while it is only 15 per cent thicker than in a monkey it has at least ten times more surface area."

John Hands, *Cosmosapiens*, p.402.

Glial cells and neurons

The glial cells essentially provide a framework and infrastructure for the neurons, which are believed to do the mental work of the brain. Neurons have a typical nucleus like other cells, but also long tentacles stretching out through the brain. Neurons are perhaps the most complex cells in the human body. Seung, *Connectome*, p.103. The human brain contains close to 100 billions neurons—a truly staggering number. *Id.*, p.xix.

The tentacles or branches are called neurites, and they come in two forms, called dendrites and axons. Dendrites are the receiving arms of the neuron. They are shorter and thicker than axons. The axons are the signaling or sending arms, and they are thin and long and can reach all the way across the brain. The neurites of various neurons intertwine and overlap—their purpose is to make contact with neurites of many, many other neurons. They have been described as like a plate of spaghetti. *Id.*, p.41. The brain appears to consist of grey matter on the surface (the cortex) and white matter inside. The grey matter is made up of the main bodies of the neurons, while the white matter consists only of axons. Id., p.205.[7]

Although neurites may be in contact or very close proximity to many other neurites, they communicate with one another only at particular points, called synapses, that occur between dendrites and axons. Where there is a synapse, one finds on the axon side transmitter molecules and on the dendrite side receptor molecules. Communications through neurons occur as electrical signals running through the neuron from a receptor on a dendrite to a transmitter on an axon. The transmitter on the axon communicates to the receptor on the neighboring dendrite (across the synapse) with a chemical signal. There are many different chemical signals (over 100 have been identified), but each neuron generally uses a few particular signals to send its message. Id., pp.43–4.

The electrical signal in a neuron will be activated by the receipt of a certain number of positive chemical signals by its dendrites across their synapses, which converge when the dendrites meet the body of the neuron. When a sufficient stimulus is received, there will be an electrical "spike" that flows through the neuron's axons,[8] causing the transmitters at the (relatively) distant synapses of the axon to emit chemical signals across to dendrites of other neurons. The signal can pass in this fashion through many neurons following a pathway until it reaches a synapse between an axon and a muscle. The axon then sends a chemical signal to the muscle. It is also possible that the signals received by a neuron are not sufficient to make it spike, in which case the neuron will be a dead-end or stopping point for the transmission of the signal. Such inhibitory actions are clearly important since it is not desirable for all stimuli to activate muscles. Id., pp.56–7. Thus, neurons can be said "to *compute*, not just communicate." Id., p.58.

Although it might be tempting to liken the electrical microcircuits of the brain to the switches in computers, there is an important difference in that the brain's "circuits are capable of more subtle responses than the simple 'on' and 'off' of a digital computer circuit." Sagan, *The Dragons of Eden*, pp.42–3. Indeed, the amount and combination of stimuli necessary to cause a reaction appears to be variable and complex.In addition, it appears that mistakes can and do occur, especially at the margins, due to chance factors. See Penrose, *Shadows of the Mind*, p.353.

Now, not all neurons are connected to all others in the brain, not nearly. Although, there are perhaps 85 billion neurons in a human brain; each axon has synapses that number, at the most, in the tens of thousands. Seung,

Connectome, p.86. That is a lot of connections, but dramatically less than the number of potential connections that could, in theory, occur given the number of neurons. Presumably, more connections would make a brain more powerful, so the limitation on the number of connections is likely to be a matter of the conservation of space inside the skull (limiting the required size of the head) and energy (reflecting the need to limit the nutritional requirements of the body in which the brain resides). *Id.* pp.120-1; Gazzaniga, *Who's In Charge?*, p.67.

Advanced imaging has revealed that all neurons are not alike. There are the normal, pyramidal neurons ("shaped like a Hershey's Kiss") that constitute "the basic building blocks of the brain." Gazzaniga, *Who's In Charge?*, p.38. But, even those types of neurons seem to vary in shape and structure across different regions of the brain and also among different species of primates. *Id.*, pp.38-9.

In addition, a different shaped neuron was described by Constantin von Economo in 1925. It is longer and thinner and about four times larger than the pyramidal neuron, with only a single dendrite at its base (receiving inputs). *Id.*, p.39. This von Economo neuron (VEN) appears only in the parts of the human brain that are involved in cognition and, among primates, are found only in the brains of humans and the great apes. *Id.* Moreover, the VEN are present in much greater quantities (both absolutely and relative to size) in the human brain. *Id.*, pp.39-40. Among mammals, VEN are found only in the brains of social animals like elephants and dolphins. *Id.*, p.40.

Some current theories postulate that there are different types of neurons in the brains of different species, some of which types may be unique to particular species, and that the structure and connections within the brains are different as well. *Id.*, pp.36-7, 40-41. Obviously, this work raises questions about the use of research on the brains of lab animals, like rats and birds and even monkeys, to draw inferences about the human brain. *Id.*

The two hemispheres

As noted, the cerebrum consists of two hemispheres—a left and a right. (The cerebellum is also in two hemispheres.) The two sides are connected by bundles of neural fibers, many of them in the corpus callosum. The connections provide some type of communication between the two hemispheres. Generally, the right hemisphere controls and receives stimuli from the left side of the body, and the left hemisphere controls and receives stimuli from the right side of the body. There are some exceptions, like the sense of smell. Sagan, *The Dragons of Eden*, pp.161-2.

However, there is also considerable evidence that different functions and different types of thinking are performed by the two hemispheres. For example, the highly developed human capacity for facial recognition appears to "reside preferentially" in the right hemisphere. *Id.*, p.158. Speech capabilities for most people are found in the left hemisphere. *Id.*, p.162. Geometric visualization generally comes from the right hemisphere. *Id.*, p.164. Again, the earliest evidence came from persons with brain lesions or injuries. *Id.*, pp.158, 161-8. For most people, information is processed sequentially by the left hemisphere and is processed simultaneously in the right. *Id.*, pp.168-9. For such people, the activity of the left hemisphere (rationality) tends to dominate one's conscious awareness, while intuitive and creative insights arise in the right hemisphere. *Id.*, p.183. The left hemisphere is generally the intellectual side, dealing with language, speech and logical reasoning. The right hemisphere for most people is perceptual and spatial; it has the distinctive human capacity for recognizing faces, and it visualizes spacial relationships. Gazzaniga, *Who's In Charge?*, pp.57-8.

Interestingly, experimentation with the so-called split-brain patients strongly suggests that when communication between the two hemispheres is severed, the patients experience a kind of double consciousness. Each hemisphere is engaged with the external world, processing information and determining appropriate responses. *Id.*, p.59-60. However, the nature of the conscious experiences seem clearly to be different, and the hemispheres are not "equally conscious, that is, it was not conscious of the same things, and not equally capable of performing tasks." *Id.*, p.63.

Verticality

There appears also to be a relevant vertical organization of the brain. See Stephen M. Kosslyn and G. Wayne Miller, "A New Map of How We Think: Top Brain/Bottom Brain," *The New York Times*, October 20, 2013. The authors present a theory based upon "another, ordinarily overlooked anatomical division—between its top and bottom parts." The top and bottom parts are largely denoted by the Sylvian fissure, visible from a side view of the brain. The top part consists of the entire parietal lobe and the larger top part of the frontal lobe.

The theory originates out of research done on rhesus monkeys. It asserts that the top part of the brain is engaged in efforts to formulate plans and then in monitoring the implementation of the plans to achieve intended goals. The bottom part organizes signals received from the senses and, using information stored in memory, interprets the signals. The two parts, like the two hemispheres, work together. However, the authors assert that the way in which the two parts interact varies among individuals, giving rise to different personality types or modes of thinking.

CONNECTIONS

While we know that certain capabilities reside generally in the left hemisphere and certain others in the right hemisphere, that certain functions are apparently performed largely in the upper part of the brain and others in the lower part and that various specific local areas of the brain seem to be associated with particular actions or feelings; it seems to be clear today that much of the brain function arises from the interactive participation of multiple areas of the brain, that is, that the brain functions largely as an integrated whole. See, e.g., Kosslyn and Miller, "A New Map of How We Think: Top Brain/Bottom Brain," *The New York Times*, October 20, 2013. Thus, the questions of how the various parts are connected and how those connections function seem to be crucial to an understanding of the mind.[9]

Seung postulates that the locations where synapses are created are, from the very early development of the brain, essentially random. *Connectome*, pp.87–8. The patterns that enable functionality arise through the elimination of various synapses and the strengthening of others. Some neuroscientists use a Darwinian analogy (*see id.*, pp.88–9), which I think is not very apt (although, not entirely inapposite). It does not seem likely to be a matter of the "survival of the fittest," and it certainly is not the result of more successful reproduction. Instead, I think that the concept is the survival of the connections that are used and, in some sense, useful, with the disappearance of the ones that are not utilized.

The idea of the strengthening of synapses through use was set forth by Donald Hebb in 1949, a process that has come to be called Hebbian plasticity. *Id.*, pp.82–3. The big picture appears to be that the genome establishes the materials and location of the brain, while life experiences fine tune the particular structure of what Seung calls the connectome (the entire structure of the connections, in and out, of all the neurons).

PLASTICITY

One of the important discoveries of the second half of the twentieth century was that the human brain (and other brains) is characterized by plasticity—the capability to change—throughout adult life. Changes in the actual anatomy of the brain occur as a result of environment and experience. See, e.g., Sagan, *The Dragons of Eden*, pp.45-6; Norman Doidge, MD, *The Brain that Changes Itself* (2007), p.xviii. New neuro-pathways are activated and used, perhaps even created, as a result of the demands placed on the brain by the new activity. Synapses are created and eliminated throughout the life of the brain.[10]

The pace of such activity is dramatically faster at the beginning of life, during the first few years for a person, but it continues through adulthood. In infancy, neurons are created at a very rapid rate, some half a million per minute, and the brain reaches about 90% of its adult size by the time the child is 5 years old. Thereafter, neurons and synapses are continuously being pruned and new ones being formed, influenced by environmental factors such as diet and experience. See, e.g., Robert Lee Hotz, "Differences in How Men and Women think are Hard-Wired," *The Wall Street Journal*, December 9, 2013.

> "Embryos develop from a clump of cells into highly organised structures. However, until now the signals orchestrating this transformation have remained hidden from observation inside the womb. Measuring gene activity in three dimensions, researchers have generated molecular maps of the second week of gestation as it has never been seen before. ...The second week of gestation is one of the most mysterious, yet critical, stages of embryo development. ... But with [pre-]existing methods [one] could not explore week two of development, after the embryo implants into the womb."

University of Cambridge, "Molecular 3D-maps unlock new ways of studying human reproduction," *Research Bulletin*, 16 June 2022.

There is some controversy about the amount of plasticity that exits, especially in the adult brain, but everyone agrees that it is highly significant in young children and continues to exist, at least to some extent, throughout adult life. A college classmate (a physicist) informs me that:

> "[T]he nerves develop organically with the rest of the body from a very early stage of the embryo, when all the cells destined to become distant parts of the body are still in close proximity. The nerve connections between those body parts and the developing brain then have a chance to reflect the geometry of the body in corresponding patterns in the brain. ...After the body parts be-

come well defined and spatially separated from the brain, the establishment of these nerve patterns, end to end, becomes impossible. ...[T]he **plastic parts are usually associated with higher-level functions, like cognition, rather than sensory/motor functions,** *which are hard-wired to the body... ."*

Richard Kellogg, email, June 21, 2022 (emphasis added).

The human brain will have its peak number of synapses and neurons in an infant or young child. As combinations are utilized, connections form between the pairs or among the groups of neurons and related synapses. Repeated usage will strengthen the connection. Over time, unused neurons and synapses will disappear. Thus, the sayings: "fire it to wire it; use it or lose it." Bruce Hood, "Developing a Sense of Self," Darwin College Lecture Series, Cambridge University, 6 February 2015.

Thus, appropriate stimulation and social interaction is highly important for children from birth, affecting not just intelligence but also the development of personality functions that are useful for constructive social interaction in later years. It seems, in fact, that while deficiencies in certain mental capabilities arising from lack of stimulation can be overcome later in life, some of the functions needed for social interaction are simply not capable of development after a certain period of neglect. Id.

This neuro-plasticity arises through what Seung calls the four Rs: reweighting, reconnection, rewiring and regeneration. *Connectome*, p.xv. Reweighting refers to the changes in the number and strength of incoming signals necessary to cause an electrical spike in the neuron. Reconnection refers to the creation of new synapses connecting different neurons. Rewiring is the changes in the pathways that can result. And, regeneration is the creation of new neurons (and the elimination of old ones). Id., pp.77–80, 116–33.

One type of consequence of these processes is the ability of the brain, at least to some extent, to compensate for injuries by relocating the "controls" for certain functions from a damaged region to another region. The brain appears to have substantial excess capacity (at least for contemporary humans) and built-in redundancies, but the interesting capability is that of being able to reconfigure connections so as to make use of that capacity. Another example is the ability of the brain to recalibrate the relationships among the senses when, for some reason, they become "out of sync," even to the extent of enabling a person to adjust to seeing things upside down (a person can begin to see things right side up after a period of wearing spectacles that inverts the images seen). Id., p.125.[11] From a mental health standpoint, there is evidence to suggest that various activities can deter or delay deterioration in brain function, somewhat like what physical exercise can do for muscle mass and bone density.[12]

Plasticity does not create genetic changes, so it would not normally be part of the story of evolution. However, genetic changes presumably occurred to give rise to plasticity. Moreover, the emergence of plasticity clearly would have had adaptational consequences. Also, it may have been coupled with other changes such as prolonged periods of dependency and longer childhoods, which could have had disadvantages for the species.

One can imagine, therefore, a complex but profound change occurring as *hominim*[13] became increasingly social and increasingly domesticated, a transition dependent upon and, therefore, reflected in genetic changes. For example, the 10–15% decrease in the size of the humanoid brain that occurred after the last ice age could be related to domestication of the human species and to the remarkable degree of plasticity we have in our brains. Bruce Hood, "Developing a Sense of Self," Darwin College Lecture Series, Cambridge University, 6 February 2015.

SOME OTHER ISSUES

Differences

Several studies find an apparent difference in the wiring of male and female brains of adults. Robert Lee Hotz, "Differences in How Men and Women Think Are Hard-Wired," *The Wall Street Journal*, December 9, 2013. A study led by Ragini Verma at the University of Pennsylvania used images of the brains of 949 people, 521 females and 428 males. The researchers created composites of images of hundreds of brains because "[in] any one person, gender patterns may be subsumed by the variations in brain shape and structure that help make every person unique." Id.

The study indicated that there were no obvious systematic differences among the brains of children. Dr. Verma stated that "Most of the changes we see start happening in adolescence. That is when most of the male–female differences come about." Id. The differences observed were that females tended to have more wiring between the hemispheres (side to side), while males tended to have more wiring within each hemisphere (front to back). Id. Importantly, we do not know what causes the differences in the structures of the adult brains, or the extent to which it is genetic and the extent to which it is environmental.

A substantial project is currently underway in Hong Kong to attempt to ascertain whether there are discernible DNA differences among people with unusually high intelligence. Gautum Naik, "A Genetic Code for Genius," *The Wall Street Journal*, February 15, 2013. This work is being done by 20-year-old Zhao Bowen in the cognitive genomics lab at BGI, a private company partly funded by the Chinese government. The facility is reportedly processing some 2200 DNA samples collected from individuals with exceptionally high measured IQs. The objective is to discover genetic material shared by the highly intelligent that is different from the genetic material in the average person, a so-called "IQ gene". *Id.*

Other recent work involves efforts to understand the nature of altruism and cooperative social behavior in humans. See Elizabeth Svoboda, "Hard-Wired for Giving," *The Wall Street Journal*, August 31, 2013. Neuroscientists in the 1980s had noticed that returning veterans with damage to the frontal lobes could have apparently full cognitive abilities and motor functions, but had become largely indifferent to the feelings and opinions of other people. *Id.* The inference was that there was a specific portion of the brain, which in the case of these veterans was a portion that had been damaged, that influenced empathy and social inclinations. Work at the National Institutes of Health utilized fMRIs to examine in real time brain functions in subjects while they were making decisions about charitable giving. The results indicated that charitable acts caused activity in the same parts of the brain "that control cravings for food and sex." *Id.* Studies by economist Bill Harbaugh and psychologist Ulrich Mayr, again using fMRI, showed similar results. *Id.* The conclusion seems to be that charitable acts frequently generate pleasurable feelings or satisfy needs or cravings. In that sense, they are selfish acts.

The question, then, is what created the conditions that altruistic acts gave rise to pleasurable feelings. Is it solely cultural? Experiments on mice that indicate that the stimulation of a specific area of a mouse's brain can make the mouse noticeably more socially inclined. *Id.* That work would suggest the possibility of some existing physical neurological arrangements giving rise to social behavior.

Genes

Scientists have searched for genetic causes of mental disorders, just as they have for physical diseases. It was recently reported that researchers had found common genetic abnormalities or variations among individuals suffering from five different types of mental disorder (schizophrenia, bipolar disorder, autism, major depression and attention deficit hyperactivity disorder). Gina Kolata, "Same Genetic Basis Found in 5 Types of Mental Disorders," *The New York Times*, February 28, 2013. The study indicated that persons with these mental disorders were likely to have these particular genetic abnormalities, but only a small percentage of persons with the abnormalities displayed symptoms of the disorders. The genetic variations could create a susceptibility to the disorders, but the causes of the actual diseases are much more complex and probably include environmental as well as genetic factors.

For example, the development of the brain is controlled by the genes, which can be activated or deactivated by various proteins that act as regulators. A protein called REST turns off certain genes during the fetal period so as to keep neurons in an immature state until the time for the development of brain functions during infancy. Recent research has indicated that the REST proteins become active in the brain again as people age and that the absence of such proteins in the relevant regions of the brain is linked to Alzheimer's disease. Pam Belluck, "Protein May Hold the Key to Who Gets Alzheimer's," *The New York Times*, March 19, 2014. In healthy adults in their 70s or older, REST levels are high. Yet, persons with advanced Alzheimer's appear to lack REST in the key areas of the brain affecting memory and cognition.

One possibility is that the protein turns off genes that promote cell death, reducing the effects of aging on the neurons. *Id.* Alzheimer's may be a result of the lack of this protein. The implications of this research work are obviously still highly speculative. The studies on the human brain are conducted post-mortem. Apart from the severe restrictions on the scope of sampling, there is the question of the extent to which death itself affects the data. Also, this limitation makes it particularly difficult to find indicators of the direction of causation. Perhaps it is Alzheimer's, triggered by something entirely different, that results in the reduction of the presence of REST.

While we would expect that our genes determine much about the structure and functioning of our brains, there is ample evidence that certain acquired instincts (such as the fear of snakes) also come to be contained within our genetic heritage, passed on to offspring automatically. In some sense, certain memories seem to be contained within our genes or, at least, in our genetic inheritance. However, we do not really know how it happens. In any event, not surprisingly, the end result that constitutes the human brain arises from a complex interaction of genetic and environmental/social factors. *See, e.g.,* Gazzaniga, *Who's In Charge?*, pp.183–6.

What the Brain Does

I am only going to talk about some of the things that the brain does, things of which we are acutely aware. These things are memory, sleep, dreams and information processing. As we shall see, these subjects present mysteries enough.

Memory

One of the more straightforward questions about the operation of a brain is how memories are stored and recollected. One theory is that ideas are represented by neurons and associations of ideas by connections among neurons. The memories are stored in that assembly of neurons, to be triggered by various stimuli. Gazzaniga, *Who's In Charge?*, p.73. This theory is at best suggestive of a possible explanatory model; it seems utterly inadequate as is.

We know, of course, that non-humans have memories or something closely resembling what we call memories in the context at least of various specific activities. Recent research conducted by Dr. Jason Bunch of the University of Chicago, published in the *Proceedings of the Royal Society B*, identified evidence of long-term memories in dolphins. Matt McGrath, "Dolphins have 'longest social memory' among non-humans," BBC *News*, 7 August 2013. Dolphins learn to recognize the distinctive whistles of other dolphins with which they have socialized, and they react when they encounter another dolphin that they know in a manner different from how they react when they encounter a stranger. Studying dolphins that were in captivity (so that it was possible to establish the social contacts of the dolphin over long periods of time), researchers were able to play recordings of the whistles of other dolphins and observe how the dolphin under study reacted. They discovered that these dolphins could recognize the whistles of friends even when they encountered them again after as long as 20 years.

There is good evidence that short-term and long-term memories reside largely in different parts of the brain, suggesting that the brain performs a sort of buffering function for short-term memories, which are then transferred, under the right circumstances, to longer term storage. See Sagan, *The Dragons of Eden*, pp.74-6. An interesting question is whether memories are physically stored or are encapsulated in on-going, dynamic processes.

Certainly, persons interested in cryonics would like to know the answer to the question of whether memories disappear entirely when the brain activity stops. Some empirical evidence can be found, for example, inf persons whose brains have temporarily stopped, say, by a sufficient lowering of the body's temperature. The answer seems to be that long-term memory can survive the cessation of brain activity. Id., p.32; Seung, Connectome, p.91. (However, there are many other issues with trying to reactivate a human body or brain that has been frozen. Seung, Connectome, pp.233–53.)

There are a number of other interesting questions about memory. For example, how do we access memories? Some memories are sharp and readily available; others are more remote and seem to require some type of trigger to become accessible. Another question concerns false memories (not a lack of recall but a clear recollection of something that did not in fact happen). How are they formed and what happens in the brain to establish a false memory? Do false memories have any adaptive function, that is, do or did they help humans to survive, and if so, how?

It has been long known that it is not difficult to generate false memories in humans. Recent experiments with mice have shed some light on how such memories might be formed. James Gorman, "Scientists Trace Memories of Things That Never Happen," *The New York Times*, July 25, 2013. Scientists at the Riken-M.I.T. Center for Neural Circuit Genetics had published in *Science* the results of a study in which they were able to cause mice to "remember" that they had received an electric shock in a particular physical location, when in fact it had occurred elsewhere.

Using new techniques called optogenetics to locate and label neurons and to make them susceptible to activation by a light transmission, the scientists explain that they were able to identify the neurons in the brains of the mice that were associated with the memory of a particular physical location. The scientists put the mice in another location and then shocked them while activating the neurons associated with the first location. On the following day, the mice exhibited signs of stress when placed in the first location, evidencing that they "remembered" the shock as having been administered when they were in that place. In fact, the mice had been shocked while remembering the first place due to the activation of the particular neurons. Id.

Gorman, in reporting on the publication of these results, quoted Edvard Moser, a neuroscientist at the Norwegian University of Science and Technology, as explaining that memory formation in mammals is of ancient origin so that the function in mice and humans is thought to be very similar, indicating that these studies could provide significant information about memory formation in the human brain. *Id.* This work is consistent with other recent evidence that physical traces of particular memories exist in particular brain cells and that stimulation of those cells can activate the memory.

The studies of memory function were among the earliest efforts to analyze the mind's capability to examine itself, that is, "the relationship between cognition about cognition, and cognition itself... ." Stephen M. Fleming, Raymond J. Dolan and Christopher D. Firth, "Metacognition: computation, biology and function," *Philosophical Transactions of The Royal Society B* (2012) 367, p.1280. In recent years, many different efforts have been undertaken to formulate empirical studies of what has been called "metacognition," including the assessment of individuals' evaluation of their own cognitive performance reflected, for example, in the degree of confidence or uncertainty that they feel about their own judgments or conscious experiences. *See, e.g., id.*, pp.1280-86; Stephen M. Fleming and Raymond J. Dolan, "The neural basis of metacognitive ability," *Philosophical Transactions of The Royal Society B* (2012) 367, pp.1338-49.

Among the questions considered is the extent to which consciousness is related to such second order awareness of first order mental experiences. Other issues involve the extent to which evaluations of one's confidence about observations or predictions can affect decision-making behavior.

SLEEP AND DREAMS

In her 2013 book, neuroscientist Penelope Lewis summarizes the current state of our knowledge about the nature and purposes of sleep, based upon extensive experiments largely using electroencephalography ("EEG"). *The Secret World of Sleep.*

Lewis describes that sleep has four stages. In the first stage, electrical activity in the brain slows down and the wave signals become bigger. In stage two, "sleep spindles" begin to appear. They are brief bursts of "frenetic activity," often from specific areas of the brain. *Id.*, p.8. In stage three, called "slow wave sleep," the brain waves are slow with high amplitudes and several areas of the brain appear to be acting in a coordinated fashion. This stage characterizes deep sleep. *Id.*, pp.8-9. The last stage is REM sleep ("rapid eye movement"), where the brain activity increases and becomes more like that of normal wakefulness, the eyes make rapid darting motions under closed lids and, in contrast, the skeletal muscles are paralyzed. *Id.*, p.10. The four stages generally last about 90 minutes total and recycle throughout the night. However, in the early cycles, the slow wave sleep will consume much of the period and there will be little if any REM sleep; in the last cycles, REM sleep will dominate. Id., pp.8–9.

Interesting questions about brain function also arise with respect to the activity of sleeping. We have been aware for some time of evidence that dreams appear to aid in the memory-storage function by which short-term memories are moved to long-term storage or by which memories are preserved and continued. See Sagan, *The Dragons of Eden*, pp.142–4. It appears that mammals and birds dream but reptiles do not. Id., p.144. On the other hand, all animals, even amoebas, sleep (using a relatively relaxed definition of sleep as a period of inactivity characterized by less than normal responsiveness to external stimuli). Lewis, *The Secret World of Sleep*, pp.1–2.

Curiously, while it is clearly established that people and animals need to sleep and that the body will force sleep when confronted with sufficient sleep deprivation, we do not know why sleep is necessary or what function it performs. Sleep deprivation in humans impairs hand–eye coordination, interferes with the performance of complex functions involving creativity or flexibility and activities like driving a car, alters the senses of taste and smell and other perceptions, impairs learning and distorts emotions. Id., pp.16–21. At the same time, it appears not to adversely affect athletic performance, critical reasoning or IQ. Id., p.15.

The absence of any evidence of a physical function of sleep (like conserving energy or resting muscles) has led many to speculate that sleep performs some other highly useful Darwinian role. At first blush, sleep seems to be a very dangerous activity that could be explained only by some biological imperative. Obviously, sleep deprivation has serious adverse consequences, but why? What is the biological purpose or beneficial function of sleep? The answer to that question has not been found.

The function(s) of sleep

Recent experimentation with babies and young children has indicated that sleep may provide a period in which the brain is able to assimilate and process information free from the constant influx of new sensory data. Alison Gopnik, "Sleeping Like a Baby, Learning at Warp Speed," *The Wall Street Journal*, March 22, 2013. Thus, learning may be enhanced through sleep, including especially dreamless sleep. Babies and young children sleep as much as 12 hours a day. Perhaps that amount of sleep reflects the needs of the exceptionally rapid learning that the young child experiences. Curiously, the use of electronic monitors of brain activity indicate that children had much longer periods of supposedly dreamless sleep than adults, suggesting that such dreamless sleep is important to the learning process. Id.

Of course, young children are not the only mammals that sleep a lot. Look at your average domestic dog. So why do so many animals sleep so much? Among possible explanations, a strong contender is the hypothesis that sleep provided highly useful defensive or protective functions in causing animals to be less active and, therefore, less visible and exposed during periods of the day when the risk of predation was greatest, enabling superior survival leading to superior reproduction. For example, prey may sleep when the relevant predators are most active and most effective (such as sleeping during the day if the principle threats are daytime predators dependent upon light to find the prey or if one's relative advantages for hunting and survival favor darkness). Perhaps, therefore, sleep patterns have changed as the nature of predation risks has changed (for example, through the extinction of the dinosaurs). Sagan, *The Dragons of Eden*, pp.129–38.

Various other strategies have appeared to mitigate the dangers of sleeping. Horses and cows regularly sleep while standing up. Their front legs are able to lock so as to keep them standing even when asleep. If danger approaches, the animals are able to respond quickly without the relatively time-consuming task of getting to their feet from a supine position. However, such animals do seem to need some brief periods of sleep lying down, presumably during which time they obtain the equivalent of REM sleep (discussed below). However, horses and cows appear to need only about 3 hours of sleep in a 24-hour period; they get much of their sleep during the day time, and it is common for some members of a group to stay awake while the others are sleeping.

Giraffes also get much of their sleep standing up (or leaning against a tree), but they also need to lie down for brief periods. Bottle-nose dolphins and many birds have developed a "split-brain" sleep in which one hemisphere at a time will sleep while the other stays awake to respond to threats and maintain certain basic functions necessary for the animal to stay afloat. See, e.g., Lewis, *The Secret World of Sleep*, pp.2–3. Thus, in Darwinian terms, it seems that many animals have developed adaptive behavior or characteristics (one is tempted to use the misleading word "strategies") that reduce the risk of sleeping. Such adaptations do suggest that sleep is important, but they again do not tell us why.

The brain controls the states of waking and sleeping. The arousal system consisting of clusters of neurons in the brainstem send signals upwards to cause wakefulness, using acetylcholine and other neurotransmitters. Id., p.39. For going to sleep, neurons in the preoptic region of the brain send signals downward to the arousal system preventing the "wake up" signals from reaching the rest of the brain. Neurons in the preoptic region also prevent the brain stem from sending signals arising from external stimuli that would interfere with sleep. Id., p.41.

Several apparent consequences of sleep have been identified, at least in theoretical models. One theory states that slow wave sleep is necessary to allow new information stored in the hippocampus to be transferred to the neocortex, a process called consolidation. The transfer is inhibited by acetylcholine, which is present when we are awake. Id., p.47. There is also evidence that slow wave sleep allows the brain to clean up the huge number of synaptic connections that occur during the day as a result of the accumulation of unimportant information and stimuli. The process may allow the synapses to be reset, so as to be ready for the new day. Id., pp.50-51. Perhaps the more important information and things that we want to remember have been encoded more strongly than have the noise and trivia, so a general weakening of the synaptic connections would eliminate the unimportant and leave sufficient, even if weakened, records of the important inputs. Id., pp.51-2.

Sleeping, then, may play a role in strengthening memory. The hippocampus and neocortex seem often to replay memories acquired during the day. The repetition of the activation of a particular set of neurons presumably strengthens the connections among them, embedding the memories. Id., pp.69-70. It is possible that the process is enabled simply because the inflow of new information is suspended during sleep, allowing the consolidation of memories to proceed without competition. It is also possible that there is something about sleep that enables the processing of memories in a way that is not possible when one is awake. Id., p.75. For example, slow wave sleep may enable the coordination of the timing of the replays of memories among disparate regions of the brain, so that they happen at the same time, thereby strengthening the connections. Id., pp.76-7.

(I have focused on the functions of sleep with respect to the operation of the brain, but sleep also appears to facilitate the healing of the body, perhaps by conserving energy for the healing process).

Dreaming

Interestingly, the replay of memories appears to occur not just during dreaming; much of it may be subconscious. Lewis, *The Secret World of Sleep*, pp.69–70. At the same time, dreaming seems to occur during all stages of sleep, not just during REM sleep. Id., p.84. Neural activity in the neocortex can produce the experience of sensory perception, without the involvement of the senses, which we would characterize as dreaming. Id., p.82. There is evidence that dreams originate in and are controlled "by the thinking part of the brain" and that there is some correlation between the complexity of dreams and the cognitive abilities of the person dreaming. Id., pp.86, 88.

So, we ask what is the purpose of dreaming? There is evidence suggesting that dreams constitute a type of rehearsal of potentially threatening events or situations, enabling us to respond more constructively to real threats; that dreams may assist us in coming to terms with traumatic experiences, by reliving them in a less emotionally vulnerable context (chemically speaking); and that dreams help regulate long-term moods. Id., pp.89–90, 131–5. Finally, it appears that sleep and probably dreaming assist in the identification of patterns in prior experiences, the connection of superficially unrelated phenomena and the integration of memories with new possibilities, all promoting creativity and flexibility. Id., pp.103–5.

The question of why we dream certainly illustrates the complexity of the mind and the limitations on what we currently understand. The challenges become even greater when one attempts to fit the theories that have been propounded for humans into the animal kingdom generally. Obviously, humans may have greater or even different needs and may have evolved variations on sleep and dreaming that meet those needs. But, the basic processes presumably must serve important purposes for most if not all animal species.

INFORMATION PROCESSING

Somewhat ironically, the feelings that we enjoy as a result of consciousness (our inner lives) are really the things that we experience most directly. The external world, in contrast, is known to us only indirectly through the interpretations by our brain of signals from our nervous system resulting from external stimuli. Thus, while the most difficult questions may have to do with consciousness and feeling, it is still quite challenging to explain how the brain makes sense out of our sensory perceptions.

As noted in the first chapter, theorizing and model building are how we understand the world. In fact, there is evidence that certain basic models central to our perception and understanding of the world seem to be in some sense built-in or hard-wired (although, the characterization "hard-wired" may be misleading).

In addition, the brain is highly selective in the filtration of sensory stimuli, so that only a small percentage of the information collected by our senses is actually processed by the brain, and how the brain somehow provides or creates interpretations of the limited data that is processed. See also, Jim Dwyer, "The Day When Neurons Go on Trial," *The New York Times*, September 17, 2013 (quoting Joshua R. Sanes, director of the Center for Brain Science at Harvard University: "The brain just throws away 90% of the information [from the retina], because the brain's job is to keep you alive"). Indeed, it seems that the brain primarily communicates with itself, that the interaction of neurons with neurons far exceeds the signals that are received from the senses. See, e.g., Harris, *The End of Faith*, p.209.

How does it work?

Much of the functioning of the brain occurs at an unconscious level. Certainly, the control of daily bodily functioning—like the beating of the heart, the digestion process and breathing—occur as if on an automated program. Other reactions to external stimuli like pain or sudden shock also clearly occur prior to and without any conscious processing. Even though we might say that we jumped because the pot was too hot or a sudden noise was too loud, we generally are aware that our body reacted to the stimulus much faster than our conscious brain was able to register the event. There is obviously a range of activities between completely unconscious and willful. Our heart beats completely independent of our conscious control (at east, for most of us). Our breathing can be controlled but also can, and very often does, proceed on a type of auto-pilot, responding appropriately to changes in the needs of our body.

"Fast and slow"?

Nobel Laureate (in Economics) Daniel Kahneman, approaching the subject from a social science rather than a biological or molecular perspective, distinguishes between two systems or levels of thinking, which he calls System 1 and System 2. *Thinking, Fast and Slow* (2011), pp.20-25.

The System 1 mode of thinking is automatic and quick, "with little or no effort and no sense of voluntary control." *Id.*, p.20. The System 2 mode involves effort and attention and takes some (processing) time. It can be (and often is) disrupted by distractions. System 2 thinking (such as daydreaming or reading for pleasure or walking at a faster than normal pace) normally operates in a relatively low-effort state until the demands of the situation require more. Interestingly, practice and repetition can turn what would be System 2 activities essentially into System 1 activities, as in the case of acquired skills or expertise (like riding a bicycle or swimming or, for the pro, hitting a golf ball). Id., pp.21-2.

The processing of information by the organism takes time. Information from external stimuli must be collected, transmitted and then interpreted. Indeed, the amount of time or the speed at which certain types of stimuli are processed varies significantly among living organisms. It is estimated that the housefly is able to process visual information about four times faster than a human, which is one reason that it is difficult to swat a fly with one's hand. For the fly, the hand movement would appear in slow motion, relative to our perception of it. See, e.g., Ian Johnson, "Q. Why is it so hard to swat a housefly? A. It sees you coming in slow motion," The Independent, 16 September 2013.

The differences in perception of time appear to correlate roughly with the size of the animal. Presumably, the differences reflect the varying abilities to perceive very small, quick movements, such as the flickering of a light (apparently dogs, which are able to perceive much more rapid movements than humans can, do not watch TV because the screen looks to them like rapidly flickering lights, rather than a continuous picture), as well as the distances that the signals need to travel within the animal (e.g., from the retina to the brain and back to the muscles that respond).

Since the transfer of information through electrical spikes and chemical reactions takes time, our brain connections may often make decisions about what we will do before we think that we consciously decide that we will do it. In such circumstances, our conscious brain will effectively create the explanation for our actions (i.e., subscribe causes) after the "decision" to so act and even the action itself has occurred—a kind of *post hoc* justification. See Gazzaniga, Who's In Charge?, pp.76-8.

Computers and Consciousness

> *"AlphaZero seemed to express insight.*
> *It played like no computer ever has,*
> *intuitively and beautifully,*
> *with a romantic, attacking style."*
>
> Steven Strogatz
> NYTimes.com
> December 26, 2018.

Mathematician Roger Penrose has written extensively about the inadequacy of science, as we know it, to explain human consciousness (or consciousness of any living being). See, e.g., Shadows of the Mind (1994); The Emperor's New Mind: Concerning Computers, Minds, and the Laws of Physics (1989).

He argues that consciousness is susceptible to a scientific explanation, but to achieve that explanation, we need new science—science with a different scope than that of the established traditions. These general assertions are very much like Nagel's, but Penrose's writings have not generated the same outrage among scientists. Presumably, the reason is not just that his books are much longer and denser, but that the outline of the solution he presents does not appear to give comfort or support to the religious communities.

In *Shadows of the Mind*, Penrose approaches the subject from the question of whether it is likely that computers can ever achieve consciousness. From the perspective of artificial intelligence, at least as be based upon general purpose computers, the issue is whether consciousness can arise from computational abilities of sufficient complexity and power alone. Penrose sets out four possibilities for the relationships among consciousness, physical processes and computability:

- First, one might assert that all thinking is computation, in which case computers will, at least theoretically, be conscious some day.
- Second, one may claim that consciousness or awareness is a result of physical activity in the brain which can be simulated through computation, but consciousness cannot be achieved through computation alone, i.e., something (like a computer) may be able to behave as if it

were conscious, when seen by an observer, but it would not necessarily experience the state of consciousness.

- Third, one can argue that consciousness results from physical activity of the brain, but such activities cannot be simulated computationally.
- Last, one might conclude that consciousness simply cannot be explained by physical activity of the brain, in which case consciousness will forever be beyond the reach of science. (For example, the views of Gottfried Wilhelm Leibniz, German philosopher and mathematician from the late seventeenth century and contemporary of Newton, fall clearly in this last category.)

Shadows of the Mind, p.12.

The middle two possibilities present the more interesting ground for further analysis. The second case would entertain computers that seem to be conscious, but in a sense that would not really be fully comparable to a living brain. The third case would not include computers that could even simulate consciousness, but would still present the possibility of a scientific explanation of consciousness with the achievement of new scientific developments, although such explanation may be outside of currently known physical laws.

Penrose favors the third case and attempts, first, to demonstrate its correctness and, second, to suggest the future developments that could give a "scientific" explanation. Id., pp.15-6. (Let me note again that the issue being discussed here concerning "computability" is related to, but is not the same question noted above as to whether the "mind" can be explained in terms of its structure, that is, the constituent matter and connections that constitute the brain.) In passing, Penrose also observes that it is possible to have completely deterministic systems that cannot be simulated computationally. Id., p.33. Thus, the conclusion that the mind performs feats that are beyond computation or computability does not itself establish that the human experience is not fully determined by scientific laws nor does it necessarily resolve the question of whether we have free will.

UNDERSTANDING WITHOUT COMPUTATION

One needs to be careful and recognize, in this context, that Penrose clearly distinguishes between mathematical explanations (or proofs) and computation. He offers clear and insightful examples of mathematical reasoning that cannot be reduced to computation. For example, there are well-known mathematical problems that appear not to be susceptible to computational solutions but that do have answers. Id., pp.28-33. Also. various creative, perhaps intuitive, solutions to mathematical questions can be achieved only through geometrical visualizations. See, e.g., id., pp.68-72. "[V]isualization involves an element of appreciation of what it is that is being visualized; that is, it involves *understanding*." Id., p.55.

With a variety of mathematical problems, the obstacle for the computer is that there is no logical basis on which it can determine when (or be instructed by the programmer to conclude that) it can "halt" its investigation. For example, if the mathematical proposition has solutions, *i.e.*, "yes" answers, then the computer can run repeated values until such an answer (say X) is reached and be instructed that the investigation is complete (and that the answer is X). In contrast, if the proposition has no solution, *i.e.*, the answer is "no," then the computer could test examples indefinitely without being able to conclude definitively that the answer is no.

Perhaps the easiest of Penrose's illustrations to understand is the proposition that there is a pair (or triplet or any other number) of even numbers the sum of which is odd. We can readily recognize that the proposition is false. We could even fashion a "proof" based upon a visualization of combinations of even numbers (every even number can be represented by two equal columns of dots of equal length; the addition of two such pairs of columns necessarily results in two longer but still equal-length columns—it cannot possibly result in two columns differing in length by one dot, which would be necessary to have an odd number). Yet, a computer could attempt to test every possible pair of even numbers forever, finding no odd-number results in the process but still without being able to reach a firm conclusion. Id., pp.26-33.

Many other examples could be constructed in which the ability to perform extensive computations very rapidly would be of limited value because an insightful human would realize (predict) what the results of such performance would be without doing the additional computations at all.

Computation without Understanding

The other issue, I think, is that the programming of computers must include various rules and "tips" for the development of tactical strategies. The programmer can instruct the computer to use certain opening moves or to respond to specified situations with particular actions. The modern computer can easily outpace the human mind in testing the consequences of various moves by simulating thousands of possible scenarios. With proper programming, it may be able to learn from experience, again using a vast and easily accessible memory store. As a result, computers today can rival and beat the best of the human chess players.

But, can a computer on its own experience insights or originate hunches, like humans seem readily able to do? Can computers be "intelligent"? See Hofstadter, *Gödel, Escher, Bach*, pp.24-7 (proposing capabilities that are inherent to intelligence, like taking advantage of unexpected opportunities, recognizing similarities between different objects or differences between similar objects, etc.).[14]

Penrose suggests that computers cannot do so because they lack understanding. In his view, computers will lack intelligence, because intelligence requires understanding and understanding requires awareness that, he argues, is not achievable only through computation. See *Shadows of the Mind*,, pp.37-41. All complicated activities "–*if* they have been understood in terms of clear-cut computational rules–are the things that modern computers are good at; but the very understanding that underlies these computational rules is something that is itself beyond computation." Id., p.48. And, "the insights available to human mathematicians–indeed, to anyone who can think logically with understanding and imagination–lie beyond anything that can be formalized as a set of rules." Id., p.72.

Again, a good programmer can incorporate insights and understanding already achieved or learned by the programmer into the software, so that the computer may appear to utilize hunches and shortcuts; but, the understanding is developed by humans and provided to the computer as an input.

This situation may be changing, however.

> "[O]n Dec. 5, 2017, the team had stunned the chess world with its announcement of AlphaZero, a machine-learning algorithm that had mastered not only chess but shogi, or Japanese chess, and Go. The algorithm started with no knowledge of the games beyond their basic rules. It then played against itself millions of times and learned from its mistakes. In a matter of hours, the algorithm became the best player, human or computer, the world has ever seen. ...It played gambits and took risks."

Steven Strogatz, "One Giant Step for a Chess-Playing Machine," NYTimes.com, December 26, 2018.

But, even if the capabilities are as described, there is an important caveat. "What is frustrating about machine learning, however, is that the algorithms can't articulate what they're thinking. We don't know why they work .... AlphaZero gives every appearance of having discovered some important principles about chess, but it can't share that understanding with us. Not yet, at least. As human beings, we want more than answers. We want insight." Id. This situation harkens back to the discussion of prediction versus explanation in the first chapter Knowledge and Understanding. Dramatic progress in artificial intelligence may offer us greater power of prediction with little gain in understanding.[15]

JUST COMPUTATION?

Stephen Pinker (who was director of the Center for Cognitive Neuroscience at MIT) disagrees with Penrose. He asserts that the brain is a combination of computational systems that process information and that the brain and intelligence are the result of the process of natural selection. Thus, the driving forces that shaped the brain were the needs of survival.

However, one consequence of that derivation of the brain and our intelligence may be that we are and will remain utterly incapable of understanding consciousness. Pinker observes that "[s]entience is not a combination of brain events or computational states.... . Our thoroughgoing perplexity about the enigmas of consciousness, self, will, and knowledge may come from a mismatch between the very nature of these problems and the computational apparatus that natural selection has fitted us with." *How the Mind Works*, pp.564, 565. So, in the end, these thoughts put Pinker in the last of Penrose's categories listed above, despite his views on computational systems.

I note that there is a substantial project underway in Europe to attempt to create a simulation of the human brain with microchips. See Tim Requarth, "Bringing a Virtual Brain to Life," *The New York Times*, March 18, 2013. The challenges of the project are awesome, even in the mere physical dimensions. As noted above, the human brain has close to 100 billion neurons existing in interconnected networks that could have up to 100 trillion connections or synapses. A virtual copy of that structure in microchips would be a staggeringly large supercomputer. There are many scientific criticisms of the project that go beyond physical feasibility, some of the more fundamental arising out of the basic lack of knowledge we have about the way in which the brain operates. Id.

However, the founder of the project, Henry Markram, reportedly asserted that "[i]f I build in enough biological detail, ...it would behave like a real brain." *Id.* Maybe, but one needs to wonder when we will ever know enough of the biological detail of the human brain and, then, whether it is even theoretically possible to build such detail into a computer.

The "Turing test"

In 1950, British mathematician and one of the founders of what we now call computer science (and artificial intelligence or "AI") Alan Turing, focused on the question of whether it would be possible ever to construct a machine (computer) that was able to think. He proposed an experiment that has come to be called the "Turing test." A panel of suitable judges would pose questions by electronic communication to two individuals, one of which would be a computer and the other a live person. The question would be whether the judges could ascertain which was which based upon five minutes of questioning. Turing predicted that by the year 2000 (fifty years later), the computer would be able to convince 30% of the judges that it was the human. The prediction did not come true; but, in 2008, the computer came close.

Brian Christian, a poet with computer science training, sought and was selected to be one of the human contestants in a 2009 "Turing test" competition in Brighton, England. He wrote a book about his experience. *The Most Human Human: What Artificial Intelligence Teaches About Being Alive* (2011), p.4. Christian describes the training in which he engaged in order to prepare himself to be perceived as human. Of most interest here, however, is what he learned about the strategies of the computer programmers designing computers to participate in the competition. The programmers strategized about what kinds of answers would fool the judges. So, for example, some programmed computers to make spelling and grammatical mistakes. Others worried about what degree of knowledge about particular subjects might be perceived by a judge as "machine like" or too good to be human. Other human characteristics identified were the appearance of *non sequiturs* and the use of jokes. *Id.*, pp.16–30. Ironically, perhaps, Christian found himself also strategizing about how he could seem most human.

The point here is that in all cases, the strategizing was being done by human beings, either for themselves as human competitors or for the machines that they were programming. The computers were not doing it.

(By the way, Christian reports the results of the 2009 competition. He won the award for the "most human" participant. More importantly, the computer contestants lost badly. As Christian reports: "Doug, Dave, and Olga were my comrades far more than they were my foes, and together we'd avenged the mistakes of 2008 in dramatic fashion. 2008's confederates had given up a total of five votes to the computers, and almost allowed one to hit Turing's 30% mark, making history. But between us, we hadn't permitted a *single* vote to go the machines' way. 2008 was a nail-biter; 2009 was a rout." Id., p.261. The machines continued to improve over time, but the human competitors improved directly in response to the challenge. Id., pp.264-5.)

In the end, the interesting philosophical/scientific question is whether a convincing simulation of the human mind would be "conscious" or, even, intelligent. It appears that Turing would have categorized a computer that could not be distinguished from a human based upon conversation as intelligent. But, as conscious? Of course, one could treat these questions as definitional. One can imagine definitions of intelligence that computers could meet to varying degrees, but do we have a grasp on a definition of consciousness? The fact that we may not be able adequately to define consciousness does not mean that it does not exist, that is, that it does not have some real meaning in the world in which we exist. (Although, the lack of an adequate definition does inhibit clear discussion and effective scientific inquiry.)

Something More?

The scientific community in the nineteenth century sought to solve this problem with the notion of a "vital force" in living things that was somehow outside of or not captured by chemistry and physics—an idea reflected in a theory referred to as "vitalism." *See, e.g.,* Seung, *Connectome*, p.240. The scientific community in the twentieth century has been rather hostile to such ideas; sometimes very hostile. But, one need not embrace "vitalism" to conclude that there is something more than what science as we know it can explain about the human mind.

Seung summarizes the debate as follows:

> *"Some philosophers believe that it's fundamentally impossible to simulate consciousness on a computer. They say that a simulation of water, no matter how accurate, isn't actually wet. Similarly, your simulation might seem adequate to your friends and family, might even proclaim its [own] satisfaction, while still*

> lacking the subjective experiences that we call consciousness. That may not seem bad, but it certainly doesn't sound like a route to immortality."

Id., p.262.

This summary combines a couple of different things that can be usefully separated. First, if there is not an imaginable way to determine whether the simulation is actually conscious or not, then how do we say it is not? If consciousness is solely a subjective experience, with no objective method of identification, then can it ever be more than a purely individual reality—I "know" that I have it; you think that you have it; but neither of us can know for certain about the other. In that case, how certain can we really be about ourselves?

A separate notion is that of immortality. Would a completely convincing simulation of our own mind give us a type of immortality that we would find satisfying or even meaningful? If you, like I, would answer that question in the negative, then our response is another piece of evidence that consciousness is something more. By the way, Seung's personal view is that "a sufficiently accurate brain simulation would be conscious. The real difficulty is not philosophical but practical: Can that level of accuracy really be achieved?" Id., p.263.

Penrose, in considering the scenario in which computers are deemed conscious, asserts that a likely consequence of that position is that computers should then come to have rights and responsibilities, like humans do. *Shadows of the Mind*, pp.35-7. I think that he fails to recognize that another potential conclusion would be that there is no good reason why people, if ultimately not more than sophisticated computing machines, should themselves have rights and responsibilities.

Beam me up Scotty? (again)

> *This might be an appropriate place for a brief digression on a science fiction topic: the transportation of things through space at the speed of light. Since all examples of each type of fundamental particle are identical (all electrons, all protons, all neutrons—with caveats not pertinent here), one could theoretically disassemble an object in one place; communicate the precise, entire design instructions to another place; and assemble the object in the distant location from electrons, protons and neutrons*

that are found in that location but that are identical in all respects to the constituent parts of the original object far, far away. One would observe the disappearance of the object in the original place and the reappearance of the object in the new place. The two iterations of the object would be indistinguishable except for the location. Have we not transported the object great distances in the blink of an eye? But, is it the same object? The argument is that if it is truly identical and indistinguishable, then how can one say that it is not the same object (now in a different place)? However, suppose that the first object were not disassembled (and it were possible to copy it without disturbing it), then we would have two identical objects in two different places. Would we say that it is the same object in both places at once? For a table or chair, this form of transport may seem quite satisfactory. For a work of art, would the mechanical reassembly of the work produce an identical, indistinguishable copy? I suppose that many scientists would say "of course", but I am not sure that all artists would. The next question is would this process work for a living organism? I suppose it may, if one were able to duplicate the "spark" of life in the reassembled organism. The next, and last, question is would it work for a person and his consciousness?

THE SOURCE OF CONSCIOUSNESS?

> "[C]onsciousness rises along the human lineage
> to the point where it becomes conscious of itself ... .
> [Humans are] the only species known to possess
> reflective consciousness."
>
> John Hands
> Cosmosapiens, p.430.

The discussion above provides a framework for attempting to explain learning and memory and the development of physical skills, such as riding a bicycle, ice-skating and skiing. But, it certainly does not go very far toward explaining self-awareness or consciousness or creative intelligence, in other words, the mind.

BEGINNINGS

Reasoning and self-awareness in other animals

Despite the enormous gap in functionality between the higher primates and modern man; there do seem to be some examples of intermediate stages of mental development. Sociologist and physician Nicholas Christakis argues that we have seriously underestimated the presence of intelligence and self-awareness in other animals. Blueprints, p.283.[16]

Work at the University of Auckland has revealed some human-like capabilities in the New Caledonian crows. The research utilizes the concept of "cumulative technological evolution," the meaning of which is pretty clear. It is assumed that such evolution is one consequence of the uses of theories by humans. The essential characteristics for such evolution to occur have been identified as consisting of "diversification of tool design", "cumulative change" and accurate "social transmission." Gavin R. Hunt and Russell D. Gray, "Diversification and cumulative evolution in New Caledonian crow tool manufacture," *Proceedings of the Royal Society*(London), vol. 270, no. 1517 (22 April 2003), pp.867-74.

In one survey, they claim to have documented cumulative evolution in tools used by the crows. *Id*. It has been subsequently reported that scientists at the University of Auckland found that the New Caledonian crows would use several tools in succession to obtain food and would fashion tools better to extract food in inaccessible nooks. Rebecca Morelle, "Clever New Caledonian crows can use three tools," BBC News, 20 April 2010. This behavior indicates problem-solving that involves some degree of reasoning, using models of the physical environment that the crows confront. See also, Alex A.S. Weir, Jackie Chappell and Alex Kacelnik, "Shaping of Hooks in New Caledonian Crows," Science Magazine, February 1, 2009, (reporting on experiments performed with two captive crows residing in the laboratory of the Department of Zoology at Oxford University).

In 2012, Auckland researchers reported on further experimental studies that they concluded demonstrated that the New Caledonian crows are capable of reasoning about "hidden causal agents," something thought to be unique to humans, because of how they reacted to events where the human agent was visible to them and when the agent was not visible. See Alex H. Taylor, Rachael Miller and Russell D. Gray, "New Caledonian crows reason about hidden causal agents," Proceedings of the National Academy of Science of the United States of America, September 17, 2012. The capabilities of these crows, as reported, are quite impressive and surprising. One might expect behavior of this type from the higher primates, but one would probably not think that crows could achieve it.

It is not currently known how much of such behavior might be found elsewhere in the animal kingdom. (Although, Nicholas Christakis argues that it is quite common. Blueprints, pp.281-331.) One obstacle to better understanding, obviously, is the difficulty of designing appropriate experiments to detect such behavior. There are challenges both to achieving an effective conceptual design for the experiment as well as to the implementation of the necessary controls actually to conduct it. Nonetheless, it appears that the more we look, the more examples we find. See, e.g., Michelle Douglass, "Incredible tool use in the animal kingdom," BBC Earth, 10 November 2014; Nicholas Bakalar, "Hawaiian Crows Join Tool-Users Club," The New York Times, September 19, 2016.

It is likely that fresh looks at the behavior of primates and birds, free of limiting prejudices, will continue to reveal much more complex and surprisingly human characteristics. Perhaps those discoveries will lead to plausible theories of how what we might loosely call intelligence has arisen through evolution or, at least, a description of the steps or stages by which it came about. However, tool use and an elemental understanding of the concept of causation are mere first steps toward what we think of as consciousness.

Nature programs often feature depictions of animals that appear to be displaying recognizable emotions and engaging in acts that appear to require imagination. Studies have shown evidence of other human-like perceptions in non-primates. See, e.g., Carl Zimmer, "Elephants Get the Point of Pointing, Study Shows," *The New York Times*, October 10, 2013 (describing a study by Richard W. Byrne, of the University of St. Andrews, published in the journal *Current Biology*, finding that elephants appear to understand the significance of pointing); University of Cambridge, "Elephants' 'body awareness' adds to increasing evidence of their intelligence," *Research Bulletin*, 21 April 2017 (describing a study by Josh Plotnik and Rachel Dale indicating elephants are aware of the distinction between their bodies and other objects in their environment).

Neuroscience professor Stuart Firestein describes what is referred to as the "mirror illusion," recognition that the reflection in the mirror is just oneself. *Ignorance*, pp.101–106. Reactions to one's reflection in a mirror range from complete disregard, to apparent perception of another, to recognition that it is oneself. He explains that humans are able to recognize the illusion by 18 months of age. Studies in the late 1970s showed chimpanzees and other great apes were able to do so. More recent studies have indicated that such recognition can also be accomplished by dolphins and elephants. *Id.* Self-awareness seems to be an important element of consciousness and something that appears to be found only among a few other species, like the great apes.

We already discussed in the chapter Darwinism the recent discovery in a cave system in South Africa of the remains of what may be a new species called *Homo naledi*. The bones of some 15 individuals were found in a chamber deep underground. The scientists who discovered the remains speculate that the location of the bones of individuals of different ages and of both sexes were placed there in some type of burial ceremony. There were no indications of the remains having been washed there by floodwaters or dragged there by animals. See, e.g., Pallab Ghosh, "New human-like species discovered in S Africa," *The BBC News*, 10 September 2015. The bones are dated from 226,00 to 335,000 years ago. Michael Greshko, "Did This Mysterious Ape-Human Once Live Alongside Our Ancestors?" *National Geographic*, May 9, 2017. But, this find may be evidence of a form of self-awareness, reflected in the practices concerning death, by a different species of genus *Homo*.

Of course, the burial of the dead, if that is what it was, may simply reflect a practical solution to the needs arising from living in settled communities (to avoid the odors of decomposition, the attraction of scavengers, and so on), which needs do not arise for nomadic species. If other primates were confronted with the same problems, they might have developed the same solution. Frans de Wall, "Who Apes Whom?," *The New York Times*, September 15, 2015.

Emergence of artistic expression

A whole other, and important, step forward for humans was the use of symbolic representation, the beginnings of artistic expression. It may be claimed that "[t]he greatest innovation of humankind was neither the stone tool nor the steel sword, but the invention of symbolic expression by the first artists." Chip Walter, "The First Artists," *National Geographic*, "The Book of Firsts," January 2015, p.33.

We are familiar with the claim that the cave paintings found in the Cave of Chauvet-Pont-d'Arc in France, dated to 35,000 years ago, are the earliest known examples of artistic creation. The earliest known work of ritual art, the wooden Shigir Idol, is now dated as 12,500 years old.

> "Skeptics argued that the statue's complex iconography was beyond the reach of the hunter-gatherer societies at the time; unlike contemporaneous works from Europe and Asia featuring straightforward depictions of animals and hunt scenes, the Shigir Idol is decorated with symbols and abstractions. ...'The idol was carved during an era of great climate change, when early forests were spreading across a warmer late glacial to post-glacial Eurasia,' Dr. Terberger [an author of the study] said. 'The landscape changed, and the art – figurative designs and naturalistic animals painted in caves and carved in rock – did, too, perhaps as a way to help people come to grips with the challenging environments they encountered.'"

Frank Lidz, "How the World's Oldest Wooden Sculpture Is Reshaping Prehistory," NYTimes.com, March 22, 2021.

Some symbolic representations seem to have been created by the Neanderthals. "Neanderthals painted caves in what is now Spain before their cousins, *Homo sapiens*, even arrived in Europe... . The finding [published in Science] suggests that the extinct hominids, once assumed to be intellectually inferior to humans, may have been artists with complex beliefs. " Emma Marris, "Neanderthal artists made oldest-known cave paintings," *Nature.com*, 22 February 2018. Examples of shell beads, engraved ocher and materials apparently used to create ocher pigments (perhaps for body painting) have been found that date to 100,000 years ago. Other examples of efforts at decorative or individualistic expressions have been found with dating to 65,000 and 75,000 years ago.

Full-blown artistic expression by *Homo sapiens* now appears to have emerged about 40,000 years ago:

> "[o]n the wall of a cave deep in the jungles of Borneo, there is an image of a thick-bodied, spindly-legged animal, drawn in reddish ocher. It may be a crude image. But it also is more than 40,000 years old ... making this the oldest figurative art in the world. Until now, the oldest known human-made figures were ivory sculptures found in Germany. Scientists have estimated that those figurines ... were at most 40,000 years old."

Carl Zimmer, "In Cave in Borneo Jungle, Scientists Find Oldest Figurative Painting in the World," NYTimes.com, November 7, 2018. *See also,* Michael Marshall, "The Four Biggest New Things We Learned About Human Evolution In 2018," *Forbes*, December 20, 2018.

More recently, on Sulawesi, a much smaller island just to the east of Borneo, more elaborate but similarly old paintings have been found:

> "Cave art depicting human-animal hybrid figures hunting warty pigs and dwarf buffaloes has been dated to nearly 44,000 years old, making it the earliest known cave art by our species. The artwork in Indonesia is nearly twice as old as any previous hunting scene and provides unprecedented insights into the earliest storytelling and the emergence of modern human cognition."

Hannah Devlin, "Earliest known cave art by modern humans found in Indonesia," *The Guardian*, 11 December 2019.

"So, we find the apparent emergence of artistic expression at roughly the same time (40,000 years ago) appearing at opposite ends of the world." *Id. See also,* "Sulawesi art: Animal painting found in cave is 44,000 years old," *BBC News*, 12 December 2019.

Interestingly, these recent finds in Spain and Indonesia were dated using a new technique. When water seeps into a cave, it can contain uranium. Since uranium has an established, uniform half-life, researchers are able to date the water "stain" or "cave popcorn" on the cave wall. If the stain or growth is over the cave drawing, then the drawing must have occurred prior to the date of the deposit, in this case, some 40,000 or more years ago. In some cases, researchers may be lucky enough to find one deposit under the drawing and one over the drawing, enabling them to bracket the date of the drawing. *See, e.g., id.*

One early December afternoon in 2014, I and visited the newly renovated Museum of Archeology and Anthropology at Cambridge University, on Downing Street. Like many museums, the institution is a testament to the incredible power of the human spirit. The urge to find solutions to the challenges of daily life through the development of tools and to enhance the experience of that life through decoration is a pervasive, maybe universal, manifestation of that spirit. The artifacts display impressive creativity and innovativeness. Clearly, the creators of the new tools and of the beautiful decorative designs were able to visualize something, imagine a means of achieving the vision and, ultimately, implement the plan to produce the results.

Two important questions arise from the evidence that has been compiled on the history of art. First, what caused or enabled such early artistic expressions—*e.g.*, was it a change in the brains of our evolving predecessors? Second, why does the record seem to be so episodic or sporadic up until about 35,000 years ago?

Then?

Attempts have been made to explain the brain in terms of evolution, and there is evidence of the evolutionary development of intelligence. In oversimplistic terms, changes in the complexity of the organism and in the sophistication of its relationships with its environment may have made additional mental functionality more useful, enabling the evolution of additional brain structures that resulted in more complicated and adaptable brains.

But, there is still the question of how (and when) the unique characteristics of human consciousness arose.

- Are they the result of the relative handful of genetic differences between us and the other great apes?
- If so, how can chemical reactions resulting from those different genes cause such incredible capacities and propensities?
- Separately, even if the answer is "in the genes", are these capabilities the product of gradual adaptation operating under the pressures of natural selection?
- If so, where are the intermediate stages?

- And, even if one could theorize about how they could evolve in small, gradual steps, how long would that take?
- Alternatively, are they the result of some dramatic leap? In fact, could they even be the ultimate example of irreducible complexity?

Admittedly, one can make the case that these capabilities have survival value in today's world, but, as philosopher Thomas Nagel has argued, humans display cognitive capabilities (like rational thought) that far exceed anything that could have been considered adaptive for primitive man. Nonetheless, the capabilities that we see as distinguishing modern humans seem to have been present in *Homo sapiens* very early, perhaps even from the beginning of the species. See David Berlinski, *The Devil's Delusion: Atheism and Its Scientific Pretensions* (2009), pp.157–64.

Suppose one took a child from a primitive, isolated community, whose ancestors never had any contact with modern civilization, and bring him or her into modern society. Presumably, the child would develop and eventually indulge and exploit the very human characteristics that one finds among modern, civilized man. There seems to be every reason to suppose that the same would be true if one could transport a very young child from 10,000 years ago to England and enroll him in public school. Berlinski attributes the foregoing observation to Alfred Russel Wallace in 1869.[17] If this speculation is correct, then when could the supposed gradual evolution of these capabilities have taken place?

The main migration of *Homo sapiens* from Africa was not until about 70,000 years ago. Sometime over the next 20,000 years (50,000 to 70,000 years ago), there occurred what Professor Harari calls the "Cognitive Revolution". Language and abstract thinking (religion, art, myths, etc.) arose, bringing to the world what we now call "culture". *Sapiens*, pp.20–9.

Simon Baron-Cohen, professor of developmental psychopathology at the University of Cambridge, also identifies a "cognitive revolution" that he says occurred between 100,000 to 70,000 years ago.*The Pattern Seekers: How Autism Drives Human Invention* (2020). Baron-Cohen hypothesizes that an important change occurred in the human mind's capabilities: a unique improvement in the ability to think systematically and logically in order to discern and manipulate causal patterns. That change, he says, underlies humans' dramatic feats of innovation over the next 70,000 years. He asserts that this ability did not exist in earlier examples of the genus *Homo* and does not exist in animals, which display only associative learning capabilities. And, he finds evidence of a genetic component.

Did the important characteristics of modern man appear in the earliest *Homo sapiens* some 200,000 years ago, or did they emerge over the period between then and 70,000 years ago? Or, did they appear only in the period between 70,000 and 40,000 years ago? Or, even, only some 10,000 years ago?

We simply do not—and may never—know. But, "it is almost certainly misguided to think we could ever specify a single, more recent point in time when Homo sapiens 'emerged'— that is , when all the various elements of the modern human condition converged, definitively, in some stupendous moment of creation." Graeber and Wengrow, *The Dawn of Everything*, p.83.

Possible triggers

In all events, what happened to enable or provoke these changes?

Climate change?

Environmental changes may have played significant roles at times in this process. There is evidence that environmental changes today are "causing" the more rapid evolution of the brains of certain animals. See Carl Zimmer, "As Humans Change Landscape, Brains of Some Animals Change, Too," *The New York Times*, August 22, 2013 (describing a study by Dr. Emile C. Snell-Rood of a handful of species of small mammals in which it was found that brain sizes were larger for individuals living in suburban and urban environment than for those living in rural habitats and that brain size increased over time with increased agriculture and other human development of the native habitats of the animals).

Appearance of empathy?

One possibility is the emergence of empathy, especially cognitive empathy through which one begins to imagine the point of view or situation of another and see oneself through the eyes of others. "[W]e each have a concept of 'my self' only because at a certain point we learn to project onto ourselves the idea of being human as an additional feature that evolution has led us to develop during the course of millennia in order to engage with other members of our group... ." Rovelli, *The Order of Time*, p.177. Does the appearance of empathy coincide with the beginnings of artistic expression, like the creation of jewelry and self decoration some 40,000 years ago? Does it have a genetic basis? See Baron-Cohen, "The Pattern Seekers," Lecture, *The Royal Institution*, 07 January 2021 (*YouTube*).

Population density?

Another possibility is that consciousness or mind appears only when a sufficient level of complexity is achieved. Maybe the brain of our predecessors finally reached the requisite complexity some 50,000 years ago and then made another leap about 12,000 years ago, when agriculture began. (When did the urge to express oneself emerge?) But, the lack of continuity in the historical record suggests that there is more involved. Maybe, instead, a certain level of population density, and the resulting interactions among the individual members of the species, was necessary. If so, then variations in density could answer the questions posed above concerning the appearances of artistic expression, i.e., the relative brief appearances of artistic expression followed by long periods without. See Chip Walter, "The First Artists," *National Geographic*, "The Book of Firsts", January 2015, pp.41-2. Indeed, "[h]uman thought is inherently dialogic. . ..Humans were only fully self-conscious when arguing with one another, trying to sway each other's views, or working out a common problem." Graeber and Wengrow, *The Dawn of Everything*, p.94.[18] So, some concentration of population would seem to be at least a necessary, if not sufficient, condition.

Process through time?

One thing that seems indisputable is that consciousness is not a state of affairs—it does not exist in a static context. One cannot take a "photo" of it (not because of the technology of photography, but because consciousness cannot be frozen in time). It must be thought of as a process, as something that exists in and through time. See, e.g., Deacon, *Incomplete Nature*, p.175. In that realization, there may be a clue as to why science has struggled to explain consciousness. Our scientific technology generally captures "states" rather than processes. We simply do not have the technological capability to capture the dynamic process that occurs.[19]

Emergence?

Consciousness may be an emergent phenomenon. It may have just appeared when a sufficient degree of complexity in the human brain was reached. (We have already discussed the issues concerning emergence above in the context of chemistry and particle physics, including the question of whether the elements of the emergent state are physically contained in the matter existing before the emergence occurs. The illustrations from chemistry are interesting, but incomplete.) Terrence Deacon cites an example of emergence used by the Nobel-prize winning neuroscientist Roger Sperry in 1980: the wheel. *Incomplete Nature*, p.160–61. The molecules constituting the material from which the wheel is made are not changed by the formation of the material into a wheel; however, the structure or organization created capabilities, such as the ability to roll, that exist only for the structured collective, not for the individual parts nor for other forms of organizations of those parts (such as a plank). So, what is the source of this emergent phenomenon? Of course, this example is "tainted" somewhat by the fact that the construction of the particular structure (a wheel) is generally carried out by design, as a result of the intention of someone to realize a perceived potentiality of the material. See *id.* Arguably, in the example of the wheel, the source of the characteristics that arise from the organization of the pieces is the imagination or intention of the designer/creator of the wheel. Obviously, that is not considered by most scientists to be a satisfactory answer with respect to naturally occurring phenomena. The answer that the "source" is the function or purpose fairs not much better.

Something else?

In contrast, Penrose, twenty years ago, observed that the traditional view of the organization of the brain and the operation of neurons and their connections was essentially a classical model, resembling a computer. *Shadows of the Mind*, pp.352–5. If, therefore, consciousness cannot be achieved through computation, that model must be inadequate. If so, a new model is needed. Therefore, rather than being an emergent phenomenon, consciousness might result from the ability of the organization of the brain to utilize non-computable action that exists in physical laws (although not yet discovered by us in our conscious, analytical capacity), which ability does not appear in matter not so organized. *Id.*, pp.216–7.

Mind & Cosmos

The introduction to this chapter used the controversy over the recently expressed views of Thomas Nagel to set up the questions to be addressed. It is time to discuss Nagel's main arguments.

He notes, as we have, the lack of apparent progress in finding a physical connection between the physics and chemistry of the brain and the characteristics that we refer to as mind. So Nagel assumes for his analysis that such a connection does not exist. Making the (initial) assumption that there is no direct connection between the mind and the physical structure of the brain is not objectionable. Unless and until it can be proven that there is such a connection, it is clearly a useful line of inquiry to try to ascertain what the consequences are of that assumption being true. (His analysis of the logical options, given the assumption, is persuasive.)

On the assumption that such a connection does not exist, he concludes that it must be the case that either:

- aspects that are mental (not physical, like particles and energy), which now feature in what we call mind, must have existed all along as part of the materials that make up the Universe, only to assume a significant or detectable role with the evolution of complex, sophisticated neurological structures in the higher beings; or
- the mental qualities must have appeared as emergent phenomena when a sufficient state of sophistication in the physical structures constituting the brain was achieved.

Mind & Cosmos, pp.54–5.

The analysis

Nagel's first heresy appears to be in the subsequent assertion that reliance on emergent phenomena is no explanation unless a case can be made as to why such phenomena may be likely to emerge from the underlying physical elements when such elements are organized in certain ways. Id., pp.55–6, 60. This viewpoint brings us back to some of the issues addressed in the opening chapters about the nature and requirements of theories and explanations.

I would note, however, that the issue that Nagel raises is not that encompassed in the contrast between explanation and prediction. Nagel would be satisfied, presumably, with a prediction of the emergence of consciousness and intelligence out of the organization of the physical elements constituting the brain, even if the basis for that prediction provided no real insight or understanding (or explanation).

The next heresy is Nagel's argument that full rationality, as he defines it, does not appear to be adaptive; that is, it does not seem to be a characteristic that could be expected to "evolve" through natural selection. Plausible arguments might be made that some aspects of consciousness and even of cognition have survival value, particularly with respect to the flexibility and capacity for responding to changing circumstances that such qualities presumably do provide. *Id.*, pp.72–3. These arguments might be extended to the capacity to generate scientific theories that generate empirical predictions, even with complex mathematics. *Id.*, pp.76–7. But, it is "the distinction between appearance and reality, and the existence of objective factual or practical truth that goes beyond what perception, appetite, and emotion tell us, the ability of creatures like us to arrive at such truth, or even to think about it" that is difficult to incorporate into a neo-Darwinian framework. *Id.*, p.73.

Nagel asserts that "[t]he distinctive thing about reason is that it connects us with the truth directly. Perception connects us with the truth only indirectly." *Id.*, p.82. He continues "[s]omething has happened that has gotten our minds into immediate contact with the rational order of the world, or at least with the basic elements of that order, which can be used to reach a great deal more." *Id.*, p.83. The argument now becomes reminiscent of the debates about the reality or independent existence of mathematics, discussed above.

Let me mention one additional argument that Nagel suggests. It is that rationality seems to require the functioning of the whole conscious mind, something greater than the sum of the individual parts. *Id.*, p.87. Nagel acknowledges that one option is to assign consciousness and rationality to the category of highly fortunate accidents. One could conclude that they are simply coincidental byproducts of other physical developments that were naturally selected over the three million years of the evolution of life on this planet. The fact that they may seem to be of such great significance to us in the scheme of things is of no real consequence. *See id.*, p.88.

Nagel properly relegates this position to the category containing similar arguments that accept the reduction of explanation to the existence of a "brute fact." These are not very satisfying as explanations; not much better than the arguments that purport to find an explanation in the fact that we could only be asking the question because such and such is so—if there were no Universe, we would not be asking why there is a Universe; if there were no life, we would not be asking why there is life; if there were no intelligence, we would not be asking why there is intelligence. See *id.*, p.95n.

Nagel declares his preference for a teleological explanation of the causation of consciousness and rationality. *Id.*, pp.91-5. He postulates the existence of some additional Laws of Nature that create a bias or tendency for evolution to move toward certain goals, through the promotion of certain values or qualities. Nagel cannot articulate those Laws and can only speculate about the values. See *id.*, pp.97–126. The affirmative case is left for future scientists and philosophers to make.

THE REACTIONS

The book has reportedly generated a minor uproar in the scientific community.

In Jennifer Schuessler's "An Author Attracts Unlikely Allies," *The New York Times*, February 6, 2013, the author (or editor) thought it more interesting to cast the article in terms of the "allies" rather than the "enemies" attracted. Nagel is a self-proclaimed atheist. So the "news" is that creationists and conservative commentators have praised Nagel's recently expressed views, because the views appear to conflict with neo-Darwinism. The article, however, goes on to describe the "deeply skeptical to scorching" reactions from parts of the scientific community. Various book reviews have been similar, leading the Guardian to name Nagel's book as "the 'most despised science book of 2012.'" *Id.*

What I find to be of interest, though hardly surprising (and, therefore, probably not "news"), is the tone of some of the critiques by scientists. The critical scientists generally do not dispute the assertion that the physical sciences have been unable to explain things like consciousness (although, some neo-Darwinians seem to argue that evolution does). Similarly, there is no consensus in the scientific community even that the established sciences will someday be able to do so. The outrage seems to be over (i) Nagel's explicit criticism of neo-Darwinism (as clearly reflected in the subtitle of the book) and (ii) his proposed answer in the form of a new scientific approach reflecting a "natural teleology" with a "cosmic predisposition" or natural tendency toward the emergence of consciousness and intelligent beings. Darwinism has long been the sacred cow of the anti-religion scientist. Public criticisms of the theory are viewed as a form of treason because of the aid such criticisms allegedly provides to the enemy, *i.e.*, to the creationists. Indeed, some of the most emotional criticism seems to be based almost exclusively on that feature, not the substance, of Nagel's arguments.

The New York Times does discuss the attacks by the critics, but only after highlighting the support of the "allies," which is the emphasis. The column notes that "even one of the more tolerant responses" is titled "Thomas Nagel is not Crazy." *Id.* The title of the referenced review by Malcolm Thorndike Nicholson may seem to offer only faint praise, but Nicholson clearly believes that the book is noteworthy and worth reading. He challenges philosophers like Daniel Dennett who assert that there is no more to consciousness than physical phenomena that can be explained by science. "To us it really does *feel* as if there is something 'it-is-like' to be conscious. Besides their strange account of consciousness, Nagel's opponents also face the classic problem of how something physical like a brain can produce something like a mind." *Prospect*, February 7, 2013. (Nicholson does conclude, however, that "his positive arguments for a natural teleology end up looking every bit as intuitively implausible as a description of reality that leaves out consciousness.")

The reference to the Guardian review is even more misleading. Mark Vernon titles his column "The Most Despised Science Book of 2012 is …worth reading" and the byline is: "Philosophers that break scientific taboos, such as Thomas Nagel with *Mind and Cosmos*, risk much, but we need them." *The Guardian*, 4 January 2013. Vernon has a theme. His awards of the Most Despised Science Book of the Year in 2010 and 2011 also went to books that challenged the correctness of current Darwinian theory. (The award in 2010 went to Jerry Fodor and Massimo Piattelli-Palmarini for *What Darwin Got Wrong*, already discussed above. His runner-up for 2012 was Rupert Sheldrake's *The Science Delusion*, which critiques scientific "dogmatism" and has been cited several times in prior chapters.)

As for Nagel's book, Vernon observes that "[d]isparagement is particularly unfair … because the book is a model of carefulness, sobriety and reason." *Id.* He observes that Nagel's apparent embrace of teleological arguments is what inflames the evolutionary biologists. It is the suggestion that evolution may have a tendency toward certain outcomes, like consciousness, that is the "taboo."

A CONCLUDING NOTE

Nagel's arguments are, in fact, neither radical nor confrontational. To a large extent, he explicitly sets forth his personal assumptions and opinions on certain issues concerning the requirements of an adequate explanation of human consciousness, assumptions and opinions with which one is free to differ. Yet, those assumptions and opinions do reflect an apparent rejection of the prevailing orthodoxy, which has been met with cries of outrage from the purported "true believers."

Of course, we do need to examine the assumption upon which Nagel based his analysis, that is, that there is no direct connection between the physical characteristics of the brain and the experience of consciousness. We have discussed the structure of the brain. What about the mind? One question that arises when one considers the structure of the brain is the extent to which the particular construction affects the nature of the mind. "Most of us are convinced that minds differ because brains differ"; but, as neuroscientist Sebastian Seung acknowledges; "so far, however, there is little proof." *Connectome*, p.21. If such a connection could be demonstrated (or discovered), then we could perhaps hope to develop a scientific theory of consciousness within the structure of the current physical sciences, tying intelligence and consciousness back to underlying relationships explained by chemistry and physics.

Still, one would have hoped that the response of the scientific establishment would have been a vigorous engagement with the arguments and an open-minded debate of the assumptions and opinions involved, rather than engage in what is not much more than name-calling.

One can still hope.

Endnotes

[1] Until the seventeenth century, in a tradition beginning at least with Aristotle and culminating with Thomas Acquinas in the thirteenth century, the unique qualities of humankind were identified with the soul. Makari, *Soul Machine*, pp.4–7.

[2] As described later, in the discussion of the studies of New Caledonian crows, there may be more than the mere "beginnings" found in some other species.

[3] With Thomas Hobbes, followed by Descartes, the seat of human qualities became identified as the "mind." Makari, *Soul Machine*, pp.19–24, 28. To Descartes, the mind was immaterial, in contrast with the body; while Hobbes sought to extend the mechanistic explanations of the world in terms of matter and motion to every aspect of man, including the mind. The material mind certainly seems more susceptible to scientific investigation than does the soul; but, nonetheless, a fundamental issue continues to persist.

[4] The Human Brain Project run by Henry Markram at the Swiss Federal Institute of Technology with the very ambitious goal of creating a computer simulation of a human brain received in January 2013 a European Union grant for 10 years of up to 1 billion euro. Tim Requarth, "Bringing a Virtual Brain to Life," *The New York Times*, March 18, 2013. However, the European project is encountering criticism from within the scientific community. Hundreds of neuroscientists from around the world signed an open letter to the European Commission on 7 July 2014 taking issue with the scope and focus of the project. See Gary Marcus, "The Trouble With Brain Science," *The New York Times*, July 11, 2014 (the most important shortcomings in our knowledge of the brain concern how it works, so what we need are new and better theories of the brain, not just more data).

[5] Imaging technology is highly significant for medical as well as research purposes. Old fashioned two dimensional x-rays were effective in disclosing breaks in bones, various types of cancerous growths and metal objects that had been ingested or embedded as result of a gunshot or an explosion; however, the technology had obvious investigatory limitations, in addition to the health concerns of excess exposure. Computed Tomography (CT scans) uses ionized radiation as well, but is capable of providing series of two-dimensional thin slices that approximate a three-dimensional

image. Continued advances in the computer programs utilized in the interpretation of the data resulting from the CT scan (the generation of the images) have enabled significant reductions in the amounts of radiation exposure necessary for the scans, but radiation cannot be eliminated. Positron Emission Tomography (PET scans) also generates images of a series of thin slices, but involves the use of radioactive isotopes. Ultrasound uses sound waves and can generate images of organs of differing densities. Magnetic Resonance Imaging (MRIs) uses strong magnets to excite the nuclei of hydrogen atoms (protons) in water molecules in the body.

[6] It is important to note that we are talking about size differences among modern human brains. There is very strong evidence that a major factor in the evolution of primates and then mankind was the increase in brain size. Human characteristics may require a brain of greater than 750 cubic centimeters (the modern human brain is about 1375 cubic centimeters, with a mass of around 1375 grams, almost 3 pounds). Sagan, *The Dragons of Eden*, pp.33, 85–91,102–4. Scientists have ranked much of the animal kingdom by the ratios of brain mass to body mass, with *Homo sapiens*, followed by dolphins, ranking at the top of the list. *Id.*, pp.37–9. This ranking is thought roughly to correspond to intelligence. *Id.*, p.40. Curiously, however, the Neanderthals had larger brains, both absolutely and relative to body mass, than the more successful ancestors of modern man. See Gazzaniga, *Who's In Charge?*, pp.30–31.

[7] The traditional investigation of the brain and nervous system focuses on genetics and chemistry. However, current research is also exploring the physics of the brain. Growth means the existence of motion and motion means the existence of forces. Brain cells are subject to physical constraints and the laws of physics. For example, the growth and shape of cells and even the expression of genes and of the proteins are affected by the "stiffness" of the substrate on which the cell grows. Different body tissues have different relative stiffnesses, *e.g.*, grey matter in the brain is much stiffer than white matter. Thus, there may be physical signaling through characteristics such as stiffness as well as chemical signaling. Understanding the physics of brain cell development may have important benefits for medical science. As an illustration, it is possible that the adverse foreign body reaction of the brain to an implanted electrode is the result not of chemical signaling but of relative stiffness. If so, then coating an electrode with a substance that achieves the appropriate degree of perceived stiffness could reduce the reaction. Kristian Franze (St. John's College and the Department of Physiology, Development and Neuroscience), "Physics in the Nervous System," presentation to the Trinity Science Society (Trinity College, Cambridge), 21 October 2014.

⁸ The spike has been likened to the use of short and long bursts used in Morse code to send messages long distances over transmission lines (although, the spikes are more like digital signals of different frequencies, conferring information about the intensity of the stimuli). The use of pulses of electrical charge makes the transmission over lines with some possible static or interference more accurate, with less likelihood of error than with a continuously varying signal, like the human voice. Seung, *Connectome*, p.49. This is another interesting example of a design advantage embodied in nature and also, independently, discovered by man.

⁹ Seung believes that the way forward for understanding the operations of the brain is through the study of connections. He notes that "the function of a neuron depends on its output connections, not only its input connections," that is, the places to which it sends signals as well as the sources from which it receives them. *Id.*, p.68. His "mantra" is: "The function of a neuron is defined chiefly by its connections with other neurons," a doctrine that he calls "connectionism." *Id.* The totality of these connections he calls a "connectome" (like a genome). *Id.*, p.xiii. Seung argues that the focus of research should be on efforts to map regions or parts of the connectome, rather than the established focus on mapping regions of the brain. *See, e.g., id.*, pp.177–80, 275. Nonetheless, he admits that we currently lack the experimental techniques and technologies to attempt such a map. *Id.*, p.69.

¹⁰ This plasticity was a bit of a surprise to scientists, in part because of the long and well-established concept of the brain as a mechanism, following Galileo's and Newton's theories of motion, applied to the human body by William Harvey and Rene Descartes, also in the sixteenth century. *See id.*, pp.xviii, 12–13.

¹¹ This phenomenon was discovered by George Stratton at the end of the nineteenth century, through an experiment conducted upon himself. George M. Stratton, "Some Preliminary Experiments on Vision without Inversion of the Retinal Image," Paper read at the Third International Congress for Psychology, in Munich, August 1896. The objective of the experiment was to determine whether, as was assumed, the perception of things as upright depended upon the inversion of the retinal image that occurs when the image is projected upon the retina. Stratton concluded from the experiment that that assumption was incorrect–inversion was not necessary for accurate perception. Originally, it was thought that the brain inverted the image to be upright. More recently, it has been concluded that

the visual image remains inverted, but that the brain combines the image with other sensory stimuli to create an interpretation of the image that achieves coherence with the other clues, "making sense out of chaos." See Polanyi, "The Creative Imagination," in Kransz (ed.), *The Idea of Creativity*, pp.150-151.

[12] Studies have been done to determine whether there is any relationship between level of education and the incidence of dementia on the hypothesis that education could either reduce the likelihood of the occurrence of brain pathologies that are associated with dementia ("neuroprotection") or enable the brain better to ameliorate the consequences of such pathologies ("compensation"). See C. Brayne, H.A. Keage, et al., "Education, the brain and dementia: neuroprotection or compensation?," *Brain* 133 (Pt8), 2210-6 (2010). The findings indicated that the apparent effects of education in reducing dementia were the results of compensation.

[13] *Hominin* is the currently preferred classification for the species in the *Homo* family, man and his immediate ancestors (traditionally referred to as *hominid*). *Hominoid* refers to the family that today includes humans, gibbons and the great apes. See *National Geographic Style Manual*, http://stylemanual.ngs.org/home/H/hominid.

[14] A poker-playing computer will soon be arriving at casinos around the world. Michael Kaplan, "The Steely, Headless King of Texas Hold 'Em," *The New York Times Magazine*, September 5, 2013. The box, being produced by G2 Game Design, derives from work done by Fredrik Dahl for the Norwegian Defense Research Establishment; using the equivalent of neural networks, the computer has been capable of developing strategies from very large numbers of repetitive calculations (or, more recently, from playing very large numbers of games of poker) and, thereby, "learning" how to win. This example of artificial intelligence is described as an "extremely focused, one-dimensional version of the human brain." Id.

[15] Researchers at Carnegie Mellon University and the Facebook artificial intelligence lab in New York City have designed a competitor. "Pluribus, a poker-playing algorithm can beat the world's top human players... . ...Pluribus learned the nuances of Texas Hold 'Em by playing trillions of hands against itself. After each hand was done, it would evaluate each decision, determining whether a different choice would have produced a better result." Cade Metz, "Hold 'Em or Fold 'Em? This A.I. Bluffs With the Best," *The New York Times*, July 11, 2019. Thus, while the computer may be able to "learn" and even to derive strategies that surprise its designers, a

significant (perhaps determinative) part of that capability is attributable to raw computing power—the ability to explore huge numbers of combinations of plays. Whether it could overcome or ameliorate some of Penrose's objections is open to serious question.

[16] Penrose believes that consciousness exists far beyond human beings. See *Shadows of the Mind*, pp.407–8. He asserts that consciousness "can be a matter of degree, and it is not just a matter of being 'there' or 'not there.'" Id., p.408.

[17] Wallace had concluded a review of a geological text by Sir Charles Lyell with several paragraphs expressing his doubts about the sufficiency of Darwin's theory to explain the qualities of human beings and his view that there was a need for a divine Designer (a "Higher Intelligence"). He noted, in particular, that man's intellectual, emotional and moral capacities appeared to differ very little between civilized and primitive man. "Sir Charles Lyell on Geological Climates and the Origin of Species," *Quarterly Review*, April 1869 (published anonymously).

[18] A theory that the origin of human consciousness arose from communication—a dialogue—between the two hemispheres of the brain had a brief flurry of publicity in the 1980s, before disappearing. See Julian Jaynes, *The Origin of Consciousness in the Breakdown of the Bicameral Mind* (1976). There was some additional publicity following Jaynes' death in 1997.

[19] Functional MRI (fMRI) and positron emission tomography (PET) have, as discussed above, provided detailed insights into the workings of the brain. But, the resulting images still capture processesonly at instances in time. (Of course, a series of such snapshots could be played as a video and create the illusion of observing a process; but, the information captured is still a series of states.) Today, it is recognized that it would likely be quite wrong to conclude that the illuminated area identified as associated with a particular brain function is "where" that function is performed. We now think that we understand that "[e]ven the simplest conscious cognitive act inevitably involves rapidly shifting, fleeting, and low-level metabolic changes across dozens of diverse brain regions..." Id., p.176.

Some Other Matters

The Practice of Science

Reproducibility

> "Replication,
> the ability of another lab to reproduce a finding,
> is the gold standard of science,
> reassurance that you have discovered
> something true."
>
> George Johnson
> *The New York Times*
> January 20, 2014.

The above-quoted statement reflects "common wisdom" among scientists. So, what is there to say about it?

Well, reproducibility may be the gold standard, yet various meta-studies have indicated that many of the preclinical results reported in medical publications were not reproducible. *See, e.g.*, C. Glenn Begley & Lee M. Ellis, "Drug development: Raise standards for preclinical research," *Nature*, 483, 531-1, 28 March 2012 (reporting on efforts by scientists at the biotechnology firm Amgen, including Begley, to confirm the findings in 53 published "landmark" studies were successful in only 6 (11%) of the cases); John P. A. Ioannidis, "Contradicted and Initially Stronger Effects in Highly Cited Clinical Research," *The Journal of the American Medical Association*, July 13, 2005 (reporting that of 45 highly cited clinical research studies examined, almost one third were contradicted or seriously qualified by subsequent studies). There are particular problems in preclinical medical research that do not so directly apply in the traditional physical sciences, problems mainly centering around how the human body responds to pathogens and to treatments and the obstacles to truly controlled experimentation.

And, the issue of reproducibility is not limited to medical research. It has arisen in many areas of research science. And, the issue has attracted considerable attention. The journal *Nature* has established an archive entitled "Challenges in Irreproducible Research," compiling articles addressing the issue in the life sciences generally. *Nature* also announced on April 24, 2013, new editorial policies designed to "ease the interpretation and improve the reliability of published results." *Nature*, Editorial, "Announcement: Reducing our irreproducibility."

Retractions

In a relatively unusual development, a co-author of a pair of papers on new techniques for making stem cells in a laboratory published in *Nature* in January 2014 developed concerns about possible mistakes in the research and called for the retraction of the studies. *See* Alexander Martin and Gautum Naik, "Japanese Institute Weights Retracting Stem-Cell Studies," *The Wall Street Journal*, March 10, 2014; Hiroko Tabuchi, "One Author of a Startling Stem Cell Study Calls for Its Retraction," *The New York Times*, March 10, 2014.

The co-author's actions were provoked by critical comments about certain aspects of the study and claims that the result could not be replicated. The Riken Center for Developmental Biology in Kobe, Japan, began to investigate the allegations of irregularities. By July, the decision had been made and announced to retract the two papers. *See, e.g.*, Andrew Pollack, "Stem Cell Research Papers Are Retracted," *The New York Times*, July 2, 2014. All of the authors joined in a notice of retraction published in *Nature*, setting out the mistakes that had been identified in the papers. Haruko Obokata, et al., "Retraction: Stimulus-triggered fate conversion of somatic cells into pluriotency," *Nature* 505, 641–7, July 2, 2014.

Sadly, one of the co-authors, Yoshiki Sasai, deputy director of the Riken Center, committed suicide shortly thereafter. *See, e.g.*, Alexander Martin, "Japanese Stem-Cell Scientist Yoshiki Sasai Commits Suicide," *The Wall Street Journal*, August 5, 2014. At the end of 2014, Haruko Obokata, the lead author of the retracted papers, resigned from the institute. Obokata, who continued strongly to defend the work, had been given three months in a dedicated lab to attempt to replicate her earlier findings. She failed to do so. Martin Fackler, "Scientist Who Had Claimed Stem Cell Breakthrough Resigns From Japanese Research Institute," *The New York Times*, December 19, 2014.

Flawed or fraud?

There have been suggestions that this lack of reproducibility of published results suggests that there were serious procedural flaws and biases (and, sometimes, even fraud) in the work that has been reported in the publications. Perhaps, the researchers saw what they wanted to see; perhaps, some researchers treated the empirical results selectively, reporting only the favorable outcomes; perhaps, some researchers manipulated the data. See, e.g., George Johnson, "New Truths That Only One Can See," *The New York Times*, January 20, 2014; Michael Sak-Young Chwe, "Scientific Pride and Prejudice," *The New York Times*, January 31, 2014.

But, there are also problems due to lack of adequate reporting of experimental design and of the resulting data, in some cases perhaps to maintain some secrecy or competitive advantage over other researchers. See, e.g., Hank Campbell, "The Corruption of Peer Review is Harming Scientific Credibility," *The Wall Street Journal*, July 13, 2014. And, there are concerns about tainted peer reviews, where one or more of handful of reviewers has some interest in the research or reason to favor the researcher. Id.

We do, of course, also find some quite notorious examples of scientific fraud (material misrepresentations undertaken intentionally for reputational or financial benefit), such as the false claims about the successful cloning of a human embryo by South Korean veterinary professor Hwang Woo-suk in 2005. Dr. Hwang also falsely claimed to have cloned stem cells matching the DNA of patients. He achieved international fame and multiple material benefits in South Korea before his deceptions were revealed and he was dismissed from his university position and publically disgraced. "Hwang fired for stem fraud," *USA Today*, March 20, 2006.[1]

The Hwang affair actually generated considerable attention to the issue of how scientific journals and the news media could better review scientific works to reduce mistakes and fraud. See, e.g., Nicholas Wade, "Journal Faulted in Publishing Korean's Claims," *The New York Times*, November 29, 2006; Julie Bosman, "Reporters Find Science Journals Harder to Trust, but Not Easy to Verify," *The New York Times*, February 13, 2006; David Dobbs, "Trial and Error," *The New York Times*, January 22, 2006.

The "art" of research

It is likely, of course, that, at least on occasion, mistakes and even fraud have happened in research submitted for publication and that some instances even survived peer review to be published in the most selective and prestigious journals.

I suspect, however, that it is also the case that most of the lack of reproducibility is due to the fact that the techniques and procedures are new, that documentation fails to capture every significant influence leading to the successful result and that human error permeates both the successes and the failures to reproduce the results. In other words, real life is complex and often uncooperative. Just as our most elegant and respected scientific theories can predict little about the most mundane, common aspects of our lives; it is difficult to structure experiments that are sufficiently isolated from all uncontrollable effects to be regularly repeatable.

Moreover, it should be recognized that much of contemporary research concerns the detection of quite subtle and often very small phenomena. Thus, the equipment must be highly specialized and the experiments very carefully controlled. Almost undetectable differences in the microenvironment may affect the results. Also, the techniques used may require years of practice to perfect, making them difficult to replicate. *See, e.g.*, Mina Bissell, "Reproducibility: The risks of the replication drive," *Nature*, 503, 333-4, 20 November 2013. Experimental research is often more of an art than a science.

PREDECESSORS

> "Discoveries are not in general made
> before they have been led up to
> by the previous trend of thought,
> and by that time many minds are in
> hot pursuit of the important idea."
>
> Alfred North Whitehead
> *An Introduction to Mathematics*, p.219.

This thought brings me to a related subject. Experimental science is based upon decades or centuries of accumulated know-how. It is almost inevitable that we will fail to appreciate the magnitude of the discoveries, inventions and both hard work and luck (or chance) that underlies the simplest of laboratory experiments. Many of us have probably marveled at the ready availability at very low prices of things that we could not conceivably imagine being able to produce on our own if suddenly thrust into a technology-free environment, even one with ample raw materials. In the woods, how would you ever produce a needle, let alone a simple piece of machinery? Society, with freedom of exchange, has enabled the accumulation of techniques, processes, equipment and know-how that allows the production of things that no one person or even a handful of persons could ever do alone.

A mountain of ants

Similarly, in the laboratory experiment, the scientist relies upon the immense body of prior work of others to develop and produce the lab equipment and tools, to discover and make available the chemical inputs and to develop the processes and theories that underpin the experiment that is being undertaken, including what is being measured, how it is being measured and, indeed, what it even means to measure it.

The scientist may bring to the table an idea or hypothesis that enables something new to be attempted. When successful, the scientist should be congratulated on that original contribution, but one must recognize the enormous small and large preceding achievements of others that made the experiment possible. In pure theoretical science, one may say with some justification that Newton, Einstein or Hawking stood "on the shoulders of giants" in making their contributions. But, for the empirical scientists, let alone the developers of technology, it would be much more accurate to say that they stand on a mountain of ants.

Indeed, for the reason quoted above, Alfred North Whitehead objected to the emphasis on the few individuals who happened to make even the most important discoveries of science: "In considering the history of science, it is both silly and ungrateful to confine our admiration with a gaping wonder to those men who have made the final advances towards a new epoch." An Introduction to Mathematics, p.218.

The importance of failures

Because of the importance of the work that necessarily precedes every success in the lab, it may be that the culture of academia today is interfering with progress. I have specifically in mind that lack of publication of negative results, of the failures, many times more numerous than the successes.

Science publications generally only publish the new finds or discoveries, not the processes and procedures that did not work. Academic careers, including university positions and tenure, as well as professional standing depend significantly upon being published. It may be that the system generates pressures to misrepresent results or, at least, to ignore unhelpful data. See, e.g., Brendan Nyhan, "To Get More Out of Science, Show the Rejected Research," The New York Times, September 18, 2014 (advocating the sharing of data and the publication of studies that utilize satisfactory methods regardless of the results to reduce "publication bias"). One consequence, discussed above, may be that the published papers contain errors and can mislead future researchers. But, another consequence that must certainly occur is that important learning (about what does not work) is not available. An understanding of the alternatives that have been attempted and the negative results achieved would likely be very helpful for further research.

Thanks to the internet, there have been emerging tools for the publication to the science community of a wider range of works, making the dissemination of a wider range of research results (successes and failures) more easily available. In the social sciences, a service called the Social Science Research Network, incorporated in 1994, has expanded dramatically. It is organized into networks for various subjects or specialties. The goal is to make new research available rapidly and cheaply. Most downloads are free.[2]

The Internet

> "These ... are cognitive tools,
> actively amplifying our collective intelligence,
> making us smarter and so better able to solve
> the toughest scientific problems."
>
> Michael Nielsen
> Reinventing Discovery, p.3.

There is an important and transformative role in science being played by the internet: the facilitation of collaborative endeavors on a scale and within time frames that are utterly unprecedented. The possibilities for the internet to accelerate dramatically the pace of scientific discovery and to provide greatly enhanced opportunities for the exercise of collective intelligence are subjects of a recent book by Michael Nielsen. *Reinventing Discovery: The New Era of Networked Science* (2014).

Nielsen observes that the advances have appeared in a variety of forms that are quite different in terms of what is added by the collaboration made possible by the internet. One obvious example is large pooled databases, like the GenBank operated by the US National Center for Biological Information. *Id.*, pp.4-5. Such projects require a means of overcoming the natural incentives of scientists to keep research data and discoveries to themselves for their own publication. Well-known projects of this sort include Wikipedia and the earlier (commencing in 1998) creation of the Linux operating system. *Id.*, pp.44-53. There are also efforts to make use of the vast resources of the broader community and to draw upon the supposed wisdom of the crowd (part of a broader phenomenon now known as "crowdsourcing"). Amateur volunteers can assist in the analysis of large quantities of data and images and in the collection of observational data. *See, e.g.*, Tom Feilden, "Citizen science is the new black," BBC *News*, 15 October 2013; Tim Adams, "Galaxy Zoo and the new dawn of citizen science," *The Guardian*, 18 March 2012.

The internet even allows sociologists to conduct real-time experiments on thousands (or, perhaps, millions) of participants, for example, in the form of on-line "games". *See, e.g.*, Christakis, *Blueprints*, pp.103-11. As an example, MIT neuroscientist Sebastian Seung (discussed in the prior chapter) launched EyeWire, an online game that seeks to harness the resources of thousands of people who in the course of playing the game would be marking the branches of neurons presented to them in the image of a cube. Although computers and artificial intelligence programming can be used to develop the images, the human brain is unsurpassed in its skills of pattern recognition, which is what is necessary to trace the connections. Reportedly, the game has attracted 165,000 players in 164 countries. Gareth Cook, "Sebastian Seung's Quest to Map the Human Brain," *The New York Times*, January 8, 2015. Of course, botany and astronomy have long made use of amateurs to collect observational information and field data. The key to the success of these projects is modularity, with the ability to tap the imagination and creativity of hundreds of different people, working at their own speed and at times of their own choosing.

Another type of use is illustrated by a mathematician at Cambridge University named Tim Gowers, who invited through his blog other mathematicians to help solve a sophisticated mathematical problem on which he was working. Nielsen, *Reinventing Discovery*, pp.1-3. This project, called the Polymath Project, enabled Gowers to "scale up creative conversation" (*id.*, p.2), dramatically broadening the participants, adding significant diversity, and speeding up the pace of interaction. It allowed participants to avoid conversations that they judged to be likely to be unproductive, to work when inspired (anytime, day or night), at their individual paces, formulating concise and accurate statements and contributing where they were most able to assist based upon their own interests and expertise. *Id.*, pp.2-3.

Finally, the internet is increasing public awareness of science. Public engagement helps address the apparent distrust of science and scientists. Public distrust has always existed with respect to public science. Schaffer, "Why trust public experiments?", presentation to the Trinity Science Society, 28 October 2014. Perhaps it is a legacy of the context in which public demonstrations by scientists first appeared, as a form of entertainment alongside magicians, medicine men, fortune tellers, spiritualists and other supposed charlatans. The greater potential access and transparency available today may even dramatically alter the public perception of science and scientists.

Specialization

> **"Science isn't just for scientists.
> ... As citizens, we all need a feel for how much
> confidence can be placed in science's claims."**
>
> Lord Martin Rees
> *From Here to Infinity*, p.10.

Excluding the public

The consensus view is that, in contrast to the eighteenth and nineteenth centuries, it is no longer possible for any person to be reasonably informed about all subjects of relevance to public policy or scientific advancement. Clearly, although there are many more people with the education, leisure and financial independence to pursue scientific pursuits as a hobby; it is highly unlikely that any of them would make a significant original contribution in any, let alone multiple, areas of inquiry as happened in the days of the great "gentlemen" scientists. But, at the same time, there is no reason why many such persons could not be sufficiently informed to be able to assess scientific developments and claims with a critical eye. Indeed, there are strong reasons why the public should be encouraged to do so. There is more to science than the creation of new theories and the performance of sophisticated experiments. As Hasok Chang has stressed,

> "the cutting edge is not all there is to science, nor is it necessarily the most valuable part of science. ...In a way, I am calling for a revival of an old type of science, the kind of 'natural philosophy' that was practiced by the European 'gentlemen' of the eighteenth and nineteenth centuries with such seriousness and delight."

Inventing Temperature (2004), pp.4-5.

Limiting the scientist

As our knowledge has advanced, it has been useful for particular areas to become carved off or out for focused attention by persons who increasingly specialize in the area. Obviously, that process has been an effective technique for the advancement of the studies of the individual areas. Repeatedly, ever smaller areas have been identified and investigated with great effect. However, as reflected over and over again in the discussion above, we do seem to exhaust many of these specialties, leaving us with the feeling that the real explanations lie outside or beyond the sphere of interest (or the methods) of the specialty.

Let me give you an example. My daughter Sarah read biological anthropology (evolutionary biology) at Queens' College, Cambridge. She became interested in the relationship between chronic stress and the immune response. For her MPhill and the beginning of her work on her PhD at Cambridge, she focused on a broader array of questions concerning viruses in primates. She was particularly interested in primates in the wild, for which invasive procedures such as blood samples were not a viable technique for the measurement of viral load. As she experimented with the extraction of DNA from fecal samples and the amplification of the extracted DNA in the laboratory, she concluded that biological anthropology was not sufficiently lab-oriented to support her work. So she switched to the virology section of the pathology department. Those researchers understood lab work and viruses. But, she found that their interests were often in the specific function of small sections of a single virus. They wanted to study viruses, but had little interest in the hosts. Of course, what makes viruses distinctive is that they can reproduce and "live" only within a host. To Sarah, it is the interface between the virus and the host that is key—that is where the action is. Ultimately, she found her place in the virology section of the vet school—surrounded by people who were interested in viruses but also loved animals.

There is increasing recognition of the limitations being imposed by the segmentation of research and theoretical modeling into such concentrated areas. That can be seen in the popularity of multi-disciplinary and interdisciplinary projects and programs. Too often, however, the multi-disciplinary activity consists of experts from various disciplines talking at one another, each about his or her own specialty. While there are undoubtedly areas in which input from multiple disciplines can be beneficial, there is also a need for persons who can free themselves from the blinders (or the microscope) of their own particular discipline and view the question more expansively and more creatively. To that end, I think that forays by experts of one subject into the activities of another subject hold the promise of progress.

Limiting the vision

I note that the argument has been made that paradigm-shifting ideas in the sciences are more likely to be made in books, rather than journal articles, and often in books for the trade rather than in academic texts. See Stephen C. Meyer, *Signature in the Cell: DNA and the Evidence for Intelligent Design* (2009), pp.6–7. Meyer asserts that this is true because peer-reviewed journal articles tend to be very focused and narrow (specialized), like the journals in which they appear and that the article format does not lend itself to length or detail argumentation, including the refutation of alternative theories. Meyer also asserts that paradigm-shifting theories are more likely to be cross-disciplinary in terms of the evidence used and the phenomena being explained. His reasons in favor of trade books are quite persuasive. Whether he is correct about the past or likely future source of paradigm-shifting ideas is, however, not so clear. And, at least conceptually, this claim is a factual one. (I should add that this view certainly serves Meyer's own interests in that he is a historian/philosopher of science presenting what he might hope to be a paradigm-shifting idea—Intelligent Design as the source of life—in a book for the general reader. His arguments are discussed below.)

The "Evolution" of Science

> The "ultimate good desired
> is better reached by free trade in ideas—
> that the best test of truth is the power of the thought
> to get itself accepted
> in the competition of the market."
>
> Justice Oliver Wendell Holmes
> Abrams v. United States
> November 10, 1919

Philosopher of science Stephen Toulmin described the aims of science as lying "in the field of intellectual creation." *Foresight and Understanding*, p.38. The dramatic progress in understanding that has been achieved is largely the result of the "tradition of criticism," which has been embraced by post-Enlightenment intellectuals and scientists. Criticism has enabled the differentiation between good and bad explanations. "Bad explanations" (like myths) tend to have a flexibility that accommodates new evidence and experience without changing the substance of the explanation. As a result, bad explanations thwart progress. In contrast, a good explanation is "hard to vary, because all of its details play a functional role." Deutsch, *The Beginning of Infinity*, p.24. Proper scientific methods allow for the detection and correction of errors in theories. However, scientific theories providing explanations often originate as conjectures or guesses; the discovery of theories is an act of intellectual creation by man. Id., pp.1-30. Thus, scientific progress is always (and will always be) a work in progress, as existing explanations get replaced by better explanations which will themselves be replaced by better explanations in the endless quest for the best explanation.

The role of advocacy

If one looks at scientists' efforts to defend and protect their own theories from conflicting evidence and criticism, it is often apparent that the scientists are not behaving as strictly objective, rational truth-seekers. Is that a problem? I think not, as long as one sees what is transpiring and is not misled by claims of special expertise and superior knowledge or experience.

We have models that suggest that biased advocacy, in contrast to neutral investigation, can be a powerful tool for the discovery of truth. That is the model that Popper and Lipton and many others seem to have had in mind for the proper scientific process. But, that model has been criticized by various others as naïve and misleading. For example, it has been argued that the process will work only if there is a relative balance of biases across interested groups and that there is some open-mindedness to criticism among at least some parties to the debate. See, e.g., John Stuart Mill, *On Liberty* (1859), Chapter 2. Nonetheless, the Anglo-American legal tradition is based on the belief that interested parties aggressively advancing their individual interests in an adversarial process before a passive finder of fact can dispel the false and distill the truth on average more effectively than can a magisterial process in which the inquiry is initiated and directed by an investigative official. The assumption is that the best and most creative arguments, the most probing and effective inquiries and the most damaging and incisive challenges are more likely to be inspired by self-interest than by good intentions.

Thus, there are potential benefits from having self-interested scientists, motivated by the promotion of their own ideas and own personal success. The corrective forces arise from the competing self-interests of other scientists. The vigorous competition among ideas (and theories) implemented and fueled by the personal competition among the individual participants may be expected to bring about more rapid scientific progress. At the same time, it is highly beneficial for the process to have some reference to the goals of objectivity, peer-group review and empirical testing.

(The parallels to the arguments in favor of free markets are obvious, as is the relevance of the question of whether some external regulation against fraud or misrepresentation is needed to promote the proper functioning of the marketplace for ideas, as well as the marketplace for goods and services.)

The underlying point is the benefit of competition. One can picture the development or "evolution" of science as a "struggle for survival" among opposing scientific models or theories in which, over time, the most "fit" or satisfying of which emerges. In the prior chapters, I have frequently criticized the neo-Darwinian theory of natural selection. It is an admittedly stubborn and pervasive paradigm of explanation in our "modern" thinking. However, the correct model of "evolution" here is one based on competition, not natural selection. It is a "struggle for success".

Natural selection and "the invisible hand"

The two concepts are historically related and certainly have a common feel and appeal; yet, the explanatory bases of the two paradigms are different. It is common to overstate the similarities. For example, Shermer goes so far as to claim that :

> "Charles Darwin's theory of natural selection **is precisely parallel** to Adam Smith's theory of the invisible hand. Darwin focused on showing how complex design and ecological balance were unintended consequences of initial competition among organisms. Smith focused on showing how national wealth and social harmony were unintended consequences of individual competition among people."

Why Darwin Matters, p.136 (emphasis added).

Shermer asserts that "Darwin scholars are largely in agreement that he modeled his theory of natural selection after Smith's theory of the invisible hand," citing to the "considerable literature on the connection between them." Id., p.182, n12. Similarly, Toni Vogel Carey refers to "a central domain-crossing link between Adam Smith's doctrine of the 'invisible hand' and Darwin's theory of natural selection," discussing some of the literature on the topic. "The Invisible Hand of Natural Selection, and Vice Versa," Biology and Philosophy, 13 (1998), p.427.

Curiously (to me), Carey goes on to make the following assertion: "Natural selection is the explanatory lynchpin of Darwinian evolution. But the appeal to an invisible hand seems ... conspicuously non-explanatory—more a place-holder for an explanation than an explanation in its own right." Id., p.428. As to that point, I have a very different view.

While I agree that the invisible hand was a placeholder for the economic theories that subsequently developed; through theoretical efforts over the next 150 years, economists developed a relatively comprehensive structure (neoclassical price theory and subsequent industrial organization theories) that provides a basis for explaining and understanding why those thousands of individual, selfish profit-motivated decisions can achieve a satisfactory allocation of resources, both in the organization of productive resources and in the distribution of the economic output. As noted above, however, it is not an "invisible hand" at work but a complex system of information exchange and related institutional developments that permits a more or less free market to achieve the allocative efficiency of neoclassical theory. This model has substantial explanatory force, even if it has only tenuous application to any particular real life situation.

However, I disagree that natural selection in biology was explanatory. Certainly, as originally propounded by Darwin, it was no more explanatory than the "invisible hand," since Darwin lacked the subsequently developed science necessary to provide an explanation either for how variation arose or for how, once it arose, it could be preserved.

As one scholar has observed:

> *"In proposing the theory of evolution by means of the mechanism of natural selection he was not really supplying a mechanism at all. Rather, he was providing an abstract account at a general level of how favorable variations might be preserved. He had to keep his account at a certain level of abstraction since, as he confessed, he could not specify either the laws of variation or the precise means by which variations were preserved."*

Young, *Darwin's Metaphor*, p.98

Subsequent developments of the theory of genes and, then, discoveries concerning the actual mechanisms of heredity, at least initially, seemed to supply important missing pieces. Since then, however, the theory of natural selection has not achieved a comprehensive theoretical development. If one could identify Laws of Natural Selection, then the theory would clearly have potential empirical content (whether correct or not). However, there have been no such Laws identified. One can also see, as already discussed, that it is highly unlikely that such Laws could be found. In short, the theory has not yet achieved satisfactory explanations of the observed phenomena in which we are most interested.

Replication, variation, selection: "Memes"

The concept of natural selection has been applied at many levels—genes, individuals, populations, *etc*. Fodor and Piattelli-Palmarini, *What Darwin Got Wrong*, p.57. Moreover, as previously noted, the use of this paradigm or model has not been limited to biology. But, the more generalized form of "evolution" today essentially dispenses with the driving force of scarcity. The bare requirements of replication, variation and selection allow the theory to be applied to language, culture, knowledge and, even, to the origin of life. See, *e.g.*, Nowak, *SuperCooperators*, pp.110, 117, 171–73; Deutsch, *The Beginning of Infinity*, p.93.

Dawkins has argued that wherever the necessary elements are present (replication, variation, selection), a type of evolution will occur. *The Selfish Gene*, pp.13–24. He, along with others, has even speculated that his theory of the "selfish gene," set forth as an explanation of biological development and human evolution, could be applied to the evolution of human culture.

Dawkins hypothesized the existence of something comparable to a gene that embodies units of cultural content. He called them "memes." *Id.*, p.206. These units or "memes" could compete with one another in a process that "explains" the emergence and development of human culture. *Id.*, pp.203-15. A whole literature has rather recently emerged exploring the potential nature of a "meme" and the applicability of a theory of evolution to "memes."

We have previously noted above the way in which such a model can be applied to scientific discovery and progress. Uses of "evolutionary" models can be found in the writings of Popper, Kuhn and Lipton, among others. In the language of this section, we might characterize such models as versions of the "meme" theory, where the discrete elements (or "memes") are the individual scientific theories. See also, Deutsch, *The Beginning of Infinity*, pp.369-97. However, the introduction of a new name does not mean that it is really a novel theory and does not itself establish a step forward in knowledge or understanding.

I think that it is important to note the distinction between biological or genetic evolution through natural selection and these similar-seeming concepts of the evolution of cultural and knowledge, including science, through competition among varying values, practices or ideas.

The Darwinian conception derives from the scarcity of resources relative to the rate of reproduction. Those conditions require that some substantial number of individuals not survive to reproduce. There is built into the process an imperative—a force driving or imposing selection. That is the power of the theory. Only some will survive and reproduce. As Kate Distin observed in her book, "[O]rganisms reproduce, passing their characteristics almost (but not always quite) accurately on to the next generation, and if their environment does not supply them with unlimited resources for their survival, then they will evolve; there will be a struggle for survival, and those organisms will be preserved whose traits are best fitted to the given environment." *The Selfish Meme*, p.3.

Admittedly, the development of cultures, the emergence of societies, the growth in knowledge and the increasing sophistication of ideas could be called "evolution". But, the processes are very different. One does not have reproduction, at least in the sense of mechanical replication of traits in offspring. Also, one does not have a scarcity of essential resources or the necessity of non-survival. Thus, there is often ample room for conflicting ideas to coexist indefinitely.[3] The essential elements of Darwin's process are not present. (Curiously, Distin did not seem to notice.) Instead, we have "evolution" through competition, the necessary conditions of which are that alternatives are being generated and the individuals affected have freedom of choice, some criteria by which they can assess the alternatives and certain mechanisms by which they can choose based upon their assessments. If the criteria or standards reflect values that we consider to be worthy, the process will generate progress.

The most "fit"

Also, it is not correct to say that the most "true" survives, since survival in the competition of ideas is not necessarily dependent upon truth or conformity to the real world. Instead, survival depends upon satisfying the needs and demands of the scientific community that is engaged in the competition. Presumably, the criteria of being most "fit" or satisfying should have some content beyond the mere fact of survival in the struggle. Here, the historian of science and the philosopher of science have something to contribute. They can look at how science has in fact "evolved" or developed over time and say things about why and how certain theories came to supplant other theories. We may say that "fitness" of a scientific theory certainly means at least that the theory offers the promise and then the capability of achieving, with manipulation and additional work, a consistency with observed phenomena. Hopefully, it will also achieve the ability to predict phenomena to be observed, in which case it will prove to have, perhaps, significant practical applications in the form of technology. We might also be able to say that our understanding of the world will be advanced because of the power and success of the new theory. In part, however, such success will be due to the sense or experience of insight and understanding that the theory brings to us. We "feel" that we see more clearly. The scientific relevance of that feeling, unfortunately, is unclear.

What about reality (again)?

The paradigms that are used within a society are clearly relevant to the nature of the members' understanding of the world around them and to the types of explanations that they will find satisfactory. But, what is the relationship between the changing paradigms and truth or reality? Is there an expected or likely relationship or is any relationship purely coincidental?

To the modern man or woman, it is probably not very persuasive to assert that there is no reality, that all perception is illusion. It also would not be very modern to claim that there is an ideal reality that exists separate from the world we inhabit, which world is merely an approximation of the perfect forms. It is probably more acceptable to assert that all perception is inherently subjective, with the result that we may never be able to see reality as it really is. We struggle with the belief that there is something that constitutes reality but are limited in our abilities to see it or comprehend it.

Michael Polanyi put it as follows:

> "[S]cientific discoveries are made in search of reality—a reality that is there, whether we know it or not. The search is of our own making, but reality is not. ...For the scientist's quest presupposes the existence of an external reality. ... [However,] [a]ttempts to eliminate these indeterminacies of science merely replace science by a meaningless fiction."

"The Creative Imagination," in Kransz (ed.), The Idea of Creativity, pp.160, 161.

In other words, our quest for knowledge is premised upon the assumption that there is such a thing as reality, but we would be foolish to confuse even our best scientific theories with it.

THE "EVOLUTION" OF COOPERATION

> "Selfishness beats altruism within groups.
> Altruistic groups beat selfish groups.
> Everything else is commentary."
>
> David Sloan Wilson
> This View of Life, p.77

David Sloan Wilson summarizes the argument made by Darwin in The Decent of Man about what we call multilevel or two-levels selection—that selfish behavior may often be more beneficial than cooperation to an individual within a group, but cooperative behavior within a group is clearly more beneficial to the group and its members in competition among groups. He quotes the statement above from a joint review he wrote with Edward O. Wilson in 2007. This View of Life: Completing the Darwinian Revolution (2019).

We have previously discussed theories of the evolution of altruism, kin selection and group selection. But, the relevant examples of cooperation that are all around us go well beyond what can be "explained" by these theories.[4] Indeed, what we are ultimately interested in is the appearance and development of intentional social behavior in all of its ramifications. Such behavior most likely originates in the emergence of conscious or intentional (strategic) cooperation.

Francis Fukuyama explains how Hobbes, Locke and Rousseau all began the development of their political theories on the false assumption that man initially was a solitary creature. (From that assumption, each went on to theorize how it was that society and political order came about.) Fukuyama describes that these political scientists were not attempting to set forth historically accurate portrayals of the origin of society, but were using a heuristic device to help identify what they believed to be the important underlying features of human nature. In the words of Rousseau, these studies were "not to be taken for historical truth, but merely as hypothetical and conditional reasonings, fitter to illustrate the nature of things, than to show their true origin." Quoted in *id.*, p.28.

The common view is that modern political organization appears as the end result of the transition from family to band to tribe to chiefdom to nation or state. *See, e.g.,* Fukuyama, *The Origins of Political Order* (2011), Ch. 3. However, recent archeological findings suggest that the story may be much more complicated and suggestive.

> "...[T]he earliest known evidence of human social life resembles **a carnival parade of political forms, far more than it does the drab abstractions of evolutionary theory.** ... A cosmopolitan Upper Palaeolithic is followed by a complicated period of several thousand years, beginning around 12,000 BC, in which it first becomes possible to trace the outlines of separate 'cultures' based on more than just stone tools. ...[T]he overall direction of history—at least until very recently—would seem to be the very opposite of globalization. It is one of increasingly local allegiances: **extraordinary cultural inventiveness, but much of it aimed at finding new ways for people to set themselves off against each other.**"

David Graeber and David Wengrow, *The Dawn of Everything: A New History of Humanity* (2021), pp.119,123, 124 (emphasis added).

The evidence strongly suggests that man was a social being from the beginning. In part, this conclusion is based on the observation that our closest relative, the chimpanzee, with whom we presumably descended from a common ancestor, is clearly a social animal. Fukuyama, *The Origins of Political Order*, pp.26-29. Thus, it is considered highly likely that that common ancestor was also a social animal. So, efforts to explain the origin and evolution of social and political relationships among animals, including humans, must look much earlier in the history of the planet.

The aggregation of peoples into larger groups with some social and political order seems clearly to have provided multiple advantages for survival and reproduction, particularly in competition with other groups of humans for necessary resources. Once cooperating communities appeared, those communities could succeed and evolve further through a form of natural selection (actually, the competition model); but, how did cooperation come about in the first instance?

We need to go behind group behavior to understand the emergence of cooperation. Various successful forms of cooperation require the sufficient development of mental capacity and societal organization to make such cooperation possible. In addition to the emergence of greater cognitive abilities (and larger brains), the development of emotions (like shame, guilt and pride) and the appearance of religion facilitated and promoted social relationships and political organization. See, *e.g.*, Fukuyama, *The Origins of Political Order*, pp.38-41, 59-63.

Is there a credible basis for presenting cooperation and organized society as results of evolutionary forces?

The "Green-Beard" Effect

In *The Selfish Gene* (1976), "Dawkins called gene-based discrimination 'the Green-Beard Altruism Effect'. He envisaged a gene that encoded both a phenotypic label, the green beard, and the tendency to be nice to green-bearded individuals." David Haig, "Imprinted green beards: a little less than kin and more than kind," *Biology Letters* (The Royal Society), 23 December 2013.

The hypothetical poses that certain genes have a distinguishing trait (the green beard), the capacity to recognize that trait in others and a tendency to behave cooperatively with similar genes. All of these phenotypic factors must be inheritable. If these conditions are met, then the qualifying genes might become a successful subgroup that outperforms other genes and gains dominance. Thus, we find what looks like altruistic behavior arising through natural selection. This exercise assumes that there is some mechanism by which one gene recognizes another. Such a gene was purportedly found in 1998. See, e.g., Laurent Keller, Kenneth G. Ross, "Selfish genes: a green beard in the red fire ant," Nature (394, 573–575), 6 August 1998. There is also some (limited) theoretical basis for the possibility that "compatible" genes could "recognize" one another. See, e.g., Andy Gardner and Stuart A. West, "Greenbeards," August 25, 2009.

There are two major challenges for such a theory. First, what is the genetic basis of such social behavior (selflessness) enabling it to be replicated? Second, how does one overcome the "free-rider" problem (selfish genes with green beards, enjoying the benefits without the costs)? See, e.g., Samir Okasha, "Biological Altruism," Stanford Encyclopedia of Philosophy, June 3, 2003 (substantive revision July 21, 2013).

THE PRISONER'S DILEMMA

A popular work that affected the public's (and, apparently, many scientists') perception of the issue of cooperation was the book *The Evolution of Cooperation* (1984; revised edition 2006), by Robert Axelrod, based on research work performed and published in various articles in the years shortly after the publication of Dawkins' *The Selfish Gene*.[5]

Axelrod's interest was in understanding the possibilities for cooperation between strategic entities (persons, groups or nations) where circumstances involved differing but not necessarily conflicting interests. He observed that although such entities should be assumed to pursue purely selfish ends, cooperation did appear sometimes to arise voluntarily, not just when imposed by a policing authority. Using game theory, a branch of mathematics that had been developed over the prior three decades and attributed to Johnny von Neumann and Oskar Morgenstern and, subsequently, mathematician John Nash (in which the behavior of other players was explicitly incorporated into the analysis of rational conduct—in other words, investigating the concept of strategy), he explored how such results might come about.[6]

Axelrod employed a game called the Prisoner's Dilemma to explore logically various scenarios in which cooperation could arise. Matt Ridley sets out a history of the Prisoners' Dilemma. *The Origins of Virtue: Human Instincts and the Evolution of Cooperation* (1996), pp.51–66. He says that the game was first formalized in 1950 by Merril Flood and Melvin Dresher of the RAND Corporation and "rephrased as an anecdote about prisoners" shortly thereafter by Albert Tucker of Princeton. On its terms, it appeared that the only stable outcome was that of mutual defection, which supported a rather pessimistic view of the human condition. However, when two young economists, Armen Alchian and John Williams, played the game repeatedly between themselves, cooperation to capture the greater benefits arose. The stage was set for Axelrod.

In the game, each of two accomplices in a crime held as prisoners are given the choice of refusing to talk or of confessing and informing upon the other. The game posits various possible trade-offs from the alternative courses of action, dependent upon what choice is made by the other prisoner. The decisions are made independently by each simultaneously or, at least, without knowledge of the other's decision. The dilemma arises because both prisoners would be better off if neither confesses, but each would be considerably worse off if he refused to talk while the other informed. If they both informed on each other, each would receive a result somewhat worse than the best case but substantially better than the worst case. With such rewards established, it becomes easy to show that under many various scenarios of assumed payouts, it will be the case that in any single, isolated game, the rational course for each prisoner will be to inform on the other.

The rewards system Axelrod used for his studies was based upon points (the more the better). If both cooperated (refused to talk), they would each get 3 points. If both defected (tried to turn in the other), they each get 1 point. If one defected, while the other cooperated, the defector gets 5 points and the "sucker" gets 0 points. *The Evolution of Cooperation* (Revised Edition), p.8. One's rewards clearly depend upon the strategic choice made by the other prisoner as well as on one's own. In a single game, each player will rationally recognize that if his counterpart defects, he himself will be better off defecting, and, if his counterpart cooperates, he himself will again be better off defecting. Thus, it is relatively safe to predict that both players will decide to defect. Therefore, a selection by each player made in his own best interests (the rational choice) results in a less than optimal outcome for the two combined. In other words, the pursuit of self-interest will diminish the benefits of the group.

The situation has more interesting potentials, however, if the game is to be played repeatedly, with each player informed about the prior strategic decisions of his counterparty. It is possible to show that under certain circumstances, repeated iterations of the game can lead to the emergence of cooperation between the two as the dominant strategy. A strategy of

Tit-for-Tat, commenced by cooperating on the first play, can result in the greatest combined points. In the simplest case, if both parties use that strategy, they would each cooperate on the first play and then continue to do so thereafter. In addition, there are strategies that can be identified as "best" or "better", independent of the strategy selected by the other party.

Through computer modeling, One can test a wide range of strategies against one another, through many games. Of course, there are a variety of assumptions underlying the construction of this simple model (such as there being no benefits or costs external to the game that could influence decisions, that is, nothing other than the reward points—no externalities).

One intriguing experiment is the incorporating of some "discount rate" to enable a comparison of present points with future points, establishing that a point in the future is less valuable than a point now, whether because of a "time-value" of benefits (use now versus use later) or the uncertainty of the future versus the present (including the relative likelihood of repeated games—the counterparty might die or simply not be encountered again) or both. *Id.*, pp.12-3. The discount rate, however, must not be too high (such as where the likelihood of repeated games is low), or else there will be no independent "best" strategy. *Id.*, pp.15-6.

Interestingly, although it was not the source of his interest, Axelrod, with the assistance of William D. Hamilton, applied his findings to biology and Darwinism. *Id.*, pp.88-105. They concluded that foresight or intentional strategizing was not a necessary ingredient for cooperative patterns to emerge, but that a reasonable probability of repeated encounters was.

Thus, the development of capabilities of recognizing other players is an important step in the process. The authors found that the strategy of always defecting was evolutionary stable, as one would expect, but so could be the strategy of Tit-for-Tat where repeated encounters could be expected, such as in clusters of geographically proximate individuals. (This use of game theory in biology has also suggested possible insights into cancer and other diseases, as well as into evolutionary theory.)

MODELS OF COOPERATION AND NATURAL SELECTION

Martin Nowak has recently described the work done over the past few decades to examine whether and under what circumstances cooperation could arise through the process of natural selection. *SuperCooperators: Altruism, Evolution, and Why We Need Each other to Succeed* (2011).[7]

He observes that natural selection of the individual would generally oppose cooperation and, as a result, would lead to the decline in the fitness of the population. *Id.*, p.10. Nowak argues that, instead, cooperation is an additional fundamental principle of evolution that supplements the theories of Darwin. *Id.*, p.280.

Nowak identifies five circumstances that can give rise to cooperation:

- repetition (direct reciprocity);
- reputation (indirect reciprocity);
- spatial selection;
- multilevel selection; and
- kin selection.

Id., pp.270-72.

He summarizes work that he and others have done to use mathematical modeling to explore whether and under what circumstances cooperation could survive (like and including the "Prisoner's Dilemma", discussed above).[14] In general, his methodology has been to "create[ ] idealized communities," with specified characteristics and directions for action by the individual members, and use computer simulations to run hundreds or thousands of iterations "to compress the passing of eons." *Id.*, pp.xvi, 183. A key point about much of this work is that while the experimentation is often described in terms of "strategies", his models do not assume strategies that are developed and modified by intelligent beings. The "strategies" are individuals' innate responses to particular circumstances.

So, if a cell or organism appeared that was genetically disposed to "cooperate" when presented with the opportunity, what would happen to it? Not surprisingly, the answer depends on the dispositions of the other individuals with which it interacts. If they are all "cooperators," then the situation is not very interesting—nothing changes. The useful scenarios are where there are some born "cooperators" and some born "defectors". Unfortunately, in the simplest cases, the defectors systematically eliminate the cooperators.

Nowak then introduces other factors. For example, an entity may initially cooperate but if it encounters a defector, then on the next round it will defect. There are innumerable variations, and his examples introduce "action rules" (like we have just discussed), "assessment rules" (to evaluate whether a prior action should be excused) and, even, in some cases, "update rules" (which provide for changes in strategy—that is, the action rules—in response to experience). See id., pp.65, 249. The resulting experiments or simulations demonstrate that under particular strategies, cooperators may gain the upper hand for a period, but the community would display cycles of dominance by cooperators and then by defectors. Id., pp.21–50. With the introduction of structure and spatial groupings to the community under simulation, cooperation can have particular success, because the simulation begins to revolve around group competition and selection. Id., pp.69–94.[8]

Considerable work has been done utilizing mathematical techniques in evolutionary biology. (Nowak's recent positions have included "Professor of Mathematical Biology" at Oxford and "Professor of Biology and Mathematics" at Harvard.) Nowak describes various uses of mathematics in his fields of study.[9] One is the effort to isolate and define by simple formulae fundamental relationships underlying the observed phenomena. Such relationships can be causal in concept or simply descriptive. The identification of such equations enables extensive efforts to "test" the relationship against observed data to see where and to what extent the relationship appears to be, at least, consistent with what has happened.[10]

But, Nowak repeatedly depicts the use of mathematics in evolutionary biology, as well as in immunology and virology, to all of which he has substantially contributed, as analogous to or methodologically equivalent to the use of mathematics in physics. See, e.g., SuperCooperators, pp.xvii, xviii, 12, 15–16, 160 ("a kind of Newtonian law of biology"), pp.164, 252, 266. I think that these methodological claims are fundamentally misleading. It seems highly implausible that the derived formula could constitute Laws of (Biological) Nature in the sense of Newton's or Einstein's theories of gravity. At best, such models may capture some significant "tendencies" or relationships that can be expected to be associated with certain outcomes, but they will always be subject to and overshadowed by the dominating *ceteris paribus* condition of everything else being equal. Of course, all Laws are subject to this condition; but, there is, I think, a difference in kind between Laws where the condition is or can be clearly specified and even regularly realized and hypotheses where the condition is rarely or never clearly defined (remember that the condition applies only to factors that matter to the relationship under investigation, which are what need to be defined).[11]

If the identification of an equivalent methodological use is thought to be instructive, then the analogue would be economics. The fact that a model cannot be realized in the real world does not mean that it is not of value. As seen in economics, the articulation of the relationship in a simple formula can itself give rise to a sense of insight or feeling of understanding. The ability of the formula to generate insights into masses of observed data clearly contributes to an important aspect of the scientific enterprise.

So?

To return to the observations at the beginning of this section, it is clear that the conditions that promote cooperation—recognition of the other player, recollection of past conduct and the facility to adjust actions based upon prior experience—all require a level of cognitive function that could not have been present in the earlier stages of evolution. In addition, another apparent prerequisite, predictable repeat interactions, presupposes some established community of individuals. Thus, there seem to be certain threshold requirements to be realized before cooperation could then evolve through competition.

> "...[S]ocial learning correlates with increased brain complexity, together with increased intelligence measured by the invention of more novel solutions to challenging problems." [So, it may be that ultimately] [c]ollaboration play[ed] a more significant role than competition in the survival and propagation of life, from the level of genes through unicellular organisms, organelles within eukaryotic cells, eukaryotic cells within multicellular organisms, and the behaviour of plants and animals from insects to primates."

Hands, *Cosmosapiens*, p.385 (citing Peter Kropotkin, *Mutual Aid: A Factor of Evolution*, first published in England in 1902).

One can construct a plausible evolutionary path that would lead to the attainment of the requisite cognitive capabilities. There are numerous reproductive benefits one can imagine from increasing cognitive functions, including improved memory and some flexibility in responding to stimuli. So, gradual steps might lead to increasingly social animals. Thereafter, the group competition model of "evolution" takes over, both with within a species and among species: significant changes in the lives of the members of the species occurring without underlying genetic changes.

Religion and Design

> "[W]e, a product of evolution,
> possess an overwhelming
> sense of purpose and moral identity
> yet arose by processes that
> were seemingly without meaning."
>
> Simon Conway Morris
> *Life's Solution*, p.2

Science and religion

I now want to address directly a matter that has raised its ugly head several times in the discussions above—the tensions between science and religion. Indeed, Curtis White argues that, especially with the emergence of neuroscience, "scientism" is at war not just with religion, but also with art, literature, philosophy and all of what we might call high culture, not to mention creativity and imagination. *The Science Delusion*, pp.57, 87-8, 103, 109.

I have referred on several occasions to the destructive conflicts that seem often to arise when religion is involved, especially some scientists' almost paranoid hostility. Admittedly, elements of the religious communities have exhibited emotional anti-science sentiments. However, it seems to me that open-mindedness and tolerance are not necessarily expected, and certainly are not required, characteristics of religious belief. Science, in contrast, as a matter of proper methodology, is expected to be objective, open to criticism and critique, subject to empirical testing and willing to entertain new theories where the old ones are found inadequate. Thus, I think that we can and should expect something more and different from scientists than we do from those who profess a belief in a religion.[12]

The questions that linger are (i) must science deny religion and (ii) must religion deny science? The answers seem obvious, but more can be said.

An Inevitable Conflict?

J. Wentzel Van Huyssteen set out the perceived distinction as follows:

1. scientific statements are hypothetical, fallible and tentative, while statements of religious faith are dogmatic, ideological and fideistic;
2. scientific thought is always open to critical evaluation, justification or falsification, while religious faith goes against facts and often defies empirical evidence;
3. scientific thought delights in critical dissent and constructive criticism, while faith more often than not depends on massive consensus and uncritical commitment:
4. scientific beliefs are based on evidence and rational argument, while religious beliefs are founded on 'faith' only; and
5. scientific rationality is, thus, revealed as not only a very manicured and disciplined form of human reflection, but also incommensurable with, and vastly superior to, religious faith and theological reflection.

Duet or Duel? Theology and Science in a Postmodern World (1998), p.9.

Our discussion thus far should make it clear that the image of science portrayed in assertions 1-3 is not regularly achieved in practice. Perhaps, that is due to scientists, not science itself. So, we can think of them as aspirational statements. But, for example, science does look for consensus. Assertion 4 is factual. Is it true? As for science, I think we can recognize that the methodological model represented by classical mechanics does not adequately describe modern science—certainly not all or most, perhaps not any. We have seen cosmological theories that posit universes that we can never know and with which we can never communicate. We have seen "cutting edge" theories that seem to be beyond any conceivable type of falsifiability. And, so on. Should we say that such theorizing is not science? *See* Michael Guillen, *Believing Is Seeing: A Physicist Explains How Science Shattered His Atheism and Revealed the Necessity of Faith* (2021), pp.89-155 (summarizing quite clearly many of the points made in these chapters). Thus, assertion 4 is not true as to science. And, assertion 5 is simply unjustified if assertion 4 is not true.

Are the assertions correct about religion? Again, as with science, we are not talking about the behavior of particular participants, but of religion itself. Religion faces "competition" and is subject to criticism. Religion seeks justification. And, religion is not antagonistic to evidence. See, e.g., Guillen, *Believing Is Seeing*, pp.157-200. In fact, religion may represent the best explanation of what we observe and experience.

We now have much more diversity and flexibility in what we accept as good science. As a result, there is some inevitable ambiguity in how to establish boundaries between science and other types of human knowledge, using knowledge in the broad sense. Certainly, we must conclude that the methodological boundaries between these types of theories and religion are not so clearcut. I hope that I have convinced the reader, through the extended discussion in the prior chapters, that the differences between science and religion are not so clearly a matter of kind as, perhaps, a matter of degree.[13]

ROOM FOR ACCOMMODATION?

As noted, this apparent antagonism is a bit odd, since it is obvious that science and religion fundamentally address quite different questions. Religion (generally) does not attempt to quantify physical relationships or describe the physical processes of the world. Science has not purported to try to ascertain whether there is a God and certainly has generated no proof of His non-existence. Thus, the two systems of thought belong in different, barely overlapping worlds.

But, that response is not entirely satisfying for many of us and is not strictly true.

There are areas where science and religion do overlap. There are also various fundamental questions that could have different answers than those generally provided by religion. Curiously, as much of the preceding discussion has indicated, those questions have rather stubbornly resisted the efforts of scientists to find answers. Thus, there are substantial areas of fundamental importance that have not been preempted by science in that science does not purport to provide the answers, at least, not yet.

So, at the moment anyway, there is still ample room in our understanding of reality for God or some other divine force. In addition, in the areas where spirituality has the most relevance—consciousness and the panoply of subjective experiences—traditional science has made relatively little progress and may, in fact, prove to be inadequate in the end.

David Berlinski summarized the matter as follows:

> "We do not know how the universe began. We do not know why it is there. ...We have little idea how life emerged, and cannot with assurance say that it did. We cannot reconcile our understanding of the human mind with any trivial theory about the manner in which the brain functions. Beyond the trivial, we have no other theories. We can say nothing of interest about the human soul. We do not know what impels us to right conduct or where the form of the good is found."

The Devil's Delusion: Atheism and Its Scientific Pretensions (2009), p.xv.

Moreover, as Van Huyssteen argues, "our mental capacities are constrained and shaped by the mechanism of biological evolution," so that "all of our knowledge, including our scientific and religious knowledge, is grounded in biological evolution." Id., p.32. In short, our belief systems and "understanding" of the world arise from information gathering and processing processes that have increased our fitness to survive and reproduce. Thus, there is reason to believe that religion, as well as science, is grounded in cognition-gaining processes that made man more suitably adapted to his environment. In that sense, religion is by definition rational. Van Huyssteen asserts that efforts to understand the evolutionary processes by which human scientific and religious understandings have been achieved could enable a dialogue between scientists and theologians, an interdisciplinary effort to advance human understanding. Id., pp.148-57. See also, Collins, The Language of God, pp.197-211.

"These splendid artifacts of the human imagination have made the world more mysterious than it ever was. We know better than we did what we do not know and have not grasped." David Berlinski, The Devil's Delusion, p.xv

AND, YET ...

The zeal with which several prominent contemporary Darwinists seek every opportunity to attack religious beliefs is decidedly unscientific, in that the personal agenda leaves them resistant to acknowledging the existence of shortcomings or gaps in their own scientific theories. The problem is sufficiently pervasive that Stephen Cave, as a reviewer in the Financial Times of three recent books on human evolution, explicitly praises David P. Barash, Homo Mysterious: Evolutionary Puzzles of Human Nature (2012), for his "bold[ness] in embracing the unknown...[through his] catalogue of the many unresolved riddles of our history." "Planet of the Apes," Financial Times, August 18–19, 2012, p.7. Cave writes: "Some might think it imprudent of Barash to admit to so many gaps in the evolutionary account of humanity, given that religious fundamentalists are ready to exploit them for their own ends. But it is just this willingness to admit what we do not know ... that marks out the scientific method."

A NOTE ON TEACHING EVOLUTION

> While recent scientific work has raised increasing questions about the validity of Darwin's theory of natural selection as an explanation for the origin of species, there is heated opposition to suggestions that school curricula should contain critical assessments of the theory. See, e.g., "Darwin Foes Add Warming to Targets," The New York Times, March 3, 2010. The perception among many is that anyone critical of Darwinism (and, apparently now, of global warming theories) is promoting religion, just as the perception among many is that anyone advocating Darwinism is anti-religion. These reactions presumably are related to the spurious religion/science dichotomy that seems to continue to flourish.
>
> Students should be taught that all of science is a process, and is one in which none of the current answers are definitive or likely to be permanent. Wherever critical analysis or challenges to accepted dogma can be identified and taught as part of the educational process, they should be. Although, it may seem to risk a tactical defeat in the war against religious beliefs; I think that no one could really argue that it is detrimental to science to admit uncertainty, which should encourage critical thinking and inspire creativity.

THE ISSUE OF DESIGN

Controversies about Darwinism arose almost immediately with the public dissemination of the theory. Whether Darwin's *The Origin of Species* was, in fact, so blatantly problematic for traditional religious views at the time is not entirely obvious; but, it certainly was perceived to be so. (*The Descent of Man* was certainly more challenging to the beliefs of the time.)

But, Darwin's view were promptly denounced (and subsequently applauded) as such, and much of the contemporary commentary and debate about the scientific theory came to be dominated by that issue. It is unfortunate and, perhaps, surprising that those controversies have continued to surround the theory, having distorted the understanding and scientific discussions of it for 150 years.

The essential problem faced by anyone who wanted to contest the existence of God or some other form of divine Creator was the apparent existence of purposive design in the Universe. There are numerous natural wonders that many might view as evidence of a Creator, but the most dramatic examples arise in the biological sphere. Richard Dawkins characterizes the "Argument from Design" as "the most influential of the arguments for the existence of a God." *The Blind Watchmaker* (2006) (original publication, 1986), p.4; see also, *The God Delusion*, pp.137-9.

A particularly eloquent statement of the case for the existence of God based upon the presence of apparent design was made at the beginning of the eighteenth century by theologian William Paley. His book *Natural Theology—or Evidence of the Existence and Attributes of the Deity Collected from the Appearances of Nature* was published in 1802. Paley began his book with the now famous example of the watch. If one finds a watch lying on the ground, one inevitably must conclude that there was a watchmaker. The complexity of the object and the obvious usefulness of the features achieved by that complexity cannot possibly have occurred by accident. This statement seems beyond dispute. (He starts with the example of kicking a stone, which is obviously something that occurred naturally. A decade earlier, Samuel Johnson also used the kicking of a stone as a demonstration. *See above* endnote 4 to Mathematics.)

One might grudgingly accept David Hume's skeptical assertion that such complexity did not (necessarily) prove the existence of a divine Creator, but absent an alternative explanation for its existence, most would have to find the appearance of design a rather compelling indication of an intentional act of creation that could only be accomplished by a Creator. (By the way, the theory of Intelligent Design seems to me to be curiously named. What is being postulated is purposeful or intentional design.[14]) If the choice is between accident and God, then Paley's argument would generally prevail.

But, the response is not that the apparently designed biological objects, including humans, occurred by accident. The contribution of Charles Darwin was the formulation of an alternative—natural selection. Based upon the theory of natural selection, Dawkins is comfortable referring to the appearance of design as "the illusion of design." *The Blind Watchmaker*, p.21. Yet, one of the most common arguments in support of the theory of evolution by natural selection is that there is no alternative "scientific" theory to explain what we can all see; so, if we dismiss natural selection acting on random changes, nothing can explain what we observe except for a Creator and Intelligent Design. It seems to me that that observation should not be seized upon as a reason to embrace neo-Darwinism, but as a reason to be open-minded and curious about the possibilities of alternative scientific explanations. It has not been uncommon in the history of science for one theory to be dogma until another, previously unimagined, theory appears to challenge it.

Indeed, the fact that successful new explanatory theories may be unimaginable to prior scientists has been expressly acknowledged by biologist Leonid Kruglyak in the context of evolutionary theory: "It's a possibility that there's something we just don't fundamentally understand.... That it's so different from what we are thinking about that we're not thinking about it yet." "The road to genome-wide association studies." *Nature*, vol. 456 (November 2008), p.18.

THE DESIGN ARGUMENT

The question of the origin of life has become a focus of the recent "Intelligent Design" arguments (by which I mean the hypothesis that God had some meaningful input to the process of life). Natural selection, by definition, can not explain the origin of life. Moreover, efforts to demonstrate that life could have arisen by chance have been notably unsuccessful. Contemporary cosmology rules out the possibility that life has always existed. So, if the Universe has an origin (or, at least, a rebirth), then life must also. Indeed, the current models of the early Universe are inconsistent with the existence of life forms during the first few billions of years.

A reformulation

The relatively recent science of information has given rise to the new formulation of many of the arguments we have been discussing, on issues from biology to physics. See Stephen C. Meyer, *Signature in the Cell: DNA and the Evidence for Intelligent Design* (2009); Charles Seife, *Decoding the Universe*, p.385. I personally do not find the reformulation to be all that enlightening, but others may differ. I do think, nonetheless, that it is a forceful way of presenting the arguments about design that we have already briefly discussed.

The key to life, as we understand it, is the reproducible genetic code contained in RNA and DNA. So, the need for a designer in this context arises not from the complex or sophisticated ("designed") nature of organs or organisms, but from the appearance of the coded genetic information that enables life. It is "the mystery of the origin of the information needed to build the first living organism." Meyer, *Signature in the Cell*, p.14.[15] We start with the observation that what is relevant about the genetic code is the information that is stored therein. ("'The gene is a package of information, not an object. The pattern of base pairs in a DNA molecule specifies the gene. But the DNA molecule is the medium, it's not the message.'" Id., p.17, quoting George Williams, *Natural Selection: Domains, Levels and Challenges* (1992).) It is the non-random series of data that has specific meaning, that is, that instructs the performance of other operations that constitute the integral elements of life. The particular physical form of the gene is not important; what matters is the information that it contains. The information contained in the gene is highly specific (i.e., it has substantial information content). Meyer, *Signature in the Cell*, pp.107-8. And, it is highly unlikely to have occurred by chance.[16]

As we have discussed previously in examining the concept of the "gene", the information achieves things that are not the immediate result of or predictable from the physical form of the gene. It directs processes that result in various phenomena that have a function. In that sense, it is as if foresight is reflected in the information that is stored, as if the information is purposeful. Indeed, the information is exceedingly complex and rich, and it seems to have had built-in potentialities far beyond the immediate needs or functions of the organism of which it is a part.

Meyer argues:

> "*Apart from the molecules comprising the gene-expression system and machinery of the cell, sequences exhibiting such specified complexity or specified information are not found anywhere in the natural—that is, the non-human—world. [Such] structures ... are completely unknown there apart from DNA, RNA, and proteins. ...Nevertheless, human artifacts and technology—paintings, signs, written texts, spoken language, ancient hieroglyphics, integrated circuits, machine codes, computer hardware and software—exhibit specified complexity....*"

Id., p.110.

Meyer observes that when one finds information of this type, one would immediately rule out random events or chance as the explanation. Indeed, just like with Paley's watch, one would assume intentional design and, therefore, a Designer. "[I]nformation invariably reflects the prior activity of conscious and intelligent persons." Id., p.16. (Francis Collins describes DNA as "God's Instruction Book." *The Language of God*, pp.109-42.)[17]

So, established scientific methodological principles, particularly the theory of "the best explanation" (expounded by Peter Lipton), point to design. Id., pp.324-6. I note that Behe, in his earlier book, discusses the arguments made by philosopher Elliot Sober in *Philosophy of Biology* (1993), observing that "Sober thinks that intelligent design is actually an inference to the best explanation, not an inductive argument." *Darwin's Black Box*, p.219. (Lipton's theory is discussed above in the first chapter.)

We are all familiar with the fact that intelligence combined with intention regularly leads to the existence of meaningful, non-random information. We are literally surrounded by examples, from technology to the arts to entertainment. Indeed, our first reaction when confronted with such information is to assume the existence of a purposive agent. Even our scientific search for extraterrestrial life is based upon the assumption that the discovery of meaningful information is evidence of intelligence. See, e.g., Meyer, *Signature in the Cell*, p.383. Thus, we scan the skies for data that represents the value of Pi or other meaningful, non-random information. If we are ever successful in detecting such "signals," we will conclude that we have probably detected the work of another intelligent life form.

Of course, the evidence would not be conclusive, just highly supportive.

Purpose/Intention

We are well acquainted with the products of design, reflected throughout human societies. How do we recognize design? The questions that we ask instinctively are:

- First, how likely is it that what we observe could have occurred by chance or by accident?
- Second, could the phenomenon have been the result of some natural process?

The indicia are pretty obvious. But, they can never be conclusive in and of themselves. When we see a cloud or a rock or a stain that bears a striking resemblance to something or someone we know, we generally conclude immediately that it has occurred naturally, not with design or purpose. But, the amount of detail or the degree of resemblance may gives us pause. Presumably, one stumbling upon Mount Rushmore would instantly think of design and purposive activity, even if they had never heard of the monument nor seen pictures of the presidents represented. I think that we instinctively assess the probabilities. And, increased knowledge can and should influence our assessments.

Design is clearly a possible explanation for such patterns or information. Thus, the methodological test is whether there are other hypotheses that also explain what we observe. Meyer argues that with respect to the origin of life, the answer is no.

Probabilities

Douglas Axe, a strong advocate of intelligent design, argues that "invention" cannot arise by accident and that the highly improbable, is the "practicably impossible". *Undeniable: How Biology Confirms Our intuition That Life Is Designed* (2016), p.160. Axe provides numerous illustrations of things with staggering odds against arising from a random process.

But:

- His argument depends heavily upon the definition of "invention," which biases the analysis. The concept has embedded in it the notions of imagination and intentional implementation. The more relevant concept is "improvement". Axe acknowledges that natural selection can achieve "fine-tuning" of an existing organism, effectuating "minor" adjustments that can have the effect of improving functionality. What he doubts is that natural selection could create something "new".

- Also, Axe fails to relate the probabilities to a realistic scenario of evolution. As we have already discussed in the chapter on Mathematics, any and every specific complex phenomenon (like a specific series of heads and tails) will appear to be highly improbable in advance, but such phenomena occur all the time. The relevant question then is not how likely is it that any specific existing organism could have arisen by chance (although, as noted, it would be significant if one could imagine no such possible scenario); it is how likely is it that a complex organism of any type could have arisen by chance.

- The proper calculation starts with the odds, given a certain population of organisms, that an adaptive mutation would arise (the nature of which is necessarily unspecified). If there are environmental challenges (or opportunities) for the organism, the likelihood that one of the multitude of mutations will turn out to be adaptive is pretty high; the question is how many generations is it likely to take. Then, given a certain population of this modified organism, what are the odds that another adaptive mutation would arise? (Again, probably pretty high.) And, so on, until the desired level of complexity and change is achieved. (The likelihood of each of these additional adaptive mutations probably will steadily decline.) Then, all of these individual probabilities would be multiplied to give the relevant "odds" of the evolution of a new, more complex organism evolving.

- The significant differences between this calculation and the ones presented by Axe are that there is a potentially very large population from which a single adaptive mutation could occur at each stage (greatly increasing the chances that it will occur) and the nature of the relevant mutation is wide open, the only requirement being that it must be beneficial from a reproductive standpoint in the then existing environment (which also improves the odds materially).

Other explanations

Meyer asserts that: "Intelligence is the only known cause of complex functionally integrated information-processing systems." Id., p.346. But, as Behe has commented:

> "Might there be an as-yet-undiscovered natural process that would explain biochemical complexity? No one would be foolish enough to categorically deny the possibility. Nonetheless, we can say that if there is such a process, no one has a clue how it would work. Further, it would go against all human experience, like postulating that a natural process might explain computers."

Darwin's Black Box, pp.203-4.

Of course, it is permissible—indeed, advisable—to look for alternative explanations with respect to any theory. We have already considered several of the difficulties encountered by the theory of natural selection to explain the variety of the species that have existed. We discuss in a following section the apparent evolution toward more complex and capable beings, including conscious and intelligent beings, that can be observed and, of course, in the next chapter, the question of the origin of life.

Different people will evaluate the evidence and theories differently, but, as we have seen, Dawkins has acknowledged the centrality of the issue raised by the appearance of design in the world, arguing that the greatest significance of the theory of natural selection is that it has provided the one and only alternative to Intelligent Design to explain biological facts. And, he also has recognized that there is no similar mechanism that has been identified to explain what we called the "fine-tuning" problem of the physical (non-biological) world (apart from some version of "multiverses"), discussed in the next chapter.

Meyer is careful to emphasize that the hypothesis of Intelligent Design says nothing about the nature of the designer. *Signature in the Cell*, pp.442-4; *see also*, Behe, *Darwin's Black Box*, p.197. It certainly does not require that the designer still exist nor that such designer was not part of the physical Universe. But, that question unanswered certainly leaves a big gap. If we limit our inquiry to the emergence and evolution of life on Earth, the source of design could be extraterrestrial intelligence that may, or may not, in turn be explainable by naturalistic theories. Indeed, there has been speculation on the fringes of the scientific community that life on Earth is the result of the intentional or accidental input of alien matter. How life and intelligence emerged elsewhere, if it did, is difficult to address without at least some information about what such alien life forms are or were like. But, the possibility that design arose from an immanent, rather than a transcendent, source is logically consistent with Intelligent Design. *See* Meyer, *Signature in the Cell*, p.443.

And so...?

The logical order of the relevant issues is different from that of my presentation. It goes as follows:

- There were very special conditions necessary for the existence of a Universe in which there are stars, planets and other aggregations of matter.
- There are also very special conditions with respect to our Solar System and the Earth necessary for the existence of life (as we know it or can imagine it).
- With two (at least) highly unlikely sets of necessary events that seem to have somehow occurred, how could all of that "fine-tuning" have come about?
- Even with all those special conditions present, how did life appear?
- Even given life, how did consciousness arise?

- Given consciousness, what is the source or basis of man's moral sense?

Are we, thus, driven to find Design?

It is important to be careful to avoid the fallacies that arise as a result of hindsight, so I have three comments.

First, the very special conditions for the existence of life, as we know it, arise as a result of the nature of "life as we know it." Life as we know it exists, because those special conditions existed, and *vice versa*. (As Fodor and Piattelli-Palmarini point out, organisms will live in circumstances for which their phenotypic characteristics make them suitable. In effect, the organisms define the niches by their traits. *What Darwin Got Wrong*, p.142.) Second, unless those special conditions were set out as a requirement in advance, one cannot say anything meaningful about the "probability" that those conditions would occur. All one can say is that they did occur. Third, a key question is whether the life that did appear is the only kind of life that is possible. In other words, are there alternative life forms that could exist that do not require the special conditions necessary for life on Earth?

If the answer to the third question were yes and the alternatives were manyfold, then one could say with some persuasiveness that life as we know it exists because the special conditions happened to occur and that that is not a surprising outcome. However, if there are no alternatives or if the alternatives are few and similarly restricted, then the anthropic principle is not an answer, and the combination of staggeringly incredible coincidences giving rise to life on Earth continues to demand an explanation.

Intelligent Design is certainly not incompatible with natural selection. In fact, design coupled with the process of natural selection makes a powerful explanatory framework. See *id.*, p.392; see also, Behe, *Darwin's Black Box*, pp.227, 228. Facts that seem incompatible with Intelligent Design as the sole explanation of current life forms are readily absorbed in such a combined theory. The fact of "junk" DNA can be explained by the operation of natural selection upon life that was designed and created by other processes. Such operation could actually have been contemplated by the design at the time of origin. (Recent discoveries that the amount of actual "junk" in current DNA is substantially less than was supposed 20 years ago are consistent with this theory. See Meyer, *Signature in the Cell*, pp.406-7, 454-5.) Similarly, the examples of "bad" design or Rube Goldberg structures are also accommodated by this combined theory. (We do not need to interpret "created in God's image" as referring to anatomical details. Indeed, many of us would question whether physical anatomy is even a relevant issue with respect to a transcendent God.) Finally, of course, intelligent design could provide explanations for all of the phenomena discussed above that pose difficulties for neo-Darwinism, including the origin of species, the appearance of progress, the emergence of intelligence and the existence of consciousness. Those explanations could involve periodic or sporadic intervention in the natural processes over time or merely "arrangements" embedded at the origin of the Universe (or at some later point). See also, id., pp.389-91.

It is easy to hear the common objection to this line of reasoning: Until we know the details of the Designer, how can we credit the design hypothesis? But, that hardly seems like a valid objection in the context of the history of science. We have over hundreds of pages examined well-established and widely accepted theories based upon the assumption of some unknown and unexplained force or factor.

By the way, it seems to me that the strongest case against an Intelligent Designer is, somewhat ironically, the numerous examples of flawed or bad design. How could creatures that were intentionally designed have characteristics or features that are useless or inefficient or, even, detrimental to proper functioning? It is not very satisfying to say that the Designer simply made mistakes. Otherwise, I am not sure how one can explain the imperfect, sometimes almost Rube Goldberg creations of nature without the existence of some type of incremental and experimental process of evolution. Proponents of natural selection as the superior alternative to a divine Creator make regular use of this argument. See, e.g., Dawkins, *The Blind Watchmaker*, pp.91-3. Of course, the process of natural selection could have interjected these oddities into what had been designed. Also, one must remain open to the possibility that any particular arrangement is not bad design or a "mistake", but achieves some other objective that we simply have failed to identify.

Endnotes

[1] Dr. Hwang did produce, however, (also in 2005) what is thought to be the world's first cloned dog ("Snappy") (other researchers had successfully cloned cats at the time but not dogs). Gina Kolata, "South Korean Scientists Clone Man's Best Friend, a First," *The New York Times*, August 3, 2005. As a result, he enjoyed substantial financial success. He now is seeking to reestablish himself in the scientific community. *See* Choe Sang-Hun, "Korean Scientist's New Project: Rebuild After Cloning Disgrace," *The New York Times*, February 28, 2014.

[2] The SSRN, by the end of 2013, had 242,000 authors in the database of 522,000 paper abstracts and 427,000 full text papers. Downloads have reached almost 1,000,000 a month. Some 400 institutions outsource the distribution of their papers to the SSRN. Greg Gordon, "SSRN's 2013 Year-End President's Letter," December 23, 2013.

[3] I recognize that there can be stable biological populations with genetic diversity, but such populations appear as relative anomalies in terms of the Darwinian theories.

[4] For example, in October 2013, the journal *Current Biology* reported on a study that genetically traced modern ants back to mud dauber wasps and bees. Carl Zimmer, "Key to Ants' Evolution May Have Started With a Wasp," *The New York Times*, October 20, 2013. The female wasp builds a mud cylinder in which to lay her egg and then paralyzes another insect, like a cockroach, to put in the nest before sealing it shut. The larva, when it hatches, is able to feed on the captured victim. Pretty complex.

[5] Richard Dawkins wrote the foreword to the new edition, noting the significance of Axelrod's work to his own and presenting a graph of the number of citations by year in the literature to Robert Axelrod, soaring after 1985. *Evolution of Cooperation* (2006), pp.xiv–xv.

[6] The geneticist John Maynard Smith attempted to utilize game theory in biology in the early 1970s to explain why natural selection might result in "strategic" behavior by animals, such as the fact that animals generally do not fight "to the death."

⁷ One critical observation that has been made about Nowak's book is that his presentation of the efforts to understand cooperation create the impression of continuous and harmonious progress, because he "gives little sense of the [fierce] debates that have raged" over the conflicting models of evolution. See Oren Harman, "A little Help From your Friends," The New York Times Book Review, April 10, 2011, p.18.

⁸ Nowak uses a payout scenario in which cooperation by both results in a prison term of two years each, defection by both results in prison terms of 3 years each and cooperation by one with defection by the other results in 4 years for the cooperator and 1 year for the defector. Id., p.7. With those assumed payouts, defection is always the best strategy (if his colleague cooperates, the player gets a 1-year term by defecting and a 2-year term by cooperating; if his colleague defects, he gets a 3-year term by defecting but a 4-year term by cooperating). Id. Clearly, the best strategy will depend upon the assumed payouts, and other assumptions can make the strategic choices less clear.

⁹ Examples of the kinds of useful results that can be achieved through this type of analysis and modeling include the identification of a "value" reflecting certain characteristics of the structure and grouping of population that reflects the frequency of interactions among various members of the population, which value can be determined for populations of any type and can predict the likely evolution of cooperation within the population (id., p.266), and a formula that states that if the ratio of the benefits to the costs of cooperation is greater than the average number of neighbors for each member of the population, then cooperators would prevail; if the ratio was less, then the defectors would prevail (id., pp.250-255).

¹⁰ Nowak also reports on the use of similar methodological approaches to study the dynamics of the progress of viral infections and of cancer. Id., pp.40-1, 142-50. He also argues that "natural selection" could have been involved in the origin of life itself–that is, that evolution preceded life. Id., pp.116-7; see also Deutsch, The Beginning of Infinity, p.143. Cf. Shermer, Why Darwin Matters, pp.141-2.

¹¹ Computers have allowed, with the creation of descriptive mechanical models to simulate the interaction of various phenomena or courses of conduct, the investigation of the impact of various assumptions over thousands or tens of thousands of iterations. In these applications, the scientist can specify simple rules of action and assessment and, perhaps, of revision of strategy and let the computer calculate the effects of those

rules over simulated eons, by means of very rapid but purely mechanical computation.

[12] Not that we should not expect at least civility from believers as well. White provides rather vicious but effective rebuttals to the principal anti-religion arguments of Richard Dawkins and essayist Christopher Hitchens (and physicist Lawrence Krauss), but goes on to proffer extended attacks aimed more at their persons (alleged ignorance and duplicity) than their ideas. *The Science Delusion,*, pp.13–56. White demonstrates that the over-zealous and intemperate are not all on one side of this (or most any other) argument.

[13] Not that I would go so far as to call it just a matter of degree, either. That latter characterization suggests a single scale of evaluation on which different approaches can be ranked unambiguously and sequentially as to their satisfaction of the stated goals.

[14] I suppose that intelligence is likely to be a requirement for intention, and the design of what we observe in the world around us would certainly seem to require intelligence; but, intelligence is a human construct that may have quite limited application in the arena of the divine.

[15] I note that the word "information" is used here in a much more restricted sense than the way it is used by modern physicists. See, *e.g.*, Susskind, *The Black Hole Wars* (2008).

[16] Meyer cites favorably Michael J. Behe's book *Darwin's Black Box* (2006). Behe, a biochemist, argued that the most plausible explanation for the highly complex biochemical systems found in all life forms is intelligent design.

[17] Meyer devotes many pages to the demonstration that life could not have arisen by chance. See *id.*, pp.178–88, 351–72. He discusses the work in the 1920s of statistician Ronald A. Fisher, who identified a statistical test for whether a series of events varied too greatly from the expected outcome of a process to be the result of chance. Meyer also relies heavily on the more recent work of William Dembski, who advanced a theory of information that looks at patterns after-the-fact, with the benefit of hindsight, to assess whether they occurred by chance or design, *e.g.*, *The Design Inference: Eliminating Chance Through Small Probabilities* (1998).

THE BIG QUESTIONS

LIFE

> "But if (& oh what a big if)...in some warm little pond ... a protein compound was chemically formed, ready to undergo still more complex changes..."
>
> Charles Darwin
> *Letter to J. D. Hooker*
> 1 February 1871

I turn now to an issue that logically precedes most of the things discussed in the chapters on Darwinism and Consciousness. That issue is how did life occur in the first place, giving natural selection something to work with?[1] At a technical level, this question is one for the molecular biologists, chemists and geologists. But, there is no technical answer. Lord Rees acknowledges, we do not yet "understand the actual origin of life." Yet, he predicts that we will "have a better idea of how that happened within the next 10 or 20 years". "Opinion: Aliens, very strange universes and Brexit," *University of Cambridge Research Bulletin*, 7 April 2017.

Remember, we started this book with a similar prediction by Carl Sagan made over 40 years ago.

LIFE?

Curiously, the minimally-educated modern person could probably quite readily and quickly examine a list of objects and identify which are alive and which are inanimate; yet, the definition of what constitutes life has generated considerable debate among "experts". See Luisa, *The Emergence of Life*, pp.5-10, 120-3.

Most of the proffered "definitions" consist of a list of observable characteristics that we associate with life. In some respects, that is not a very satisfactory approach to a definition. Many such definitions are of limited use when applied to an individual specimen rather than to a type or group. For example, the ability to reproduce may not even be ascertainable from examining a single individual, and it may not be an actual characteristic of particular individuals that are undoubtedly alive. Many specific physical characteristics can be common to animate entities and inanimate machines (mechanical or electrical), such as mobility and the generation of sound. Indeed, even the ability to reproduce is not sufficient to establish that something is alive (nor, for any individual entity, is it a necessary characteristic).

Here are some samples of definitions with my comments:

- Erwin Schrodinger addressed the question from the standpoint of a physicist in a lecture delivered in February 1943: "When is a piece of matter said to be alive? When it goes on 'doing something', moving, exchanging material with its environment, and so forth, and that for a much longer period than we would expect of an inanimate piece of matter to 'keep going' under similar circumstances." "What is life? The Physical Aspect of the Living Cell" (1944). Surely, this definition is rather ambiguous and, as a result, may be too inclusive.
- In 1994, The Exobiology Program within NASA propounded what is often called "the NASA definition": "Life is a self-sustained chemical system capable of undergoing Darwinian evolution." That definition was intended to provide a standard by which one could identify whether any alien specimens that might be discovered were "alive". Brief reflection should reveal that this definition could not be very useful operationally. When one encounters an unknown entity, one is not likely to be able to ascertain much of anything about its heritage or the development of its type (or "Darwinian evolution"). We need something better.
- In the words of Lord Martin Rees, "metabolizing, replicating structures" are alive. "Opinion: Aliens, very strange universes and Brexit," *University of Cambridge Research Bulletin*, 7 April 2017. Succinct. A live entity will have the capability to maintain itself (that is, to convert raw materials into energy that the entity can utilize) and will have the capability and propensity to grow and, in many cases, change over time. Also. something that is alive will typically react to the environment in other ways ("irritability"), engage in some form of respiration and use energy to create some order. Metabolizing includes the conversion of

"food" into energy and the synthesis of additional matter to become part of the "structure".

- And, a version from a chemist: "a system can be said to be living if it is able to transform external matter/energy into an internal process of self-maintenance and production of its own components." Luisa, *The Emergence of Life*, p.122.
- But, note: "It isn't true ...that life generates structures that are particularly ordered, or that locally diminish entropy: it is simply a process that degrades and consumes the low entropy of food; it is a self-structured disordering, no more and no less than in the rest of the universe." Rovelli, *The Order of Time*, pp.164-5). Well?

PHYSICAL COMPONENTS OF LIFE FORMS

We recognize that life as we know it requires the existence of highly order chemical structures, the ability of those structures to self-replicate and a sufficient concentration of those structures and their constituent chemicals to permit and sustain activity. The concentration is achieved through membranes that encapsulate the organic material (creating cells). See Luisa, *The Emergence of Life*, Chapters 3 and 4; Erin O'Donnell, "How Life Began: Jack Szostak's pursuit of the biggest questions on Earth," *Harvard Magazine*, July-August 2019 (describing the work of Nobel laureate Jack Szostak).

DNA and RNA

We now know that the key to life on Earth is found in the double helix structure of DNA. Proteins are built from twenty amino acids, each of which is coded for by a triplet of three nucleotide base pairs. The base pairs are contained in the DNA set out in combinations of the four bases adenine (A), cytosine (C), guanine (G) and thymine (T), but the creation of the amino acids is carried out through RNA, which also consists of four bases A, C, G and uracil (U) rather than T. The resulting genetic code is very effective at "reliably providing the amino acids that underpin protein construction," with what has been called an "eerie perfection." Simon Conway Morris, *Life's Solution*, at pp.15, 13–19.

DNA is central in the process of reproduction. Generally, DNA is vastly longer than RNA. DNA carries the genetic code for the organism. Cells reproduce through mitosis. The DNA strand divides, each half is reconstructed into a new identical strand of DNA, and the cell divides into two daughter cells. Within the nucleus, relevant pieces of genetic material (pieces of DNA) are "transcripted", forming pieces of RNA. Those pieces of messenger RNA can leave the nucleus and are then "translated" into proteins by transfer RNA that brings the proper amino acids to the messenger RNA to create the protein. It has turned out, however, that most of the DNA strand is not "normal" genetic material—it is not copied by RNA and translated into proteins. In fact, this non-coding DNA constitutes about 98.8% of the human genome.

There are vegetables, like the onion, with DNA five times the length of human DNA. Thus, it is highly unlikely that the length of DNA bears any relationship to the complexity or sophistication of the organism at issue. See Carl Zimmer, "Is Most of Our DNA Garbage?" *The New York Times*, March 5, 2015. Also, a conclusion has been reached that there are huge amounts of "excess" DNA, based upon mathematical calculations of the amount of coding needed to produce the number of proteins that appeared to be used by the organism.

However, there is much more involved than the mere creation of the proteins. For example, DNA outside of the coding of the 20,000 human genes appears actually to contain information integral to development of the body. So mutations in these areas can and do have devastating effects, contrary to what was thought to be the case just a few years past. In addition, some of the noncoding DNA produces matching RNA. Scientists are investigating what purposes, if any, are served by that RNA. *Id.*

So, there is increasing evidence that at least some of this noncoding DNA is not "junk". Some commentators suggest that continuing discoveries will establish that much, if not most, of the human genome performs tasks, including various functions that coordinate and direct the use of other information contained in the DNA. See, e.g., Stephen C. Meyer, *Signature in the Cell* (2009), pp.406–7. Many of these functions are newly discovered as a result of increased awareness of the variety, complexity and interconnected nature of genetic information. (Yet, many scientists still reject the idea that most of the genome is functional. See Zimmer, "Is Most of Our DNA Garbage?", *The New York Times*, March 5, 2015. Zimmer, a reporter and science writer, speculates that the preference of these scientists for the messier world of the junk DNA may be influenced by their prejudices against anything that suggests Design.)

By the way, there is some reason to think that RNA appeared before DNA. It is, in fact, more ubiquitous. It appears in both "left-handed" and "right-handed" forms, which forms are capable of working together. In DNA, thymine appears instead of (in place of) uracil, which makes DNA more stable. So, RNA may have been the first self-replicating molecule.

Moreover, the sugars necessary for RNA have recently been detected in ancient meteorites. See NASA, "First Detection of Sugars in Meteorites Gives Clues to Origin of Life," *Release*, October 18, 2019.

Nonetheless, "[a]ny explanation for the origin of life must show how this complex trinity – DNA, RNA and ribosome protein – came into existence and started working." Micheal Marshall, "The Secret of How Life on Earth Began," *BBC.com*, 31 October 2016.

Membranes

It is also likely that life requires some type of envelope or packet to contain the genetic material. That envelope would serve to allow a sufficient concentration of the correct molecules for replication, to keep out other compounds and to protect the material from interactions taking place in the surrounding environment. Something like a cell needed to emerge, somehow. As to the satisfaction of this requirement, the theories become highly speculative. Apparently, complex hydrocarbons can form droplets complete with a membrane analogous to that of a cell. Yet, such long hydrocarbons were unlikely to develop on early Earth. At least, they cannot be synthesized in any process that recreates the early conditions on Earth. There are other theories about how membranes or simple cells (without DNA) might have arisen, but there is little evidence supporting any of them. See Luisa, *The Emergence of Life*, pp.85-8, 108-11 (*e.g.*,"primordial cells without DNA", "spontaneous overcrowding", "prebiotic metabolism"); Gray, *Molecules* p.67 (soap); O'Donnell, "How Life Began," *Harvard Magazine*, July-August 2019 (fatty acids).

Metabolism

A cell must be able to convert energy from an external source to its own use; it must be capable of metabolizing. The "process of harnessing energy is so utterly essential, many researchers believe it must have been the first thing life ever did." Marshall, "The Secret of How Life on Earth Began," *BBC.com*, 31 October 2016. There are various possible mechanisms for effecting such conversions. *See, e.g., id.* But, for life as we know it, the process is photosynthesis. We have already discussed how it may be that chloroplasts and mitochondria came to be incorporated into all more complex life forms, but we have not discussed how the ability to create energy from sunlight came about in the first place. The key is in the molecule chlorophyll and the process of photosynthesis. Chlorophyll seems to be very ancient; it can be found in many bacteria. The chlorophyll molecule is able to capture the energy of sunlight, but the details of how the molecule takes photons from the Sun and turns them into usable energy are not known. The process, by which carbon from carbon dioxide is combined with water to produce sugars, releasing oxygen, in a process fueled by the energy derived from the Sun, is complicated. Morris Conway, *Life's Solution*, pp.108-9. But, that process "effectively underpins the entire biosphere." Id.

And, first?

What element appeared first?

> We assume that "one crucial aspect of life – its ability to reproduce itself – appeared before all the others. ...But there are other features of life that seem equally essential. The most obvious is metabolism: the ability to extract energy from your surroundings and use it to keep yourself alive. For many biologists, metabolism must have been the original defining feature of life, with replication emerging later."

Marshall, "The Secret of How Life on Earth Began," BBC.com, 31 October 2016.

Perhaps, it was not the ability to reproduce, but the ability to metabolize. Actually, perhaps, the membrane came first. Otherwise, the chemical processes would not be condensed and protected. And, there is also the question of how all of the "ingredients" got together in a single cell.

> "In order for life to have gotten started, there must have been a genetic molecule—something like DNA or RNA—capable of passing along blueprints for making proteins, the workhorse molecules of life. But ...none of these molecules can do their jobs without fatty lipids, which provide the membranes that cells need to hold their contents inside ... [while} protein-based enzymes (encoded by genetic molecules) are needed to synthesize lipids."

Robert F. Service, "Researchers may have solved origin-of-life conundrum," *sciencemag.org*, March 16, 2015.

THE APPEARANCE OF LIFE

Up until 4 billion years ago, the Earth was subject to a regular bombardment of cosmic materials like comets and meteorites, causing a level of disruption on the surface that must have been incompatible with the development of life. The intensity of some of the collisions would have evaporated the oceans and effectively sterilized the surface. Conway Morris,

Life's Solution, pp.69–72. A significant collision probably occurred about 4.0 billion years ago, giving Earth its orbit, its wobble and its moon—all of which should have contributed to the stability of the environment on Earth.

About 3.8 billion years ago, the situation appears to have improved, as evidenced by remnants of relatively well-preserved geological sediments. *Id.*, p.72. Curiously, the first evidence of possible microbial life dates from "shortly" after that period. (around 3.4 billion years ago). See Luisa, *The Emergence of Life*, pp.10-2. Indeed, the first life on land (mere microbes) appears to have arisen about 3.2 billion years ago (in water, a bit earlier). See Martin Homann, et al.,"Microbial life and biogeochemical cycling on land 3,220 million years ago," *Nature Geoscience*, 23 July 2018; Luisa, *The Emergence of Life*, p.12.

There has been a recent discovery of fossil-like structures in rocks that maybe as much as 4.3 billion years old. If these are indeed the remains of bacteria that lived at that time, they would be the oldest fossils ever discovered and evidence of the earliest known life on Earth. (There is, however, considerable controversy about whether these structures are actually the results of early life and, even if they are, whether that life could have survived the subsequent disruptions.) See Carl Zimmer, "Scientists Say Canadian Bacteria Fossils May Be Earth's Oldest," *The New York Times*, March 1, 2017. In addition, researchers believe that they have found fossils of fungi that may be around a billion years old. Carl Zimmer, "A Billion-Year-Old Fungus May Hold Clues to Life's Arrival on Land," *The New York Times*, May 22, 2019. This discovery also potentially solves an exiting puzzle. It was believed that fungi were only about 500 million years old, but DNA testing of current fungi species suggested that the common ancestor exit a billion years ago. *Id.*

POSSIBLE SOURCES OF LIFE

How were the complex molecules that constitute the genetic material enabling life to exist and, thereafter, evolve (such as RNA, DNA and the many proteins) and the necessary membrane created?

Spontaneous appearance?

Charles Darwin included in the sentence quoted in part to introduce this section this qualification: "oh what a big if". (*Darwin Correspondence Project*, University of Cambridge). In fact, extensive experimentation in the lab indicates that the genetic materials cannot be generated by adding energy (such as heat or light or electricity) to the "soup" of carbon-based molecules that appears to have made up the pre-biotic soup on Earth some three million years ago, as Darwin hoped. See Simon Redfern, "Earth life 'may have come from Mars,'" BBC News, 29 August 2012. Attempts to create RNA in the lab from some combination of molecules and environmental stimuli have been notably unsuccessful. See Luisa, The Emergence of Life, pp.72-96, 111.

It is possible that RNA can be shaped out of simpler molecules by atoms at the crystalline surface of certain minerals that effectively create templates for the more complex molecules. See Conway Morris, Life's Solution, pp.53-6. But, Professor Steven Benner has argued that the minerals most likely to give rise to such templates were likely to have been largely dissolved in the oceans on Earth well before the appearance of RNA. See Redfern, "Earth life 'may have come from Mars,'" BBC News, 29 August 2012. In addition, apparently key elements (boron and molybdenum) were (and are) scarce on Earth. Furthermore, the templating process requires oxidation, and the necessary oxygen for that process was not sufficiently present in the early days on Earth (not until about 2.4 billion years ago).

"Visitors" from space?

The key minerals[2] and the necessary oxygen, however, were abundant on Mars. Benner argues that the RNA on Earth may have originated on Mars and been transported to Earth on rocks and other debris. Id. He notes that while Mars may have been the better environment for the origination of the genetic building blocks, Earth was to become the better environment for sustaining life and giving the opportunity for evolution to occur. Id.[3] Thus, it is possible that all of these necessary molecules came to Earth on carbonaceous meteorites. Conway Morris, Life's Solution, p.38. See also Luisa, The Emergence of Life, pp.32-3, 109. At least, it seems likely that various things that we would consider building blocks of life were created long before the Solar System itself was formed. Conway Morris, Life's Solution, p.39. And, it appears that the materials such as carbon, nitrogen and water had extraterrestrial origins and arrived here via meteors. Id., p.43.

In 2019, scientists found sugars essential to life in ancient meteorites that struck the Earth in the last 50 years:

> "'Other important building blocks of life have been found in meteorites previously, including amino acids (components of proteins) and nucleobases (components of DNA and RNA), but sugars

> have been a missing piece ...,' said Yoshihiro Furukawa, lead author of the study... . 'The research provides the first direct evidence of ribose in space and the delivery of the sugar to Earth. The extraterrestrial sugar might have contributed to the formation of RNA on the prebiotic Earth which possibly led to the origin of life.'"

NASA, "First Detection of Sugars in Meteorites Gives Clues to Origin of Life," *Release*, October 18, 2019.

Of course, this possibility only would explain how life got to Earth.

(Interestingly, there is a position at NASA with the title "planetary protection officer" with the responsibility is to ensure that places like Mars are not infected with life or evidence of life by man's efforts to explore outer space. *See* Kenneth Chang, "Mars Is Pretty Clean. Her Job at NASA Is to Keep It That Way," *The New York Times*, October 5, 2015.)

Something else?

There has been some excitement over recent theoretical work being done by Jeremy England at MIT, which work has been glamorized by author Dan Brown in *Origin* (2017), pp.396-8. One reporter has written:

> "The biophysicist Jeremy England made waves in 2013 with a new theory **that cast the origin of life as an inevitable outcome of thermodynamics.** His equations suggested that under certain conditions, groups of atoms will naturally restructure themselves so as to burn more and more energy, facilitating the incessant dispersal of energy and the rise of 'entropy' or disorder in the universe. England said this restructuring effect, which he calls dissipation-driven adaptation, fosters the growth of complex structures, including living things."

Natalie Walchover, "First Support for a Physics Theory of Life," *Quanta Magazine*, July 26, 2017 (emphasis added).

There are at least two problems with the reporter's characterization of England's work (and with Dan Brown's extrapolation of it):

First, while "[in one] paper, England and his coauthors ... [at] MIT simulated a system of interacting particles [in which] they found that the system increases its energy absorption over time by forming and breaking bonds in order to better resonate with a driving frequency" (*id.*); the scientists did not initiate the energy absorption spontaneously. As they say, "in many cases where external driving brings about a strong shift in a system's state of organization, the outcome we observe at long times should reflect that the system had to be specially configured at some time in its past in order to absorb and dissipate large amounts of work from the drive." Nikolay Perunov, Robert A. Marsland, and Jeremy L. England, "Statistical Physics of Adaptation," *Physical Review*, X 6, 021036 (2016).

Second, even though, these studies do show some response of innate matter to an initial energy source, this responsiveness is still a far cry from what we mean by life. It might come to meet Schrodinger's definition noted above (an organization of matter that resists entropy by utilizing external energy and dissipating it), but the process singularly lacks the capability of reproduction. As the MIT group states: "...even a system devoid of any recognizable self-replicator might still end up seeming 'well adapted' to environmental drives in light of the marks on its current configuration left by its exceptional dissipative history." *Id.*

Thus, we may leave this work aside in this section.

So...

As Conway Morris has explained, there are three possibilities consistent with the timing suggested by the geological and fossil records:

- Life appeared very rapidly (over maybe 20 million years) after the conditions were suitable;
- Life had appeared earlier and some organisms managed to survive the intervening intense heat and trauma;
- Or, life (bacteria) arrived from some extra-terrestrial source, like Mars, and took hold when the conditions on Earth became sufficiently hospitable.

Life's Solution, p.73.

But, whatever the actual story on timing, we still have no explanation of how life came to be, of how it could emerge even under the most perfect of conditions.

Fine Tuning

> "... [T]these numbers
> seem to have been very finely adjusted
> to make possible the development of life."
>
> Stephen Hawking
> A Brief History of Time, p.125

Leaving aside the questions of how and where life emerged, there is also the curious fact that Earth happens to have a rather special specific set of conditions hospitable to life as we know it. However, there is recent evidence that there are life forms on Earth that developed and survived under very different conditions.[4]

Conditions for Life

A "Goldilocks" planet

Earth was the right size, so that gravity was effective to stabilize things, including the atmosphere and oceans, but not so strong as to make the evolution of large creatures physically impossible. It was the right distance from the Sun, given the Sun's particular characteristics, to permit liquid water and other features of a temperate environment. (If the Sun were hotter, the Earth could have been further away; if the Sun were cooler, Earth could have been closer.) The large Moon provided stability to the Earth's axis of rotation. The large mass of Jupiter provided some protection from cosmic debris, and its position may have assisted in directing comets containing water and other necessary elements toward the Earth. See Morris Conway, *Life's Solution*, pp.77–105.

The specialness of the Solar System and the Earth raises the question of whether there are comparable planets elsewhere in the Universe, a question that was very difficult to investigate empirically until the last few years because of technological limitations. Planets around other stars ("exoplanets") were just too far away to be seen by Earth-based telescopes. Their presence was identifiable only from the effects of gravity on the wavelengths of light emanating from the star around which they orbit. Not surprisingly, the early discoveries were all of very large planets, generally orbiting very close to their particular star. Such planets, being large and close, would create relatively significant greater gravitation perturbations. Relatively small planets, like Earth, orbiting far from their suns would not generally produce a detectable effect. See, e.g., id., pp.77–83.

The Search

Interestingly, the "discovery" of an Earth-like planet was announced with considerable excitement in 2010. Gliese 581g supposedly orbited a star in the constellation Libra, about 20 light years from Earth. The planet appeared to be several times the size of Earth, but was believed to be the first planet discovered that appeared to be potentially inhabitable. See Dennis Overbye, "New Planet May Be Able to Nurture Organisms," The New York Times, September 29, 2010. The planet appeared to orbit a dim red dwarf star at a distance that falls within the so-called "Goldilocks zone", where the temperature is neither too hot nor too cold for life to exist. R. Paul Butler, one of the leaders of the team that made the discovery, was quoted as saying "This is really the first Goldilocks planet." Steven S. Vogt, the team co-leader, was quoted as saying, with the qualification that he is an astronomer and not a biologist, "the chances of life on this planet are almost 100 percent." Id.

Yet, almost four years later, scientists from Pennsylvania State University announced that Gliese 581g (and its presumed companion Gliese 581d) may not actually even exist. Douglas Quenqua, "Earthlike Planets May Be Merely an Illusion," The New York Times, July 7, 2014. The astronomers involved collected and analyzed data that they believe indicate that these two planets may be illusions, caused by intense magnetic activity on the star that they were thought to orbit. Id. Other planets thought to orbit the star were clearly confirmed by the data. Id. So, we must be careful about premature conclusions.

Recently, using the orbiting Kepler space telescope, NASA scientists collected data on 150,000 stars. In 2014, NASA announced the discovery of another 715 new planets orbiting distant stars (bringing the total number then discovered to 1700). In those new discoveries, four that were of a size sufficiently similar to Earth and possibly in the right temperature range to be potentially suitable for life. Robert Lee Hotz, "NASA Scientists Discover 715 New Planets," The Wall Street Journal, February 26, 2014. Shortly thereafter, NASA announced the discovery of an Earth-size planet only 500 light years from Earth. Kenneth Chang, "Scientists Find an 'Earth Twin,' or Perhaps a Cousin," The New York Times, April 17, 2014. The planet is bigger than Earth and probably somewhat colder. However, as of now, the mass of the planet cannot be determined, scientists cannot tell whether it has oceans and they do not know what type of atmosphere it has. Its orbit is considerably closer to its star than is Earth's, but it still may fall within the "habitable zone" where water may be liquid (not gas) since its star is cooler. Id. (The planet is circling a "red dwarf," which is much smaller and cooler and probably older than our Sun. That fact is interesting, because red dwarfs are the most common type of stars in the Universe.)

At the beginning of 2015, astronomers announced the discovery of eight new planets in orbits potentially compatible with liquid water. Dennis Overbye, "So Many Earth-Like Planets, So Few Telescopes," *The New York Times*, January 6, 2015. In July 2015, another potentially Earth-like planet only some 1,400 light years away was identified, with a radius of 1½ that of Earth and an orbit of 385 days (Kepler 45). It, also, is too far away for meaningful investigation with current technology. Nonetheless, Jenkins of NASA's Ames Research Center was quoted as saying: "It's awe-inspiring to consider that this planet has spent six billion years in the habitable zone of its star, longer than Earth. That's substantial opportunity for life to arise, should all the necessary ingredients and conditions for life to exist on this planet." Dennis Overbye, "NASA Says Data Reveals an Earth-Like Planet, Kepler 452b," *The New York Times*, July 23, 2015.

During 2016, the pace of new discoveries of such planets accelerated, largely due to improvements in detection techniques and new facilities for observation. *See, e.g.*, Dennis Overbye, "Kepler Finds 1,284 New Planets," *The New York Times*, May 10, 2016; Chris Jones, "The Woman Who Might Find Us Another Earth," *The New York Times Magazine*, December 7, 2016.

In early 2017, scientists reported the remarkable discovery of a small, weak star (another red dwarf) only 39 light-years from Earth with seven planets all roughly the size of the Earth (actually, all smaller, between the sizes of Venus and Earth) and orbiting within a distance from the star that is believed to generated temperatures compatible with liquid water. The star is called Trappist-1. University of Cambridge, "Newly discovered planets could have water on their surfaces," *Research Bulletin*, 24 February 2017.

Then, the Transiting Exoplanet Survey Satellite or TESS replaced the Kepler space station:[5]

> *"After nine and a half years in orbit, 530,506 stars observed and 2,662 planets discovered around other stars, the little spacecraft [Kepler] will be left to drift forever around the sun. ...NASA's new space observatory ... [TESS] has already taken up the search for planets in the nearby cosmos, and giant telescopes both on the*

> ground and in space are being designed to detect and observe exoplanets".

Dennis Overbye, "Kepler, the Little NASA Spacecraft That Could, No Longer Can," NYTimes.com, Oct. 30, 2018.

In mid-2019, NASA made the following announcement:

> "NASA's *Exoplanet Archive announces 31 newly confirmed exoplanets – planets beyond our solar system – discovered by ground and space-based telescopes. Five were detected by the recently launched TESS space telescope. They push the official planet count past the 4,000 mark for the first time. ...They include: A planet almost exactly as big around as Earth, called EPIC 201833600 c, orbiting a star some 840 light-years away. An entirely new planetary system with at least three planets, L 98-59 b, c and d, in Earth's size range... ."*

NASA, "Discovery Alert: A Record Haul--Planet Count Hits 4,000," *Exoplanet Exploration*, June 13, 2019.

And, two months later: "[A] planet, dubbed GJ 357D, orbits a star approximately 31 light years distance from Earth. The super-Earth is in so-called habitable zone, an area far enough from its star to not be too hot but close enough to not be too cold." Tom Fish, "NASA exoplanet discovery: Space agency spots closest 'habitable' Earth-like planet yet," *Express*, August 2, 2019. Of course, most of these potential planets are simply too far away for us to do any investigation of them. Even for the relatively close planets, new technology will be needed to determine whether they are, in fact, inhabitable.

The "Fulton gap"

There is also a curious fact that not many planets between one and a half and two times the size of Earth have been found. "For some reason, planets with radii between 1.5 and two times that of Earth are rare. "The paucity of planets in that range, known as the 'Fulton gap' after the lead author of the paper that pointed it out, first appeared in the findings of the Kepler Space Telescope... ." Rebecca Boyle, "As Planet Discoveries Pile Up, a Gap Appears in the Pattern," *Quanta Magazine*, May 16, 2019.

The first roughly Earth-sized planet found in the TESS survey is 90% of the size of Earth. The others are more than twice as large. *Id.* Perhaps, the problem is one of detection. Or, there may be physical reasons making it difficult for the "Fulton gap" size planets to exist.

For example:

> "These worlds are harder to see because of simple geometry. TESS detects a planet's presence by studying blips in a star's brightness, which indicate something passing in front of it. Planets orbiting at great distances from their star take a long time to cross in front, creating a prolonged blip that is harder to pick up, and they dim the starlight less."

Id.

Also, a planet's size includes its atmosphere. Planets too close to their star may have their atmosphere burnt away, making them "smaller". Perhaps, in addition, only larger planets farther from their star are currently detectable. However, some very recent discoveries may begin to challenge the supposed "Fulton gap".[6]

AND, ALSO...

Water

Of course, life as we know it requires water.

In the last few years, astronomers believe that they have found substantial bodies of water in several places within the Solar System. See, e.g., Kenneth Chang, "Suddenly, It Seems, Water Is Everywhere in Solar System," *The New York Times*, March 12, 2015. More precisely, the scientists have found what they consider to be evidence of the presence of water. NASA's Galileo spacecraft had detected moving magnetic fields on Ganymede, the largest moon of Jupiter, which could be caused by a conductive free flowing saltwater ocean beneath a large layer of ice on the surface (which layer is not a good conductor). Unfortunately, there are other possible causes for the measurements that could be hypothesized. The new research done using the Hubble space telescope studied the swaying of the auroras over Ganymede caused by the changes in magnetic fields that occur during the 10-hour rotation of Jupiter. Computer simulations were used to predict the degree of sway that would occur if the moon were covered in ice with what would occur if there were a saltwater ocean underneath a layer of ice. The actual sway (2 degrees) matched the predicted sway if there was an ocean (versus 6 degrees for all ice). Id.

A recent article in *Nature* reported on the discovery of evidence of hydrothermal vents on Enceladus, a moon of Saturn. *Id.* These vents indicate that the water underground on Enceladus is quite hot (90 Celsius), which would be an unusual phenomenon and in a surprising location, since the moon is not nearly large enough to have a hot radioactive core. One hypothesis is that the strength of the tides caused by the orbit around Saturn could generate enough energy to produce the heat. The evidence for the presence of the hot water is the discovery of tiny dust particles that contain silicon but not sodium or magnesium. The particles appear to be silica, which experiments demonstrate can be produced when alkaline water is heated to at least 90 degrees Celsius. The dust appears to originate in the plumes emanating from Enceladus' south pole. Thus, the supposition that the moon harbors some very hot water. *Id.* The existence of plumes of ice crystals shooting out of geysers on Enceladus were observed some years ago by NASA spacecraft Cassini, and various measurements of gravitational fields suggested the presence of a substantial body of something like water under the surface. *See, e.g.*, Kenneth Chang, "Saturn Moon Has Geysers, Hinting Life Is a Possibility," *The New York Times*, March 10, 2006; "Under Icy Surface of a Saturn Moon Lies a Sea of Water," *The New York Times*, April 3, 2014.

There have even been recent discoveries of further evidence that there might still be water on Mars. *See, e.g.*, Keith Cooper, "Astrobio Top 10: Liquid Water Discovered on Mars," *Astrobiology Magazine*, January 1, 2019 (originally published July 25, 2018). And, evidence of water has been detected on some exoplanets. For example, "[t]wo scientific teams have announced their independent discovery of water—the foundation of biology as we know it—in the atmosphere of a transiting planet dubbed K2-18 b. The planet orbits in the habitable zone of its star, the sweet spot in which starlight may sufficiently warm a world to allow water to pool and flow on its surface." Lee Billings, "Astronomers Find Water on an Exoplanet Twice the Size of Earth," *Scientific American*, September 11, 2019.

Oxygen?

Animals on Earth require oxygen, but life itself does not:

> "The challenge for all life on Earth is the same...[:] to find both a source of electrons and a place to dump them to complete the circuit. Animals get their electrons from the sugar in the food ... . In a series of chemical reactions that happen inside animal cells, these electrons are released and bind to oxygen. That flow of electrons is what powers animal bodies. ...[O]lder, more prim-

itive lifeforms evolved to use other elements as their electron dumps. Many of these lifeforms – such as bacteria and archaea – are still living happily without oxygen today."

Jasmin Fox-Skelly, "There is one animal that seems to survive without oxygen: It is microscopic, looks a bit like a jellyfish, and survives in a place that would kill every other known animal species," BBC.com, 26 January 2017.

There are some very small "animals" that do not require oxygen for metabolization. (We may just be dealing with a problem of categorization, rather than an exception. We have no evidence that these alternative life structures can reach large sizes or complex forms.) These organisms can also live at high temperatures and under very high pressures. Id.[7]

However,

"Life started to develop the ability to metabolise oxygen ... sometime over 1.45 billion years ago. A larger archaeon engulfed a smaller bacterium, and somehow the bacterium's new home was beneficial to both parties, and the two stayed together.That symbiotic relationship resulted in the two organisms evolving together, and eventually those bacteria ensconced within became organelles called mitochondria. Every cell in your body except red blood cells has large numbers of mitochondria, and these are essential for the respiration process. They break down oxygen to produce a molecule called adenosine triphosphate, which multicellular organisms use to power cellular processes."

Michelle Star, "Scientists Find The First-Ever Animal That Doesn't Need Oxygen to Survive," *Sciencealert.com*, 25 February 2020.

Originally, Earth had essentially no oxygen, and the level of oxygen remained quite low for billions of years. It suddenly appears to have surged between 2.2 and 2.0 billion years ago. See Luisa, *The Emergence of Life*, p12. This event is referred to as The Great Oxidization Event. Shortly thereafter, however, "oxygen levels fell dramatically. Researchers say conditions for life on Earth went from 'feast to famine'–and that these conditions persisted for about one billion years." Hannah Osborne, "Two Billion Years Ago, Up To 99 Percent of Life on Earth Died," *Newsweek*, September 2, 2019.

Early bacteria acquired the ability to carry out photosynthesis from sunlight, absorbing carbon dioxide and producing oxygen. One possibility is that over two billion years, these microbial entities produced enough oxygen that the accumulated level became sufficient to support the evolution of animal life. A new theory argues that much of the credit actually goes to sponges. *See* Carl Zimmer, "Take a Breath and Thank a Sponge," *The New York Times*, March 13, 2014. It has been discovered that sponges, technically animals, are able to survive with very low levels of oxygen. Perhaps, millions of years ago, sponges released an abundance of oxygen that transformed the planet. *Id.* While bacteria in the oceans produced oxygen, the dead bacteria floated to the surface and were eaten by microbes, which also consumed much of the oxygen that was being produced, leaving the depths of the ocean oxygen-free. Then algae and other larger organisms appeared. They tended to sink when they died, allowing oxygen to accumulate in the deeper waters. Sponges then appeared. Acting as seawater filters, sponges took nutrients from the water that would otherwise have fed the bacteria and also trapped bacteria, which became food for the sponges. *Id.*

In all events, somehow, over time, a high enough level of oxygen accumulated to permit the evolution of more complex animals in the oceans.

VERY, VERY SPECIAL INITIAL CONDITIONS

Life requires not just a very special Earth, but also a very special Universe.

Lord Rees identified six empirical numerical relationships (dimensionless Constants), which he asserts reflect fundamental facts about the Universe that are crucial to the existence of large, complex entities ranging from galaxies and stars and planets to intelligent life. *Just Six Numbers*, pp.1-4. If the actual values of any of these numbers (or ratios) were very much different than what they in fact are, then various events would have occurred that would have prevented the world we know from existing:

- matter would not have clustered together to form galaxies and stars and planets;
- the creation of more complex chemicals would have been altered so as to result in only a few elements (rather than the 92 that we have found occurring in nature);

- the creation of new elements may have progressed so rapidly as to have largely eliminated all of the hydrogen, preventing the subsequent appearance of water;
- the Universe would have collapsed too soon for evolution to have generated sophisticated life; and/or
- gravity would have been too great for the survival of large animal forms.

These are several numerical relationships that have been identified in the studies of the Universe. They may not actually be constants, but in the span of several billion years they have had significant consequences for the Universe. The numerical values of these characteristics are found, not in theory, but in experimental observations. Moreover, the numbers are not integral parts of scientific theories; they are just empirical relationships that seem to exist.[8]

So, one must conclude that our existence is a highly improbably event. See, e.g., Barrow, *The Constants of Nature*, pp.xiii–xiv, 141–56.

Obviously, we must be careful not to be misled by a *post hoc* examination. Any specific set of circumstances can seem highly improbable in retrospect. For example, many of us have probably been haunted by a sense of fate or predestination arising when we realize that our current circumstances are largely the result of a number of unpredictable and often unlikely events that occurred in our past. But for all or most of those events, our present lives would be quite different. Yet, that fact is simply an inevitable characteristic of anything that results from a series of discrete past events, like a large number of flips of a coin: one may predict the average results (the percentage of heads) with some precision, but any particular sequence that in fact occurs (whether 100 heads in a row, or 100 tosses that regularly alternate between heads and tails, or absolutely any other sequence that is specified) was highly improbable in advance.

However, Lord Rees argues that his six numbers are in fact preconditions of any universe in which life as we understand it and, particularly, intelligent beings could exist, because of basic requirements such as the formation of stars and planets, relatively complex chemistry and a sufficiently stable physical habitat for hundreds of millions of years. In addition, the traditional view is that the assumed Laws of Nature exist independently, so the unique and improbable characteristics of the Universe are the result of the very specific initial conditions upon which those Laws have operated over the life of the Universe.[9]

Thus, the highly unlikely circumstances that facilitated the appearance of life and of intelligent beings must be attributable to and dependent upon those initial conditions. However,

> "[y]ou'd really have to consider independent modifications of all parameters to be able to conclude there is only one combination supportive of life. But this is not presently a feasible calculation. ... [But,] we now think there is more than one combination of parameters that will create a universe hospitable to life. [Recent work indicates that] a chemistry complex enough to support life can arise under circumstances that are not anything like the ones we experience... ."

Sabine Hossenfelder, *Lost in Math: How Beauty Leads Physics Astray* (2018), pp.113-4.

The "Problem"

These observations have been viewed as very bothersome, in large part, because people may be tempted to view them as evidence of the existence of a Creator and, in part, because the situation is almost inconceivable. Like a miracle.

A "brute fact"?

One possibility is that there is no scientific explanation for the special characteristics of the Universe; that the Universe is the way it is may just be what cosmologists tend to refer to as a "brute fact." See, e.g., Rees, *Just Six Numbers*, p.174; Paul Davis, *The Mind of God: The Scientific Basis for a Rational World* (1992), pp.60, 69, 88. It is this highly improbable fact that invites or, at least, allows the alternative of a Creator. The assertion that the Universe is just a "brute fact" for which there is no explanation is attributed to Bertrand Russell. In a 1947 BBC radio debate with Father Frederick C. Copleston about the existence of God, Russell responded to Copleston's presentation of what is called the cosmological argument (everything has a cause; the only imaginable cause of the Universe is God; therefore, God must exist) by denying the necessity of a cause: "the universe is just there, and that's all." See, e.g., Bruce Reichenbach, "Cosmological Argument," *The Stanford Encyclopedia of Philosophy*, Spring 2019. Russell used the phrase "brute fact" to refer to something that "just is" in several of his writings. See, e.g., *The Writings of Bertrand Russell* (1961), pp.614, 616.

THE ANTHROPIC PRINCIPLE?

In the end, one cannot help but ask whether the six numbers are just the accidental preconditions of the Universe in which we find ourselves today or have more profound significance. It may be that these circumstances arose by chance, against extremely long odds. One could say that we should not be surprised, because if such circumstances had not occurred, we would not be here to observe them. It is hard to dispute that statement, but most of us probably cannot help but wonder whether there is something more satisfying to say about the matter. Some physicists and cosmologists have tried to provide something more.

In an article in 1957 and then four years later in a letter published in *Nature*, physicist Robert Dicke made the observation that the great size and age of the Universe that we observed was not a coincidence, because there was a certain amount of time required for the formation of the chemicals of which the life forms with which we are familiar, including ourselves, are made (about 10 billion years). Thus, we could not possibly be able to observe a smaller or younger Universe, since we could not have existed at the time of such a Universe. Similarly, after some period, when the relevant stars have burnt out, we could not be here to observe the Universe. See Barrows, *The Constants of Nature*, pp.106-8, 109-10.

One might think of this phenomenon as a kind of observer bias. There are all sorts of constraints on what an observer can observe, as a result of the place and time in which the observer exists. And, it would be rather foolish to draw implications from the fact that the specific circumstances of the observation might have been viewed as highly unlikely in advance. Like the specific sequence of results from a coin toss, many of the things that do happen were highly improbable in advance.

The underlying concept here was given the name the "anthropic principle" in 1974 by the mathematician and astro-physicist Brandon Carter. Dawkins, *The God Delusion*, p.162; Barrow, *The Constants of Nature*, pp.160-2. Carter's use of the word "anthropic" is somewhat misleading, because it suggests that the principle depends upon or is related to some human characteristic, while it is much more general. It would apply to any type of observer and observation. Carter himself later preferred the phrase "self-selection principle." See Barrow, *The Constants of Nature*, pp.162-3; Dawkins, *The God Delusion*, p.440 n67. In short, the argument is that "because this series of events happened, you are here"; or "because you are here, this series of events happened". Either of these statements probably seems rather odd as an answer to the fine-tuning problem.

A more personalized example might help illuminate what is at issue. Many of you have probably had at some time the strange sensation that some event or situation in your life was "preordained". When you looked back at the series of events leading up to your then current situation, it would appear that each of the events in that series was a contributing factor to who or where you then were, which is true—you are what and where you are because of your entire history of experiences. You recognize that the series of events in question was highly unlikely to have occurred, which may also be true. It is the next step that raises the issue. The realization then may lead to the conclusion that that those events must have been planned or preordained, because they are highly unlikely to have occurred by chance. But, that conclusion is not justified. Remember our discussion of probabilities. In advance, the probability that the series of events leading to your current situation maybe was extremely unlikely; but, after the fact, what occurred, did occur. It is that simple. Indeed, anything that actually occurs was highly unlikely in advance. In the end, there is no support here for a belief that that your life was planned or preordained. Of course, there is also no support for a belief that it was not.

Dawkins sets out a weaker yet more plausible form of the anthropic argument as an explanation of the relative uniqueness of Earth and its inhabitability: There are many billions of planets in the Universe. (Dawkins asserts that the number of planets is likely to be 30 billion per galaxy and the number of galaxies to be 100 billion; however, our efforts to locate planets outside of our Solar System have not generated numbers suggesting anything like that magnitude, as discussed in a later chapter.) So, despite the long odds against the appearance of a planet like Earth, it is not so unlikely that many such planets would appear by chance. Similarly, with enough inhabitable planets in the Universe, even a very highly unlikely (but possible) event like the spontaneous appearance of life would still be likely to occur in at least one instance. Moreover, if there is one such planet on which life arose, that is the very planet on which we would find ourselves, by definition. Therefore, he says, we should not be surprised to find ourselves here—it is the only place we could be. *The God Delusion*, pp.16-9.

Dawkins acknowledges, however, the potential difficulties of applying this version of the argument to the uniqueness of the Universe itself. *Id.*, pp.169-73. We would need to conclude that there are or have been a huge number of universes with differing characteristics (such as different initial conditions and, perhaps, different Laws of Nature) out of which at least one very special case has not surprisingly happened to appear. And, of course, that is the one in which we as the observers exist.

So, how do we get millions of universes? Not surprisingly, such a theory appeared.

MULTI- AND PARALLEL UNIVERSES?

In the twenty-first century, some cosmologists (including Lord Rees) have proposed that there are multiple, perhaps an infinite number of, universes that exist simultaneously (if that concept even applies in this context) in parallel, whether in the same space or in adjacent spaces (if the concept of space even applies). See, e.g., Rees, *Just Six Numbers*, pp.164–74. We have already noted that some of the puzzles that emerged and were highlighted in the development of quantum theory led to theories of multiple parallel universes. Thus, the multiple universes for this purpose could be many parallel, simultaneous universes; a series of different, sequential universes; even different regions of one universe that are vast enough to appear to us to be universes in themselves or something else. See, e.g., Krauss, *A Universe from Nothing*, at pp.128–9, 134–5; Rees, *Just Six Numbers*, pp.166–71; Tegmark, *Our Mathematical Universe*, pp.119–3, 184, 233. Another possibility is that the Universe is a cycling phenomenon: creation, expansion, collapse, followed by a new creation, etc. If time is infinite, then there may have been essentially an infinite number of differing universes, of which our Universe is just one.

In all cases, if it is possible to assert that if there are infinitely many universes (in any of these forms), then it is supposedly almost guaranteed that at least one would be suitable for human life (since the one thing that we know for certain is that such a universe is possible) and that it is obviously logically necessary that we find ourselves in that one rather than in some other, inhospitable one. So, we should not be surprised.The fact that we find ourselves in one such universe is entirely explainable by the fact that we could not exist in the ones that are not suitable, so if we are to exist, then it must be in a universe like ours. See, e.g., Hawking, *A Brief History of Time*, p.124; Krauss, *A Universe from Nothing*, pp.125–7.

OR?

Of course, there may be other solutions to the requirements for intelligent life that would be compatible with other values for some or all of the six numbers. Certainly, in the biological context, the process of nature has evidenced remarkable ingenuity and diversity in the solutions that have emerged to address similar challenges. Yet, as discussed above, there is considerable controversy about the apparent drive toward complexity and sophistication that seems present in the biological world.

It is also possible that there is an underlying, more fundamental theory that we have not yet discovered that mandates the apparent fine-tuning of the Universe, for example, by tying together the numerical relationships that permit the Universe to exist. In other words, there may be a connection among the numerical relationships and between them and the appearance of a universe (a real final theory of everything).

Or, there may be some physical process that would cause the development of these numerical relationships over time, regardless of the actual initial conditions at the beginning of a universe (such as the inflationary hypothesis, discussed above). In either of these two cases, the numerical relationships (fine-tuning) would not just be accidents arising by chance (or through divine intervention), they would be an inevitable part of the physical process by which a universe is created.

Either explanation would be scientific, but neither would serve to rebut an argument for the existence of God based upon design. We still face the question of the source of the process. *See id.* Indeed, it is difficult to imagine how science can avoid that regress. Not surprisingly, some contemporary scientists have been searching for a mechanistic, self-implementing explanation like natural selection that could be applied to cosmology.

INFINITY (AGAIN)

It is somewhat strange that scientists have been promoting the hypotheses of multiple universes as a possible answer to the fine-tuning problem, since no one has postulated a meaningful empirical test of any of these hypotheses. (Tegmark argues that falsifiability exists because these multiverses are the implications of theories that do generate testable propositions, *albeit* of things that are predicted to occur within our Universe. *Our Mathematical Universe*, pp.124-5. *See also*, Hawking and Mlodinow, *The Grand Design*, pp. 117-9. As I have noted, the fact that certain elements of a theory generate testable propositions does not constitute much verification for other, more remote aspects of the theory.)

Indeed, most of the theories themselves specify that the various universes can have no impact on each other. If the universes appear sequentially, then each big bang will have eliminated all evidence and all "memories" of each prior universe; so each new universe will, for all practical, scientific purposes be new. If the universes exist simultaneously in parallel, it is assumed that there is no conceivable communication among them. So, for all practical purposes, each one is independent and alone. Even if they do exist in different regions of a single Universe but are so distant from one another as to prevent any type of communication; then again, as a practical matter, each one stands alone. In other words, these hypotheses are not subject to testing or falsification, because the other universes are, by definition, irrelevant to and have no impact on us or on our Universe.

Leave aside "multiple universes". The argument is that the Universe is infinite. As a result, our mathematical definition of infinite tells us that anything that could possibly occur anywhere does occur somewhere. Thus, there must be numerous (even infinite) worlds more or less just like ours out in the Universe somewhere. See, e.g., Deutsch, *The Beginning of Infinity*, p.98, n*; Greene, *The Hidden Reality*, pp.11-2, 29-42.

But, is this really an "explanation" for—the "why" of—our existence ("anything that can happen, does happen")?

I do not find this "answer" to be satisfying or satisfactory. The alleged existence of all of those parallel worlds may cause some excitation of the imagination, but otherwise what does it really mean to say that such worlds exist but can have no influence on us, or we on them, other than that stimulation of the imagination? *Cf. id.*, pp.189-216. It seems to me that this use of infinity is a good example of what David Deutsch and other physicists, when speaking in a philosophical frame of mind, call a "bad explanation," since it is inherently capable of proving anything, it appears to prove everything. *The Beginning of Infinity*, p.22. See also, Krauss, *A Universe from Nothing*, p.134 (referring to string theory); Rees, *Just Six Numbers*, p.76 (definition of "a 'bad' theory").

At the heart of this argument, we find the concept of "infinity". As previously suggested, this concept—foreign to any actual human experience—is problematic and troublesome. It seems to me that the concept is a significant distraction. Of course, apart from the mathematical complexities, the concept of infinity can be very convenient for, after all, if we are truly dealing with infinities, then anything that is "not impossible is virtually guaranteed to occur somewhere...." Deutsch, *The Beginning of Infinity*, p.126.

But, is that really science?

And, so...

If hypotheses have no empirical meaning or consequence for our world, then is it not the accepted scientific method to disregard them?[10] Or, in this case, does the possible use to rebut the religious inference give these hypotheses "meaning" that matters anyway (even if the meaning cannot be a scientific one)? There is, perhaps, considerable irony in the use by scientists of untestable hypotheses to rebut the presumably untestable hypothesis of the presence of a Designer for the Universe. The strong form of the so-called anthropic principle (and multi-universes) is not to me a persuasive or credible response. (Yet, the argument has also been criticized by some scientists as being too suggestive of or supporting Intelligent Design, rather than being an alternative to it. See, e.g., Susskind, The Cosmic Landscape, pp.7–12; Luisa, The Emergence of Life, pp.24-6.) Thus, to date, we can only say that the finely-balanced structure of the Laws of Nature, the numerous Constants of Nature and various other fundamental phenomena like the Cosmological Constant (or, in today's terminology, dark energy), which, in the aggregate, allow the appearance of intelligent life, are explainable by science only as a staggeringly unlikely set of coincidences—we, our world, the Universe are all just an extraordinarily accident.

Progress

> "...[F]rom the war of nature,
> from famine and death,
> the most exalted object
> which we are capable of conceiving,
> namely, the production of the higher animals,
> directly follows."
>
> Charles Darwin
> *The Origin of Species*
> (1859)

> "People who see
> a direction in human history,
> or in biological evolution, or both,
> have often been dismissed
> as mystics or flakes."
>
> Robert Wright
> *Nonzero*, p.1

Many of us cannot help but see the appearance of progress in biological evolution–toward bigger, smarter, more sophisticated and more adaptable organisms–even if we have doubts about the direction of human history.

It is accepted dogma that "not all evolution constitutes progress, and no (genetic) evolution optimizes progress. ...It has optimized not the functional adaptation of a variant gene to its environment...but the relative ability of the surviving variant to *spread through the population*." Deutsch, *The Beginning of Infinity*, p.91. To make the point rather dramatically, "it is *fittest* for nothing except preventing variants of itself from procreating," that is, for being able to deprive its competing genetic variations from surviving. Id. Thus, "[e]volution can even favor genes that are not just suboptimal, but wholly harmful to the species and all of its individuals." Id. In other words, in such situations, everyone (of that species) is worse off.Indeed, in anthropology, sociology and other cultural studies, the 1920s saw a backlash against the Spenserian theories of the nineteenth century that characterized social evolution in normative terms that were later seen to be racist and prejudiced. See, e.g., Wilson, *This View of Life*, pp.16-35. (This reaction, however, never fully overcame vocabulary usages like "developed" and "developing" in reference to societies or economies.)

Yet, the sense that societal evolution, at least, is progressive is hard to overcome. It seems indisputable that such evolution leads to more complex, more powerful and more productive forms of social organization, even if one studiously avoids the use of more emotive normative terms. See Fukuyama, *The Origins of Political Order*, pp. 50-3. In fact, even Richard Dawkins expressed a belief in "the evolution of evolvability," that the ability to achieve more adept and rapid evolution could itself be captured in reproducible traits and could provide an advantage that would be favored by natural selection. See *The Greatest Show on Earth*, pp.423-5.

"[M]ost evolutionary biologists today deny any pattern of evidence that implies progress." John Hands, *Cosmosapiens: Human Evolution from the Origin of the Universe*, p.349. But, "the overall pattern of the fossil record over time is from simple to more complex; the available evidence for animals indicates that complexity increases along each lineage, and this is clearly the case for the lineage leading to humans." *Id.*, p.348. Indeed, "[n]early all physical things in the universe are complex and form nested hierarchies of increasing complexity, from subatomic particles through atoms, molecules, unicellular organisms, multicellular organisms, to societies of organisms." *Id.*, p.349.

EVOLUTION AND COMPETITION

DARWIN'S VIEW

To start, however, it is useful to note that Darwin used expressions that strongly suggested both the concept of "progress" from evolution and its likely inevitability. In the closing words of the original (1859) edition of *The Origin*, Darwin wrote:

> "Thus, from the war of nature, from famine and death, the most exalted object which we are capable of conceiving, namely, the production of the higher animals, directly follows. ...[W]hilst this planet has gone cycling on according to the fixed law of gravity, from so simple a beginning endless forms most beautiful and most wonderful have been, and are being, evolved."

Quoted in Dawkins, *The Greatest Show on Earth*, p.399.

Indeed, one can also find some suggestions of the elements of a more comprehensive and explanatory theory of progress in Darwin's own writings, ideas that resonate with Adam Smith's work. See Carey, "The Invisible Hand of Natural Selection, and Vice Versa," Biology and Philosophy, pp.427–42. Darwin wrote that as a result of natural selection,

> "each creature tends to become more and more improved in relations to its conditions. This improvement inevitably leads to the gradual advancement of the organization of the greater number of living beings throughout the world. [Where advancement means] the amount of differentiation of the parts of the same organic being ... and their specialisation for different functions... ."

Id., p.436 (quoting Darwin, The Origin of Species (1859), p.92).

The concepts suggested in this passage of Darwin are specialization, or Smith's division of labor, and order (organization). There is a further concept underlying the works of both Darwin and Smith that in modern terms would be called efficiency. There appears to be some natural tendency toward structure and organization that promotes the efficient use of resources and enables more to be obtained from less. Another word that has been suggested is "parsimony." See id., pp.438–9. Indeed, all that is missing from Darwin's theory compared to Smith's is the benefit of free trade (which does not fit very well in the biological context).

An additional parallel between the theories of free market economies and evolution was highlighted by the work of the twentieth century Austrian school of economists, represented by Frederick Hayek (who was discussed in the chapter on Social Science). The recognition that complex interrelated economies can arise and function based upon, not central planning, but on thousands or millions of individual parochial or selfish decisions strongly supports the claim that the staggering complexity of the animal or human body can arise through thousands or millions of small, incremental "local" events. In both contexts, results are determined not by some grand plan or over-riding structure, but in response to factors of a decidedly narrow, localized nature.

As Dawkins observed:

> "The key point is that there is no choreographer and no leader. Order, organization, structure—these all emerge as by-products of rules which are obeyed locally and many times over, not globally. And that is how embryology works. It is done by local rules, at various levels but especially the level of the single cell. ...The

> beautifully 'designed' body emerges as a consequence of rules being locally obeyed by individual cells, with no reference to anything that could be called an overall global plan."

The Greatest Show on Earth, p.220; see also, pp.213-43.

I understand that there are a variety of related ideas and concepts that can emerge from the intellectual ethos of particular periods of human history and that the enormous potency of competition is clearly an intellectual discovery of eighteenth and nineteenth century England. It is the investigation of the implications of competition that tie together Darwin and Smith. To some extent, interestingly, both of these theories contain an implicit appeal to a notion of a "natural order" of things, a concept that can be thought to be suggestive of design and, therefore, a Designer. See Id, p.390.

VERSATILITY

There are limits to the benefits of specialization. Vesatility and adaptability have survival value too. So, evolution may take alternate directions.

> "[T]he bacteria and archaea we observe today are not those that existed 2-3 billion years ago when the Earth was very different and lacked oxygen. Those prokaryotes evolved by forming new species to cope with each new environmental stress as a whole range of environmental niches developed on the evolving Earth, resulting in the current astronomical number of such species. Animals, on the other hand, evolved as more complex species, each able to cope with several environmental stresses. One branching lineage produced the most complex species we know. Homo sapiens is capable of surviving more and different environmental stresses than any less complex species. Uniquely, it has made the whole planet its habitat."

Hands, Cosmosapiens, pp.353-354.

TRADE AND EXCHANGE

Science writer Matt Ridley argues that the past successes of human civilization (and a reason for optimism about the future) lay in the innate drive of humans to engage in trade and exchange. The drive applies not only to material goods but also to ideas. Ridley compares the exchange and intermingling of ideas with the intermingling of genes in reproduction, a process that leads to the selection of the "best". *The Rational Optimist: How Prosperity Evolves* (2010), pp.5–7.

Again, this characterization utilizes the competition model of evolution. In fact, although Ridley introduces Dawkins' concept of memes and the process of natural selection at the beginning (a theory I criticize above), the preponderance of the book is an elaboration of the successes attributable to the processes identified by Adam Smith and David Ricardo: division of labor, specialization and trade (including, the theory of comparative advantage).

Given the apparent propensity of humans, unlike any other species, to engage in trade with non-family members (the cause of which is not explained), the logic of comparative advantage seems to lead inevitably to productivity-enhancing exchange when and where ever the institutional structure allows it (*e.g.*, relative individual economic freedom and stable laws) and the population density is sufficient to permit significant specialization or division of labor.

Ridley presents what I consider to be a meaningful explanatory paradigm or theory:

> "*Somewhere in Africa more than 100,000 years ago, a phenomenon new to the planet was born. A Species began to add to its habits, generation by generation, without (much) changing its genes. ...This [trade and exchange] gave the Species **an external, collective intelligence far greater than anything it could hold in its admittedly capacious brain.**"*

Id., p.350 (emphasis added).

"NON-ZERO-SUMNESS"

In modern parlance, the phenomenon of interest is the "non-zero-sumness" of human economic interaction. *Id.*, p.49 (a phrase attributable to Robert Wright, as set out in Wright's 2001 book, *Nonzero: The Logic of Human Destiny*, from which I quote at he beginning of this section). The point is that economic activity, like many other community phenomena, is not limited to how to divide up a given "pie." To the contrary, the activity can increase the size of the pie, and do so very dramatically. As a result, there is more for everyone. This concept may apply to biological developments far below the level of human economic and political organization.

In his highly popular and influential book, Wright emphasized the importance of the concepts of game theory and the insights to be gained from the recognition that various situations of choice offered not just an opportunity to win or lose (a zero-sum game) but the chance for both (or all) players to win. Robert Wright, Nonzero, pp.4–5. His thesis is that there is "a kind of force–the non-zero-sum dynamic–that has crucially shaped the unfolding of life on Earth so far." Id., p.5.

Wright developed the thesis through an examination of human history, which he characterized as evidencing a trend or tendency toward cooperative solutions that have dramatically increased the benefits available for mankind collectively. He then turns to the evolution of life on Earth:

> "How did life begin? Beats me. But from the beginning, one of its driving forces was non-zero-sum logic. ...From alpha to omega, from the first primordial chromosome on up to the first human beings, natural selection has smiled on the expansion of non-zero-sumness."

Id., p.252.

Wright's argument is based upon the fact noted above that much of life and most of the life forms consist of individual pieces or entities that were once or could once have been independent but that exist now in coalitions of collectives that constitute the more advanced forms of life, including even the cells that make up those organisms (with mitochondria and various organelles). The key assertion is that these cooperative or parasitic relationships can and do promote reproduction of genes.

Just as the survival and replication of a particular gene will depend upon the survival and success of its host (actually, some of its hosts), arrangements that promote the survival and success and proliferation of a host organism can be highly beneficial for the gene that seeks only to reproduce. (I apologize for the anthropomorphism, but it seems to be the easiest way to make the point. Obviously, the gene does not "seek" anything, but the theory of evolution by natural selection assumes that the only relevant driving force is the successful reproduction of the selfish gene. Wright argues that there is something more.)

Wright made his point very graphically with the example of the benefits to the genes of domesticated animals that result from the fact of domestication: "There is today a lot more pig DNA around than its undomesticated kin, wild boar DNA. In that sense–in the Darwinian sense–getting eaten is the best thing that ever happened to pork." Id., p.256.

Now, it is important to note the difference between the issue being introduced here and the concept of altruism as discussed in the chapter on Darwinism. There, cooperation was defined as the process in which individuals do things that would appear to be irrational from a purely individual, selfish standpoint. As we discussed, from the standpoint of the selfish gene, such self-sacrificing behavior by an individual organism can promote the successful reproduction of the gene where genetic relatedness is involved. However, for organisms with sufficient capacities, the concept of reciprocity and an extended time line for measuring welfare can promote real cooperation (the selection of an alternative that seems contrary to self-interest in the short term or isolated context). The point here is that competition among groups can give rise to cooperative behavior that promotes the interests or advantages of one's own group relative to the others. (In his book, Wright incorporates both types of cooperation—kin selection and altruistic reciprocity, as well as benefits from cooperation such as in the division of labor and specialization—as part of the full story. Id., pp.261-4.)

Specialization of functions in the biological sphere could offer non-zero-sum advantages to the organisms, analogous to those attainable by human organization. Ridley cites the obvious specialization that occurs in the body of any complex organism, from the readily observable examples of organs, bones and muscles to the incredibly complicated and effective operation of the immune system or the division of functions between DNA and RNA. Id., pp.41-6. To the extent that the opportunity to realize benefits from specialization can "guide" biological evolution, there would be a tendency toward larger organisms in which greater specialization and the related economies of scale can be realized.

Coincidence

Are there other processes that tend to lead to more and more complex organisms? How about coincidental or opportunistic interactions of organisms or features?

Accidental circumstances have apparently lead to parasitical, symbiotic relationships among types of cells or bacteria. Similarly, the reproductive success of particular genes can be enhanced by variations that enable the host organism to do more with less—to achieve greater success in the competition with all other organisms over the scarce resources that support life, not by getting relatively more of the resources, but by making better use of the share that is obtained. (As Ridley has observed: "Civilisation, like life itself, has always been about capturing energy. That is to say, just as a successful species is one that converts the sun's energy into offspring more rapidly than another species, so the same is true of a nation." *The Rational Optimist*, p.244.)

Moreover, changes that occur for one "reason" or no reason (that is, adaptive or not) may inadvertently offer opportunities for new functions or features not "anticipated" by the initial changes (excuse, again, the anthropomorphism). We have already mentioned feathers and flight (in the chapter Darwinism). A similar example might be the arrangement of multiple wings on an insect. See Christakis, *Blueprints*, pp.86-9. One theory is that the incipient wings developed as thermal regulatory devices, to help control the body temperature of the insect. *Id*. At some point, the appendages that were presumably favored traits because of that use may have began to serve another, totally unrelated function in connection with the mobility of the insect. If this scenario is correct, then the usefulness of the appendages for different purposes led to the ultimate emergence of wings that enabled flight. Or, the *bacterial flagellum* that enables bacteria to move (or "swim") may have started out as shorter "stingers" used as weapons. See Collins, *The Language of God*, pp.185-6,191-3. (Of course, we are necessarily involved here in *post hoc* speculation.)

Not only can physical changes arise that take advantage of environmental opportunities, but opportunities can arise that provide advantages to prior physical changes. The saying that "the organ precedes the function" has an ancient origin. It was expressed over 2,000 years ago by Aristotle. Or, more poetically:

> *"Since naught is born in body so that we*
> *May use the same, but birth engenders use:*
> *No seeing ere the lights of eyes were born,*
> *No speaking ere the tongue created was;*
> *But origin of tongue came long before*
> *Discourse of words, and ears created were*
> *Much earlier than any sound was heard;*
> *And all the members, so meseems, were there*
> *Before they got their use: and therefore, they*
> *Could not be gendered for the sake of use."*

Lucretius
De Rerun Natura, Book 4 (BCE 50)[11]

The proposition has a certain logical appeal. To the extent that it reflects actual history, it poses interesting questions for the Intelligent Design advocates, as well as for the neo-Darwinians.

PHYSICAL PROCESSES

A "CONSTRUCTAL LAW"

We have noted the potential importance of the concept of efficiency above. A theory based on a notion of efficiency was systematically developed by Adrian Bejan, a professor of mechanical engineering. He describes discovering what he calls "the constructal law," which he asserts applies in biology as well as in the physics of natural phenomena. *Design in Nature* (2012) (with J. Peder Zane). His theory is based upon the characterization of most phenomena as "flows" governed by his law, which dictates that "flow systems should evolve over time, acquiring better and better configurations to provide more access for the currents that flow through them." Id., p.5.

Bejan explains that he "realized that the world was not formed by random accidents, chance, and fate but that behind the dizzying diversity is a seamless stream of predictable patterns." Id., p.2. He characterizes this law as a law of physics that affects "biology, hydrology, geology, geophysics, [and] engineering" (id., p.6) and as a "first principle" that cannot be deduced from other laws (id., p.14).[12]

Bejan applies the flow analogy to all movement that will necessarily encounter resistance, arguing that in all of these instances a design will appear over time that will maximize the flow (that is, minimize the impact of the resistance). He claims that the patterns of design that will occur are predictable and can be predicted by calculations and computer simulations given the conditions of the flow and the resistance. He applies the theory to map the patterns of the erosion from rain, the tributaries of streams, the pattern of lightening, the basins or deltas of rivers, the growth of trees, the circulatory systems of living organisms, the locomotion of animals and birds, and so on. He also derives various "scaling" relationships that are predictable as size increases.

His argument is that movement, whether "driven" by the Second Law of Thermodynamics (whereby temperatures, the levels of liquids, the density of matter all tend to equalize over time) or by the needs of living organisms, there will be a tendency over time for the patterns of movement to conform more and more closely to designs that enable the movement at the least cost (in energy) or toward the balanced tradeoffs of competing needs (such as the ability of a heart to pump blood balanced against the additional demands that a larger heart places on the body, in terms of weight, space, etc.). Id., pp.96-8. He also finds designs for skeletons, tree trunks and the like as reflecting basic principles of physics. See, e.g., id., pp.120-1.

I admit that my first reaction to Bejan's law was that it was an obvious and empty truism: like taking the path of least resistance. Certainly, the concepts of faster and easier are value-laden and lend themselves to circular reasoning. Continuing to read, however, I realized that the proposition was different—the flow does not take the path of least resistance; instead, flow patterns or designs will arise over time that enable the flow to reduce the impact of the resistance that is inevitably there, the impact on the flow, facilitating the attainment of more movement from the available energy (or of the same movement with less energy).

As is obvious from the examples cited, Bejan's principles (or laws) apply to all aspects of nature, not just to living things, thereby presenting a connection between biology and various other sciences, including mechanical engineering. "The *design* we see in nature—the shape and structure of rivers, animals, cities, *etc.*—is a manifestation of this tendency in nature to generate shape and structure to facilitate flow." Id., p.31.

An interesting aspect of this theory is that, as Bejan claims, it may provide an objective, measurable standard against which alternative designs can be assessed to identify which is better. That standard then could establish one means by which one can judge the results of natural development. Bejan asserts:

> "The constructal law also challenges another idea that has become dogma since Darwin—**that there is no overarching direction to evolution**. ...The closest they come is through a piece of circular logic that says: A change is better if it aids survival; any change that aids survival is better. The constructal law, by contrast, predicts that **evolution should occur because of the tendency of all flow systems to generate better and better designs** for the currents that flow through them. It expresses 'better' in unambiguous physics terms—change that facilitates faster, easier movement."

Id., p.23 (emphasis added).

There are numerous examples of phenomena in nature that appear to have realized optimal physical solutions, maximizing the benefits obtained from particular resources. Many of these phenomena reflect what would be calculated by skilled engineers to be the optimal. Examples include the design of leaves on plants and trees, allowing for the laws of evaporation and the rules of scaling with great precision; the foraging strategies of bees, achieving great efficiency in achieving the necessary objectives of searching for and collecting pollen; and, the wing strokes of various flying insects. Fodor and Piattelli-Palmarini, *What Darwin Got Wrong*, pp.83–9. The sophistication and complexity of the air-borne maneuvers of certain insects are essentially unique. See, e.g., Michael Dickinson and Karl G. Gotz, "Unsteady Aerodynamic Performance of Model Wings at Low Reynolds Numbers," *The Journal of Experimental Biology*, 174 (1993), pp.45–64.

There are also examples of engineers finding solutions by looking at what nature "produces," that is, solutions found in nature. *See, e.g.*, Jacqueline Garget, "Suction cups that don't fall off: Insects in torrential rivers may inspire engineering solutions," *University of Cambridge Newsletter*, 20 December 2019; Dawkins, *The Blind Watchmaker*, pp.21–37 (bats have developed multiple different solutions to the same basic technical problems of designing an effective and workable sonar navigation system, many of which solutions were also developed in the twentieth century through the ingenuity of human engineers). This process has been called "biomimicry." *See* Courtney Linder, "This Badass Beetle Could Keep Your Car Cool: See how scientists are inspired by the bug, which survives at extreme temperatures," *Popular Mechanics*, June 14, 2020.

I note that the concepts of maximization or minimization (more generally, optimization) have a long history. "Optimization problems arise naturally, and were well known to mathematicians in antiquity." Julian Barbour, *The End of Time* (1999), p.109. This history includes various mathematical techniques to determine the shortest paths between points on various types of surfaces and the derivation of the "principle of least action." *See id.*, pp.109–20. As Barbour observes: "A rich body of mathematical and physical theory has developed out of [such] problems. It cannot explain why the universe is, but given that the universe does exist it goes a long way to explain why it is as it is and not otherwise." *Id.*, p.110.

The "law of forms"

Bejan is actually is a recent contributor to a line of thought that has a substantial history (a history of which Bejan shows little awareness): it is the concept of a "law of forms."

Fodor and Piattelli-Palmarini describe a recent reemergence of attention to that concept. *What Darwin Got Wrong*, pp.72-92. The concept refers to the appearance of optimal solutions in nature: "[C]ases when optimal structures and processes have been found in biological systems. These are naturally occurring optimizations, probably originating in the laws of physics and chemistry." Id., p.xix. They predict that "other self-organization processes by autocatalytic collective forces are almost sure to be elucidated in the near future." Id., pp.xix-xx. They note that certain very specifically defined (mathematically) forms appear in a wide range of natural phenomena, such as Fibonacci series and spirals appearing in spiral nebulae, the arrangement of magnetically charged droplets, seashells, the alteration of leafs on stalks of plants and the placement of seeds in a sunflower. Id., p.74. The inference would seem to be that these arrangements are imposed by laws of physics and chemistry that create "constraints on possible biological forms, more particularly on stable and reproducible biological forms." Id. The more efficient forms will tend to exist where there is some force or limitation that gives an advantage to such efficiency.

Fodor and Piattelli-Palmarini attribute the phrase "law of forms" to D'Arcy Wentworth Thompson, who argued in the early twentieth century that biologists needed to pay more attention to the physical laws that affect the form and structure of organisms. Id., p.75. They observe that work reflecting this concept was undertaken by mathematicians Vito Volterra, Alfred J. Lotka, Alan Turing and Rene Thom, and by chemists Boris P. Belousov, Anastol M. Zhabotinsky and Ilya Prigogine. In the second half of the twentieth century, several biologists continued to stress the importance of the law of forms, including Conrad Hal Waddington, Stephen Jay Gould and Richard C. Lewontin. Id., pp.76-7. The authors also comment on the work done by Bejan and others on the apparent laws of scaling based upon multiples of quarters, not thirds. Id., pp.78-9

Michael Denton and Craig Marshall asked in an article in *Nature* whether "some of the higher architecture of life might be determined by physical law? ...[If so, then] all the diversity of life is a finite set of natural forms that will recur over and over again anywhere in the cosmos where there is carbon-based life." "Law of forms revisited," *Nature*, vol. 410, 2001, p.417. This theme is also pursued by Simon Conway Morris in his theory of convergence in evolution.

The "forms" that seem to be dictated by chemistry and physics clearly cannot be the result of or caused by natural selection; however, in the biological sphere, it is possible that natural selection plays a role in the frequent appearance of these "forms" among living organisms. Perhaps, various physical laws establish the optimum structures and arrangements of elements, and the process of natural selection through random variation (or mutation) tends to "discover" those forms over time. Once discovered, the optimum structures and arrangements would tend to be stable and persistent.

Thus, we may have another example of constraints or limitations imposed on the process of natural selection by other Laws of Nature.

A ROLE FOR EFFICIENCY

Curiously, Bejan's thesis asserts a type of directionality, but he does not clearly articulate the potential mechanism through which that directionality is achieved. in the natural world, an inherent parsimony is the directing force, but how does it operate? Bejan merely suggests that, like in the Second Law of Thermodynamics, nature moves toward uniformity. Equilibrium arises when no change can increase the degree of uniformity or improve the flow.

Economics presumes that man's greed and self-interest propel the adjustments leading to equilibrium; Darwinian theories propose that scarcity drives natural selection operating on random variations (but cannot itself explain the apparent directionality). Since many of Bejan's examples are not biological, subject to natural selection, such as the course of a stream or the contour of a coastline, what could be the mechanism that propels the changes toward equilibrium? My guess is that the inefficient patterns of organization are inherently less stable than the most efficient, optimal patterns; thus, disruptions, whether caused by wind, rain, the tides or other phenomena, will cause the inefficient patterns to change until, by chance, the optimal pattern is achieved. Further disruptions will often not imbalance the system.

A similar idea is reflected in Jeremy England's theory of "dissipative adaptation," already discussed in the prior section:

> *"while any given change in shape for the system is mostly random, the most durable and irreversible of these shifts in configuration occur when the system happens to be momentarily better at absorbing and dissipating work. With the passage of time, the 'memory' of these less erasable changes accumulates preferentially, and the system increasingly adopts shapes that resemble*

> those in its history where dissipation occurred. Looking backward at the likely history of a product of this non-equilibrium process, the structure will appear to us like it has self-organized into a state that is 'well adapted' to the environmental conditions. This is the phenomenon of dissipative adaptation."

"Dissipative adaptation in driven self-assembly," *Nature Nanotechnology*, November 2015.

England is exploring systems that are being "fed" energy (are "driven") and are then dissipating heat or some lower form of energy. His modeling suggests that the more efficient organizations for the conversion of energy can be favored. "[E]ven a system devoid of any recognizable self-replicator [like present in life forms] might still end up seeming 'well adapted' to environmental drives in light of the marks on its current configuration left by its exceptional dissipative history." Nikolay Perunov, Robert A. Marsland, and Jeremy L. England, "Statistical Physics of Adaptation," *Physical Review*, X 6, 021036 (2016).

So...

We are considering theories other than natural selection that help explain evolution and that suggest non-circular means of detecting or defining progress. Importantly, some of these forces, if they exist, apply to inanimate matter, as well as to living things. Thus, they may help explain even more of the world that we experience than can natural selection. The foregoing discussions suggests the importance of cooperation and possible theories of multi-level "natural selection" (that is, of natural selection and competition working simultaneously at different levels, from the cell to the nation-state, with differing and potentially conflicting influences on the course of "evolution"), combined with endosymbiosis and horizontal gene transfer, non-adaptive sexual selection and the various types of forces suggested by Bejan's "constructal law", the law of forms and England's "dissipative adaptation". And, we have Nagel's argument for teleology. In the end, I present all of these ideas as suggestive of significant revisions to our theory of evolution that could supplement neo-Darwinian natural selection. Such revisions could, in fact, represent a partial return to the actual ideas of Darwin, which were considerably more expansive and flexible than the more formalistic neo-Darwinism that emerged in the twentieth century. See Luisa, *The Emergence of Life*, pp.182-3 ("Conversation with Denis Noble").

Intelligence

> **"Homo sapiens is an entity,
> not a tendency."**
>
> Stephen Jay Gould
> Wonderful Life, p.320

> **"... [O]nce it has emerged from the inorganic,
> life continues naturally ...
> to become both [more complex] externally
> and more conscious internally;
> ...up to the ... emergence of reflection."**
>
> Pierre Teilhard de Chardin
> Christianity and Evolution, p.230

This brings us to a particularly contentious debate, not just as to whether evolution by natural selection is directional (discussed in the prior chapter), but, more specifically, whether the end result (in our case, human beings) is in some sense preordained or, at least, highly likely. This issue is part of a broader controversy over contingency (its role and importance) and determinism. See, e.g., Luisa, *The Emergence of Life*, pp.14-9.

So, we start with this question: Are there certain characteristics of life on Earth that are inherent in the original conditions and Laws of Nature, so that one would expect similarities to appear in the results if the process could be started over and rerun again? The consensus answer seems to be yes, for reasons discussed in a prior chapter. Even paleontologist Stephen Jay Gould, a denier of the inevitability of the evolution of human-type life, agreed.

As Gould explained:

> "Am I really arguing that nothing about life's history could be predicted, or might follow directly from general laws of nature? Of course not.... **Life exhibits a structure obedient to physical principles. ...I suspect that the origin of life on Earth was virtually inevitable**, given the chemical composition of early oceans

and atmospheres, and **the physical principles of self-organizing systems**. Much about the basic form of multicellular organisms must be constrained by the rules of construction and good design. The **laws of surfaces and volumes**, first recognized by Galileo, require that large organisms evolve different shapes from smaller relatives in order to maintain the same relative surface area. Similarly, **bilateral symmetry** can be expected in mobile organisms built by cellular division."

Wonderful Life (1989), p.289 (emphasis added). (The title of the book is based upon the movie It's a Wonderful Life, with Jimmy Stewart.)

CONTINGENT?

Gould, however, emphasized the significance of contingency or chance in the evolutionary process, expressing "amazement ... at the fact that humans ever evolved at all." Id., p.289. He identifies that numerous occasions on which the human evolutionary path could have been wiped out by events, and he noted the examples of earlier branches of the Homo sapien line that did in fact become extinct. Thus, his conclusion quoted above. Gould's thesis is as follows:

> "**The divine tape player holds a million scenarios**, each perfectly sensible. Little quirks at the outset, occurring for no particular reason, unleash cascades of consequences that make a particular future seem inevitable in retrospect. But the **slightest early nudge contacts a different grove, and history veers into another plausible channel, diverging continually from its original pathways**."

Id., pp.320-21 (emphasis added).

Robert Wright, whose views are discussed in the preceding section, responds to Gould asserting that the question of interest is not whether Homo sapiens would emerge if we were to rewind the tape and start over again, but whether there is reason to expect that sophisticated intelligence, even self-consciousness, would tend to emerge from the process: "whether the evolution of something as smart and complex as us was very likely." Nonzero, p.269. Without ignoring the impact of chance and accident, Wright asks whether there are non-random factors that could be expected to lead toward intelligence and complexity. Id., pp.270-2. He suggests that a key is "positive feedback," that is, that complexity leads to more complexity, similar to the "arms race" metaphor (more sophisticated armaments lead to more sophisticated defenses, *etc.*). He notes that increases in intelligence have happened innumerable times, as have increases in flexibility and adaptability—all characteristics with potential survival value, like eyesight. *Id.*, pp.274–5. Wright is not responding to the fact of randomness, which relates to the types of mutations or changes that arise. He is making a point about the mutations that are more likely to survive, the concept of natural selection.

Gould's view has also been challenged by a fellow paleontologist, Simon Conway Morris, who had been involved with Gould in the work on which Gould's book was based. See *The Crucible of Creation* (1998). Both books examine the fossil discoveries in the Burgess Shale in the Canadian Rockies, discovered and investigated by Charles D. Wolcott in the early twentieth century. That discovery provided an unusual insight into the state of life on Earth some 520 million years ago, shortly after the Cambrian "explosion".

The issue is the following: Life (bacteria) existed on Earth for some 3.5 billion years and multi-celled organisms (eukaryotes) for almost a billion years. Then, suddenly (over a period of only 5 to 10 million years), the variety and complexity of life forms on Earth exploded. Essentially all of the major phyla in existence today appeared during the Cambrian explosion. (And, this complex life has persisted over 520 million years, despite the five major mass extinctions.) Why so late and why so suddenly? A short "challenge" and "reply" by the two appeared in the magazine *Natural History* in 1998. Conway Morris and Gould disagree over the novelty of the life forms that are found in the Burgess Shale and over the changes that have occurred since. That debate certainly illustrates the difficulties of interpreting a fossil record.

But, the dispute of interest here is between Gould's conclusion that intelligent, conscious life like humans arose only by reason of a serendipitous, chance event and Conway Morris' argument that given the life, "a creature with intelligence and self-awareness on a level with our own would surely have evolved—although perhaps not from a tailless, upright ape." Simon Conway Morris and Stephen Jay Gould, "Showdown on the Burgess Shale," *Natural History*, 107(10): 48–55 (1998).

CONVERGENT?

Conway Morris relied on the evidence of convergence in evolution, that is, the appearance of the same solution to an environment challenge (or opportunity) on multiple occasions and from various different ancestral organisms. The examples of the eye and of flight are perhaps the best known. Gould responded by suggesting that Conway Morris overstates the extent and significance of convergence. Id., pp.48–55.

Conway Morris has expanded his argument in later books. See *Life's Solution: Inevitable Humans in a Lonely Universe* (2003); *The Runes of Evolution* (2015). He has also argued that evidence of convergence can be found in much earlier developments and in lower parts of the taxonomic hierarchy. "Evolution: like any other science it is predictable," *Philosophical Transactions of The Royal Society*, Biological Sciences, 24 November 2009, pp.133–45. Some part of the excitement that his scientific ideas have generated undoubtedly relates to the fact of his religious beliefs. However, Conway Morris criticizes Intelligent Design and asserts the incontestability of the Darwinian mechanism of evolution, noting that the fact of convergence supports adaptational explanations. But, he recognizes that his argument that "something very like a human is an evolutionary inevitability ... hardly sits comfortably with neo-Darwinian orthodoxy." Id.

(Interestingly to me as an aside, Gould noted during the debate discussed above that "Conway Morris charges that my arguments for contingency arise 'not from the evidence of paleontology but from [my personal credo about the nature of the evolutionary process].' This claim, however ungenerously stated, is—and must be—true, for any general view of life must read evidence in the light of a favored theory... ." "Showdown on the Burgess Shale," *Natural History*, pp.48–55. In short, no one can be truly objective, as discussed in the chapter on Knowledge and Understanding.)

In the years since the debate, much more evidence has been collected. See Losos, *Improbable Destinies* (2017). With the increasing availability of DNA sequencing in the twenty-first century, researchers obtained some startling results. It appears that many, if not most, of the widespread adaptive physical and behavioral characteristics that constitute solutions to common environmental challenges or opportunities have arisen independently several times, that is, are examples of convergent evolution. Id., pp.28-78. At the same time, we know of many examples in Madagascar, Australia and, especially, New Zealand of apparently unique species occupying the type of environmental niches that are home to similar species in other regions of the world. Id., pp.81-105.

In his latest book, Conway Morris argues for a "map of life" that sets out the rules by which all living things are likely to develop, once life appears. *The Runes of Evolution*. This "map" (the outcome of convergence) sets out the expected or preferred forms by which the necessary functions of living organisms—eating, moving, sensing the environment, reproducing, and so on—are achieved. Where there are a finite number of solutions for a particular environmental challenge or opportunity and one of the solutions is noticeably better (in terms of net benefits), then one may expect to see convergence on that best solution. This process of convergence extends to the emergence of intelligence and even consciousness.

This hypothetical, however, is quite incomplete. For example, it assumes that many of the existing organisms at the beginning have available a genetic path from their current positions to that best solution. For many or even all, that may not be the case—i.e., it may be that you simply cannot get there from here. But, even where such a path is available (a series of mutations constituting the steps from one place to another), each step along that path must be an adaptive one—i.e., one that in and of itself brings reproductive benefit. That may be a very tall order. Moreover, if at any one of these steps there are alternative adaptive mutations and those alternatives are inconsistent with reaching the final goal, then the organism may (or, will, where such an alternative arises) reach an evolutionary dead-end relative to that best solution. Thus, the number of steps required can make a big difference to the likely outcome.

Of course, there also may be significantly different solutions for any particular environment challenge or opportunity. For example, in response to a new, large predator, some organisms might develop the ability to hide or disguise themselves, some may develop increased speed (*e.g.*, longer legs) to escape the predator, some might develop defensive features like armor or spikes (like the armadillo and the porcupine) to discourage the predator, and so on. And, each such alternative can come in a multitude of forms. Similarly, Dawkins discusses of the different means by which various varieties of bats manage to navigate in the dark at night using sonar. *The Blind Watchmaker*, pp.21–37.

However, some of our observations may be the result of "confirmation bias"—if one looks for examples of convergence and finds them, one may tend to conclude that convergence is quite common. See Nick Longrich, "Evolution tells us we might be the only intelligent life in the universe," *The Conversation*, October 18, 2019. What if one looks instead for examples of uniqueness (non-convergence)? Paleontologist Nick Longrich writes:

> "...when you look for non-convergence, it's everywhere, and critical, complex adaptations seem to be the least repeatable, and therefore improbable.What's more, these events depended on one another. Humans couldn't evolve until fish evolved bones that let them crawl onto land. Bones couldn't evolve until complex

> animals appeared. Complex animals needed complex cells, and complex cells needed oxygen, made by photosynthesis. ... [M]any critical events in our evolutionary history are unique and, probably, improbable. One is the bony skeleton of vertebrates, which let large animals move onto land. The complex, eukaryotic cells that all animals and plants are built from, containing nuclei and mitochondria, evolved only once. Sex evolved just once. Photosynthesis, which increased the energy available to life and produced oxygen, is a one-off. For that matter, so is human-level intelligence. ...All this convergence happened within one lineage, the Eumetazoa. Eumetazoans are complex animals with symmetry, mouths, guts, muscles, a nervous system. Different eumetazoans evolved similar solutions to similar problems, but the complex body plan that made it all possible is unique. **Complex animals evolved once in life's history... .**"

Id. (emphasis added).

Moreover, it appears that the human brain, the most remarkable and adaptational development of all, other than life itself, has appeared only once on Earth.

> "[L]arge, compressed and more complex than the brain of any other species. It has tripled in size over the last six million years, with most of this transformation occurring 200,000-800,000 years ago, largely before the emergence of Homo sapiens. ...If a brain resembling the human one is indeed so unambiguously beneficial for survival, why has no other species developed a similar brain over billions of years of evolution?"

Oded Galor, Humanity: The Origins of Wealth and Inequality (2022), p.14.

Of course, the human brain comes with a cost—it requires considerable energy and its need to develop after birth leaves human infants helpless for years. Id., p.15. But, no alternative means of achieving the same capabilities and functions has appeared.

PREDICTABLE?

Not surprisingly, it appears that most examples of convergent evolution have arisen where the starting point is a set of similar or, perhaps, closely related species, most of whom would be likely to have available genetic paths to the best solution; although, it appears that often the genetic changes to get there are different (the same phenotype is achieved with different genetic structures). Where the original organisms are significant different, in contrast, one would expect that the evolutionary results will be quite different, even where similar functions are achieved.

What does this all mean for predictability? The new evidence suggests that some significant predictability can be expected in the short run if the starting conditions are known, but chance is still important. It seems to me that in terms of "rerunning the tape", it is likely to make a considerable difference how far back one rewinds. Millions of years ago, it is likely that the range of possible directions for evolution was substantially greater than it was hundreds of thousands of years ago; and the role of chance, therefore, was more significant. That would suggest that as evolution progresses, the range of ultimate possible outcomes narrows and does so significantly. This concept makes sense and comports with many of our own personal experiences in life. And, it suggests that evolution will eventually take a direction and that that direction will increasingly become more focused.

I also think that the evidence of convergence is, at least, modest support for the proposition that some of the broader characteristics are likely to arise over time. Some of these broader characteristics presumably have many alternative genetic paths for their realization; others may have only one. Where we can say that the characteristic has clear and widely applicable reproductive benefits (such as intelligence, giving flexibility and adaptability and enabling innovation in behavioral patterns), then I am tempted to say that overtime, it is likely to appear, even if it is not "inevitable".

Similarly, Bejan asserts based upon his "construal law" (discussed in the prior section) that:

> "[I]f the tape of evolution were rewound and if swimmers, runners, and fliers appeared again, their shapes and structures should produce the same types of speeds, stroke-stride frequencies, and force outputs of these forms of locomotion as exist today. Their circulatory systems would still have a tree-shaped design; their organs would still have characteristic sizes; and, when useful, they would follow movement and migration patterns."

Design in Nature, p.100.

On one point made forcefully by Gould, there seems to be little disagreement. The timing by which these various things may happen is quite unpredictable. It might be a billion years or 10 billion years between important steps of evolution. See Luisa, *The Emergence of Life*, p.15.

Perhaps the strongest statement that evolution, very broadly defined, has a decided, inevitable direction comes from Pierre Teilhard de Chardin (1881-1955), a Jesuit priest and paleontologist who studied chemistry, physics, botany, and zoology and received his doctorate in geology. He concluded:

> "*Man used to be regarded as an anomaly in the universe; from now on he is tending to be seen **as the extreme point attained at this moment**, in the field of our experience, by the [process of] 'negative entropy' ...–or, more simply, evolution. ... Left to itself, under the influence of chance, **matter tends to group itself into as large molecules as possible. And, experientially, life stands as the natural and normal continuation of this 'moleculization'** process.*"

Christianity and Evolution: Reflections on Science and Religion (1969), p.221 (emphasis added).

But, Alone?

Intelligent life should emerge repeatedly throughout the Universe. See Conway Morris, "Map Of Life' predicts ET. (So where is he?)", *Cambridge University Research Magazine*, 2 July 2015. However, "[t]he almost-certainty of ET being out there means that something does not add up, and badly. ...We should not be alone, but we are."

In their earlier debate, however, Conway Morris acknowledged that we do not know how life arose or how probable that event was, even given the most hospitable environmental conditions. He concluded, "we must admit the real possibility that life arose but once and that we are alone and unique in the cosmos... ." "Showdown on the Burgess Shale," *Natural History*, pp.48-55. Similarly, Longrich (quoted above) asserts: "None of this happens without the evolution of life, a singular event among singular events. All organisms come from a single ancestor; as far as we can tell, life only happened once." "Evolution tells us we might be the only intelligent life in the universe," *The Conversation*, October 18, 2019.

In contrast, Gould concluded: "The origin [of] life seems reasonably predictable on planets of earthlike composition, while any particular pathway, including consciousness at our level, seems highly contingent and chancy." "Showdown on the Burgess Shale," *Natural History*, pp.48–55. And, Teilhard de Chardin, in further contrast, declared that self-reflective, conscious life must exist across the Universe. *Christianity and Evolution*, pp.230-1 ("the idea of a single hominized planet in the universe has already become in fact ... almost as inconceivable as that of a man who appeared with no genetic relationship to the rest of the earth's animal population").

In the preceding two sections, we looked at the string of circumstances that had to be realized for an inhabitable planet like Earth to exist. The accumulating series of apparently highly unlikely events suggest that the appearance of life is a "happy accident" or "near miracle." Conway Morris, *Life's Solution*, p.67. Indeed, it may be that we are alone, not because evolution would not often lead to the appearance of intelligence and consciousness, but because life has not appeared elsewhere or, at least, not where the conditions were suitable for the evolution of complex organisms. This possibility creates disquiet among some scientists, because of the opening it seems to provide for a religious explanation. See *id.*, pp.67-8.

Of course, we do not know that we are alone—we just can find no evidence of other intelligent life (so far). In addition, convergence would predict other intelligent life forms only if life has arisen multiple times.

And, even if we are alone, we face two quite different possibilities:

- Life has arisen in many places, but convergence is not as powerful as Conway Morris asserts; or
- The appearance of life is extremely rare in the Universe, perhaps even unique to Earth.

It would be interesting to know which proposition is true. Or, if the truth is something else entirely.

And, Us

> "...[F]ace to face with
> something which does not exist,
> to which it alone
> can give reality and substance,
> which it alone
> can bring into the light of day."
>
> Marcel Proust
> *Swann's Way* (Overture)

> "'Thou great star!
> What would be thy happiness
> if thou hadst not
> those for whom thou shinest!'"
>
> Friedrich Nietzsche,
> *Thus Spake Zarathustra*
> (1883, 1891)

Free Will

I now want to address directly the matter of free will. Philosophical debates about the existent of man's freedom to choose and act have been around for a long time, but important scientific advances and the full emergence of the scientific method and philosophy of science in the twentieth century arguably altered the nature of the debate. The concept of determinism dates back at least to the time of Newton; but a significant feature of the intellectual life of the mid-twentieth century was the reassertion of new, serious, scientific arguments for determinism. I believe in free will. I also think that the determinist argument is one of the more negative influences that science has injected into the world. The damage, however, is largely superficial, since the philosophy is in such contrast to the human spirit.

French mathematician, astronomer, and physicist, Pierre-Simon, Marquis de Laplace, (1749 –1827) presented a deterministic theory of the physical world which excluded supernatural events of divine intervention. Continued progress in physics gave rise to increasingly plausible support for the belief there were laws of nature that determined the outcome of everything—that with sufficient knowledge of the initial conditions and the applicable Natural Laws, the entire history of the Universe would be predictable. That belief raised issues about the meaningfulness of the concept of free will for human beings. See, e.g., Carnap, *Philosophical Foundations of Physics* (1966), pp.216-22.

The appeal of the determinist philosophy undoubtedly owes a great deal to the intellectual achievements of the natural sciences and the remarkable successes of applied science and technology, especially during the nineteenth and early twentieth centuries. As we have discussed in various contexts, the paradigm of the scientific explanation for the physical sciences was the model of Newtonian mechanics. The model postulates the existence of universal Laws that are invariable and define relationships that are mandatory. Thus, given a state of initial conditions, the application of the universal Laws to those conditions will generate predictions that are logically necessary, subject to the condition of *ceteris paribus*.

We may never be able easily to determine the full state of initial conditions necessary to predict the future of the world, and we may fail to achieve an explanation for those initial conditions (where they came from and how they arose). But, the Laws predict the future course of events given those initial conditions. The indeterminacy created by the *ceteris paribus* condition could be further and further reduced by broadening the theory (and model) to include more and more facts about the Universe, even though many of those facts may have very minor short-term impacts on the matters under investigation. As the theory encompasses more and more, the relevance of the *ceteris paribus* condition diminishes.

But, developments in quantum mechanics introduced uncertainty and random events into mainstream physics; other developments, like chaos theory, further complicated the picture of a completely predictable theory of the Universe. It seems to me, however, that these developments present challenges, but are not themselves fatal, to determinism.

But, the need to rely on some version of the Darwinian theory of natural selection, in addition to the laws of physics and chemistry, in order to explain biological phenomena does present a fundamental problem. Acceptance of any such theory undermines determinism, since all recognized versions are premised upon random variations. By definition, it is not a predictive theory and, so, cannot support a deterministic view of the world.

There is still the question of whether we can ever reduce the entire scope of what happens in the Universe to a set of universal Laws. Again, science does not provide an unequivocal answer.

Choice

The issue here is not whether genetic or chemical factors can affect a person's behavior. Clearly they can. It is not whether in certain circumstances a person should not ethically or legally be held accountable for his actions. The law has long recognized defenses base upon diminished capacity or insanity. The issue is whether as a matter of fact we actually make any choices in a meaningful way, not just think that we do so.

On this question, the exponents of determinism confront a serious obstacle in the form of our personal subjective experience. Do we decide when to get out of bed, whether to stand up or sit down, what to eat for lunch, whether to forgive a wrong, whether to get married? It is inconceivable that all of these discretionary actions could be programmed into the mind so that the actions are mechanically determined by what is already present. The decision to eat another potato chip, even if strongly influenced by the creative ingenuity of the manufacturer and its advertising agency, is ours and is made differently from time to time.

We even have the experience of exercising free will when we strain to assert willpower in order to resist temptation. I do not see how one could argue that the simple decisions are matters of free will, while the big things are not. That is not to say, at the same time, that there are not serious constraints and influences on our decision-making.

The twentieth-century physicist and philosopher of science Rudolf Carnap clearly fits in the determinist school. *See, e.g., Philosophical Foundations of Physics*, pp.216–22. He tried to soften the blow, as have others, by arguing that the concept of free will and choice was still meaningful even though the choices that were made by people were, at least theoretically, completely predictable, because the choices were not being compelled by external forces. *Id.* The absence of compulsion supposedly means that the perception of choice is not a mere illusion.

As a method of reviving the concept of human dignity, the argument has some value. However, it seems to me that it fails badly in the context of moral dilemmas or crises of conscience. The nobility of that dark night in the Garden of Gethsemane or when one struggles painfully during 40 days of temptation in the desert seems to evaporate if one is assured that the outcome was always inevitable.

Consciousness and intent

As discussed above, many would say that a science that cannot provide a satisfactory explanation of consciousness must be seriously inadequate, because consciousness is one of the most undeniable and relevant features of intelligence. Indeed, the issue of free will puts a spotlight on the nature of consciousness.

We have considered the arguments about whether consciousness is susceptible to a full explanation in physical terms, that is, through the processes described by chemistry and physics. We have also considered the issue of whether consciousness is reducible to computation so that it could be achieved by a computer. And we have discussed the introduction of probabilistic theories into physics. This development presents its own challenges to the deterministic view of the Universe. But, none of these matters really addresses the issue of free will. We are interested not just in whether future events are not predictable, but also in whether our experiences of making free and meaningful choices are real.

I grew up at the time when the debate over nature versus nurture was raging. Was our conduct primarily the result of our genetic inheritance or of our upbringing (family and community)? Studies of identical twins raised separately in different families were thought to hold the answer. The studies were inconclusive. But, the question here is whether there is something else of importance, beyond "nature" and "nurture". As Penrose put it: "[M]ight there be something that is beyond our inheritance, beyond environmental factors, and beyond chance influences—a separate 'self' that has a profound role in controlling our actions?" *Shadows of the Mind*, p.401.

Terrence Deacon uses an illustration that I find particularly provocative. A boy picks up a flat, smooth stone and throws it so that it "skips" across the surface of the lake. *Incomplete Nature*, pp.18–22. The boy presumably was taught how to accomplish this feat by someone else. Because of that instruction or example, he was able to visualize the possibility of the event and take steps to realize it. The result was a stone "skipping" on water. "[P]rior to the evolution of humans, the probability that any stone on any beach on Earth might exhibit this behavior was astronomically minute." *Id.*, p.20. Something happened that changed the way in which matter behaved and interacted.

What was the "cause" of the stone's unusual behavior? Various people could describe in painful detail the bases in classical mechanics for the fact that the stone "skipped," *given* its shape, size, speed and trajectory. But, what "caused" the phenomenon? Certainly, the boy's decision to try to skip the stone was a "cause," and one that "explains" the shape, size, speed and trajectory of the stone. And, that decision appears to have been the result of his ability to visualize or imagine a future result and his understanding of how to bring about that result. See *id.*, pp.21-2.

It is difficult to imagine the "deterministic" version of the boy skipping the stone.

Implications

I cannot avoid commenting briefly on another dimension to the belief in free will.

We have discussed altruistic behavior in a biological sense, But, for a conscious being with free will, altruism raises a qualitative different issue. Why consciously choose self-sacrifice or a selfless act? Not as a matter of a biological imperative, but as a volitional act, as a decision? We have discussed above the perception of moral laws and morality, the "awareness" of right and wrong. Those concepts provide a context in which people act. Do they actually make choices or is right and wrong merely an illusion?

In a prior chapter, I began the section The End of Time with a quotation from the Dolly Parton song "In the Meantime." She notes how people apparently have been anticipating the end of time since at least the beginning of recorded history, but that we do not know when that end will come: "It might be today, it might be tomorrow, or in a million years or two." The essence of free will is captured in the refrain: "In the meantime, in between time, let us make time to make it right."

Time matters not because it is change, but because it provides the opportunity for change. Perhaps we cannot ever make it "right", but maybe we can make it better. That is the promise of free will. And, perhaps, also a hint of the Divine.

IMAGINATION

What is missing in the deterministic account of nature is a place for imagination—"thoughts of things that are not present, or not yet present—or perhaps never to be present..." Polanyi, "The Creative Imagination," in *The Creative Idea*, p.155. It is the urge set forth in the quotation from Proust with which I begin this sections: the need to create.

I previously mentioned the human reactions to art and music as challenges to science. but what about the creative impulse giving rise to those works? Many thousands of words ago I referenced Ely Cathedral, rising improbably above the fens in East Anglia. The cathedral itself is a magnificent monument to the human spirit: vision, inspiration, creativity, aesthetic sensitivity, persistence, will—the creation from something imagined of something concrete and, from a human perspective, something almost eternal. The construction of the current cathedral, occurring during the twelfth through the fifteenth centuries (with some modest "modern" additions), involved thousands of craftsmen and laborers, most of whom, as noted in the Introduction (quoting Lord Rees, who was presumably borrowing from John Ruskin), never lived to see—and never expected to live to see—anything close to the finished product.

I admit that these overly dramatic examples somewhat miss the point at issue. The question is not whether imagination is explained in the determinist worldview, but whether it is real, that is, a part of the reality that is to be explained. However, I think that the examples speak to that question as well.

Something more has bothered me and, I now realize, has bothered others too.

SOMETHING MORE?

We have noted the enormous similarities in the genomes of humans and apes (and pigs and many, many other mammals and even notable similarities with other types of life form). One can also observe substantial similarities in physical structure, bodily processes and biological functions. Yet, look at the differences between man and all other known life forms. Morality, science, art, music, literature, politics, passions, guilt, ambition. The list could go on and on, with each of us having our favorite examples. Not only are these things big, dramatic differences between us and the rest, they encapsulate most of what we would think of as the defining and meaningful characteristics of our species.

"No nonhuman animal behaves in ways other than to acquire food and shelter, escape predators, find a mate, and rear progeny or to reinforce collaboration within its group to increase its own chances of survival and reproduction." Hands, *Cosmosapiens*,.p 537.

Let me note three caveats. First, distinguishing between animals and humans is complicated by the tendency to anthropomorphize animal behavior. We may misconstrue the nature of some animal activities and mannerisms. Second, I exclude social, altruistic and cooperative inclinations, all of which can be explained by natural selection (as discussed in the chapters Darwinism and Consciousness). Third, I recognize that many of the distinctive human capabilities could cumulatively develop and progress very rapidly through competition once started.

Art and music

While working on this book, I frequently browsed through the collection of examples of the pinnacle of human artistic expression contained in the Fitzwilliam Museum in Cambridge. Of course, many of the paintings exhibited exceptional skill of execution, but they also set out an array of differing artistic visions. Perhaps most importantly, the art works reflected individuality and even, sometimes, uniqueness. Can one really believe that the stunning blues of Sassoferrato's seventeenth-century painting of The Holy Family were the outcome of mere genetic material responding to subsequent physical stimuli? Perhaps, the painting itself was predictable (the subject matter was well-known and frequently depicted in the visual arts), but the color that was achieved is what fills me with awe. In that December, I also attended Christmas concerts in the Trinity Chapel and in the magnificent Kings College Chapel. The sounds were made even more beautiful by the astounding architectural settings.

Morality and evil

Then, there is the matter of morality or moral laws (right and wrong). Many philosophers, like Immanuel Kant and contemporary political theorist Hadley Arkes, have believed that humans have an innate sense of morality, an awareness of right and wrong. Recent research by sociologists has led to a similar conclusion. For example, Nicholas Christakis cites studies that indicate that children as young as three months have a sense of fair play and an ability to sense what characters are good and which are bad. *Blueprint: The Evolutionary Origins of a Good Society* (2019), pp.4-5.

Similarly, based on the results of a host of empirical studies, sociologist Jonathan Haidt asserts that the sense of morality appears to be an innate human characteristic, but the precise details of what is right and wrong are ultimately derived from the culture in which the individual is raised.

Thus, e.g., the scope of morality is narrower in Western, individualistic societies than it is in Eastern, family and community-centered societies. As he says: "We're born to be righteous, but we have to learn what, exactly, people like us should be righteous about." *The Righteous Mind: Why Good People are Divided by Politics and Religion.* (2012), p. 31. However, there is a strong common core across the human race. See also, Christakis, *Blueprint*, pp.17-8.

Finally, what about evil? Not bad, but evil. I think that we all know that evil exists, even if the idea offends our materialistic inclinations. Of course, it is a unique hallmark of *Homo sapiens*, if not of the genus *Homo*, at least on this world. It is curious that the concept exists in our minds; it is more curious that so much evil (violence, cruelty, greed, dishonesty, anger, vengefulness) appears among humans. How "adaptive" is evil? (Or, alternatively, a sense of morality? Or, altruism? Or, even, free will?)

Evolution may explain man's sense of morality and strong social instincts. Christakis argues that "natural experiments" which can be observed and studied (a methodology discussed in the chapter on Social Science) establish the "evolutionary origins" of the bases for "a good society." *Blueprints*. I find his evidence somewhat unsatisfying. In examining historical events, I think he tends to see what he wants to see—or, more accurately, he tends to see what he is looking for. Like us all. But, in any event, his argument does not attempt to answer what adaptive function is served by evil or immoral acts.

Awe and wonder

Do other animals stop and stare at a beautiful vista or a sunset? Do they experience awe and wonder?

What can current science tell us about these feelings, these experiences? Anything even remotely relevant? The appreciation of music might have some adaptational (or competitive) value in promoting community and group activity. The same may be true for the feelings of sympathy and empathy (and the shedding of tears during sad movies). But, as adaptive traits, do not these things seem like obvious overkill? And, what about awe, wonder, joy, or the apprehension of beauty? What adaptive functions could they have served? Neo-Darwinism can offer explanations of altruism and sacrifice in evolutionary terms; but, art, music and other aesthetic expressions or, even, the passionate search for truth and understanding? I think not. We know in our hearts that something more is going on. We just seem unable as yet to explain precisely what it is or how it comes about.[13]

You know my answers to these questions, for whatever they are worth. The relevant question for the reader is what are your answers?

Our Role in the Universe

We have discussed the anthropic principle that, in some forms, postulates that the very unusual features of the Universe are as they are so that intelligent, conscious beings can exist. Of course, most scientist-proponents of that principle deny that such a relationship implies an intentional causality. Instead, the "so that" merely means that intelligent beings would not have existed "but for" those extraordinary conditions. Those proponents would contend that conscious intelligence is just a highly unlikely, accidental, but fortuitous, phenomenon in our Universe (or a highly probable event in one or more of very, very many universes).

A necessary component?

At the other extreme, one might speculate that the Universe simply would not exist if we were not there to be conscious of it. Leaving aside the possibility that reality is a mere illusion or some type of hologram, most of us would probably conclude that there is a physical reality that exists in the absence of the conscious observer. But, does that mean that reality is independent of consciousness? Or, could it be that reality is more or different in the presence of consciousness?

As noted at the very beginning, what we perceive of our world is a result of the functioning of our brains. Music (or even sound) does not exist without someone who "hears" it; color, without someone who "sees" it; flavor, without someone who "tastes" it; odor, without someone who "smells" it; and hardness, softness, texture (silkiness or roughness), hot or cold, without someone who "feels" it. *See, e.g.*, Levitin, *This is Your Brain on Music* (2006), pp.22-4. The only things that independently exit are the physical phenomena (frequencies, vibrations, molecules) that have the potential of being much, much more—that have the potential of being perceived as beauty.

Curtis White marvels (sarcastically) that even staunch materialists like Richard Dawkins and Leonard Krauss enthusiastically use words like "awe", "amazement" and "beauty", despite the fact that their scientific views would seem to deny the relevance or even the reality of such concepts. He notes that what they fail to "imagine" is "that the existence of our universe was always dependent on our consciousness of it. After consciousness is gone, the stars can burn all they want, whether near or far, but not a single star will 'shine', and nothing will think that's 'tragic'." *The Science Delusion*, pp.48-9 n**. White's comment borrows from the Prologue of the epic poem by Friedrich Nietzsche quoted above to introduce this section. That statement follows Zarathustra's appearance: "...[R]ising one morning with the rosy dawn, he went before the sun, and spake thus unto it... ." [14]

What have we learned?

Can we say anything more analytical with respect to these matters? Let me try. I start with a very brief summary of some things we have already discussed:

- We observed in the chapter on Darwinism how the ways humans choose to live and their life experiences appear to impact some of the inheritable characteristics of their offspring. We also discussed in the chapter on Neuroscience how the brain is shaped and changed by experiences and influenced by aspects of what we call culture, *i.e.*, that one might say the "mind" helps shape the "brain."
- We have noted in various contexts how conscious intelligence is much more than a mere, passive observer of the natural world. We, certainly, as the best-known example of conscious intelligence, are very much participants in the world. Our unique capabilities have made our participation particularly pervasive and significant. As Jacob Bronowski begins *The Ascent of Man*: "Man is a singular creature. He has a set of gifts which make him unique among the animals: so that, unlike them, he is not a figure in the landscape—he is a shaper of the landscape. In body and in mind he is the explorer of nature, the ubiquitous animal, who did not find but has made his home in every continent" (p.19).
- Traditional science portrays the scientist as an objective, detached observer of nature. Yet, we know that that picture of science is quite misleading. It is obvious that the acts of observation and analysis do influence the subject being examined in the social sciences, like social anthropology and economics. See, *e.g.*, Byers, *The Blind Spot*, p.100. In an important sense, the same is true for history (*e.g.*, "those who cannot learn from history are doomed to repeat it"[15]). What about the natural sciences? Are they also dependent upon the participation of the observer? The following two points suggest that the answer is yes.
- We have seen that in quantum mechanics it is believed that it is the act of observation that causes the collapse of the wave function and

enables the "particle" to act like (or to "be") a particle rather than a wave. Similarly, the Uncertainty Principle suggests that position and momentum do not exist simultaneously until a particle is observed. Does the act of observation change the reality that exists (or that would otherwise exist, if unobserved)? Or could it be that the underlying reality does not change but that the act of observation creates a new "reality," the reality of the observer? (And, I do not mean that in the sense of alternative or parallel realities.)

- We discussed in the first chapter how our observations about reality generated through our senses are not independent of our minds. In fact, in an important sense, we see or observe with the mind, not with the senses. The mind fills gaps, provides extensions or extrapolations, adds things that are not actually there. That does not make these things arbitrary or meaningless, but what reality do they have absent the observer? *See* Byers, *The Blind Spot*, pp.128-9.
- As also discussed in that chapter, pattern and order "exist" only because of—or from the perspective of—the observer. We comprehend things through the use of models and theories—very much human constructs. Indeed, reality would be meaningless to us in the absence of such constructs and the mind's ability to utilize them.
- Finally, our ability to predict results may continue to improve, but there may be inherent and insurmountable limitations to our ability to understand.

Do these conclusions point to any answers?

Another reality?

To me, these observations suggest that, in some important sense, there is a "reality" that exists separate from the natural world itself, a reality that would not exist without the presence of conscious intelligence. Could it be that we (and any other form of intelligent consciousness that may exist in the Universe) play some fundamental role in the "scheme of things"? Perhaps, there is a "reality" dependent upon and resulting from conscious intelligence (even if there is also "something" that existed before consciousness and that will continue to exist when consciousness is gone).

For example:

> "The distinguished theoretical physicist John Wheeler... took **the conscious-dependent view of physical reality to its logical conclusion.** He argued that the universe depends for its existence on

the presence of conscious observers to make it real, not only today but also retrospectively to the Big Bang. **The universe existed in a kind of indeterminate probabilistic ghost state until conscious beings observed it,** *thus collapsing the wave function for the entire universe and bringing it into physical existence."*

John Hands, Cosmosapiens, pp. 87-8 (emphasis added).

What might this mean for us and for science?

- First, the traditional concept that there is an objective reality (one independent of the human mind) that we have been able more and more closely to approximate by our scientific theories may simply be inadequate. It certainly seems incomplete.
- Second, if there is an objective physical reality independent of but including conscious intelligence, we still have the questions about the nature and extent of the relationship between that world and our perceptions and understanding of it. Does our progress in science brings the two closer and closer together? Or, do they inevitably remain forever separate?
- Finally, it is possible that there is, for all practical purposes, another "reality" of which we are an integral part. if so, what can or does science say about that "reality"? If there are two different "realities", then one is and always will be meaningless to us and all that we really can comprehend is the other, the one that is intertwined with consciousness.

Physicist Paul Davies has concluded that there must be "a deeper level of explanation" than one that portrays man as a mere accident of nature:

> *"I have come to the point of view that mind—i.e., conscious awareness of the world—is not a meaningless and incidental quirk of nature, but an absolutely fundamental facet of reality. That is not to say that we are the purpose for which the universe exists. Far from it. I do, however, believe that we human beings are built into the scheme of things in a very basic way."*

The Mind of God, p.16.

Is it just human pretense and vanity to suggest that the stars are given "meaning" by the presence of intelligent and conscious beings who can observe them? In all events, we certainly must recognize that it is the existence of consciousness that makes this world such a wonder.

**"And I think to myself,
What a wonderful world."**

Louis Armstrong

Endnotes

[1] It may seem curious to us now, but serious questions about the origin of life were not asked until the mid-nineteenth century. Of course, for all of those in the Hebraic tradition, the answer was contained in the Book of Genesis. Moreover, there was also widespread belief in spontaneous generation. It was thought that lower life forms, like worms, insects and, even, rats, arose spontaneously out of their environments. See Luisa, *The Emergence of Life*, pp.3-4.)

[2] The elements generally required for life on Earth are carbon, hydrogen, oxygen, nitrogen and phosphorous (phosphorous is essential to the molecule adenosine triphosphate, or ATP, used in the transfer and storage of energy in the cell). Conway Morris, *Life's Solution*, pp.24-5.

[3] This speculation presumes that life existed on Mars. At the beginning of the twentieth century, observations of the apparent "canals" on Mars led to the conclusion that Mars was or had been home to a technologically-sophisticated civilization of intelligent beings. Later observational evidence suggested, at least, that Mars had once had flowing water, raising the possibility that it had at some stage harbored life. Most recently, observational evidence obtained through satellite photographs from the Mars Reconnaissance Orbiter has led some scientists to conclude that flowing water currently appears on Mars in the form of leaks from what may be underground oceans. See, e.g., Robert Lee Hotz, "Mars Shows Signs of Flowing Salt Water," *The Wall Street Journal*, September 28, 2015.

[4] Corey S. Powell, "Strange life-forms found deep in a mine point to vast 'underground Galapagos'," nbcnews.com, September 7, 2019:

> "[T]he deepest spot ever explored on land and the reservoir of the oldest known water[,] ...7,900 feet below the surface, in perpetual darkness and in waters that have remained undisturbed for up to two billion years, ... is teeming with life. ...The single-celled organisms don't need oxygen because they breathe sulfur compounds. Nor do they need sunlight. Instead, they live off chemicals in the surrounding rock... ."

[5] The Kepler project was ended in 2013 because of technical problems with the telescope, but the analysis of the data already collected will take

years more. As of the end of 2014, the Kepler project had identified 4,175 potential planets and 1,004 of those have been confirmed as real. Dennis Overbye, "So Many Earth-Like Planets, So Few Telescopes," *The New York Times*, January 6, 2015.

[6] See, NASA Exoplanet Exploration, "Discovery Alert: New Twin Planets Prompt Comparisons to Earth", June 20, 2019 ("Two new planets were detected orbiting Teegarden's Star, an ultracool, red dwarf less than 13 light-years away. At a minimum, the new planets both weigh in at about 1.1 times the mass of Earth"); Jeanette Kazmierczak, NASA Exoplanet Exploration, "TESS Finds Its Smallest Planet Yet", June 28, 2019 ("The planet ... marks the tiniest discovered by TESS to date [about 80% of the size of Earth]. ...The two other worlds in the system, L 98-59c and L 98-59d, are respectively around 1.4 and 1.6 times Earth's size"). See also, e.g., Kenneth Chang, "7 Earth-Size Planets Orbit Dwarf Star, NASA and European Astronomers Say," *The New York Times*, February 22, 2017.

[7] Kerr Than, "Oxygen-Free Animals Discovered—A First: Deep in the Mediterranean, scientists have discovered the first complex animals known to live without oxygen." *National Geographic*, April 17, 2010.

> *It was previously thought that only viruses and single-celled microbes could survive without oxygen long-term. But three new species of multicellular animals found during recent research expeditions live comfortably in oxygen-free depths... The new animals appear to have modified versions of mitochondria called hydrogenosomes, which can produce ATP without oxygen. Hydrogenosomes were previously known only in single-celled organisms."*

Also, Michelle Star, "Scientists Find The First-Ever Animal That Doesn't Need Oxygen to Survive," *Sciencealert.com*, 25 February 2020 ("a jellyfish-like parasite [living in salmon muscle] doesn't have a mitochondrial genome—the first multicellular organism known to have this absence. ...[I]t lives its life completely free of oxygen dependency").

[8] No one Constant in any particular Law needs to be any specific value: it could be anything and the Law would still be the same (the Laws simply specify that there is a constant value as part of the relationship). However, when all of the different Laws are put together to explain the world (rather than to explain just isolated phenomena or relationships), many of the values do matter. In combination, they create the conditions necessary for life. See Barrow, *The Constants of Nature*, pp.158-9.

⁹ It may be that the Laws themselves are effectively part of those initial conditions, established in or by the Big Bang or, as Martin Rees puts it, "were imprinted into [the Universe] at the time of the initial Big Bang." *Just Six Numbers*, p.1. Similarly, the asymmetry (the flow) of time may have been introduced by or been part of these initial conditions. See Smolin, *Time Reborn*, p.206; Penrose, *Shadows of the Mind*, p.232.

¹⁰ Leonard Susskind wrote: "There is a philosophy that says that if something is not observable—unobservable in principle—it is not part of science. If there is no way to falsify or confirm a hypothesis, it belongs to the realm of metaphysical speculation, together with astrology and spiritualism. By that standard, most of the universe has no scientific reality—it's just a figment of our imaginations." *The Black Hole War*, p.438. This statement has been quoted approvingly. See, e.g., Curtis White, *The Science Delusion*, p.151. However, from his other writings, it is obviously that Susskind has a very poor opinion of the "philosophy" to which he refers and of the usefulness of philosophy of science in general. See, e.g., *The Cosmic Landscape*, pp.192-7.

¹¹ Translation by William Ellery Leonard (*The Internet Classic Archive*). Quoted, with a different translation, in Luisa, *The Emergence of Life*, p.62.

¹² Bejan initially stated his new law of physics as follows: "For a finite-size flow system to persist in time (to live), its configuration must evolve in such a way that provides easier access to the currents that flow through it." *Id.*, p.3.

¹³ These types of observations, plus the transcendent experiences of awe and joy, are cited by Francis Collins as powerful evidence of the existence of a theistic God. *The Language of God*, pp.33-54. He has a point.

¹⁴ Richard Strauss set the spirit of Nietzsche's poem to music. Strauss' "Also spake Zarathustra" premiered in 1896. The opening section, corresponding to this quotation, was used in the "dawn" sequence of Stanley Kubrick's film *2001: A Space Odyssey*.

¹⁵ Although, regularly attributed to Edmund Burke; the original version appears to be from George Santayana, *The Life of Reason*, Vol. I, "Reason in Common Sense" (1905-6): "Those who cannot remember the past are condemned to repeat it."

Conclusion

There are a variety of themes that appear again and again in the preceding pages. I shall not repeat them wholesale. But, I do want to emphasize certain things, some of which have been discussed extensively and some only addressed by implication.

As a place to start, I think that it is useful to distinguish again among technology (or applied science), scientific theory and the empirical work of scientists.

The progress of technology has been the result of the interaction among theory, intuition or hunch and pragmatics. It is almost always incremental, building often through very small additions or adjustments to existing practice. Of course, sometimes, there have been large leaps. Those leaps have generally come about not as a result of a new scientific insight, but as an imaginative visualization of a possible way to accomplish something—sometimes a better way to do something we were already doing, sometimes the idea of doing something that no one had previously even thought to do. In all events, technology is about predictability. We do not need to know how or why, but we need to be reasonably certain that the results of an action will be as predicted.

As we have discussed at length, pure scientific theory in contrast is often in the form of deductive logical constructs that are necessarily true within the context of their own assumptions and scope. The difficulty is that these models do not necessarily bear any particular resemblance to reality or to the world of our senses and experiences. Some scientists are engaged in purely theoretical work. By definition, they work in a world of logic.

The more complex and nuanced work occurs in the laboratories of the empirical and experimental scientists. They may be engaged in attempting to falsify or support an existing theory (maybe one that is well established, maybe one that is new and radical) or attempting to gather empirical data from which theories may be derived. The important fact about this empirical work is that it is not mechanical or purely objective; it is individualized and personal. In some ways, the empirical work can be called more of an art. It takes practice, intuition and, often, a special touch. That is why certain experiments require hundreds or thousands of hours to perfect. It is also one of the reasons why many results are not readily reproducible.

However, other than the large array of experiments that can be performed in relative isolation, science has great difficulty modeling the real world that we inhabit. We have reviewed the fact that many of the most fundamental questions about our world and our position in it have no answers from science. The open questions include the origin of the Universe, the content of the Universe, the source of or explanation for the incredible fine-tuning of the Laws and Constants of Nature, the emergence or origin of life and the origin and nature of consciousness.

We focus on the simple (and elegant) equations like "$E=mc^2$" or "the force of gravity between two objects is equal to the Gravitational constant times the product of the masses of the two objects divided by the square of the distance between them." We might assume that the mathematics underlying so much of contemporary physics is similarly straightforward. It is not.

These equations are really the exception rather than the rule. They were discovered for certain simple relationships that have some meaning and uses in isolation from the complexities of the rest of the natural world. Einstein's field equations, for example, do not engage the imagination of the layperson in the same way as does $E=mc^2$. Moreover, the performance of actual calculations with these equations using data reflecting the real world is beyond the capabilities of the most capable people and even powerful computers. (Before the advent of supercomputers, it was not possible to investigate fully the implications of the equations of General Relativity. Smolin, *Time Reborn*, p.74.)

One thing that I find particularly striking arising from this review of the state of many of the most significant areas of contemporary science is the extent of the deeper methodological similarities of the issues in the sciences, as well as in economics and, to some extent, the other social sciences.

We saw, in the context of the debates among economists concerning the theory of the firm that models we use can come in very different forms. The assumption that firms act so as to maximize profits has been defended as a definition of a "firm", in terms of a role that may or may not be played by any particular actual person or organization, but which is an integral part of a model of a competitive economy. Others asserted that real-life firms generally attempt to maximize profits and over time will tend to do so. Others argued that because of competitive forces, actual firms behave as if they were seeking to maximize profits or accidentally manage to do so; otherwise, they would have been driven out of business. In other words, the assumptions underlying a model may be characterized in very different ways. They may be a definition (no real content), an empirical or contingent assertion (that may be true or false) or just a simplifying assumption (that is more or less true, but clearly false).

The models based upon a defined "role" rather than contingent assumptions about specific actual entities may be useful because they can be predictive of what actually takes place. They mat appear to reflect the fundamental tendencies at work and the results toward which real-life will tend to move despite repeated short-term disruptions, effectively predicting, at least, long-term trends. These models may be perceived to be useful even if there is little expectation that they will ever be meaningfully predictive of actual outcomes. What could be useful about a model or theory that cannot be used to predict outcomes or events? It may be perceived to provide understanding, to help explain what is happening. They may be said to capture or reflect fundamental causal relationships at work.

Alternatively, a useful model may enable highly accurate predictions of certain events, even where that model may bear scant relationship to reality in anything other than its predictions. In such a case, none of the axioms or propositions within the model necessarily needs to be true. The model needs only to have some operative terms that are tied to observable or measurable physical elements so predictions can be made.

At the beginning, we discussed the issue of prediction versus explanation—whether all we should and can expect from a good scientific theory is accurate prediction or whether something more expected? As we noted, the indefiniteness of the concept of explanation and the difficulties of formulating criteria beyond falsifiability and the related element of predictive power may suggest that those scientists who said that prediction was enough were probably right. In the subsequent chapters, however, we have seen that accurate forecasting is not as pervasive among established scientific theories as one might at first have assumed.

Some sciences, because the nature of the subjects that they address and the nature of the models involved, are not normally even expected to be predictive. In addition, more complex phenomena, especially dynamic processes, appear to have inherent chaotic elements or, at least, are beyond our capabilities to capture, so predictive models are not realizable. In fact, most aspects of the world that occupy our everyday attentions fall into this last category. The more common examples of the predictive power of science involve, in effect, controlled experiments in which the relevant variables are very limited and external influences are effectively excluded. These successes are taken to support the applicability of the theories or models more generally, but there is a large leap of faith necessarily involved.

Indeed, ultimately, unless the world is deterministic, complete predictability cannot be required, let alone be sufficient, since it can never be achieved.

I suppose that a standard based upon the creation of a "feeling" of understanding is inherently problematic for scientists. It certainly has a strong subjective flavor, and a personal one. How does one assess whether a theory actually generates a sense of understanding, apart from the testimony of the audience? And whose experience or understanding is relevant and sufficient? Clearly, the view of any one person is not very significant; on the other hand, we would not want to define it as subject to majority rule.

However, processes over time seem to ameliorate the difficulties significantly. A good theory will be fruitful or productive, leading if not to accurate predictions, then at least to broader or additional theories that facilitate our understanding of a wider range of phenomena. And, the stronger theories will, with time, succeed in competition with weaker alternatives, becoming more widely accepted. The best new theories may even become the accepted theories of a future generation. And, achieving a better understanding of our world or of reality should, with time, have beneficial consequences, consequences that are observable and enduring.

So, I conclude that science must require something other than prediction. Indeed, one could say that successful predictive models that are merely predictive are really more properly categorized as technology or applied techniques. We should expect real science to provide understanding and insight. For example, even where our models are inadequate to make accurate predictions, because of the complexity of or inherent randomness in the phenomena under investigation; good scientific theories can enable us to identify and appreciate the factors that contribute to or influence the outcome and, thus, to begin to understand what has happened. In that sense, there can be a kind of foresight achievable even where we cannot accurately predict the precise consequences: A type of "foresight" that includes some "insight", as a companion of understanding.

Postscript

The thoughts that I am going to be expressing here did not occur to me, at least not fully formed, until some four years after I finished writing the original book and three years after the publication of the first edition.

Shortly after publication, a college classmate of mine read the first edition and wrote to me with his comments. After the obligatory compliments, he stated that he thought that "an implicit assumption underlying the book was a need to attribute meaning." I was mildly offended, perceiving a smugness lurking behind the comment. I thought I understood what he meant—I was being accused of needing to find purpose to life. In our era of the late 1960s in the elite Eastern academic establishment, such a need was not cool. It was inconsistent with our pretense of sophistication, our self-image as intellectuals. Indeed, I certainly had not believed, nor would I admit, that such a "need" was an implicit assumption underlying my work. (Of course, I may also have misinterpreted his intended meaning.)

Over the following three years, I thought repeatedly about his comment and my reaction to it. I also reread the book several times (resulting, by the way, for unrelated reasons, in new, heavily revised versions). I think that I now see clearly, from my (obviously biased) perspective, what had happened while I was originally writing it.

I began the project with the simple goals stated in the preface—to understand what it is about the world that we actually know and what the open questions are, with the prejudices disclosed. I believe that I had, in fact, made no assumptions about meaning when I started. However, as the work progressed, I found myself regularly stumbling over, bumping into or accidentally uncovering glimpses of things that suggested the existence of meaning in our Universe.

In rereading what I wrote, I noticed that the tone gradually changed from the first to the last chapter, going from relatively objective, analytical and dry to more expansive, speculative and, what I hope I can call, almost lyrical. The changes in tone reflect what I now view as the my "progress" on a journey of discovery and awakening—the gradual, growing realization of the possibility of something more; something that transcends our normal reality; something, in fact, that might make sense of the existence (and consequent experience) of anticipation, wonder, joy, curiosity, aspiration, integrity and artistic expression; something that might even explain hope, laughter, tears and love.

What it is, I cannot say. But, its presence is dogged and pervasive. And, my increasing awareness of it affected, and was reflected in, my writing.

For a new version, I contemplated revising the entire approach of the book (and the title) to be explicitly a story of "the discovery of the possibility of purpose and meaning". But, I rejected that approach as weakening what I actually already had. The book as originally written documented and revealed, both in content and in tone, the stages of my journey. That story, I did not want to lose. My revisions may have heightened what was happening but have not change it.

With the book continuing in this form, I hope that you can experience this journey more or less as I did.

REFERENCES

BOOKS

Aczel A.D. *Chance*. Thunder's Mouth Press: New York, 2004.

Arkes H. *First Things: An Inquiry into the First Principles of Morals and Justice*. Princeton University Press: Princeton, 1986.

Atkins P. *Four Laws That Drive the Universe*. Oxford University Press: Oxford, 2007.

Axe D. *Undeniable: How Biology Confirms Our intuition That Life Is Designed*. HarperOne: New York, 2016.

Axelrod R. *The Evolution of Cooperation*. Penguin: London, 1984; revised edition, Basic Books: New York, 2006.

Barash D.P. *Homo Mysterious: Evolutionary Puzzles of Human Nature*. Oxford University Press: Oxford, 2012.

Barbour J. *The End of Time: The Next Revolution in Physics*. Oxford University Press: Oxford, 1999.

Baron-Cohen S. *The Pattern Seekers: How Autism Drives Human Invention*. Basic Books: New York, 2020.

Barrow J.D. and Tipler F. *The Anthropic Cosmological Principle*. Oxford University Press: Oxford, 1988.

Barrow J.D. *The Constants of Nature*. Vintage: London, 2003.

Barrow J.D. *Theories of Everything: The Quest for Ultimate Explanation*. Oxford University Press: New York; 1991/Vintage: London, 2005.

Baumol W.J. *Business Behavior, Value and Growth*. Macmillan: New York, 1959.

Behe M.J. *Darwin's Black Box: The Biochemical Challenge to Evolution*. Simon & Schuster: London, 1996.

Bejan A. and Zane J.P. *Design in Nature*. Doubleday: New York, 2012.

Berle A. and Means G. *The Modern Corporation and Private Property*. Macmillan: New York, 1932.

Berlinski D. *The Devil's Delusion: Atheism and Its Scientific Pretensions*. Basic Books: New York, 2009.

Boulding K.E. *A Reconstruction of Economics*. J. Wiley: Chichester, 1950.

Braithwaite R.B. *Scientific Explanation*. Cambridge University Press: Cambridge, 1968.

Bronowski J. *The Ascent of Man.* BBC Books: London, 1973 [paperback 2011].
Brown D. *Origin.* Doubleday: New York, 2017.
Bryson B. *A Short History of Nearly Everything.* Transworld Publishers: London, 2003.
Byers W. *The Blind Spot: Science and the Crisis of Uncertainty.* Princeton University Press: Princeton/Oxford, 2011.
Carnap R. *Philosophical Foundations of Physics.* Basic Books: New York, 1966.
Chamberlain N. *A General Theory of Economic Process.* Harper: New York, 1950.
Chamberlin E.H. *The Theory of Monopolistic Competition.* Harvard University Press: Cambridge, MA, 1933.
Chang H. *Inventing Temperature: Measurement and Scientific Progress.* Oxford University Press: Oxford, 2004.
Christakis N.A. *Blueprint: The Evolutionary Origins of a Good Society.* Little, Brown Spark: New York, 2019.
Christian B. *The Most Human: What Artificial Intelligence Teaches About Being Alive.* Anchor Books: London, 2011.
Cohen K.J. and Cyert R.M. *Theory of the Firm.* Prentice Hall: New Jersey, 1965.
Collins F.S. *The Language of God: A Scientist Presents Evidence for Belief.* Free Press: New York, 2006.
Colyvan M. *An Introduction to the Philosophy of Mathematics.* Cambridge University Press: Cambridge, 2012.
Conway Morris S. *Life's Solution: Inevitable Humans in a Lonely Universe.* Cambridge University Press: Cambridge, 2003.
Conway Morris S. *The Runes of Evolution: How the Universe Became Self-Aware.* Templeton Press: West Conshohocken, PA, 2015.
Cox B. and Forshaw J. *Why Does E=mc2? (And why should we care?).* Da Capo Press: Cambridge, MA, 2009.
Cox B. and Forshaw J. *The Quantum Universe: Everything that Can Happen Does Happen.* Allen Lane: London, 2011.
Coyne J. *Why Evolution Is True.* Oxford University Press: Oxford, 2009.
Darwin C. *The Origin of Species. On the Origin of Species by Means of Natural Selection, or the Preservation of Favored Races in the Struggle for Life.* Collier Books: New York, 1962 [from the sixth edition of 1872, the last revised by Darwin].
Davies P. *The Mind of God: The Scientific Basis for a Rational World.* Simon & Schuster: London, 1992.
Dawkins R. *The Selfish Gene.* Oxford University Press: Oxford, 1976.
Dawkins R. *The Extended Phenotype.* Oxford University Press: Oxford, 1982.

Dawkins R. *The Blind Watchmaker*. Norton: New York, 1986/ Penguin: London, 2006.
Dawkins R. *The God Delusion*. Black Swan: New York, 2006.
Dawkins R. *The Greatest Show on Earth*. Free Press: New York, 2009.
Deacon T.W. *Incomplete Nature: How Mind Emerged from Matter*. W.W. Norton: New York, 2012.
Dembski W.A. *The Design Inference: Eliminating Chance Through Small Probabilities*. Cambridge University Press: Cambridge, 1998.
Dennett D. *Darwin's Dangerous Idea*. Simon & Schuster: London, 1995.
Deutsch D. *The Beginning of Infinity*. Viking: New York, 2011.
Diamond J. and Robinson J.A. (eds) *Natural Experiments of History*. Harvard University Press: Cambridge, MA, 2010.
Distin K. *The Selfish Meme*. University Cambridge Press: Cambridge, 2005.
Doidge N. *The Brain that Changes Itself*. Penguin Books: London, 2007.
Doyle R.O. *Free Will: The Scandal in Philosophy*. I-PHI Press: Cambridge, MA, 2011.
Eddington A.S. *The Philosophy of Physical Science*. Cambridge University Press: Cambridge, 1939.
Einstein A. and Infeld L. *The Evolution of Physics*. Simon & Schuster: London, 1938.
Feyerabend P. *The Tyranny of Science*. Polity Press: Cambridge, 2011 [original published in Italian, 1996].
Feynman R. *The Character of Physical Law*. The Modern Library: New York, 1994 [originally published by The BBC in 1965, based upon a series of lectures given at Cornell University in 1964].
Feynman R.P. *QED: The Strange Theory of Light and Matter*. Princeton University Press: Princeton, 1988.
Firestein S. *Ignorance: How It Drives Science*. Oxford University Press: Oxford, 2012.
Fodor J. and Piattelli-Palmarini M. *What Darwin Got Wrong*. Picador: New York, 2011.
Frank A. *About Time: Cosmology and Culture at the Twilight of the Big Bang*. The Free Press: New York, 2011.
Fukuyama F. *The Origins of Political Order*. Farrar, Straus and Giroux: New York, 2011.
Galor O. *The Journey of Humanity: The Origins of Wealth and Inequality*. Dutton: New York, 2022,
Gazzaniga M.S. *Who's In Charge? Free Will and the Science of the Brain*. HarperCollins Publishers: New York, 2011.
Gott J.R. III. *Time Travel in Einstein's Universe: The Physical Possibilities of Travel through Time*. Houghton Mifflin: New York, 2001.
Gould, S.J. *Wonderful Life*. W.W. Norton & Co.: New York, 1989.

Gould S.J. *The Crucible of Creation.* Oxford University Press: Oxford, 2000.

Goulson D. *A Sting in the Tail: My Adventures with Bumblebees.* Picador: New York, 2014.

Gray T. *The Elements: A Visual Exploration of Every Known Atom in the Universe.* Black Dog & Leventhal: New York, 2009.

Gray T. *Molecules: The Elements and the Architecture of Everything.* Black Dog & Leventhal: New York, 2014.

Gray T. *Reactions: An Illustrated Exploration of Elements, Molecules, and Change in the Universe.* Black Dog & Leventhal: New York, 2017.

Graeber D. and Wengrow D. *The Dawn of Everything: A New History of Humanity.* Farrar, Straus and Giroux: New York, 2021.

Greene B. *The Fabric of the Cosmos.* Penguin: Melbourne, Australia, 2004.

Greene B. *The Elegant Universe.* Vintage: London, 2005.

Greene B. *The Hidden Reality.* Vintage Books (Alfred A. Knopf): New York, 2011.

Guillen M. *Believing Is Seeing: A Physicist Explains How Science Shattered His Atheism and Revealed the Necessity of Faith.* Tyndale Refresh: Carol Stream, Illinois, 2021.

Guillen M. *Five Equations that Changed the World: The Power and Poetry of Mathematics.* Hyperon: New York, 1996.

Hacking I. *The Emergence of Probability.* Cambridge University Press: Cambridge, 1975.

Hacking I. *Why Is There Philosophy of Mathematics at All?* Cambridge University Press: Cambridge, 2014.

Haidt J. *The Righteous Mind: Why Good People are Divided by Politics and Religion.* Vintage Books: New York, 2012.

Hall R.L. and Hitch C.J. *Oxford Economic Papers.* Oxford University Press: Oxford, 1939.

Hands J. *Cosmosapiens: Human Evolution from the Origin of the Universe.* Overlook Duckworth, Peter Mayer Publishers: New York, 2016.

Hardy G.H. *A Mathematician's Apology.* Cambridge University Press: Cambridge, 1940.

Harari, Y.N. *Sapiens: A Brief History of Humankind.* HarperCollins Publications: New York, 2015.

Harris S. *The End of Faith: Religion, Terror, and the Future of Reason.* W.W. Norton: New York, 2004, 2005.

Hawking S.W. *A Brief History of Time: From the Big Bang to Black Holes.* Bantam Books, New York, 1988, 2010.

Hawking S.W. and Mlodinow L. *The Grand Design.* Bantam Books: New York, 2010.

Hempel C.G. *Philosophy of Natural Science*. Prentice Hall: Englewood Cliffs, NJ, 1966.

Hicks, J.R. *Value and Capital*. Oxford University Press: Oxford, 1941.

Hossenfelder S. *Lost in Math: How Beauty Leads Physics Astray*. Basic Books: New York, 2018.

Hofstadter D. *Gödel, Escher, Bach: An Eternal Golden Braid*. Basic Books: New York, 1979; Vintage Books Edition, September 1980.

Holt J. *Why Does the World Exist? An Existential Detective Story*. Liveright Publishing: New York, 2012.

Horgan J. *The End of Science: Facing the Limits of Knowledge in the Twilight of the Scientific Age*. W.H. Freeman: San Francisco, 1977.

Hume D. *A Treatise on Human Nature*. D.F Norton and M.J. Norton (eds) Oxford Philosophical Texts. OUP: Oxford, 2000 [original published in 1739].

Hume, D. *An Enquiry Concerning Human Understanding and Other Writings*. S. Buckle (ed.) Cambridge University Press: Cambridge, 2007.

Hutchinson T.W. *The Significance and Basic Postulates of Economic Theory*. Macmillan: London, 1938.

James T. *Elemental: How the Periodic Table Can Now Explain (Nearly) Everything*. Abrams Press: New York, 2019.

Jaynes J. *The Origin of Consciousness in the Breakdown of the Bicameral Mind*. Houghton Mifflin Harcourt: Boston, 1976.

Kahneman D. *Thinking, Fast and Slow*. Farrar, Straus and Giroux (Macmillan): New York, 2011.

Kant I. *Critique of Pure Reason*. Macmillan: London, 1922 (translation). [Original published in 1781.]

Karasu S.R. and Byram Karasu T. *The Art of Marriage Maintenance*. Jason Aronson: Northvale, NJ, 2005.

Kean S. *The Bastard Brigade: The True story of the Renegade Scientists and Spies who Sabotaged the Nazi Atomic Bomb*. Little, Brown and Company: New York, 2019.

Keynes J.N. *The Scope and Method of Political Economy*. Macmillan: London, 1891.

Kline M. *Mathematics for the Nonmathematician*. Dover Publications: New York, 1967.

Knight F. *Risk, Uncertainty, and Profit*. Houghton Mifflin: Boston and New York, 1921.

Krauss L.M. *A Universe from Nothing: Why There is Something Rather Than Nothing*. The Free Press (Atria Paperback). Simon & Schuster: New York, 2012.

Kuhn T.S. *The Structure of Scientific Revolutions*. University of Chicago Press: Chicago, 1962.

Lane N. *Life Ascending: The Ten Great Inventions of Evolution*. W.W. Norton: New York, 2009.

Laughlin R. *A Different Universe (Reinventing Physics from the Bottom Down)*. Basic Books: New York, 2005.

Lewis P.A. *The Secret World of Sleep: The Surprising Science of the Mind at Rest*. Palgrave Macmillan: London, 2013.

Lindley D. *Uncertainty*. Doubleday: New York, 2007.

Lipton P. *Inference to the Best Explanation* (Second Edition). Routledge: London, 2004.

Levitin D.J. *This is Your Brain on Music: The Science of a Human Obsession*. The Penguin Group: New York, 2006.

Losos J.B. *Improbable Destinies: Fate, Chance, and the Future of Evolution*. Riverhead Books: New York, 2017.

Luisa P.L. *The Emergence of Life: From Chemical Origins to Synthetic Biology*. Cambridge University Press: Cambridge, 2016 (Second Edition).

Macbeth N. *Darwin Retried*. Dell Publishing Company: New York, 1971.

Macknik S.L. and Martinez-Conde S. *Sleights of Mind: What the Neuroscience of Magic Reveals About Our Everyday Deceptions*. Henry Holt: New York, 2010.

Makari G. *Soul Machine: The Invention of the Modern Mind*. W.W. Norton: New York, 2015.

Mangueijo J. *Faster than the Speed of Light: The Story of a Scientific Speculation*. Perseus Publishing: Cambridge MA, 2003, 2011.

Mazur J. *The Motion Paradox*. Dutton: New York, 2007.

McGuire J.W. *Theories of Business Behavior*. Prentice Hall: New Jersey, 1964.

Meyer S.C. *Signature in the Cell: DNA and the Evidence for Intelligent Design*. Harper One: New York, 2009.

Mill J.S. *A System of Logic*. Longmans: London, 1904.

Nagel E. *The Structure of Science*. Harcourt, Brace & World: New York, 1961.

Nagel T. *Mind & Cosmos: Why the Materialist Neo-Darwinian Conception of Nature is Almost Certainly False*. Oxford University Press: Oxford, 2012.

Nielsen M. *Reinventing Discovery: The New Era of Networked Science*. Princeton University Press: Princeton, 2014.

Nowak M.A. and Highfield R. *Super Cooperators* (Altruism, Evolution, and Why We Need Each other to Succeed). Free Press: New York, 2011.

Penrose R. *The Emperor's New Mind: Concerning Computers, Minds, and the Laws of Physics*. Penguin Books: London, 1989.

Penrose R. *Shadows of the Mind: A Search for the Missing Science of Human Consciousness*. Oxford University Press: Oxford. 1994, 1995.

Penrose R. *The Road to Reality: A Complete Guide to the Laws of the Universe*. Alfred A. Knopf: New York, 2005.

Penrose R. *Cycles of Time. An Extraordinary New View of the Universe*. The Bodley Head: London, 2010.

Pinker S. *How the Mind Works*. W.W. Norton & Company, New York, 1997.
Polan M. *In Defense of Food*. The Penguin Press: London, 2008.
Popper K.R. *Objective Knowledge: An Evolutionary Approach*. Oxford University Press: Oxford, 1972.
Popper K.R. *The Logic of Scientific Inquiry*. Basic Books: New York, 1959.
Prum R.O. *The Evolution of Beauty: How Darwin's Forgotten Theory of Mate Choice Shapes the Animal World – and Us*. Doubleday: New York, 2017.
Rees M. *Just Six Numbers: The Deep Forces That Shape The Universe*. Basic Books: New York, 2000.
Rees M. *From Here to Infinity* (Reith Lectures 2010). WW Norton: New York/Profile Books: London, 2011.
Ridley M. *The Origins of Virtue: Human Instincts and the Evolution of Co-operation*. Penguin Books: London, 1996.
Ridley M. *The Agile Gene*. Harper-Collins: London, 2004.
Ridley M. *The Rational Optimist: How Prosperity Evolves*. Harper-Collins: London, 2010.
Robbins L. *The Nature and Significance of Economic Science*. Macmillan: London, 1935.
Robinson J. *The Economics of Imperfect Competition*. Macmillan: London, 1933.
Rovelli C. *The Order of Time*. Penguin Publishing Group, Kindle Section: New York, 2018.
Russell B. *The Writings of Bertrand Russell*. George Allen Unwin: London, 1961.
Ryan F. *Virolution*. Collins: Glasgow, 2009.
Ryle G. *Dilemmas*. Cambridge University Press: Cambridge, 1954.
Sagan C. *Broca's Brain: Reflections on The Romance of Science*. Random House: New York, 1979.
Sagan C. *The Dragons of Eden*. Random House: New York, 1977.
Schumpeter J.A. *A History of Economic Analysis*. Oxford University Press: New York, 1954.
Seife C. *Zero: The Biography of a Dangerous Idea*. Penguin Books: London, 2000.
Seife C. *Decoding the Universe. How the new science of information is explaining everything in the cosmos, from our brains to black holes*. Viking: New York, 2006.
Senior N.W. *Political Economy*. London: George Allen and Unwin: London, 1836.
Seung S. *Connectome: How the Brain's Wiring Makes Us Who We Are*. Allen Lane: London, 2012.
Sheldrake R. *The Science Delusion: Freeing the Spirit of Enquiry*. Coronet: London, 2012.

Shermer M. *Why Darwin Matters: The Case Against Intelligent Design.* Henry Holt: New York, 2007.

Smolin L. *Three Roads to Quantum Gravity.* Basic Book: New York, 2001.

Smolin L. *Time Reborn: From the Crisis of Physics to the Future of the Universe.* Allen Lane: London, 2013.

Sober E. *Philosophy of Biology.* Westview Press: Oxford, 1993.

Stigler J.G. *Five Lectures on Economic Problems.* Longmans, Green: New York, 1950.

Susskind L. *The Cosmic Landscape: String Theory and the Illusion of Intelligent Design.* Little, Brown & Co.: New York, 2006.

Susskind L. *The Black Hole War: My Battle with Stephen Hawking to Make the World Safe for Quantum Mechanics.* Little, Brown & Co.: New York, 2008.

Susskind L. and Hrabovsky G. *The Theoretical Minimum.* Allen Lane: London, 2013.

Tarski A. *Introduction to Logic.* Oxford University Press: New York, 1946.

Tegmark M. *Our Mathematical Universe: My Quest for the Ultimate Nature of Reality.* Allen Lane: London, 2014.

Teilhard de Chardin P. *Christianity and Evolution: Reflections on Science and Religion.* Houghton Mifflin Harcourt: New York, 1969.

Toulmin S. *Foresight and Understanding.* Indiana University Press: Bloomington, 1961.

Toulmin S. and Goodfield J. *The Discovery of Time.* University of Chicago Press: Chicago, 1965/ Phoenix Edition,1982.

Unzicker A. and Jones S. *Bankrupting Physics: How Today's Top Scientists Are Gambling Away Their Credibility.* Palgrave Macmillan: Basingstoke, 2013.

Weinberg S. *To Explain the World.* Harper: New York, 2014.

Wengrow D. and Graeber D. *The Dawn of Everything: A New History of Humanity.* Farrar, Straus and Giroux: New York, 2021.

Wentzel van Huyssteen J. *Duet or Duel? Theology and Science in a Postmodern World.* SCP Press: Norwich, 2012 [original published by Trinity Press International, 1998].

White C. *The Science Delusion: Asking the Big Questions in a Culture of Easy Answers.* Melville House: New York, 2013.

Whitehead A.N. *An Introduction to Mathematics.* Watchmaker Publishing: Seaside, OR, 1911.

Wilczek F.*Fundamentals: Ten Keys to Reality.* Penguin Press: New York, 2021.

Williams G. *Natural Selection: Domains, Levels and Challenges.* Oxford University Press: Oxford, 1992.

Williamson C.E. *The Economics of Discretionary Behavior.* Prentice-Hall: Englewood Cliffs, NJ, 1964.

Wilson D.S. *This View of Life: Completing the Darwinian Revolution.* Pantheon Books, New York, 2019.

Wilson E.O. and Hoelldobler B. *Journey to the Ants: A Story of Scientific Exploration.* Belknap Press: Cambridge, MA, 1994.

Wright R. *Nonzero: The Logic of Human Destiny.* Vintage Books: New York, 2001.

Young R.M. *Darwin's Metaphor: Nature's Place in Victorian Culture.* Cambridge University Press: Cambridge, 1985.

ARTICLES, CHAPTERS, LECTURES

Adams T. "Galaxy Zoo and the new dawn of citizen science." *The Guardian*, 18 March 2012.

Amos J. "Plank satellite: Maps detail Universe's ancient light." *BBC News*, Science & Environment, 21 March 2012.

Amos J. "Cosmos may be 'inherently unstable.'" *BBC News*, 18 February 2013.

Amos J. "Diamond to shine light on infections." *BBC News*, 17 February 2013.

Angier N. "William Hamilton, 63, Dies; An Evolutionary Biologist,." *The New York Times*, March 10, 2000.

Anon. "Foundations of Logic and Mathematics." *International Encyclopedia of Unified Science*, No. 3, 1939.

Anon. "Hwang fired for stem fraud." *USA Today*, March 20, 2006.

Anon. "Neutrinos Not Faster Than Light in Retest, Proving Einstein Right." *The New York Times International*, 17 March 2012, p.A7.

Anon. "How many ants are there alive on the planet?" *AntBlog*, April 8, 2013.

ATLAS Collaboration: Aad G., Abajyan T., Abbott B., et al. "Observation of a new particle in the search for the Standard Model Higgs boson with the ATLAS detector at the LHC" (reporting on the data collected through the use of ATLAS). *Physics Letters B*, 716(1): 17 September 2012.

Bach-y-Rita P., et al. "Vision Substitution by Tactile Image Projection." *Nature*, 221: 963–4, 1969.

Bakalar N. "First Mention/DNA, 1947." *The New York Times*, February 28, 2012.

Bacalar N. "Hawaiian Crows Join Tool-Users Club." *The New York Times*, September 19, 2016.

Barrow J.D. "Beyond numbers—the role of infinity in the understanding of the universe." *University of Cambridge Alumni News*, May 2015.

BBC News, 1 August 2012. [Many questions remain as to whether the particle is indeed the long-sought Higgs boson. AND The formal threshold for claiming the discovery of a new particle.]

BBC News, 12 November 2012. [The results of recent experiments with the Large Hadron Collider appear to conflict with the predictions of supersymmetry.]

BBC News, 19 December 2012. [Hugh Jones and colleagues detect five planets orbiting Tau Ceti.]

BBC News, 12 December 2019.[Animal painting found in cave is 44,000 years old.

Begley C.G. and Ellis L.M. "Drug development: Raise standards for preclinical research." *Nature*, 483, March 28, 2012.

Belluck P. "Protein May Hold the Key to Who Gets Alzheimer's." *The New York Times*, March 19, 2014.

Billings L. "Astronomers Find Water on an Exoplanet Twice the Size of Earth: Water vapor in the skies of the world K2-18 b may make it 'the best candidate for habitability' presently known beyond our solar system." *Scientific American*, September 11, 2019.

Bissell M. "Reproducibility: The risks of the replication drive." *Nature*, 503: 333-4, November 20, 2013.

Blythman J. "Butter is bad—a myth we've been fed by the 'healthy eating' industry." *The Guardian*, 23 October 2012.

Boffey P.M. "The Next Frontier Is Inside Your Brain." *The New York Times*, February 23, 2013.

Bosman J. "Reporters Find Science Journals Harder to Trust, but Not Easy to Verify." *The New York Times*, February 13, 2006.

Bostwick D.G. and Meiers I. "Neoplasms of the Prostate." Chapter 9, in D.G. Bostwick and L. Cheng, *Urologic Surgical Pathology*. Mosby Elsevier: Oxford, 2008, pp.443–580.

Bosveld J. "Isaac Newton, World's Most Famous Alchemist: For centuries some of the world's greatest geniuses struggled in secret to turn base metals into gold. In a sense they succeeded: In their restless quest, they unlocked some of nature's greatest secrets." *Discover*, December 27, 2010.

Botkin D.B. "Absolute Certainty is Not Scientific." Opinion, *The Wall Street Journal*, December 2, 2011.

Boyle R. "As Planet Discoveries Pile Up, a Gap Appears in the Pattern." *Quanta Magazine*, May 16, 2019.

Brainard C. "The Archaeology of the Stars." *The New York Times*, Space & Cosmos, February 10, 2004.

Brayne C., Keage H.A., et al. "Education, the brain and dementia: neuroprotection or compensation?" *Brain* 133 (Pt8), 2010.

Burba J. "'Unknown Dark Energy'- A Fifth Force or New Form of Matter?" *The Daily Galaxy*, August 19, 2019.

Campbell H. "The Corruption of Peer Review is Harming Scientific Credibility." *The Wall Street Journal*, Opinion, July 13, 2014.

Cappellini E., Welker F., ...Willerslev E. "Early Pleistocene enamel proteome from Dmanisi resolves Stephanorhinus phylogeny." *Nature*, 11 September 2019.

Carey B. "Can We Really Inherit Trauma? Headlines suggest that the epigenetic marks of trauma can be passed from one generation to the next. But the evidence, at least in humans, is circumstantial at best." NYTimes.com, December 10, 2018.

Carey T. "The invisible hand of natural selection, and vice versa." *Biology and Philosophy*, 13, 1998.

Castelvecchi D. "The forgotten quantum pioneer who turned wartime spy: Davide Castelvecchi enjoys the extraordinary life of Sam Goudsmit, the atomic sleuth nominated for a Nobel prize 48 times." *Nature* 563: 320-321, Books and Arts, 2 November 2018.

Castelvecchi D. "Gravitational waves hint at detection of black hole eating star: LIGO and Virgo observatories have spotted ripples from what could be the first-ever detection of this long-sought event." *Nature* 569: 15-16, 26 April 2019.

Cave S. "Planet of the Apes." *Financial Times*, Books in Life & Style, 18–19 August 2012.

Center for Genomic Regulation. "Artificial intelligence applied to the genome identifies an unknown human ancestor." *Science News*, January 16, 2019.

Chang K. "Saturn Moon Has Geysers, Hinting Life Is a Possibility." *The New York Times*, Space & Cosmos, March 10, 2006.

Chang K. "Going, Going, Still Going? Voyager 1 at Solar System's Edge." *The New York Times*, June 27, 2013.

Chang K. "Under Icy Surface of a Saturn Moon Lies a Sea of Water." *The New York Times*, Space & Cosmos, April 3, 2014.

Chang K. "Scientists Find an 'Earth Twin,' or Perhaps a Cousin." *The New York Times*, April 17, 2014.

Chang K. "Neanderthals in Europe Died Out Thousands of Years Sooner Than Some Had Thought, Study Says." *The New York Times*, Science, August 20, 2014.

Chang K. "Suddenly, It Seems, Water Is Everywhere in Solar System." *The New York Times*, Space & Cosmos, March 12, 2015.

Chang K. "Mars Is Pretty Clean. Her Job at NASA Is to Keep It That Way." *The New York Times*, October 5, 2015.

Chang K. "7 Earth -Size Planets Orbit Dwarf Star, NASA and European Astronomers Say." *The New York Times*, February 22, 2017.

Chang K. "New Form of Water, Both Liquid and Solid, Is 'Really Strange.'" *New York Times*, February 5, 2018.

Choi C.Q. "Ancient Neutron-Star Crash Made Enough Gold and Uranium to Fill Earth's Oceans." *Space.com*, May 8, 2019.

Clarke K.A. and Primo D.M. "Overcoming 'Physics Envy.'" *The New York Times*, April 1, 2012.

CMS Collaboration: Chatrchyan S., Khachatryan V., Sirunyan A.M., et al. "Observations of a new boson at a mass of 125 GeV with the CMS experiment at the LHC." *Physics Letters B*, 716(1), 17 September 2012.

Conniff R. "Alchemy May Not Have Been the Pseudoscience We All Thought It Was: Although scientists never could quite turn lead into gold, they did attempt some noteworthy experiments." *Smithsonian Magazine*, February 2014.

Conway Morris S. and Gould S.J. "Showdown on the Burgess Shale." *Natural History*, 107(10), 1998.

Conway Morris S. "Evolution: like any other science it is predictable." *Philosophical Transactions of The Royal Society, Biological Sciences*, 24 November 2009.

Conway Morris S. "'Map Of Life' predicts ET. (So where is he?)." *Cambridge University Research Magazine*, 2 July 2015.

Cook G. "Sebastian Seung's Quest to Map the Human Brain." *The New York Times Magazine*, January 8, 2015.

Cooper K. "Astrobio Top 10: Liquid Water Discovered on Mars." *Astrobiology Magazine*, January 1, 2019 (originally published July 25, 2018).

de Wall F. "Who Apes Whom?" *The New York Times*, Opinion Pages, September 15, 2015.

Denton M. and Marshall C. "Law of forms revisited." *Nature*, 410, 2001.

Crisp A. "Human genome includes 'foreign' genes not from our ancestors." *Research Bulletin*, University of Cambridge, 12 March 2015.

De Rham C. "Massive Gravity."*Living Reviews in Relativity*, 25 August 2014.

Devlin H. "'Spectacular' jawbone discovery sheds light on ancient Denisovans." *The Guardian*, 1 May 2019.

Devlin H. "Earliest known cave art by modern humans found in Indonesia: Pictures of human-like hunters and fleeing mammals dated to nearly 44,000 years old." *The Guardian*, 11 December 2019.

Devlin H. "Has physicist's gravity theory solved 'impossible' dark energy riddle?" *The Guardian*, 25 January 2020.

Dickinson M. and Karl G. Gotz K.G. "Unsteady aerodynamic performance of model wings at low Reynolds numbers." *Journal of Experimental Biology*, 174, 1993.

Diggle S.P., et al. "Evolutionary theory of bacterial quorum sensing; when is a signal not a signal?" *Philosophical Transactions of the Royal Society B*, 2007.

Dockrill P. "It's Official: Atmospheric CO2 Just Exceeded 415 ppm For The First Time in Human History." *ScienceAlert*, 13 May 2019.

Douglass M. "Incredible tool use in the animal kingdom." *BBC Earth*, 10 November 2014.

Dwyer J. "The Day When Neurons Go on Trial." *The New York Times*, September 17, 2013.

England J.L. "Dissipative adaptation in driven self-assembly." *Nature Nanotechnology*, Vol. 10, November 2015.

Eyres H. "How to Cultivate a Growth Industry." *FT.com*, 15 October 2010.

Fackler M. "Scientist Who Had Claimed Stem Cell Breakthrough Resigns From Japanese Research Institute." *The New York Times*, December 19, 2014.

Faluger R., Hill J.C. and Spergel D.N. "Toward an Understanding of Foreground Emission in the BICEP2 Region." arXiv:1405.7351v2 [astro-ph.CO]. Cornell University Library, May 28, 2014, (revised June 20, 2014).

Feilden T. "Citizen science is the new black." *BBC News*, 15 October 2013.

Feilden T. "Gas guzzler," *BBC News*, 13 November 2012.

Feyerabend P. "Explanation, Reduction and Empiricism." In: H. Feigl and G. Maxwell (eds). *Scientific Explanation, Space and Time*. University of Minneapolis Press, 1962, pp.28-97.

Fish T. "NASA exoplanet discovery: Space agency spots closest 'habitable' Earth-like planet yet." *Express*, August 2, 2019.

Fleming S.M. and Dolan R.J. "The neural basis of metacognitive ability." *Philosophical Transactions of The Royal Society B*, 2012; 367: 1338-49.

Fleming S.M., Dolan R.J. and Firth C.D. "Metacognition: computation, biology and function." *Philosophical Transactions of The Royal Society B*, 367, 2012.

Fox-Skelly J. "There is one animal that seems to survive without oxygen: It is microscopic, looks a bit like a jellyfish, and survives in a place that would kill every other known animal species." *BBC.com*, 26 January 2017.

Frank A. "Cracking the Quantum Safe." *The New York Times*, Sunday Review, The Opinion Pages, October 14, 2012.

French L. "Disability history month: John Goodricke the deaf astronomer." *BBC News Magazine*, 18 December 2012.

Frenkel E. "Is the Universe a Simulation?" *The New York Times*, Sunday Review, February 14, 2014.

Friedman M. "The Methodology of Positive Economics." *Essays in Positive Economics*. University of Chicago Press: Chicago, 1953.

Gabbatiss J. "Interbreeding with Neanderthals gave humans ability to fight off dangerous diseases." *The Independent*, 4 October 2018.

Gaisler-Salomon I. "Inheriting Stress." *The New York Times*, Sunday Review, March 7, 2014.

Gardner A. and West S.A. "GREENBEARDS." EVOLUTION, January 2010.

Garget J. "Suction cups that don't fall off: Insects in torrential rivers may inspire engineering solutions." *University of Cambridge Newsletter*, Research News, 20 December 2019.

Gary M. "The Trouble With Brain Science." *The New York Times*, Opinion Pages, July 11, 2014.

Greshko M. "What are mass extinctions, and what causes them?" *nationalgeographic.com*, September 26, 2019.

Ghosh P. "New human-like species discovered in S Africa." *The BBC News*, 10 September 2015.

Gleick J. "Introduction," Feynman R. *The Character of Physical Law*. The Modern Library Edition, 1994 [originally published by The British Broadcasting Corporation in 1965].

Goldberg M. "'The Massive Elephant in the Room' –New Theory Says 'No Need' for Dark Energy." *The Daily Galaxy*, January 28, 2020.

Goldstein R.M. "The hard problem of consciousness and the solitude of the poet." *Tin House*, 2012, 13(3): Issue 51.

Gopnik A. "Sleeping Like a Baby, Learning at Warp Speed." *The Wall Street Journal*, March 22, 2013.

Gordon G. "SSRN's 2013 Year-End President's Letter," December 23, 2013.

Gorman J. "N.H.I. Seeks $4.5 Billion to Try to Crack the Code of How Brains Function." *The New York Times*, Science, June 5, 2014.

Gorman J. "Scientists Trace Memories of Things That Never Happen." *The New York Times*, July 25, 2013.

Gorman J. "The Brain, in Exquisite Detail." *The New York Times*, January 6, 2014.

Gray J. "Book Review." *Notices of the AMS*, Vol. 4, No. 9, p.1080, October 2000.

Greshko M. "Did This Mysterious Ape-Human Once Live Alongside Our Ancestors?" *National Geographic*, May 9, 2017.

Harman O. "A little Help From your Friends." *The New York Times Book Review*, 10 April 2011, p.18.

Hayek F.A. "The Use of Knowledge in Society." *American Economic Review*, XXXV, No. 4, p.519, September, 1945.

Helmenstine A.M. "What Is the Chemical Formula of Sugar?" *ThoughtCo*, June 22, 2018.

Helmenstine T. "What Are the Radioactive Elements?" *sciencenotes.org*, September 9, 2019.

Hempel C.G. and Oppenheim P. "Studies in the Logic of Explanation." *Philosophy of Science*, 1948.

Hicks J.R. "Annual Survey of Economic Theory: The Theory of Monopoly," *Econometrica*, Volume 3, Issue 1, January 1935.

Hicks J.R. "A Suggestion for Simplifying the Theory of Money." *Econometrica*, February 1935.

Hodgkinson J. "Learning the chemical language of bacteria." Master's Fellows' Research Talks, Trinity College (Cambridge), 4 February 2014.

Hogan C.M. "Archaea." *Encyclopedia of Life*, 2012 [online].

Hogan J. "To Err Is Progress." *The Wall Street Journal*, July 20, 2011 [reviewing Deutsch's book].

Hogenboom M. "Is this a new species of human?" *BBC Discovery*, 25 January 2015.

Holmes J. "The Case for Teaching Ignorance." *The New York Times*, August 24, 2015.

Homann M., et al.,"Microbial life and biogeochemical cycling on land 3,220 million years ago." *Nature Geoscience*, 23 July 2018.

Hood B. "Developing a Sense of Self." Darwin College Lecture Series, Cambridge University, 6 February 2015.

Hood M. "Scientists set to unveil first picture of a black hole." *Phs.Org.com*, April 6, 2019.

Horsten L. "Philosophy of Mathematics." *The Stanford Encyclopedia of Philosophy*, Spring 2019.

Hotz R.L. "Differences in How Men and Women Think are Hard-Wired." *The Wall Street Journal*, Health & Wellness, December 9, 2013.

Hotz R.L. "NASA Scientists Discover 715 New Planets." *The Wall Street Journal*, February 26, 2014.

Hotz R.L. and Naik G. "Discovery Bolsters Big-Bang Theory." *The Wall Street Journal*, March 17, 2014.

Hotz R.L. "Mars Shows Signs of Flowing Salt Water." *The Wall Street Journal*, September 28, 2015.

Hotz R.L. "Gravitational Waves Detected, Verifying Part of Albert Einstein's Theory of General Relativity," *The Wall Street Journal*, Environment & Science, February 11, 2016.

Hunt G.R. and Gray R.D. "Diversification and cumulative evolution in New Caledonian crow tool manufacture." *Proceedings of the Royal Society*, 270 (1517), 2003.

Hurley D. "Grandma's Experiences Leave a Mark on Your Genes." *Discover*, June 11, 2013.

Ioannidis J.P.A. "Contradicted and Initially Stronger Effects in Highly Cited Clinical Research." *The Journal of the American Medical Association*, July 13, 2005.

IceCube Collaboration (University of Wisconsin-Madison). "Ice Cube - South Pole Neutrino Detector." *icecube.wisc.edu*.

Jabr F. "It's Buggy Out There." *The New York Times Magazine*, February 13, 2015.

Jabr F. "Why Soap Works: At the molecular level, soap breaks things apart. At the level of society, it helps hold everything together." *The New York Times*, March 13, 2020.

Johnson G. "New Truths That Only One Can See." *The New York Times*, January 20, 2014.

Johnson I. "Q. Why is it so hard to swat a housefly? A. It sees you coming in slow motion." *The Independent*, 16 September 2013.

Jones C. "The Woman Who Might Find Us Another Earth," *The New York Times Magazine*, December 7, 2016.

Kaku M. "No, You Still Can't Go Faster Than Light." *The Wall Street Journal*, March 21, 2012.

Kaplan M. "The Steely, Headless King of Texas Hold 'Em," *The New York Times Magazine*, September 5, 2013.

Kaufman L. "Darwin Foes Add Warming to Targets." *The New York Times*, March 3, 2010. Kazmierczak J. "TESS Finds Its Smallest Planet Yet." NASA Exoplanet Exploration. June 28, 2019.

Keller L. and Ross K.G. "Selfish genes: a green beard in the red fire ant." *Nature*, 394:573–575, 6 August 1998.

Kilner R. Of the Department of Zoology, Cambridge University. Lecture entitled "Social evolution in a grave," 4 March 2014, sponsored by the Trinity College Science Society.

Kolata G. "South Korean Scientists Clone Man's Best Friend." *The New York Times*, August 3, 2005.

Kolata G. "Same Genetic Basis Found in 5 Types of Mental Disorders." *The New York Times*, February 28, 2013.

Kolata G. "In Gut Research's Latest Advance, Bacteria From Humans Can Slim Mice Down." *The New York Times*, Health, September 5, 2013.

Konstantinovsky M. "Primary Colors Are Red, Yellow and Blue, Right? Well, Not Exactly." science.howstuffworks.com, July 2, 2019.

Kosslyn S.M. and Miller W.G. "A New Map of How We Think: Top Brain/Bottom Brain." *The New York Times*, The Saturday Essay, October 20, 2013.

Kragh H. "Max Planck: the reluctant revolutionary." *physicsworld.com*, 1 December 2000.

Kristian F. "Physics in the Nervous System." Presentation to the Trinity College Science Society (Trinity College, Cambridge), 21 October 2014.

Kruglyak L. "The road to genome-wide association studies." *Nature*, November 2008; 456: p.18.

Krupp S.R. "Theoretical explanation and the nature of the firm." *Western Economic Journal*, 1, 1963.

Krupp S.R. "Types of Controversy in Economics." In Krupp S.R. (ed.) *The Structure of Economic Science*. Prentice-Hall: New Jersey, 1966.

Lane N. "Energetic Constraints on the Evolution of Life." Lecture, Cambridge University, 2 March 2014, sponsored by the Cambridge Biological Society.

Laland K.N., et al., "How culture shaped the human genome: bringing genetics and the human sciences together," *Nature Reviews Genetics*, 11, February 2010.

Leaf C. "Do Clinical Trials Work?" *The New York Times*, Sunday Review, The Opinion Pages, July 14, 2013.

Lee H.R. "Remains of Humanlike Ancestors Found in South Africa." *The Wall Street Journal*, September 10, 2015.

Leman J. "The Biggest Quantum Breakthrough Yet—Literally." *Popular Mechanics*, October 2, 2019.

Lester R.A. "Shortcomings of marginal analyses for wage-employment problems." *American Economic Review*, XXXVI, 1946.

Lester R.A. "Marginalism, minimum wages and labor markets." *American Economic Review*, 1947, XXXVII: 135–48.

Lezard N. "Explaining nothing, brilliantly," *The Guardian*, 22 March 2003.

Lidz F. "How the World's Oldest Wooden Sculpture Is Reshaping Prehistory." NYTimes.com, March 22, 2021.

Linder C. "This Badass Beetle Could Keep Your Car Cool: See how scientists are inspired by the bug, which survives at extreme temperatures." *Popular Mechanics*, June 14, 2020.

Linsky B. and Irvine A.D. "Principia Mathematica." *The Stanford Encyclopedia of Philosophy*. Summer 2019 Edition.

Longrich N. "Evolution tells us we might be the only intelligent life in the universe." *The Conversation*, October 18, 2019.

Luisi P.L. "Emergence in Chemistry: Chemistry as the Embodiment of Emergence." *Foundations of Chemistry*, 10-2002, 4(3).

Machlup F. "Marginal analysis and empirical research." *American Economic Review*, XXXVI, 1946.

Machlup F. "Rejoinder to an anti-marginalist." *American Economic Review*, XXXVII, 1947..

Machlup F. "Operationalism and Pure Theory in Economics." In S.R. Krupp (ed) *The Structure of Economic Science*. Prentice-Hall: New Jersey, 1966.

Manke K. "Heat energy leaps through empty space, thanks to quantum weirdness." Phys.org, 11 December 2019.

Marcus G. "The Trouble With Brain Science." *The New York Times*, Opinion Pages, July 11, 2014.

Margolis J. "The analysis of the firm: rationalism, conventionalism, and behavioralism." *Journal of Business*, 31, 1958.

Margulis L. [see also Sagan L.] "Endosymbiosis." *learn.genetics.utah.edu*.

Markoff J. "Obama Seeking to Boost Study of Human Brain." *The New York Times*, February 17, 2012.

Markoff J. "IBM Develops a New Chip That Functions Like a Brain." *The New York Times*, Science, August 7, 2014.

Markoff J. "Expansion Microscopy Stretches Limits of Conventional Microscopes." *The New York Times*, Science, January 19, 2015.

Marris E. "Neanderthal artists made oldest-known cave paintings: Designs at three Spanish sites are thought to predate human arrival in Europe by at least 20,000 years." *Nature.com*, 22 February 2018.

Marshall M. "Human and Neanderthal interbreeding questioned." *New Scientist*, August 13, 2012.

Marshall M. "The Secret of How Life on Earth Began." *BBC.com*, 31 October 2016.

Marshall M. "The Four Biggest New Things We Learned About Human Evolution In 2018." *Forbes*, December 20, 2018.

Martin A. "Japanese Stem-Cell Scientist Yoshiki Sasai Commits Suicide." *The Wall Street Journal*, August 5, 2014.

Martin A. and Naik G. "Japanese Institute Weights Retracting Stem-Cell Studies." *The Wall Street Journal*, March 10, 2014.

McGrath M. "Dolphins have 'longest social memory' among non-humans." *BBC News*, Science & Environment, 7 August 2013.

McKay B. "Study Links Eggs to Higher Cholesterol and Risk of Heart Disease." *WSJ.com*, March 15, 2019.

Meaney M. and Szyf M. "Epigenetic programming by maternal behavior." *Nature Neuroscience*, June 2004.

Metz C. "Hold 'Em or Fold 'Em? This A.I. Bluffs With the Best." *The New York Times*, July 11, 2019.

Miller M.C. "Gravitational waves: A golden binary." *Nature*, 551, 02 November 2017.

Mirazón Lahr M., et al. "Inter-group violence among early Holocene hunter-gatherers of West Turkana, Kenya." *Nature*, 529: 394-8, 21 January 2016.

Modigliani F. and Miller M.H. "The Cost of Capital, Corporation Finance, and the Theory of Investment." *American Economic Review*, XLVIII, 1958.

Moore H. "Are all the ants as heavy as all the humans?" *BBC News Magazine*, 22 September 2014.

Morell V. "Hummingbirds see colors we can't even imagine." *National Geographic*, June 15, 2020.

Morelle R. "Clever New Caledonian crows can use three tools." *BBC News*, 20 April 2010.

Morgenau H. "What is a Theory?" In S.R. Krupp (ed) *The Structure of Economic Science*. Prentice-Hall: Englewood Cliffs, NJ, 1966, p.26.

Morris S.C. "Evolution: Like any other science it is predictable." *Philosophical Transactions of The Royal Society, Biological Sciences*, 24 November 2009, 133–45.

Morris S.C. "Darwin was right. Up to a point." *The Guardian*, 12 February 2009.

Murphy, H. "You Will Never Smell My World the Way I Do: Scientists find that whiskey's smokiness, the smell of beets and lily of the valley perfume can be utterly different depending on your genetic wiring." NYTimes.com, May 3, 2019.

Naik G. "A Genetic Code for Genius." *The Wall Street Journal*, February 15, 2013.

Naik G. "Hint of Dark Matter Found." *The Wall Street Journal*, April 3, 2013.

Naik G. "How a Black Hole Really Works." *The Wall Street Journal*, August 16, 2013.

Naik G. "Francois Englert and Peter Higgs Win Nobel Prize in Physics." *The Wall Street Journal*, October 8, 2013.

NASA. "First Detection of Sugars in Meteorites Gives Clues to Origin of Life." *Release*, October 18, 2019.

NASA. "NASA's NuSTAR Sees Rare Blurring of Black Hole Light." *Release* 14-210, August 12, 2014.

NASA. "Discovery Alert: A Record Haul–Planet Count Hits 4,000." *Exoplanet Exploration*, June 13, 2019.

NASA. "Discovery Alert: New Twin Planets Prompt Comparisons to Earth." *Exoplanet Exploration*, June 20, 2019.

NASA. "NASA's Webb Telescope Launches to See First Galaxies, Distant Worlds." *Release* 21-175, December 25, 2021.

Natarajan P. (Professor of Astronomy & Physics, Yale University). Lecture, March 9, 2013, New York City (Cambridge in America Day).

National Geographic, *Daily News*, May 6, 2010. [It appears that a few "genes" of the Neanderthal can be found in the DNA of modern man.]

National Geographic, *Style Manual*, 2017.

Nicholson M.T. [reviewing Nagel] *Prospect*. 7 February 2013.

Nield D. "Plants May Not Have Ears, But They Can 'Hear' Way Better Than We Thought." sciencealert.com, 19 January 2019.

Noble W.J. "New Species of Human Ancestor Is Found in a South Africa Cave." *The New York Times*, September 10, 2015.

Nyhan B. "To Get More Out of Science, Show the Rejected Research." *The New York Times*, September 18, 2014.

Oberheim E. and Hoyningen-Huene P., "The Incommensurability of Scientific Theories." *The Stanford Encyclopedia of Philosophy*, Fall 2018.

O'Connor A. "Nutritional Panel Calls for Less Sugar and Eases Cholesterol and Fat Restrictions." *The New York Times*, February 19, 2015.

O'Connor T. and Wong H.Y. "Emergent Properties." *The Stanford Encyclopedia of Philosophy*, Summer 2015.

Obokata H., et al. "Retraction: Stimulus-Triggered fate conversion of somatic cells into pluriotency." *Nature*, 505, 2 July 2014.

O'Donnell E. "How Life Began: Jack Szostak's pursuit of the biggest questions on Earth." *Harvard Magazine*, July-August 2019.

Okasha S. "Biological Altruism." *Stanford Encyclopedia of Philosophy*, June 3, 2003 (substantive revision July 21, 2013).

Osborne H. "Two Billion Years Ago, Up To 99 Percent of Life on Earth Died ...." *Newsweek*, September 2, 2019.

Overbye D. "New Planet May Be Able to Nurture Organisms." *The New York Times*, September 29, 2010.

Overbye D. "Tiny Neutrinos May Have Broken Cosmic Speed Limit." *The New York Times*, September 22, 2011.

Overbye D. "The Trouble with Data That Outpaces a Theory." *The New York Times*, March 27, 2012.

Overbye D. "Particle Physicists in U.S. Worry About Being Left Behind." *The New York Times*, March 4, 2013.

Overbye D. "All Signs Point to Higgs, but Scientific Certainty Is a Waiting Game." *The New York Times*, March 4, 2013.

Overbye D. "CERN Physicists See Higgs Boson in New Particle." *The New York Times*, March 17, 2013.

Overbye D. "For Noble, They Can Thank the 'God Particle.'" *The New York Times*, October 8, 2013.

Overbye D. "Dark Matter Experiment Has Detected Nothing, Researchers Say Proudly." *The New York Times*, October 30, 2013.

Overbye D. "Space Ripples Reveal Bid Bang's Smoking Gun." *The New York Times*, March 17, 2014.

Overbye D. "Astronomers Hedge on Big Bang Detection Claim." *The New York Times*, June 19, 2014.

Overbye D. "So Many Earth-Like Planets, So Few Telescopes." *The New York Times*, January 6, 2015.

Overbye D. "NASA Says Data Reveals an Earth-Like Planet, Kepler 452b." *The New York Times*, July 23, 2015.

Overbye D. "Gravitational Waves Detected, Confirming Einstein's Theory." *The New York Times*, February 11, 2016.

Overbye D. "Kepler Finds 1,284 New Planets," *The New York Times*, May 10, 2016.

Overbye D. "Vera Rubin, 88, Dies; Opened Doors in Astronomy, and for Women," *The New York Times*, December 27, 2016.

Overbye D. "Kepler, the Little NASA Spacecraft That Could, No Longer Can," NYTimes.com, October 30, 2018.

Overbye D. "Black Hole Picture Revealed for the First Time: Astronomers at last have captured an image of the darkest entities in the cosmos." NYTimes.com, April 10. 2019.

Overbye D. "Vera Rubin Gets a Telescope of Her Own: The astronomer missed her Nobel Prize. But she now has a whole new observatory to her name." *The New York Times*, January 11, 2020.

Pagel M. "DNA-based evolutionary trees." *Nature*, DOI: 10.1038/nature08630.

Palmer J. "Protein flaws responsible for complex life." BBC News, Science & Environment, 19 May 2011.

Palmer J. BBC News, 29 November 2012. [The second largest "black hole" ever identified. AND Black holes exist in the center of large galaxies.]

Parker A. Presentation to the Trinity College Science Society (Trinity College, Cambridge), 11 March 2014

Perunov N., Marsland R.A., and England J.L.. "Statistical Physics of Adaptation ," *Physical Review*, X 6, 021036, 2016.

Plumer B. "There have been five mass extinctions in Earth's history. Now we're facing a sixth." *The Washington Post*, February 11, 2014 (interview with Elizabeth Kolbat, author of *The Sixth Extinction: An Unnatural History*).

Polanyi M. "The Creative Imagination." In: M. Kransz (ed.) *The Idea of Creativity*. Brill: Leiden, Netherlands, 2009 [original published in *Chemical and Engineering News*, 1966].

Pollack A. "Stem Cell Research Papers Are Retracted." *The New York Times*, July 2, 2014.

Porter C. "Physicists Mine Cosmic Answers Deep Underground." *The Wall Street Journal*, July 30, 2012.

Powell Corey S. "Strange life-forms found deep in a mine point to vast 'underground Galapagos'." nbcnews.com, September 7, 2019.

Pruitt S. "Early Humans Slept Around with More than Just Neanderthals," History.com, Mar.16, 2018.

Quenqua D. "Earthlike Planets May Be Merely an Illusion." *The New York Times*, July 7, 2014.

Quenqua D. "Unchanged for More Than Two Billion Years." *The New York Times*, February 9, 2015.

Rachman G. "Economists would do well to learn the modesty of historians...." NYTimes.com., September 9, 2010, and FT.com, 6 September 2010.

Redfern S. "Earth life 'may have come from Mars.'" BBC News, Science & Environment, 29 August 2012.

Redfern S. "Resurrected protein's clue to origins of life." BBC News, Science & Environment, 8 August 2013.

Rees, M. "Opinion: Aliens, very strange universes and Brexit," *University of Cambridge Research Bulletin*, 7 April 2017.

Reichenbach B. "Cosmological Argument." *The Stanford Encyclopedia of Philosophy*, Spring 2019.

Reichert C. "Mass extinction 2 billion years ago reportedly killed 99% of life on Earth." *Cnet.com*, September 3, 2019.

Rennie J. "How Complex Wholes Emerge From Simple Parts." *Quanta Magazine*, January 28, 2019.

Rheinberger H-J., Müller-Wille S. and Meunier R. "Gene." *The Stanford Encyclopedia of Philosophy*, Spring 2015.

Requarth T. "Bringing a Virtual Brain to Life." *The New York Times*, March 18, 2013.

Richtel M. and Jacobs A. "A Mysterious Infection, Spanning the Globe in a Climate of Secrecy, The rise of Candida auris embodies a serious and growing public health threat: drug-resistant germs." *The New York Times*, April 6, 2019.

Ridley M. "Experts have been feeding us a big fat myth." *The Times*, Opinion, 30 June 2014.

Ross H. "Where Did the Universe Come From? New Scientific Evidence for the Existence of God." Presentation in South Barrington, Illinois, 16 April 1994.

Russell Wallace A. "Sir Charles Lyell on Geological Climates and the Origin of Species." *Quarterly Review*, April 1869 [published anonymously].

Sagan L. [Margulis]. "On the origin of the mitosing cells." *Journal of Theoretical Biology*, 1967; 14: 225–74.

Sak-Young Chwe M. "Scientific Pride and Prejudice." *The New York Times, Sunday Review*, January 31, 2014.

Sang-Hun C. "Korean Scientist's New Project: Rebuild After Cloning Disgrace." *The New York Times*, February 28, 2014.

Schaffer S. "Why trust public experiments?" Presentation to the Trinity College Science Society (Trinity College, Cambridge), 28 October 2014.

Schrodinger E. "What is life? The Physical Aspect of the Living Cell." First published in 1944, based on lectures delivered under the auspices of the Dublin Institute for Advanced Studies at Trinity College, Dublin, in February 1943.

Schuessler J. "An Author Attracts Unlikely Allies." *The New York Times*, February 6, 2013.

Service R.F. "Researchers may have solved origin-of-life conundrum." *sciencemag.org*, March 16, 2015.

Shapin S. "Why Scientists Shouldn't Write History." *The Wall Street Journal*, February 13, 2015 [review of *To Explain the World* by Steven Weinberg].

Siegel E. "Water In Space: Does It Freeze Or Boil?" *Forbes*, December 23, 2016.

Siegel E. "This Is Why The Event Horizon Telescope Still Doesn't Have An Image Of A Black Hole: The data has been taken, collected, and analyzed.

So where is the first image of an event horizon, already?" *Medium.com*, June 11, 2018.

Starr M. "Scientists Just Created a Bizarre Form of Ice That's Half as Hot as The Sun." *ScienceAlert*, 9 May 2019.

Star M. "Scientists Find The First-Ever Animal That Doesn't Need Oxygen to Survive." *Sciencealert.com*, 25 February 2020.

Stigler G.J. "Professor Lester and the Marginalists." *American Economic Review*, XXXVII, 1947.

Stratton G.M. "Some Preliminary Experiments on Vision without Inversion of the Retinal Image." Paper read at the Third International Congress for Psychology, in Munich, August 1896.

Strogatz S. "One Giant Step for a Chess-Playing Machine." *NYTimes.com*, December 26, 2018.

Svoboda E. "Hard-Wired for Giving." *The Wall Street Journal*, The Saturday Essay, August 31, 2013.

Tabuchi H. "One Author of a Startling Stem Cell Study Calls for Its Retraction." *The New York Times*, Science, March 10, 2014.

Taylor A.H., Miller R. and Gray R.D. "New Caledonian crows reason about hidden causal agents." *Proceedings of the National Academy of Science of the United States of America*, September 17, 2012.

Teicholz N. "The Government's Bad Diet Advice." *The New York Times*, The Opinion Pages, February 20, 2015.

Than K. "Oxygen-Free Animals Discovered—A First:Deep in the Mediterranean, scientists have discovered the first complex animals known to live without oxygen." *National Geographic*, April 17, 2010.

Turner E. "Biological Sciences: the story of four billion years of evolution in seven hours." Institute of Continuing Education, University of Cambridge, 8 March 2015.

University of Cambridge. "The super-resolution revolution." *Research Bulletin*, 27 February 2015.

University of Cambridge. "Welcome to the neighbourhood: new dwarf galaxies discovered in orbit around the Milky Way." *Research Bulletin*, 10 March 2015.

University of Cambridge. "Earliest evidence of human warfare." *Research Bulletin*, 22 January 2016.

University of Cambridge, "Newly discovered planets could have water on their surfaces." *Research Bulletin*, 24 February 2017.

University of Cambridge. "Elephants' 'body awareness' adds to increasing evidence of their intelligence." *Research Bulletin*, 21 April 2017.

University of Cambridge. "Studies raise questions over how epigenetic information is inherited." *Research Bulletin*, 2 November 2018.

University of Cambridge. "'Game-changing' research could solve evolution mysteries," *Research Bulletin*, 11 September 2019.

University of Cambridge."Molecular 3D-maps unlock new ways of studying human reproduction," *Research Bulletin*, 16 June 2022.

University of California - Santa Cruz. "Study documents paternal transmission of epigenetic memory via sperm." October 17, 2018.

Wade N. "Journal Faulted in Publishing Korean's Claims." *The New York Times*, November 29, 2006.

Wade N. "Human Culture, An Evolutionary Force." *The New York Times*, March 1, 2010.

Wade N. "Anthropology a Science? Statement Deepens a Rift," *The New York Times*, Science, December 9, 2010.

Walter C. "The First Artists." *National Geographic* ("The Book of Firsts"), January 2015.

Webb J. "Evolution 'favours bigger sea creatures.'" *BBC News*, 19 February 2015.

Wei-Haas M. "Surprising leap in ancient human technology tied to environmental upheaval: Sediment core evidence reveals the critical factors that may have given rise to strikingly complex behaviors some 320,000 years ago, around the time the first members of our species appeared." *National Geographic*, October 21, 2020.

Weiner J. "Human Cells Make Up Only Have Our Bodies. A New Book Explains Why." *The New York Times*, August 15, 2016 (book review).

Weir A.A.S., Chappell J. and Kacelnik A. "Shaping of Hooks in New Caledonian Crows." *Science Magazine*, February 1, 2009.

Wigner E.P. "The Unreasonable Effectiveness of Mathematics in the Natural Sciences." *Communications of Pure and Applied Mathematics*, 1960; XIII: 1–14.

Wilford J.N. "Skull Fossil Suggests Simpler Human Lineage." *The New York Times*, October 17, 2013.

Wilford J.N. "Jawbone Fossil Fills a Gap in Early Human Evolution." *The New York Times*, March 4, 2015.

Wilson E.O. "The Riddle of the Human Species." *The New York Times*, Opinionator, February 24, 2013.

Wong K. "Ancient Fossils from Morocco Mess Up Modern Human Origins." *Scientific American*, June 8, 2017.

Woodward A. "First Image of Black Hole at Center of Milky Way Galaxy Revealed: Sagittarius A* is 4 million times as massive as the sun and some 26,000 light-years from Earth." *WSJ.com*, May 12, 2022.

Wu K.J. "Dads Pass On More Than Genetics in Their Sperm." *Smithsonian.com*, July 26, 2018.

Wu K.J. "Plants May Let Out Ultrasonic Squeals When Stressed: Human ears can't hear them, but other plants or animals might." *Smithsonian.com*, December 9, 2019.

Yardley W. "Ian Barbour, Who Found a Balance Between Faith and Science, Dies at 90." *The New York Times*, January 12, 2014.

Yin S. "Cold Tolerance Among Inuit May Come From Extinct Human Relatives." *The New York Times*, December 23, 2016.

Zimmer C. "Watching Bacteria Evolve, With Predictable Results." *The New York Times*, August 15, 2013.

Zimmer C. "As Humans Change Landscape, Brains of Some Animals Change, Too." *The New York Times*, August 22, 2013.

Zimmer C. "DNA Double Take." The New York Times, September 16, 2013.

Zimmer C. "Elephants Get the Point of Pointing, Study Shows." *The New York Times*, October 10, 2013.

Zimmer C. "Key to Ants' Evolution May Have Started With a Wasp." *The New York Times*, October 20, 2013.

Zimmer C. "Christening the Earliest Members of Our Genus." *The New York Times*, October 24, 2013.

Zimmer C. "Neanderthals Leave Their Mark on Us." *The New York Times*, January 29, 2014.

Zimmer C. "Out of Siberian Ice, a Virus Revived." *The New York Times*, March 3, 2014.

Zimmer C. "Take a Breath and Thank a Sponge." *The New York Times*, March 13, 2014.

Zimmer C. "Is Most of Our DNA Garbage?" *The New York Times Magazine*, March 5, 2015.

Zimmer C. "Scientists Say Canadian Bacteria Fossils May Be Earth's Oldest." *The New York Times*, March 1, 2017.

Zimmer C. "In Cave in Borneo Jungle, Scientists Find Oldest Figurative Painting in the World." NYTimes.com, November 7, 2018.

Zimmer C. "A Billion-Year-Old Fungus May Hold Clues to Life's Arrival on Land: A cache of microscopic fossils from the Arctic hints that fungi evolved long before plants." *The New York Times*, May 22, 2019.

www.ingramcontent.com/pod-product-compliance
Lightning Source LLC
Chambersburg PA
CBHW052005070526
44584CB00016B/1623

Praise for *Unsettling the Great White North:*

"How did Canada become white? Dispossession, erasure, and a sham multiculturalism. This extraordinary volume of essays exposes settler violence and fraudulent claims of 'inclusion' and offers instead a long, deep, and often hidden history of Black struggles for freedom, power, and self-determination. It should be required reading, not only for Canadians but for all of us on occupied Turtle Island and around the world."

Robin D. G. Kelley,
Author of *Freedom Dreams: The Black Radical Imagination*

"This timely collection challenges any remaining conception of the Great White North as a refuge from anti-Black violence and exclusion. Documenting the long and varied histories of endurance, negotiation, and resistance against racism among persons of African descent, this volume not only challenges any conception of the Black Canadian experience as 'linear, unchanging, homogenous, and recent,' it also offers a vital corrective to the erasure of Black Canadians from the national myths about who 'we' are, and who 'we' can become. Truly a must read!"

Beverley Mullings,
Professor of Geography and Planning, Queen's University

"This collection, by locating the past as a force shadowing the present and informing Black people's ongoing search for freedom, overwrites narratives of a redemptive Canadian multicultural citizenship to narrate the possibilities by which Black people have survived, and continue to survive, conquest and subjugation. This comprehensive cartography of Black life in Canada is a stunning achievement and essential reading in Black Canadian history and thought."

Andrea A. Davis, Associate Professor of Humanities,
York University, and co-editor of *The Journal of Canadian Studies*

"*Unsettling the Great White North* is a vital collection of historical scholarship featuring bold and accomplished voices in the field of Black Canadian Studies, and illuminating a striking diversity of periods, regions, communities, institutions, art forms, and emergent frames of inquiry. Expertly curated, it offers precious knowledge and critical insight to scholars and general readers alike."

David Chariandy, Professor of English,
Simon Fraser University, and author of *Brother*

"This is an important and comprehensive contribution to Black Canadian historical studies. It offers critical analyses regarding the diversity, complexity, and creativity of Black Canadian lives and new understandings about the racialized history of African Canadians as well as the ways in which Black communities have overcome systemic barriers. *Unsettling the Great White North* indeed unsettles dominant historical practices and centres the very people who have been left out of history."

Annette Henry, David Lam Chair in Multicultural Education,
Professor of Language and Literacy Education, and Professor at the Institute for Race, Gender, Sexuality and Social Justice, University of British Columbia

"*Unsettling the Great White North* is a major contribution. Encompassing essays on topics ranging from slavery to community studies about African immigrants, this indispensable book should be read by all students of Canadian history."

Harvey Amani Whitfield,
Professor of Black North American History, University of Calgary

"*Unsettling the Great White North*, the first volume of its kind, is an impressive collection in depth, scope, and quality. Twenty-one authors centre and make visible the experiences of African Canadians, including recent migrants (Rwandans and other continental Africans). Collectively, albeit with different emphases, these authors bear witness to histories of exclusion and marginalization while simultaneously underscoring African Canadians as agents of their own lives. *Unsettling the Great White North* clears up any misconception regarding African Canadians' contributions to Canada's nation-building enterprise. An authoritative text that is accessible to and suitable for both academic and general audiences interested in Black Canadian Studies and history."

Karen Flynn, Associate Professor of Gender and Women's Studies,
University of Illinois, Urbana-Champaign, and author of *Moving beyond Borders: A History of Black Canadian and Caribbean Women in the Diaspora*

"This important collection of essays engages with the presence of Black people (people of African descent) in this region colonially known as Canada. But more than this, the compelling essays demonstrate the importance of Black Studies in addressing the historical and current conditions and experiences of Black life. This is a book for all interested in Black Studies!"

OmiSoore H. Dryden, James R. Johnston (JRJ)
Chair in Black Canadian Studies, and Associate Professor,
Faculty of Medicine, Dalhousie University

"This is essential reading for all Canadians. Michele A. Johnson and Funké Aladejebi, two formidable scholars, have collated and edited a perfect mix of essays focused on Black Canadian history. These documented Black Canadian stories can now be shared and debated, bringing about awareness of the Black Canadian presence – indeed *Unsettling the Great White North*."

Honourable Dr. Jean Augustine P.C., C.M., O.Ont., C.B.E.,
First Black Canadian woman elected to the Parliament of Canada,
Proposed the motion to designate February as Black History Month in Canada

"*Unsettling the Great White North* finds a path through the hazy four centuries of Black history in Canada to elucidate the racial oppression of Blacks, and its attendant resistance-resilience dialectic. The book paints an unpleasant, albeit truthful, portrait of the procedures of *Othering* used to sustain the exclusion of Blacks in many spheres of Canadian life. The enduring desire of the majority to extract racial subsidies from Blacks in pursuit of profit is illuminated in a voice that is often deeply poetic. This is an indispensable resource for scholars of Black Studies in Canada."

Joseph Mensah, Professor,
Faculty of Environmental and Urban Change, York University